Condensed Matter
Optical Spectroscopy
An Illustrated Introduction

Condensed Matter Optical Spectroscopy

An Illustrated Introduction

Iulian Ionita

CRC Press is an imprint of the
Taylor & Francis Group, an **informa** business

CRC Press
Taylor & Francis Group
6000 Broken Sound Parkway NW, Suite 300
Boca Raton, FL 33487-2742

Printed on acid-free paper
Version Date: 20140612

International Standard Book Number-13: 978-1-4665-6956-0 (Pack - Book and Ebook)

Visit the Taylor & Francis Web site at
http://www.taylorandfrancis.com

and the CRC Press Web site at
http://www.crcpress.com

To my wife, Maria

Contents

Preface

After a long history, optical spectroscopy has become a common method of analysis in a large field of applications. The new instruments are computer controlled for faster, automated performance; this feature also offers the opportunity to quickly compare the obtained spectrum to a previous one or to a spectra library. The automatic laboratories mounted on board space probes are very good examples. Nowadays, many people use such instruments without an idea of how the machine makes an analysis, so they do not know how reliable the results are. This kind of work could be efficient in a repetitive process of analysis such as in the industry, agriculture, environment, pharma, or medicine, but it is not recommended in research. So I strongly recommend that you gather knowledge about the method you are using to characterize a material or collaborate with people skilled in that field. I often meet young researchers who choose the wrong path in their research because of some artifacts appearing in the recorded spectra. You should also carefully consider the importance of the sample history (its origin and processing) before analysis. The computer never knows this. The sample does not always have exactly the composition or structure that you are interested in. Sometimes, a correct result depends on a small peak obscured by the tail of a big peak.

Absorption spectroscopy in ultraviolet, visible and infrared, Raman spectroscopy, and fluorescence spectroscopy are powerful methods that you can use to characterize the composition and especially the structure of a sample. The same elements can be arranged in different ways and these dramatically change the properties of the compound. To understand how the arrangement of atoms in a spatial structure such as a molecule or crystal can influence the properties and spectra, the study of molecular symmetry becomes fundamental. It is vital that students in chemistry, physics, biology, materials science, and so on have a background on these topics from the undergraduate level.

There are several books about how group theory can be applied to spectroscopy. They are old but are still considered as fundamental contributions, including those written by G. Herzberg, F. A. Cotton, C. K. Jorgensen, C. G. Ballhausen, J. S. Griffith, J. C. Slater, and M. Tinkham. Other books have appeared more recently on the same subject for many reasons, including the need to adapt teaching to the level of students attending the course, showcase new technical opportunities to present a spatial arrangement, explain a subject in another way, and so on. Compressing a classical book such as Cotton's into one semester of teaching is a real challenge and requires it be reformulated.

My first encounter with this subject was many years ago through Professor A. Trutia, to whom I am grateful for the way he presented a clear overview of the field. He introduced us to the work of Cotton, McClure, and Griffith. Indeed, I'm not surprised to see that Cotton's book is still recommended for supplementary study at many universities throughout the world. Thus, in certain chapters, I follow a similar order of topics as in his book.

A very important technological achievement of the last few decades was the development of computer programs for molecular modeling and visualization. There are many programs, but I find the Jmol program, which was developed by the contributions of several enthusiastic people, including Robert M. Hanson, the most useful. Some important features are the software is freely available for any operation system; it works very well from the Internet and therefore does not require installation; it is good for molecules, crystallography, biopolymers, and proteins; at the same time, to construct your own molecules, you can easily install it and work offline with it. The ability to visualize colors and movements of images is very important nowadays and helps students to feel the pleasure of working in the field of molecular symmetry.

The book is designed as an introduction to condensed matter optical spectroscopy. After using the book, students should be able to pass three steps:

1. Choose a method or several methods for material characterization.
2. Measure a correct spectrum.
3. Interpret the spectrum or correlate the spectra obtained using different methods.

To choose suitable method(s), they have to know a few preliminary details about the sample and some important features (advantages and disadvantages) of the characterization methods (not only optical methods). Using the modeling software helps to obtain a direction of the work. In the second step, we decide what type of instrument is suitable, for example, long or short focal length, single or many detectors. Then the computer of the measuring instrument will ask you to decide about the value of some main parameters such as spectral range, slit width, single- or multiple-spectra acquisition, methods of spectra processing, and so on. The third step is the most difficult because the sample is not pure and perfect. It is the step where an understanding of group theory helps us attribute the bands of a spectrum.

The book is divided into five chapters dedicated to answering several practical questions:

Chapter 1. How can we classify the molecules according to their symmetry?
Chapter 2. What happens when an ion of an transition metal enters an environment with a given symmetry?
Chapter 3. How are atomic orbitals involved in molecular bonding?
Chapter 4. Is the molecule a rigid construction or a dynamic structure, which can either interact with light or not at all?
Chapter 5. How do we perform a reliable spectrum measurement?

The appendix (Selected Character Tables) includes several of the most important character tables, while the complete tables will be available at the book's individual webpage on www.crcpress.com, under the Supplementary Materials tab.

The book is intended for undergraduate students of different majors. Readers without a strong mathematical or quantum mechanics background may bypass the integrals or long tables with matrix multiplication, as long as they can understand the importance of the character tables in spectra interpretation.

This book differs from others because of its heavy use of illustrations (238!) to aid in understanding the main content. At a time when young people spend much time before the computer, phone, or tablet displays, the author and the publisher have made available all illustrations in electronic format from the book's website, so that the young reader can download and access them anywhere and anytime to enjoy the color and animation. Yes, you can move molecules as in a computer game. Enjoy them on your device because symmetry is beautiful but depends on the angle view!

I am indebted to Professor Alain Bulou for helpful comments on several sections and to my editor Luna Han for the idea of writing a book based on my teaching experience in the field and for her excellent cooperation in preparing the manuscript. I would like to thank to the people from different departments of Taylor & Francis Group for guidance to complete this project. I also appreciate the quality of work of the production team from Techset Composition directed by Syed Mohamad Shajahan, project manager, and I would like to thank them.

Also, thanks to Google for helping me feel in touch with the world's scientific teaching community from a long distance....

Iulian Ionita
University of Bucharest
Iulian.ionita@g.unibuc.ro

Author

Dr. Iulian Ionita is a member of the Faculty of Physics at the University of Bucharest, where he has been a lecturing professor since 1991 and is currently the director of the Research Center in Photonics–Spectroscopy–Plasma–Lasers. He is responsible for laboratories of biophotonics, atomic and molecular spectroscopy, and femtonics.

Dr. Ionita earned his PhD in physics, specializing in optics–spectroscopy–lasers, from the University of Bucharest. The title of his thesis was "Photoconductivity of Doped Ionic Crystals." His extensive and varied experience includes expertise in the characterization and processing of inorganic materials and biomedical tissues using ultrashort pulsed lasers; biomedical applications of optics, spectroscopy, and lasers; and optical and spectral properties of pure and doped crystals. He has been an invited professor at the University of Le Mans (France), the University of Heidelberg (Germany), the University of Umea (Sweden), and the University of Grenoble (France) and has published (as author or coauthor) 37 articles related to either biophotonics or spectral characterization of both inorganic materials and human tissues (Raman, fluorescence, IR and UV–Vis absorption, harmonic generation). His published books (in Romanian) include

- *Spectral Methods for Medical Analyses*
- *Condensed Matter Spectroscopy—Practical Works*
- *Atomic and Molecular Spectroscopy—Practical Works*
- *Geometrical Optics—Practical Works*
- *Wave Optics—Practical Works*

He is also the coauthor of five textbooks on physics for high school students.

Dr. Ionita currently teaches courses on wave optics (course and laboratory works), spectroscopy and lasers (course and laboratory works), condensed matter spectroscopy (course and laboratory works), and management of research projects. He is a member of the European Society Photonics21 (Work Group 3, Life Sciences and Health) and SPIE, and was the chairman at the International Conference "Photonics Europe 2010," conference 7715, "Biophotonics: Photonic Solutions for Better Health Care."

Molecular Symmetry and the Symmetry Groups

1

1.1 SYMMETRY ELEMENTS AND SYMMETRY OPERATIONS

1.1.1 INTRODUCTION

The word "symmetry" comes from the old Greek word συμμετρεῖν (symmetría) and can be roughly translated as "the same measure." Apart from its practical importance, symmetry was also a "measure" of beauty. We should also reflect on the symmetry in life and nature, which is determined by our interaction with the physical environment. Most small bacteria (archaebacteria) had a spherical shape because of the isotropy of primitive ocean water, where they evolved. A tree must grow vertically in the direction of sunlight, but its crown must be symmetrical in order to remain stable under the force of gravity. Is it possible to make a connection between the symmetry of the animal body and the symmetry of the water molecule?

Symmetry elements and operations are strongly related to each other but they must be well defined from the beginning. Looking in the mirror, we can see a virtual image that is exactly like us. We say the image is symmetrical to the object. The reflection of light on the mirror surface is the phenomenon that produces this symmetric image. The reflection in the mirror (mirror symmetry) is the most common meaning of symmetry. Other types of symmetry will also be discussed in this book. The reflection is the *symmetry operation* and the mirror is the plane of reflection, that is, it is the *symmetry element.*

Symmetry operations related to symmetry elements are defined by whether, after their application to a body, we can obtain its image, also called an *equivalent configuration.* This means that the obtained image is similar to the object, namely, it cannot be distinguished from the object, but it is not identical. Returning to the first example, our image in the mirror also has two arms but our right arm is the left arm of the image and our left arm is the right arm of the image. The man in front of us is equivalent to us regarding body structures. One performed a certain geometric operation (reflection, rotation, translation) on an object and during that transformation the geometric properties of that object did not change (Darvas 2012).

Looking inside the microworld, we take the simple molecule of water, which has two identical hydrogen atoms. By convention, we

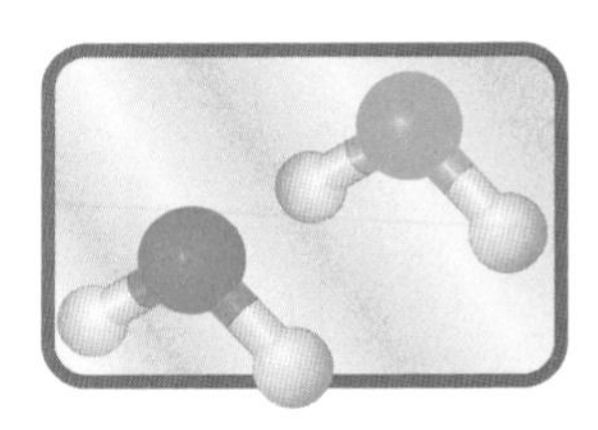

FIGURE 1.1 Water molecule "looking" in the mirror.

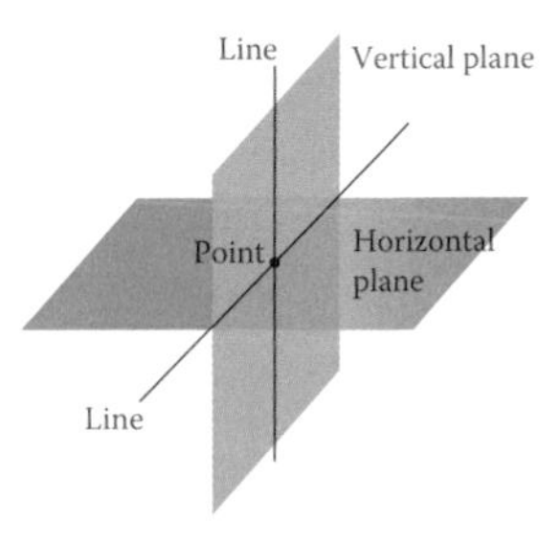

FIGURE 1.2 Symmetry elements are geometrical entities.

call them H1 and H2. The properties of the water molecule do not depend on which atom is on the left side of oxygen and which atom is on the right side. Left and right are only a convention. Looking at the image of the molecule in the mirror, atom H2 is equivalent to atom H1 and atom H1 is equivalent to atom H2 (Figure 1.1). There is no relation between the position of the mirror plane and the molecule in this example. In the following examples, we will use only mirror plane and other symmetry elements related to the molecule or the box containing the molecule. Any symmetry element will be located inside the object. The "mirror" will always be located inside the object and not outside, as in the previous example.

Symmetry elements are geometric entities such as **plane, line, and point** (Figure 1.2). Every symmetry operation is defined with respect to an element, but at the same time, every symmetry element exists only if the appropriate operation (which leaves the object unchanged) exists. Thus, we will discuss operations and elements together.

Symmetry operations are also geometric transformations, which, applied to a molecule, change only the positions of some atoms (move an atom in place of another) *without changing the size, shape, and position of the molecule.* In other words, the final state of the molecule is the same as the initial state except the labeled numbering of atoms.

There are four types of symmetry elements and related operations as shown in Table 1.1.

1.1.2 SYMMETRY PLANES AND REFLECTIONS

As we have already mentioned, the symmetry plane (mirror plane) should be chosen for the purpose of passing through the molecule, or through the box containing the molecule. Thus, the reflection in the mirror plane must transform the molecule in itself, that is, the image will be located in the same position as the original molecule. We use the symbol σ for the

TABLE 1.1 Symmetry Elements and Appropriate Symmetry Operations in Molecular Symmetry

	Symmetry Element	Element Symbol	Symmetry Operation	Operation Symbol
1	Plane, can be horizontal, vertical, dihedral	σ	Reflection	Σ
2	Center of inversion	*i*	Inversion of all atom coordinates through the center	*I*
3	Proper axis	*C*	Rotation around the axis	C_n^k
4	Improper axis	*S*	Rotation around the axis followed by reflection in the plane perpendicular to the axis	S_n^k
5			Identity	*E*

Source: Adapted from Cotton, A. F. 1990. *Chemical Applications of Group Theory*, 3rd edition. New York: John Wiley & Sons. Inc.

reflection plane and the symbol Σ for the reflection operation. In the case of a water molecule, the reflection plane could be the plane containing all three atoms (σ_v) or the secondary plane (σ'_v) containing only the oxygen atom and is perpendicular to the main plane (Figure 1.3).

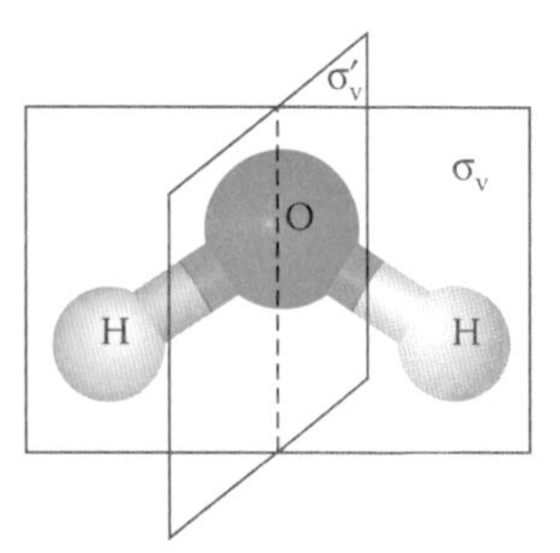

FIGURE 1.3 Reflection planes in the case of a water molecule.

There seems to be no difference between the two reflection planes. But we must remember that the first one passes through three atoms (σ_v) and the second (σ'_v) through only one atom. It is as though the first plane is "heavier" than the second plane. Thus, they do not have equal importance for the characterization of molecular symmetry.

Let us imagine a molecule with square planar geometry as represented in Figure 1.4. There are four vertical reflection planes (more exactly perpendicular to the molecular plane): two vertical planes containing three atoms and two vertical planes containing only one (central) atom. Although the reflection operation appears to be the same for all cases, there is a difference: in the first case, the reflection only moves two opposite atoms, while in the second case, the reflection moves four neighboring atoms in order to obtain the equivalent configuration. Thus, the four reflection operations form two distinct classes. Now, we have a reason to change the notation to differentiate between them. We will denote by σ_v the vertical planes containing three atoms and by σ_d the vertical planes containing one atom. The lower index *d* comes from a dihedral that, in geometry, is the angle between two planes. Finally, the square planar molecule has three distinct classes of reflection: one horizontal, two vertical, and two dihedral. Examples of square planar molecules are some transition metal complexes such as $[PtCl_4]^{2-}$ and noble gas compounds such as $[XeF_4]$.

What happens if we apply the same reflection operation twice? The last reflection brings all atoms to their original positions. The final configuration is not only equivalent but is also identical to the original configuration. The operation that leaves the configuration unchanged is called

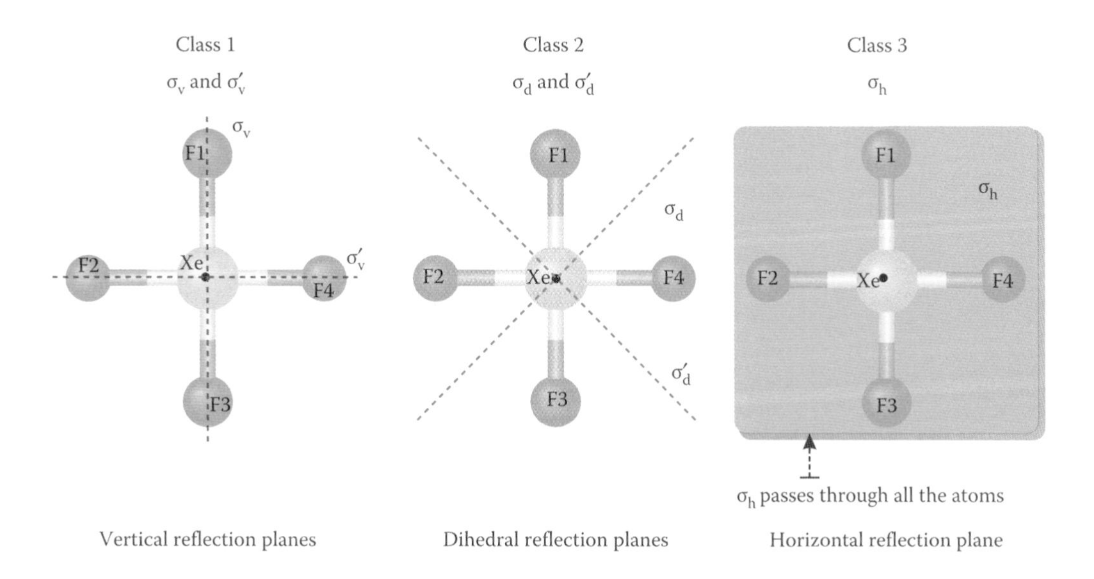

FIGURE 1.4 Reflection planes of a molecule with square planar geometry: two vertical planes (left), two dihedral planes (center), and one horizontal plane (right).

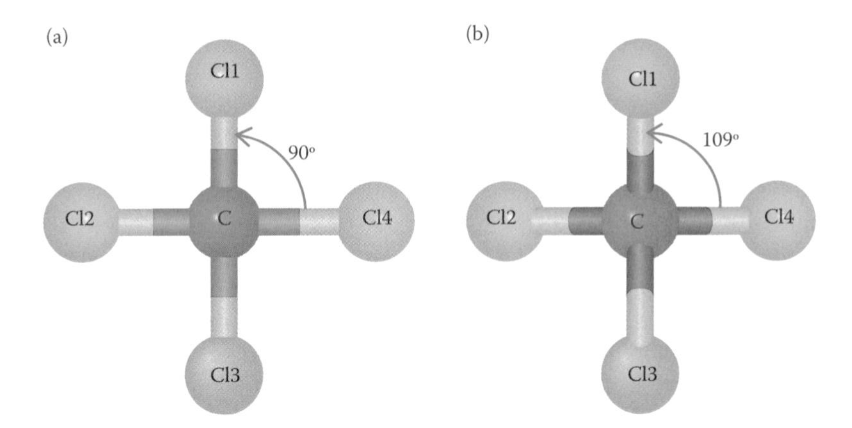

FIGURE 1.5 Electrostatic repulsion between chlorine atoms reduces the molecular symmetry by increasing the angle between the C–Cl bonds: (a) planar geometry; (b) tetrahedral geometry.

the *identity operation* and is represented by the symbol E. The identity operation is present in the symmetry of all molecules. Usually, the identity operation is the result of a combination of other symmetry operations, which brings the molecule to a configuration identical to the initial configuration. Therefore, *the symmetry plane always generates only one operation of reflection.*

$$\Sigma^n = \begin{cases} E \text{ (when } n \text{ is even)} \\ \Sigma \text{ (when } n \text{ is odd)} \end{cases} \tag{1.1}$$

There are similar molecular formulas, for example, CCl_4. What are its reflection planes? The carbon atom is central and surrounded by four chlorine atoms. What does this molecule look like? It could be as shown in Figure 1.5a. This is a flat representation of the molecule, where Cl–C–Cl bond angles appear to be 90°. But the actual value of the bond angles is around 109°. Then the real geometry is not flat but rather tetrahedral, as in Figure 1.5b. Because of the planar shape of paper, we have to imagine the three-dimensional shape of the molecule. The three-dimensional arrangement of the chlorine atoms is a direct consequence of the electrostatic repulsion of bonding electrons. In this actual case, there are only two reflection planes. Every plane passes through three atoms: the central carbon atom and two outer chlorine atoms. In the three-dimensional representation, the molecule has a reduced symmetry. We denote reflection planes without a lower index because horizontal and vertical are concepts that have no meaning now.

1.1.3 CENTER OF INVERSION AND INVERSION OPERATION

The center of inversion is a point located in the center of the body. If we attach a coordinate system where the origin lies in the center of the body, then the position of every atom will be determined by coordinates (x, y, z). Inversion changes the coordinates of a point into $(-x, -y, -z)$. Inversion is a symmetry operation if the final configuration of the molecule is equivalent to the initial one. The point related to the coordinate origin becomes the center of symmetry or the *center of inversion*. An inversion center generates only one symmetry operation. We use the symbol i for inversion center and the symbol I for inversion operation. Repeated inversion brings the molecule into a configuration identical to the initial one, which has the same effect as the identity operation E. Therefore, *the inversion center always generates only one operation of inversion.*

$$I^n = \begin{cases} E \text{ (when } n \text{ is even)} \\ I \text{ (when } n \text{ is odd)} \end{cases} \quad (1.2)$$

What is the difference between the two structures drawn in Figure 1.6? They are XeF_4 molecule in Figure 1.6a, and a pure geometrical structure, without any relation to the real molecules, in Figure 1.6b. We are looking for the inversion operation present in both structures. Also, the inversion center is located in the center of the square in both structures. In the structure in Figure 1.6a one atom is placed in the inversion center and is the only atom that is not shifted when the inversion is performed. In the structure in Figure 1.6b the inversion center still exists but has no physical substance (no atom is related to it).

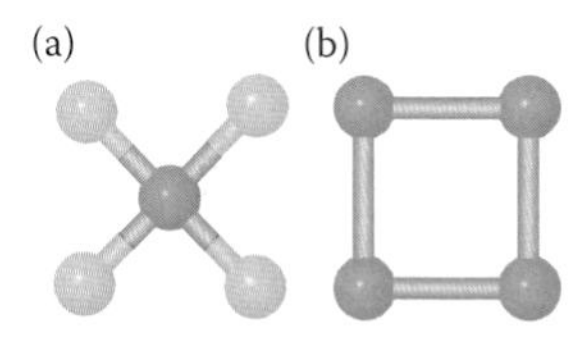

FIGURE 1.6 Two different geometrical structures (square planar) with the same symmetry. (a) The inversion center is located on the central atom and (b) the inversion center has no physical substance.

Examples of molecules with inversion center not atom-based are cyclobutadiene and benzene, as illustrated in Figure 1.7.

Could a linear molecule have an inversion center that is not atom-based? Yes, for example, C_2H_2.

1.1.4 PROPER AXES AND ROTATIONS

Let us look again at the water molecule and view it from the top (Figure 1.8). Upon rotating the molecule 180° about a vertical line (the intersection

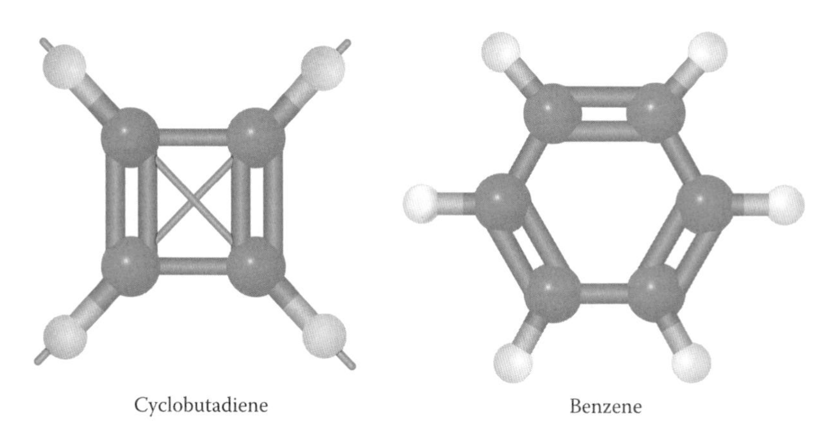

FIGURE 1.7 Examples of molecules with inversion center not based on the atom.

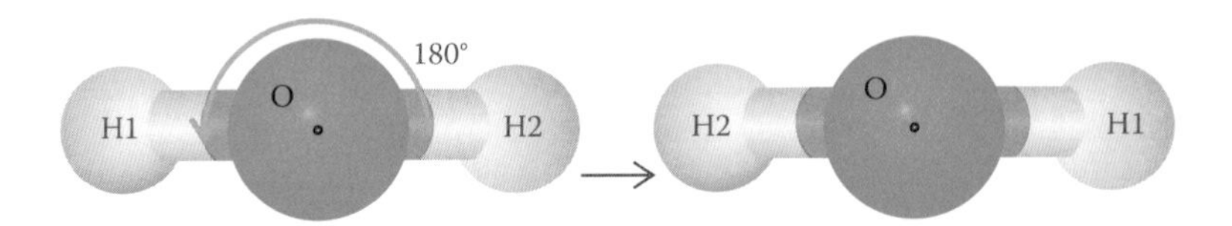

FIGURE 1.8 Top view of a water molecule. Upon rotating the molecule by 180° around a vertical line, the molecule is brought into an equivalent configuration.

line of the vertical planes σ_v and σ'_v), the molecule is brought into an equivalent configuration. The hydrogen atoms interchange their positions. This difference between the two configurations does not affect any properties of the molecule. The rotation axis passes through the oxygen atom and the rotation is the symmetry operation.

We shall use the symbol c_n for the rotation axis and the symbol C_n for the rotation operation. The subscript n shows the order of the rotation axis and rotation operation, respectively. The subscript n is the number from the denominator of the expression $2\pi/n$, which represents the value in radians of the rotation angle. In the previous example, the rotation angle is 180°, which means $2\pi/2$ radian. Thus, the rotation axis has the label c_2 and the corresponding rotation operation is C_2. If we repeat the rotation again, we bring the molecule in a position that is identical to the initial one; therefore, we can write $C_2 \cdot C_2 = C_2^2 = E$.

Let us take a more complicated case of the trigonal planar molecule, such as boron trifluoride (BF_3), sulfur trioxide (SO_3), or nitrate ion (NO^{3-}) and carbonate ion (CO_3^{2-}). The structure of any of the named molecules can be represented as in Figure 1.9 by a central atom and three outer identical atoms. The molecule can be inscribed in an equilateral triangle. A line drawn through the center of the molecule (center of the triangle) and perpendicular to the molecular plane is the principal rotation axis named c_3. Why is the order of this axis 3? It is because of the order of the three possible rotations of the BF_3 molecule around the axis c_3, as shown in Figure 1.9. After the rotation of the molecule by 120° around this axis, the final molecular configuration is equivalent to the initial one. The order of rotation is given by the denominator of the rotation angle measured in radians, that is, $2\pi/3$. Thus, the rotation label is C_3. After the rotation of the molecule by 240°, the final molecular configuration is again equivalent to the initial one. Thus, the rotation label is given by $2 \times (2\pi/3)$, that is, C_3^2, which means C_3 repeated twice. The last rotation by 360° brings the molecule into a configuration identical (not only equivalent) to the initial one. The rotation label is given by $3 \times (2\pi/3)$, that is, C_3^3, which means that C_3 is repeated thrice. Therefore, the order of the rotation axis is 3 because all possible rotations around this axis are of order 3. We note the rotation $C_3^3 = E$, so it is the identity operation. The C_3^3 rotation is not unique and can be substituted by E. The identity operation exists obviously for any molecular symmetry.

Are there other axes of rotation in the BF_3 molecule? Figure 1.10 shows the presence of three other axes of rotation. All of them are twofold axes, meaning that a 180° rotation around c'_2 axis is possible and brings the molecule into an equivalent configuration. All three c_2 axes form a

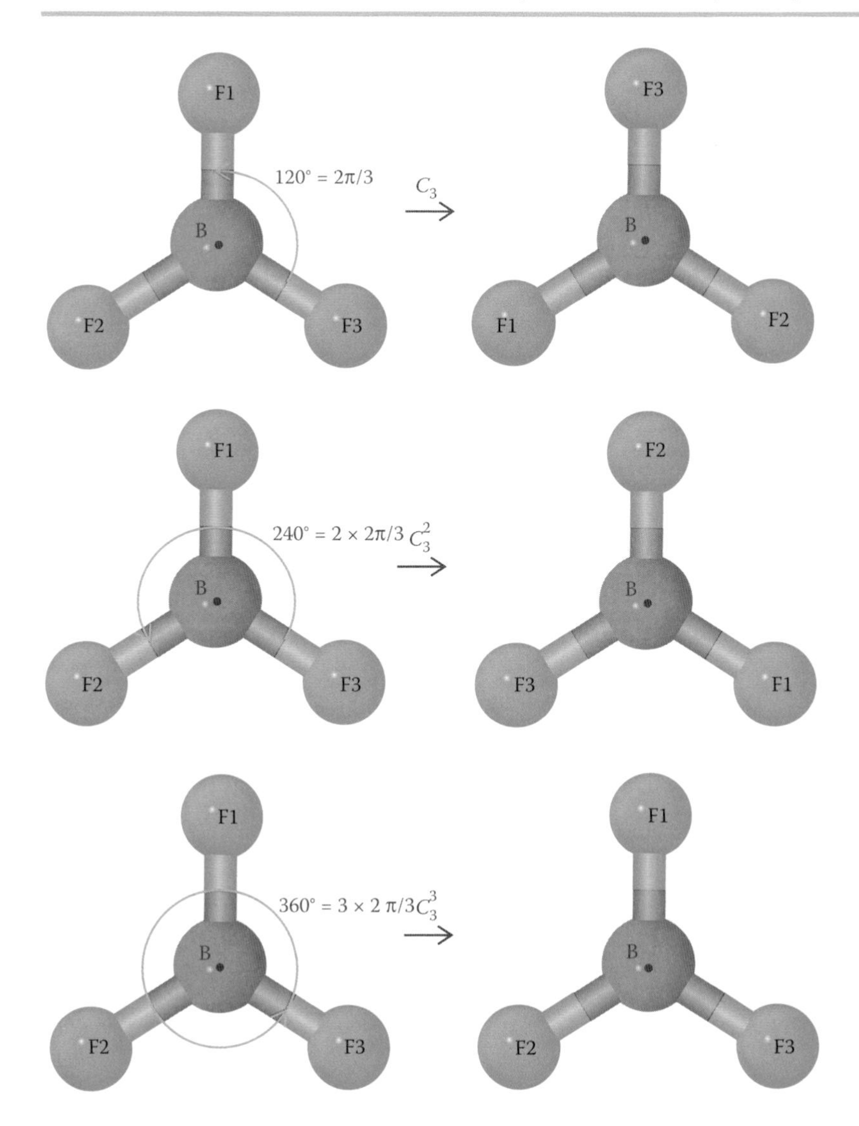

FIGURE 1.9 Three possible rotations of the BF_3 molecule around the principal axis c_3, which passes through the center of the molecule and is perpendicular to the molecular plane.

set of equivalent axes because a C_3 rotation around the c_3 principal axis generates one c_2 axis from another.

What is the order of the rotation axis when there are two types of rotation around the same axis? For the answer, we should look at the case of a square planar molecule. The presence or absence of the central atom is not important. The principal rotation axis is perpendicular to the molecular plane and passes through to the square center. It is obvious that the 90° rotation around this axis will bring the molecule into an equivalent configuration (Figure 1.11). The repeated 90° rotation will have a similar result. Therefore, the symmetry operations generated by rotation around this axis are C_4, C_4^2, C_4^3, and $C_4^4 = E$. But the 180° rotation around the same axis also brings the molecule into an equivalent configuration (Figure 1.12).

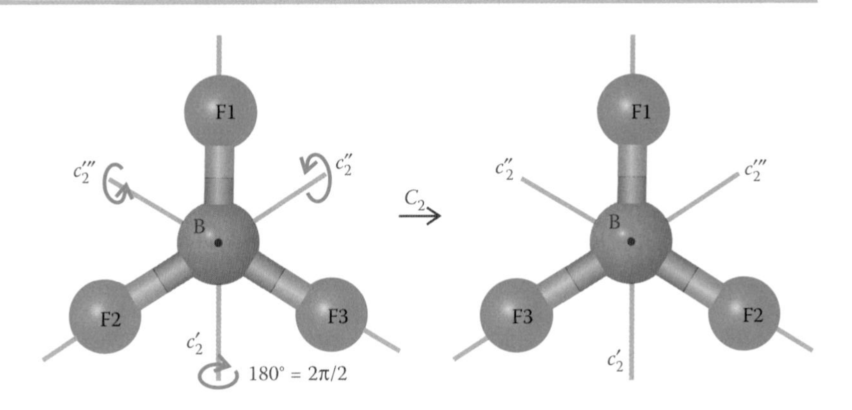

FIGURE 1.10 The trigonal planar molecule has three equivalent twofold axes perpendicular to the principal rotation axis. 180° rotation around c_2' axis brings the molecule into a configuration equivalent to the initial one.

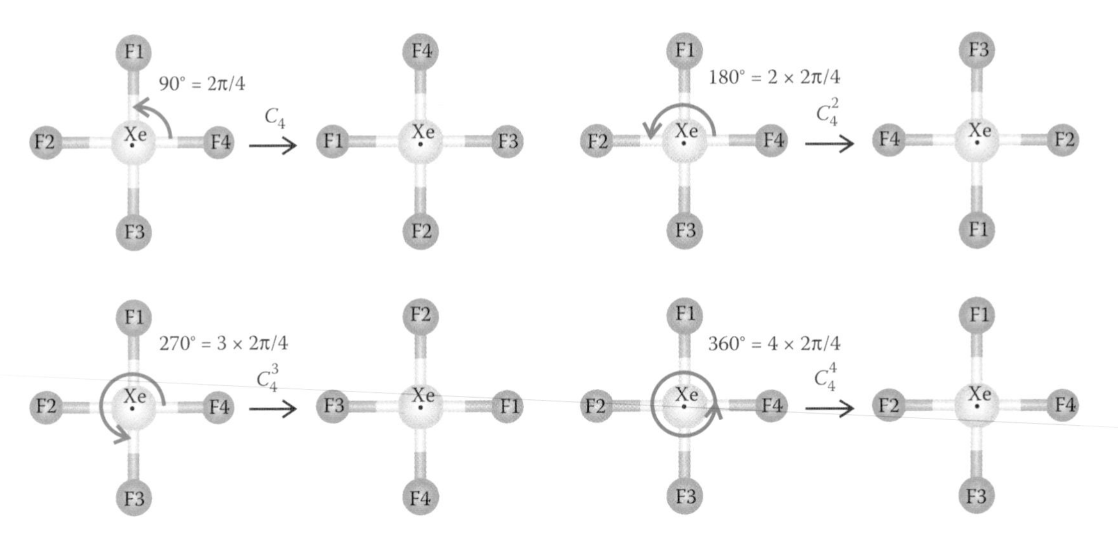

FIGURE 1.11 Any of the $m \times 90°$ rotation, where $m = 1, 2, 3, 4$, around the central axis brings the molecule into an equivalent configuration.

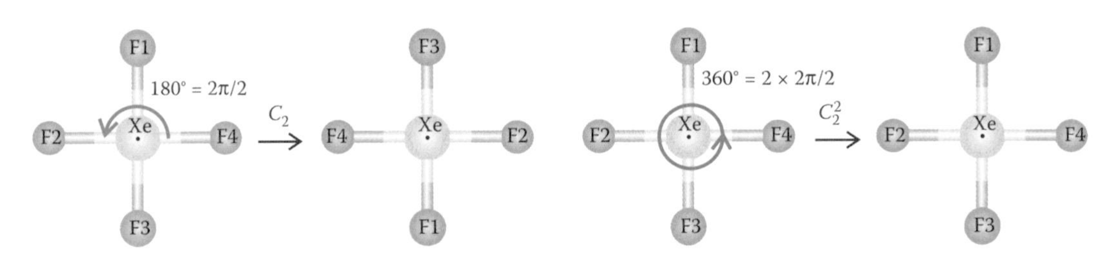

FIGURE 1.12 180° rotation around the principal axis brings the molecule into an equivalent configuration.

Thus, the generated symmetry operations are C_2 and $C_2^2 = E$. What is the order of rotation axis: 2 or 4? *The order of the rotation axis is always the highest value of n* such that the rotation $2\pi/n$ *brings the molecule into an equivalent position.* Therefore, the order of the principal rotation axis of square planar molecule is 4, that is, we may write c_4 (Figure 1.13). To conclude, the symmetry operations generated by c_4 axis are C_4, C_2, C_4^3, and E. C_4^2 is equivalent to C_2 and was replaced by the last because they have the same effect.

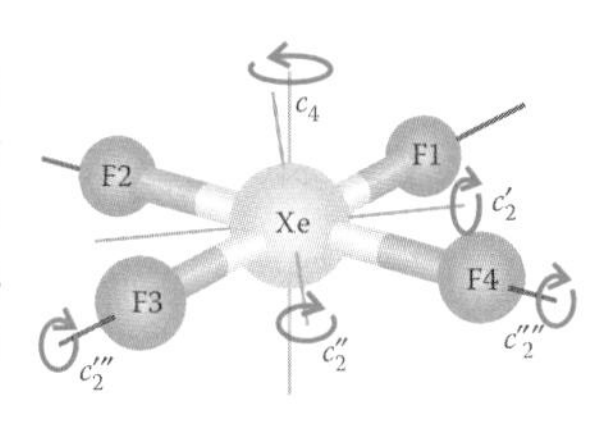

FIGURE 1.13 All possible rotation axes of the planar square molecule. c_4 is the principal rotation axis.

To complete our discussion about the rotation axis of square planar molecules, Figure 1.13 shows all the rotation axes, namely, c_4, the principal rotation axis, two c_2 rotation axes that pass through three atoms (e.g., F1–Xe–F3), and two c_2 rotation axes that pass through only the central atom. All c_2 axes lie in the molecular plane and a 180° rotation around any of them can bring the molecule into an equivalent configuration. We note that 90° rotation of the molecule around the c_4 axis changes c_2' into c_2'' and vice versa. This means that c_2' and c_2'' axes form a set of equivalent axes. From the same reasoning, it follows that c_2''' and c_2'''' axes form another set of equivalent axes.

Therefore, *an n-fold proper rotation axis generates n – 1 distinct rotations*, namely, C_n, C_n^2, $C_n^3, \ldots, C_n^{n-1}$. C_n^n rotation is always replaced by E.

A new question arises: Can finding the principal axis of rotation help us further in searching for other elements of symmetry? Or in other words, is it a way to short a search for symmetry elements? We have already seen in Figure 1.10 that one c_2 axis perpendicular to principal axis c_3 can generate two similar axes if the body is rotated around the principal axis. Figure 1.14 shows two examples in which a twofold axis is rotated around the principal axis and generates other equivalent twofold axes. Figure 1.14a, representing a regular pentagon body, shows that a twofold axis c_2 perpendicular to the fivefold axis c_5 can generate other four equivalent c_2 axes by a rotation C_5 and equivalent around the principal axis. Figure 1.14b, representing a regular heptagon body, also shows that a twofold axis c_2 perpendicular to the sevenfold axis c_7 can generate other six equivalent c_2 axes by a rotation C_7 and equivalent around the principal axis. Now, we can generalize *for any c_n principal axis (where n is odd) that if there is one c_2 axis perpendicular to it, there are in total n equivalent c_2 axes.*

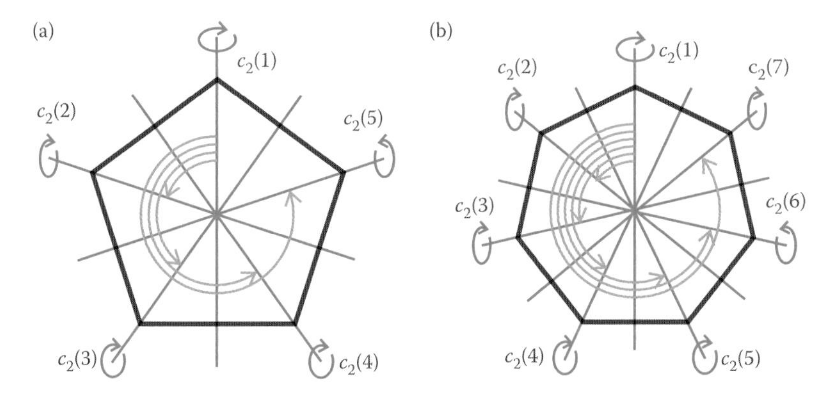

FIGURE 1.14 Rotation of twofold axis around the principal rotation axis c_n (where n is odd) can generate other equivalent twofold axes.

The results are different where n is even. As we have seen in the case illustrated in Figure 1.13, there are two sets of two c_2 equivalent axes generated by a C_4 rotation and equivalent. In case of a regular hexagon body, there are two sets of three c_2 equivalent axes generated by a C_6 rotation and equivalent.

1.1.5 IMPROPER AXES AND ROTATIONS

An *improper rotation* is the combination of a rotation about an axis and a reflection in a plane perpendicular to the axis. The order in which they are performed is not important. The final result is the same. It is recommended to perform symmetry operations in the mentioned order because it is easier to recognize the position of the rotation axis. The n-fold improper rotation axis is denoted by the symbol s_n and the corresponding improper rotation is denoted by S_n. It means the improper rotation S_n is a compound operation combining a rotation C_n followed by a reflection Σ through a plane perpendicular to the rotation axis. It results in a configuration that is indistinguishable from the original.

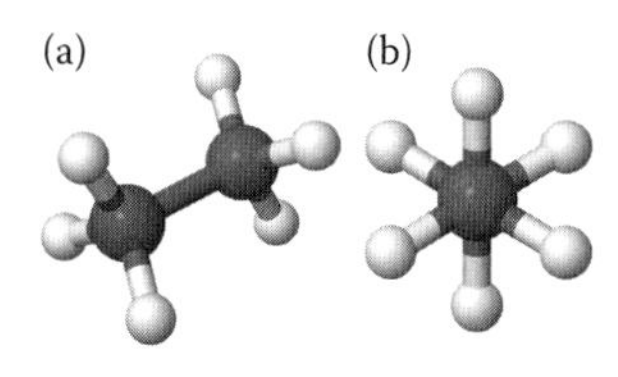

FIGURE 1.15 Molecular configuration of ethane: spatial illustration (a) and top view (b).

Common examples of molecular structures presenting improper rotation axis are methane (CH_4), ethane (C_2H_6), boron trifluoride (BF_3), and benzene (C_6H_6). Methane molecule, which is a regular tetrahedral molecule, will be discussed extensively later. Boron trifluoride and benzene are planar molecules and are not complicated.

We shall consider ethane as a first exercise. The molecular configuration of ethane is shown in Figure 1.15a. The second image (Figure 1.15b) is more suggestive as it is easy to view the improper rotation axis. This image suggests the existence of a principal proper axis of rotation c_3 but also of an improper axis s_6 on the same direction.

The symmetry operations are described in detail, step by step, in Figure 1.16. First of all, the 60° rotation (C_6) around the principal axis, which passes through the two carbon atoms, brings the molecule into an intermediate configuration nonequivalent to the initial one. Then, the

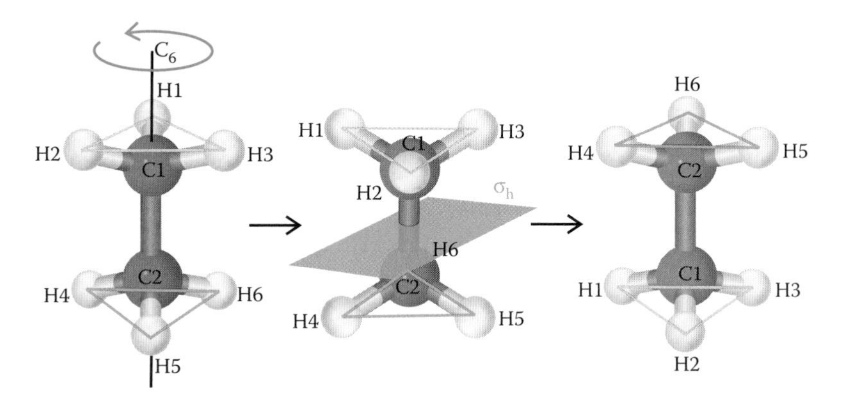

FIGURE 1.16 Two steps action of improper rotation on ethane molecule: 60° rotation around the vertical c_6 axis followed by reflection Σ through a horizontal plane σ_h.

reflection (Σ_h) through a plane (σ_h) perpendicular to the rotation axis brings the molecule into an equivalent configuration to the initial one. As we can see in Figure 1.16, all atoms (including carbon atoms) change their positions compared to the initial configuration but the molecule (as a body) appears to look the same. *Important notice: Both rotation C_6 and reflection Σ_h are not symmetry operations of the ethane molecule.* But they form together an improper rotation S_6, which is a symmetry operation of the molecule. This is a general property of the improper rotation. Reflection followed by rotation give the same result as rotation followed by reflection, including the position of atoms.

How many improper rotations are generated by an improper rotation axis? We shall start with an *improper axis of even order*. The improper rotation axis s_6 of ethane molecule generates a set of improper rotations: S_6^1, S_6^2, S_6^3, S_6^4, S_6^5, S_6^6 (see Table 1.2). *The number of improper rotations generated by an even-order improper axis s_n is n.* Do all these operations exist? We just showed that S_6 exists. The improper rotation S_6^2 means that we must perform the pair of operations $C_6^1 \cdot \Sigma_h$ twice, as you can see in Table 1.2. The reflection Σ_h carried out twice gives identity operation. The rotation C_6^1 carried out twice gives rotation C_3^1, which is still a proper rotation of the molecule. Therefore, the improper rotation S_6^2 does not exist as a distinct operation. The improper rotation S_6^3 means that we must perform the pair of operations $C_6^1 \cdot \Sigma_h$ thrice. From Table 1.2, we can see that the result is the improper rotation S_2^1. Actually, this improper rotation brings the molecule in the same configuration as inversion I, which is a simpler symmetry operation and still exists as an irreducible operation. The results of both symmetry operations can be seen in Figure 1.17. The operation S_6^4 can be reduced to a proper rotation C_3^2. The improper rotation S_6^5 remains as a distinct operation while S_6^6

TABLE 1.2 Decomposition of Improper Rotations Set Generated by Improper Axis s_6 in Simpler Operations

Improper Rotation	Decomposition in Simpler Operations	Final Result
S_6^1	$S_6^1 = C_6^1 \cdot \Sigma_h$	S_6^1
S_6^2	$S_6^2 = S_6^1 \cdot S_6^1 = C_6^1 \cdot \Sigma_h \cdot C_6^1 \cdot \Sigma_h = C_6^2 \cdot \Sigma_h^2 = C_6^2 \cdot E$	C_3^1
S_6^3	$S_6^3 = C_6^1 \cdot \Sigma_h \cdot C_6^1 \cdot \Sigma_h \cdot C_6^1 \cdot \Sigma_h = C_6^3 \cdot \Sigma_h = C_2^1 \cdot \Sigma_h = S_2^1$	I
S_6^4	$S_6^4 = C_6^1 \cdot \Sigma_h \cdot C_6^1 \cdot \Sigma_h \cdot C_6^1 \cdot \Sigma_h \cdot C_6^1 \cdot \Sigma_h = C_6^4 \cdot \Sigma_h^4 = C_6^4$	C_3^2
S_6^5	$S_6^5 = C_6^1 \cdot \Sigma_h \cdot C_6^1 \cdot \Sigma_h \cdot C_6^1 \cdot \Sigma_h \cdot C_6^1 \cdot \Sigma_h \cdot C_6^1 \cdot \Sigma_h = C_6^5 \cdot \Sigma_h^5$	S_6^5
	$= C_6^5 \cdot \Sigma_h$	
S_6^6	$S_6^6 = C_6^1 \cdot \Sigma_h \cdot C_6^1 \cdot \Sigma_h \cdot C_6^1 \cdot \Sigma_h \cdot C_6^1 \cdot \Sigma_h \cdot C_6^1 \cdot \Sigma_h \cdot C_6^1 \cdot \Sigma_h$	E
	$= C_6^6 \cdot \Sigma_h^6 = E \cdot E$	
S_6^7	$S_6^7 = C_6^7 \cdot \Sigma_h^7 = C_6^1 \cdot \Sigma_h^1$	S_6^1
S_6^8	$S_6^8 = C_6^8 \cdot \Sigma_h^8 = C_6^2 \cdot E$	C_3^1

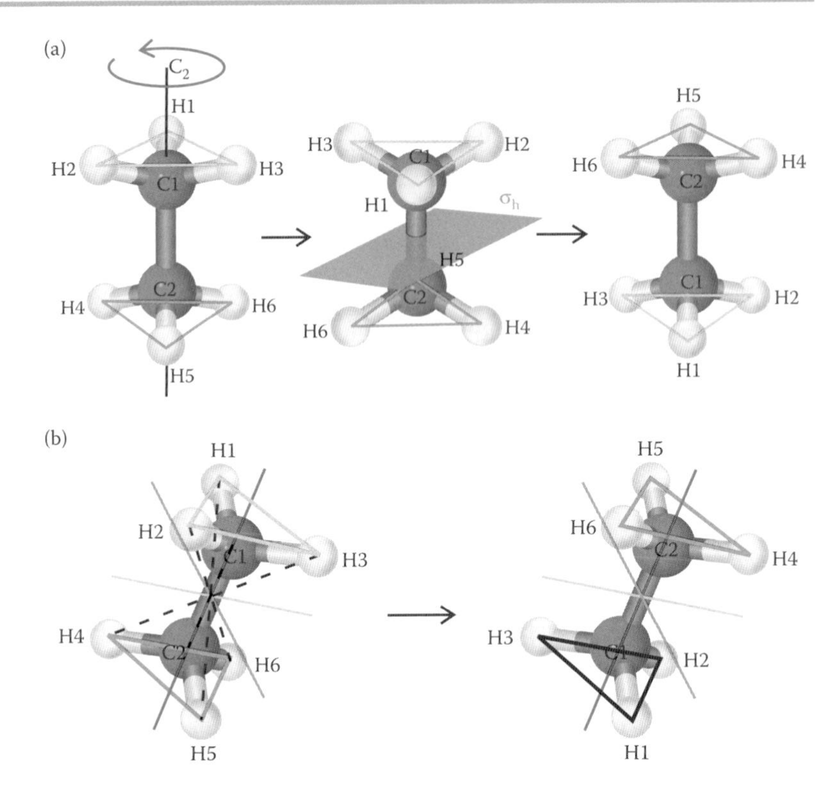

FIGURE 1.17 The action of improper rotation S_2^1 on ethane molecule gives the same configuration (a) as that directly given by inversion I (b). The continuous lines are coordinate axes. The dotted lines show the action of inversion on related atoms.

is the identity operation E. Therefore, the set of improper rotations generated by improper axis s_6 can be restricted to only two distinct improper rotations: S_6^1 and S_6^5. The operations C_3^1 and C_3^2 are generated by the proper rotation axis c_3. We can conclude that an improper axis s_n of even order requires the existence of a proper rotation axis $c_n/2$.

How many distinct improper rotations will be generated by an improper axis of odd order? Let us examine Table 1.3, where we reduced the improper rotations generated by improper axis s_5 to simpler operations as in the previous table. First, we must observe that *the number of improper rotations generated by an odd-order improper axis s_n is 2n!* In our particular case, the repetition of sequence starts at S_5^{11}. Second, from 10 generated operations, only five are fifth-order improper rotations (including E) and five are pure rotations or reflections.

We can finally summarize:

$$(1.3)\quad S_n^m = \begin{cases} C_n^m = C_{\frac{n}{2}}^{\frac{m}{2}} & \text{(when } m \text{ is even)} \\ \Sigma C_n^m & \text{(when } m \text{ is odd)} \\ I \text{ for } n/2 \text{ odd or } C_2 \text{ for } n/2 \text{ even (when } m = \frac{n}{2}) & \\ E & \text{(when } m = n) \end{cases} \quad \text{for } n \text{ even}$$

TABLE 1.3 Operations Generated by s_5 Improper Axis and Their Reduction to Simpler Operations

Improper Rotation	Decomposition in Simpler Operations	Final Result
S_5^1	$S_5^1 = C_5^1 \cdot \Sigma_h$	S_5^1
S_5^2	$S_5^2 = S_1^1 \cdot S_5^1 = C_5^1 \cdot \Sigma_h \cdot C_5^1 \cdot \Sigma_h = C_5^2 \cdot \Sigma_h^2 = C_5^2 \cdot E$	C_5^2
S_5^3	$S_5^3 = C_1^1 \cdot \Sigma_h \cdot C_5^1 \cdot \Sigma_h \cdot C_5^1 \cdot \Sigma_h = C_5^3 \cdot \Sigma_h = S_5^3$	S_5^3
S_5^4	$S_5^4 = C_5^1 \cdot \Sigma_h \cdot C_5^1 \cdot \Sigma_h \cdot C_5^1 \cdot \Sigma_h \cdot C_5^1 \cdot \Sigma_h = C_5^4 \cdot \Sigma_h^4 = C_5^4$	C_5^4
S_5^5	$S_5^5 = C_5^1 \cdot \Sigma_h \cdot C_5^1 \cdot \Sigma_h \cdot C_5^1 \cdot \Sigma_h \cdot C_5^1 \cdot \Sigma_h \cdot C_5^1 \cdot \Sigma_h$ $= C_5^5 \cdot \Sigma_h^5 = E \cdot \Sigma_h$	Σ_h
S_5^6	$S_5^6 = C_5^1 \cdot \Sigma_h \cdot C_5^1 \cdot \Sigma_h \cdot C_5^1 \cdot \Sigma_h \cdot C_5^1 \cdot \Sigma_h \cdot C_5^1 \cdot \Sigma_h \cdot C_5^1 \cdot \Sigma_h$ $= C_5^6 \cdot \Sigma_h^6 = C_5^1 \cdot E$	C_5^1
S_5^7	$S_5^7 = C_5^7 \cdot \Sigma_h^7 = C_5^2 \cdot \Sigma_h$	S_5^2
S_5^8	$S_5^8 = C_5^8 \cdot \Sigma_h^8 = C_5^3 \cdot E$	C_5^3
S_5^9	$S_5^9 = C_5^9 \cdot \Sigma_h^9 = C_5^4 \cdot \Sigma_h$	S_5^4
S_5^{10}	$S_5^{10} = C_5^{10} \cdot \Sigma_h^{10} = E \cdot E$	E
S_5^{11}	$S_5^{11} = C_5^{11} \cdot \Sigma_h^{11} = C_5^1 \cdot \Sigma_h$	S_5^1
S_5^{12}	$S_5^{12} = C_5^{12} \cdot \Sigma_h^{12} = C_5^2 \cdot E$	C_5^2
S_5^{13}	$S_5^{13} = C_5^{13} \cdot \Sigma_h^{13} = C_5^3 \cdot \Sigma_h$	S_5^3

$$(1.4)\quad S_n^m = \begin{cases} S_n^m \text{ for } m < n \text{ or } C_n^{m-n} \text{ for } n < m < 2n \text{ (when } m \text{ is odd)} \\ C_n^m \text{ for } m < n \text{ or } C_n^{m-n} \text{ for } n < m < 2n \text{ (when } m \text{ is even)} \\ S_n^n = \Sigma_h \\ S_n^{2n} = E \end{cases}$$

for n odd

1.2 POINT GROUPS AND MOLECULAR SYMMETRY

1.2.1 POINT GROUPS

A mathematical group is a *collection of elements* (whatever their nature) interrelated by an *operation*. Examples of elements that can form a group are numbers, functions, vectors, matrices, or symmetry operations.

In order to define a group, we shall use the operation (seen as group operation) called "*product*," which is not the multiplication of arithmetic. In case of molecular symmetry, the product could be named combination of two symmetry operations and it means to apply symmetry operations in a certain order. Therefore, if we write the product as $B \cdot A$, it means to apply to the molecule first the symmetry operation A and then the operation B.

In order to form a group, a collection of elements must satisfy four criteria: *closure, associativity, existence of identity element, and inverses for each element.*

We are now able to make a list of symmetry operations for a given molecule. If the set of the symmetry operations is complete, these operations form a mathematical group. Therefore, the symmetry operations satisfy the previous criteria:

1. *Closure.* Any combination of two symmetry operations of the set must generate another operation of the set. We can write it as $B \cdot A = C$, where A, B, and C are members of the group.
2. *Identity.* It must be one operation that leaves any other operation of the set unchanged. We can write it as $E \cdot A = A$, where E is the identity operation and A is any other group operation.
3. *Associativity.* The symmetry operations of the set must obey the associative law. We can write as $A \cdot (B \cdot C) = (A \cdot B) \cdot C$, where A, B, and C are members of the group.
4. *Inverses.* Any operation must have an inverse operation among the elements of the set. We can write as $A \cdot A^{-1} = E$, where A and $B = A^{-1}$ are members of the group.

For some, but not for all groups, a supplementary condition for product is to commute, that is, $B \cdot A = A \cdot B$.

A symmetry group is called a *point group if there is a point that remains fixed during all symmetry operations,* because *all the symmetry elements* (points, lines, and planes) *will intersect at this single point.*

The point should not have physical substance but it must be inside the virtual box including the molecule. In some cases, the fixed point is located at a central atom of a molecule as in the studied cases of BF_3, XeF_4, H_2O, and CH_4. If the number of symmetry operation is finite, the group is called *finite group.* The number of elements gives the *order* of the group. There are also two *infinite groups.*

1.2.2 PRODUCT OF SYMMETRY OPERATIONS

In the fulfillment of the conditions for the existence of the group, we shall discuss in detail the product of symmetry operations. It is essential to do it before checking whether the operations meet the requirements of the group because the product is involved in all requirements.

If we find two symmetry operations of a molecule, we shall always find the third operation as a product of these operations. In the following, we present two examples.

The existence of two twofold axes perpendicular to each other implies the existence of the third twofold axis perpendicular to both axes. Figure 1.18 shows the action of rotation $C_2(x)$, that is, 180° rotation around the Ox axis, followed by rotation $C_2(y)$, that is, 180° rotation around the Oy axis,

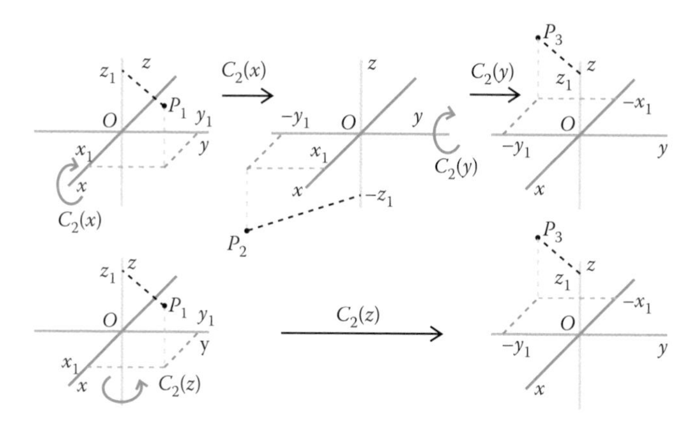

FIGURE 1.18 The product of rotations $C_2(y) \cdot C_2(x)$ has the same result as the rotation $C_2(z)$.

bring the point P_1 of coordinates (x_1, y_1, z_1) into final position P_3 of coordinates $(-x_1, -y_1, z_1)$ exactly as does the rotation $C_2(z)$. Therefore, we can write $C_2(y) \cdot C_2(x) = C_2(z)$. We note here that the result of twofold rotation does not depend on the rotation direction. It is not the case of other order rotations.

The existence of a fourfold axis and a plane containing this axis also implies the existence of a second reflection plane at an angle of 45° to the first one. Figure 1.19 illustrates the action of rotation $C_4(z)$ in a counterclockwise direction, followed by reflection $\Sigma_v(xz)$, bringing the point P_1 of coordinates (x_1, y_1, z_1) into final position P_3 of coordinates $(-y_1, -x_1, z_1)$ as reflection Σ_d can do. Therefore, we can write $\Sigma_v(xz) \cdot C_4(z) = \Sigma_d(x, -y)$.

How can we study the result of a symmetry operation in a manner simpler than drawing three-dimensional illustrations? We can use a mathematical approach to write the result of the product previously described in Figure 1.18. We can simply write the shift of general point P_1 by the product of two twofold rotations as follows:

$$C_2(y) \cdot C_2(x)[x_1, y_1, z_1] = C_2(y)[x_1, -y_1, -z_1] = [-x_1, -y_1, +z_1] \tag{1.5}$$

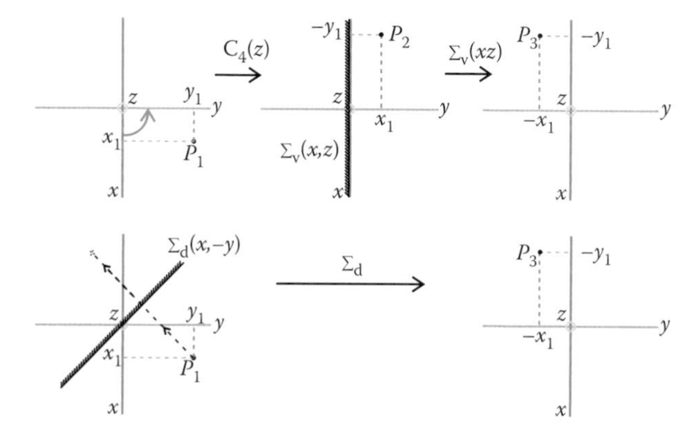

FIGURE 1.19 The product of operations $C_4(z)$ and $\Sigma_v(xz)$ has the same result as the reflection $\Sigma_d(x, -y)$.

The rotation $C_2(x)$ leaves the x coordinate unchanged and changes the sign of the y and z coordinates. The rotation $C_2(y)$ leaves the y coordinate unchanged and changes the sign of the x and z coordinates and the rotation $C_2(z)$ leaves the z coordinate unchanged and changes the sign of the x and y coordinates.

The effect of rotation $C_4(z)$ is to change the x axis into the y axis and vice versa. A supplementary condition is to take into account the effect of the rotation direction. In the case described in Figure 1.19, the counterclockwise rotation changes the sign of the x coordinate. Thus, the effect of the rotation $C_4(z)$ can be written as

$$C_4(z)[x_1,y_1,z_1] = [-y_1,x_1,+z_1] \tag{1.6}$$

In a similar way, we can write the shift produced by the reflection in the vertical plane containing the vertical axis z and x axis as in the following relationship:

$$\Sigma_v(xz)[x_1,y_1,z_1] = [x_1,-y_1,+z_1] \tag{1.7}$$

Concerning the dihedral plane containing the vertical axis z and the bisector of angle between the horizontal axes $+x$ and $-y$, the coordinate axes reflect into it:

$$\Sigma_d[x_1,y_1,z_1] = [-y_1,-x_1,+z_1] \tag{1.8}$$

Thus, the dihedral plane Σ_d leaves the z coordinate unchanged and interchanges the x and y coordinates changing also their sign. Therefore, the example illustrated in Figure 1.19 can be written as

$$\Sigma_v(xz)\cdot C_4(z)[x_1,y_1,z_1] = \Sigma_v(xz)[-y_1,x_1,z_1] = [-y_1,-x_1,z_1] = \Sigma_d(x,-y)$$

We continue in this manner of searching the effect of symmetry operations. We shall now study whether the product of two operations commutes. We take as an example the product of vertical fourfold rotation C_4 and twofold rotation C_2 around the x axis, which passes through atoms XeF_1F_3 for molecule XeF_4 (see Figure 1.13).

$$C_2(x)\cdot C_4(z)[x_1,y_1,z_1] = C_2(x)[-y_1,x_1,z_1] = [y_1,x_1,-z_1] \tag{1.9}$$

and

$$C_4(z)\cdot C_2(x)\cdot[x_1,y_1,z_1] = C_4(z)[x_1,-y_1,-z_1] = [y_1,x_1,-z_1] \tag{1.10}$$

The results are sensitive to the rotation direction of C_4. We chose the trigonometric (counterclockwise) direction.

In the examples, it was evident how to obtain the coordinates transformation produced either by a reflection through a plane containing two coordinate axes or by a twofold rotation (180°) around one coordinate axis.

The transformation produced by a fourfold rotation around one coordinate axis was a little bit more complicated but a careful inspection can give correct results. The inversion operation is also easy to apply, only by changing the sign of all coordinates. But the operation could be more complicated, such as threefold or fivefold rotation or more and reflections in planes other than those determined by coordinate axes. Therefore, we need a general formula of transformation, which will be studied later.

1.2.3 MOLECULAR SYMMETRY OPERATIONS AS GROUP

We should know how to find symmetry operations of a given molecule. We also know how to make a product of them and also how to find new operations that could exist. We must now verify whether the list of symmetry operations forms a group. This means to check that the list satisfies the four mentioned criteria.

We take as an example the allene molecule illustrated in Figure 1.20. The main proper rotation axis is a two-fold one c_2 that passes through three carbon atoms. Observing the molecule from the top, the presence of an improper axis c_4 along with c_2 is to be noticed. We can also see two vertical planes perpendicular to each other. Every plane contains three carbon atoms and two hydrogen atoms. Thus we can build Table 1.4, where the second row cells containing known operations are white while the

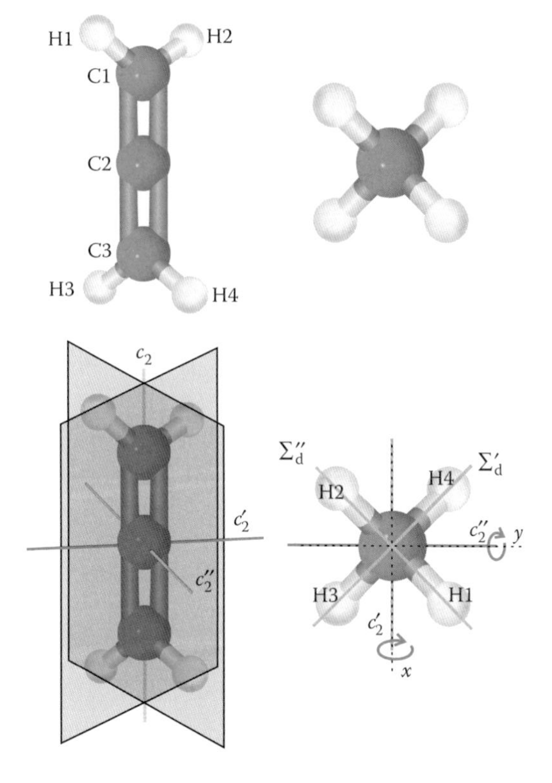

FIGURE 1.20 Configuration of allene molecule and its symmetry elements: front view (left) and top view (right).

TABLE 1.4 Set of Symmetry Operations of Allene Molecule

Symmetry Elements	C_2		σ'_d		σ''_d		S_4				C'_2		C''_2	
Symmetry operations	C_2^1	C_2^2	Σ'_d	Σ'^2_d	Σ''_d	Σ''^2_d	S_4^1	S_4^2	S_4^3	S_4^4	C'_2	C'^2_2	C''_2	C''^2_2
Irreducible operations	C_2^1	E	Σ'_d	E	Σ''_d	E	S_4^1	C_2^1	S_4^3	E	C'_2	E	C''_2	E
Point group	$E, C_2^1, \Sigma'_d, \Sigma''_d, S_4^1, S_4^3, C'_2, C''_2$													

contents of gray cells are to be discovered. It is not easy to see other two twofold proper rotation axes perpendicular to the principal rotation axis. Both axes pass only through the central carbon atom.

Is this set of operations a point group? A few operations will be reduced to identity operation. Then, we apply the product of known operations to discover new operations. We will exemplify this by calculating the product of principal rotation and all other operations.

The product of C_2^1 (about z axis) and Σ'_d (the plane passes through all carbon atoms and hydrogen atoms 3 and 4) is as follows:

$$\Sigma'_d \cdot C_2(z)[x_1,y_1,z_1] = \Sigma(3,4)[-x_1,-y_1,z_1] = [x_1,y_1,+z_1] \tag{1.11}$$

where we denoted the reflection by $\Sigma(3,4)$. This leaves the z coordinate unchanged, and interchanges the x and y coordinates (i.e., $x \leftrightarrow -y$ and $y \leftrightarrow -x$).

This is equivalent to apply one reflection Σ''_d through the plane that passes through all carbon atoms and hydrogen atoms 1 and 2, where reflection $\Sigma(1,2)$ leaves the z coordinate unchanged, and interchanges the x and y coordinates (i.e., $x \leftrightarrow y$ and $y \leftrightarrow x$):

$$\Sigma(1,2)[x_1,y_1,z_1] = [x_1,y_1,+z_1] \tag{1.12}$$

The product of C_2^1 and Σ''_d is

$$\begin{aligned}\Sigma''_d \cdot C_2(z)[x_1,y_1,z_1] &= \Sigma(1,2)[-x_1,-y_1,z_1] = [-x_1,-y_1,+z_1] \\ &= \Sigma(3,4) = \Sigma'_d\end{aligned} \tag{1.13}$$

In a similar manner, we can check the product between the principal twofold rotation and the fourfold improper rotation:

$$\begin{aligned}S_4^1(z) \cdot C_2(z)[x_1,y_1,z_1] &= S_4^1(z)[-x_1,-y_1,z_1] = \Sigma_h(xy)C_4^1(z)[-x_1,-y_1,z_1] \\ &= \Sigma_h(xy)[y_1,-x_1,+z_1] = [y_1,-x_1,-z_1]\end{aligned} \tag{1.14}$$

The result is the same as given by the improper rotation S_4^3:

$$\begin{aligned}S_4^3(z)[x_1,y_1,z_1] &= \Sigma_h(xy)C_4^3(z)[x_1,y_1,z_1] \\ &= \Sigma_h(xy)[y_1,-x_1,+z_1] = [y_1,-x_1,-z_1]\end{aligned} \tag{1.15}$$

The product of C_2^1 and S_4^3 is

$$S_4^3(z) \cdot C_2(z)[x_1, y_1, z_1] = \Sigma_h(xy)C_4^3(z)[-x_1, -y_1, z_1]$$

$$= \Sigma_h(xy)[-y_1, x_1, +z_1] = [-y_1, x_1, -z_1] = S_4^1(z) \quad (1.16)$$

The twofold rotations perpendicular to the principal rotation are not easy to be seen but also impossible to be generated by the product of other operations. They will generate each other by reporting to any known operation:

$$C_2'(x) \cdot C_2(z)[x_1, y_1, z_1] = C_2'(x)[-x_1, -y_1, z_1] = [-x_1, y_1, -z_1] \quad (1.17)$$

The result is the same as given by applying only C_2'':

$$C_2''(y)[x_1, y_1, z_1] = [-x_1, y_1, -z_1] \quad (1.18)$$

And the product of C_2^1 and C_2'':

$$C_2''(y) \cdot C_2(z)[x_1, y_1, z_1] = C_2''(y)[-x_1, -y_1, z_1] = [x_1, -y_1, -z_1] \quad (1.19)$$

The result is the same as given by applying C_2':

$$C_2'(x)[x_1, y_1, z_1] = [x_1, -y_1, -z_1] \quad (1.20)$$

Using the new operations, we shall complete Table 1.5 (the multiplication table of symmetry operations of the allene molecule). From the multiplication table, we can see that the list of symmetry operations satisfies the four conditions (closure, associativity, identity presence, and inverses

TABLE 1.5 Multiplication Table of Allene Point Group

Right \ Left	E	$C_2^1(z)$	$C_2'(x)$	$C_2''(y)$	Σ_d'	Σ_d''	S_4^1	S_4^3
E	E	$C_2^1(z)$	$C_2'(x)$	$C_2''(y)$	Σ_d'	Σ_d''	S_4^1	S_4^3
$C_2^1(z)$	$C_2^1(z)$	E	$C_2''(y)$	$C_2'(x)$	Σ_d''	Σ_d'	S_4^3	S_4^1
$C_2'(x)$	$C_2'(x)$	$C_2''(y)$	E	$C_2^1(z)$	S_4^3	S_4^1	Σ_d''	Σ_d'
$C_2''(y)$	$C_2''(y)$	$C_2'(x)$	$C_2^1(z)$	E	S_4^1	S_4^3	Σ_d'	Σ_d''
Σ_d'	Σ_d'	Σ_d''	S_4^3	S_4^1	E	$C_2^1(z)$	$C_2'(x)$	$C_2''(y)$
Σ_d''	Σ_d''	Σ_d'	S_4^1	S_4^3	$C_2^1(z)$	E	$C_2''(y)$	$C_2'(x)$
S_4^1	S_4^1	S_4^3	Σ_d''	Σ_d'	$C_2'(x)$	$C_2''(y)$	$C_2^1(z)$	E
S_4^3	S_4^3	S_4^1	Σ_d'	Σ_d''	$C_2''(y)$	$C_2'(x)$	E	$C_2^1(z)$

presence) and therefore form a group. Attention to the rule for the order of multiplication: in the product $B \cdot A$, A is the right term and will be found in the row, and B is the left term and will be found in the column.

In this manner, we have also completed the second column of the multiplication table. Then, we check if these operations also obey the commutative condition. The first example was the product of C_2^1 and Σ_d' (Equation 1.5) and the result of Equation 1.5, that is, $[x_1, y_1, +z_1]$, is the same as what is obtained in reverse order:

$$(1.21) \quad C_2(z) \cdot \Sigma(3,4)[x_1, y_1, z_1] = C_2(z)[-x_1, -y_1, z_1] = [x_1, y_1, +z_1]$$

The commutativity helps us to complete the second row of the multiplication table.

The reader should exercise all other products by making the effort to manipulate symmetry operations and compare to the results given as follows.

$$C_2''(y) \cdot C_2'(x)[x_1, y_1, z_1] = C_2''(y)[x_1, -y_1, -z_1] = [-x_1, -y_1, z_1] \equiv C_2(z)[x_1, y_1, z_1]$$

$$C_2'(x) \cdot C_2''(y)[x_1, y_1, z_1] = C_2'(x)[-x_1, y_1, -z_1] = [-x_1, -y_1, z_1] \equiv C_2(z)[x_1, y_1, z_1]$$

$$\Sigma_d' \cdot C_2'(x)[x_1, y_1, z_1] = \Sigma(3,4)[x_1, -y_1, -z_1] = [y_1, -x_1, -z_1] \equiv S_4^3(z)[x_1, y_1, z_1]$$

$$\Sigma_d'' \cdot C_2'(x)[x_1, y_1, z_1] = \Sigma(1,2)[x_1, -y_1, -z_1] = [-y_1, x_1, -z_1] \equiv S_4^1(z)[x_1, y_1, z_1]$$

$$\begin{aligned} S_4^1(z) \cdot C_2'(x)[x_1, y_1, z_1] &= S_4^1(z)[x_1, -y_1, -z_1] = \Sigma_h(xy) C_4^1(z)[x_1, -y_1, -z_1] \\ &= \Sigma_h(xy)[y_1, x_1, -z_1] = [y_1, x_1, z_1] = \Sigma(1,2)[x_1, y_1, z_1] \end{aligned}$$

$$\begin{aligned} S_4^3(z) \cdot C_2'(x)[x_1, y_1, z_1] &= S_4^3(z)[x_1, -y_1, -z_1] = \Sigma_h(xy) C_4^3(z) \\ [x_1, -y_1, -z_1] &= \Sigma_h(xy)[-y_1, -x_1, -z_1] = [-y_1, -x_1, z_1] = \Sigma(3,4)[x_1, y_1, z_1] \end{aligned}$$

$$\Sigma_d' \cdot C_2''(y)[x_1, y_1, z_1] = \Sigma(3,4)[-x_1, y_1, -z_1] = [-y_1, x_1, -z_1] \equiv S_4^1(z)[x_1, y_1, z_1]$$

$$\Sigma_d'' \cdot C_2''(y)[x_1, y_1, z_1] = \Sigma(1,2)[-x_1, y_1, -z_1] = [y_1, -x_1, -z_1] \equiv S_4^3(z)[x_1, y_1, z_1]$$

$$\begin{aligned} S_4^1(z) \cdot C_2''(y)[x_1, y_1, z_1] &= S_4^1(z)[-x_1, y_1, -z_1] = \Sigma_h(xy) C_4^1(z)[-x_1, y_1, -z_1] \\ &= \Sigma_h(xy)[-y_1, -x_1, -z_1] = [-y_1, -x_1, z_1] \\ &= \Sigma(3,4)[x_1, y_1, z_1] \end{aligned}$$

$$\begin{aligned} S_4^3(z) \cdot C_2''(y)[x_1, y_1, z_1] &= S_4^3(z)[-x_1, y_1, -z_1] = \Sigma_h(xy) C_4^3(z)[-x_1, y_1, -z_1] \\ &= \Sigma_h(xy)[y_1, x_1, -z_1] = [y_1, x_1, z_1] = \Sigma(1,2)[x_1, y_1, z_1] \end{aligned}$$

$$\Sigma_d'' \cdot \Sigma_d'[x_1, y_1, z_1] = \Sigma(1,2)[-y_1, -x_1, z_1] = [-x_1, -y_1, z_1] \equiv C_2^1(z)[x_1, y_1, z_1]$$

$$S_4^1(z) \cdot \Sigma_d'[x_1, y_1, z_1] = S_4^1(z)[-y_1, -x_1, z_1] = \Sigma_h(xy)C_4^1(z)[-y_1, -x_1, z_1]$$

$$= \Sigma_h(xy)[x_1, -y_1, z_1] = [x_1, -y_1, -z_1] = C_2'(x)[x_1, y_1, z_1]$$

$$S_4^3(z) \cdot \Sigma_d'[x_1, y_1, z_1] = S_4^3(z)[-y_1, -x_1, z_1] = \Sigma_h(xy)C_4^3(z)[-y_1, -x_1, z_1]$$

$$= \Sigma_h(xy)[-x_1, y_1, z_1] = [-x_1, y_1, -z_1] = C_2''(y)[x_1, y_1, z_1]$$

$$S_4^1(z) \cdot \Sigma_d''(1,2)[x_1, y_1, z_1] = S_4^1(z)[y_1, x_1, z_1] = \Sigma_h(xy)C_4^1(z)[y_1, x_1, z_1]$$

$$= \Sigma_h(xy)[-x_1, y_1, z_1] = [-x_1, y_1, -z_1]$$

$$= C_2''(y)[x_1, y_1, z_1]$$

$$S_4^3(z) \cdot \Sigma_d''(1,2)[x_1, y_1, z_1] = S_4^3(z)[y_1, x_1, z_1] = \Sigma_h(xy)C_4^3(z)[y_1, x_1, z_1]$$

$$= \Sigma_h(xy)[x_1, -y_1, z_1] = [x_1, -y_1, -z_1]$$

$$= C_2'(x)[x_1, y_1, z_1]$$

$$S_4^1(z) \cdot S_4^1(z)[x_1, y_1, z_1] = S_4^1(z) \cdot \Sigma_h(xy)C_4^1(z)[x_1, y_1, z_1]$$

$$= \Sigma_h(xy)C_4^1(z)\Sigma_h(xy)[-y_1, x_1, z_1]$$

$$= \Sigma_h(xy)C_4^1(z)[-y_1, x_1, -z_1] = \Sigma_h(xy)[-x_1, -y_1, -z_1]$$

$$= [-x_1, -y_1, z_1] = C_2^1(z)[x_1, y_1, z_1]$$

Examining the multiplication table previously shown, we can summarize that the distinct symmetry operations satisfy the necessary group criteria as presented in Table 1.6.

At this time, it is hard to build the multiplication table, but the exercise will develop your ability to find the correct symmetry groups of molecules.

TABLE 1.6 Group Criteria That Symmetry Operations Satisfy

Criterion	
Closure	Every line or every column contains all distinct symmetry operations of allene molecule.
	Every line or every column contains each operation only once.
	Line *n* contains all operations in exactly the same order as column *n*, where *n* takes the values 1, 2,. . ., 8 (commutativity).
Associativity	It is obvious from earlier sentences too.
Identity	Identity operation *E* is present once in every line or column.
Inverse	Each element has an inverse. Improper rotations are inverse of each other. Any other operation is the inverse of itself.

1.2.4 CLASSES OF SYMMETRY OPERATIONS

We have seen in the model of the allene molecule that some operations are equivalent operations in the group. *A set of equivalent operations is a class of group.* Two operations A and B are from the same class if they can be converted one into the other by a third operation C of the group. Therefore, we can write $C \cdot A = B$ and $C \cdot B = A$.

From the multiplication table of the allene molecule, we can see that the proper rotations C_2' and C_2'' can be converted one into the other by the third operation C_2^1 of the group. Thus, the two rotations form a class. The reflection planes Σ_d' and Σ_d'' can also be converted one into the other by the same rotation C_2^1 and they can be collected in another class. In a similar manner, the same rotation C_2^1 interchanges improper rotations S_4^1 and S_4^3 when applied to them. Thus, the two improper rotations form the third class. Finally, how many classes are in the point group of allene molecule? The identity forms a class by itself, the rotation C_2^1 forms a class by itself too, and the other three classes have already been mentioned. There are five classes in total.

The inversion operation always forms one class by itself. A reflection through a horizontal plane will always form a class by itself. There is only one exception to this rule: the molecules with octahedral symmetry (you will see further in Section 1.3.5). These molecules have three axes that can be considered as principal and, thus, they can have three horizontal planes. The horizontal plane is defined by reporting to the vertical axis and such molecule can have several equivalent horizontal planes, which form a class. When there are n equivalent vertical reflection planes, we will simply write the class as $n\ \Sigma_v$. We will proceed in the same way for equivalent dihedral planes. Concerning rotations in most of the cases, the rotations C_n^{n-m} and C_n^m are equivalent and form a class. This means that the effect of one rotation will be the same as the effect of another rotation in the reverse direction, and we write the class as $2C_n^m$. If there are three equivalent axes c_n, the class is denoted as $6C_n^m$. The improper rotations of one class will be treated as only one improper rotation; similar to proper rotations. As was seen earlier, S_4^1 and S_4^3 are equivalent and we write a class formed by $2S_4^1$ rotations.

The use of the class has the advantage that all operations of a class are represented by only one operation. Therefore, if we have a group containing tens of symmetry operations while one class has eight equivalent operations, you can imagine how much we can reduce the work of testing the effect of these operations if we test only one.

1.3 SYMMETRY CLASSIFICATION OF MOLECULES

1.3.1 NOTATIONS

All molecules may be categorized by their symmetry. There are molecules very different in terms of spatial structure but that can have the same set of symmetry operations. This means that they belong to the same point group. All molecules can be characterized by 32 different combinations of

TABLE 1.7 All Crystallographic Point Groups

Cubic	$\boldsymbol{T}\ \boldsymbol{T}_h\ \boldsymbol{O}\ \boldsymbol{T}_d\ \boldsymbol{O}_h$
Tetragonal	$\boldsymbol{C}_4\ \boldsymbol{S}_4\ \boldsymbol{C}_{4h}\ \boldsymbol{D}_4\ \boldsymbol{C}_{4v}\ \boldsymbol{D}_{2d}\ \boldsymbol{D}_{4h}$
Orthorhombic	$\boldsymbol{D}_2\ \boldsymbol{C}_{2v}\ \boldsymbol{D}_{2h}$
Monoclinic	$\boldsymbol{C}_2\ \boldsymbol{C}_s\ \boldsymbol{C}_{2h}$
Triclinic	$\boldsymbol{C}_1\ \boldsymbol{C}_i$
Trigonal	$\boldsymbol{C}_3\ \boldsymbol{S}_6\ \boldsymbol{D}_3\ \boldsymbol{C}_{3v}\ \boldsymbol{D}_{3d}$
Hexagonal	$\boldsymbol{C}_6\ \boldsymbol{C}_{3h}\ \boldsymbol{C}_{6h}\ \boldsymbol{D}_6\ \boldsymbol{C}_{6v}\ \boldsymbol{D}_{3h}\ \boldsymbol{D}_{6h}$
Noncrystallographic	$\boldsymbol{C}_\infty\ \boldsymbol{C}_{\infty h}\ \boldsymbol{C}_\infty\mathrm{V}\ \boldsymbol{D}_{\infty h}\ \boldsymbol{C}_5\ \boldsymbol{S}_8\ \boldsymbol{D}_5\ \boldsymbol{C}_{5v}\ \boldsymbol{C}_{5h}\ \boldsymbol{D}_{4d}\ \boldsymbol{D}_{5d}\ \boldsymbol{D}_{5h}\ \boldsymbol{D}_{6d}\ \boldsymbol{I}\ \boldsymbol{I}_h$

symmetry elements. Therefore, there are 32 crystallographic point groups and 17 noncrystallographic point groups, as summarized in Table 1.7.

There are two naming systems commonly used when describing the symmetry operations of the point groups:

1. The *Schönflies notation* is used extensively in spectroscopy.
2. The *Hermann–Mauguin* or *international notation* commonly used in space groups is preferred in crystallography.

In this book, we use the Schönflies notation. The name of the group is given by a capital letter: ***C*** or ***D*** in most of the groups. The symbol ***C*** is used because the molecule must have at least a rotation axis as symmetry element and this is named the principal axis. The symbol ***D*** is used when the molecule has at least one twofold rotation axis perpendicular to the principal axis. The subscripts 1, 2, 3, 4, 5, 6, 8, and ∞ denote the order of the principal rotation axis. We have already shown that in some cases, the principal axis can generate rotations of different orders (C_2 and C_4, C_3 and C_6). The order of the principal rotation axis is the highest value of n. The second subscripts v, d, and h denote that the molecule has reflection planes vertical, dihedral, and horizontal, respectively. The symbol ***S*** is used for molecules whose principal axis is improper. The symbols ***T***, ***O***, and ***I*** are used for molecules whose shape is tetrahedral, octahedral, or icosahedral. For example, the group named $\boldsymbol{D}_{4h}$ means that the molecule has a fourfold rotation axis as principal axis, which generates C_4 and C_2 rotations (coming from C_4^2 rotation), and a horizontal plane perpendicular to the principal axis, which generates a reflection Σ_h. ***D*** means the molecule also has twofold rotation axes perpendicular to the principal axis, which generate C_2 rotations nonequivalent to the vertical C_2 rotation. By direct product of known operations or by inspection of the molecule, the other symmetry operations of the group can be obtained. Examples of such molecules are XeF_4 (which is planar) and C_4H_8 (cyclobutane).

1.3.2 SYSTEMATIC METHOD

How will we make the decision about the group type after we have drawn the molecule? We will follow a sequence of steps that will lead us or a computer software to identify the point group of any molecule by a systematic method. Such a flow chart is presented in Figure 1.21 and others can be

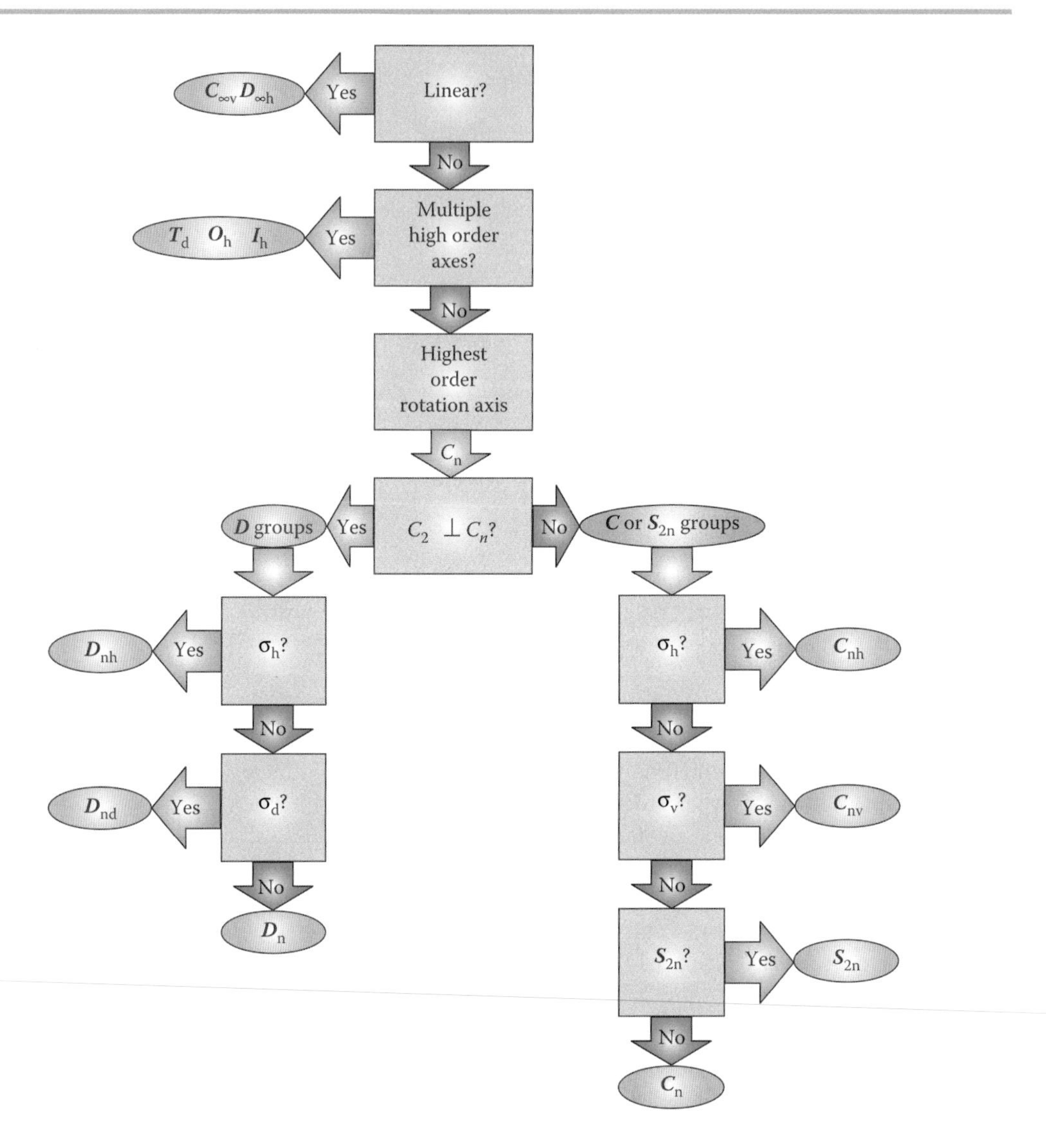

FIGURE 1.21 Flowchart of the systematic method useful to identify the point group of any molecule.

imagined. This flowchart will lead us systematically to the proper assignment of a molecule to a point group.

The first question is "Does the molecule belong to a special point group?" This question is referring to highly symmetric molecules that are easily recognizable. Here, we include two classes of molecules:

- Linear molecules
- Multiple high-order axes molecules

1.3.3 LINEAR MOLECULES

Linear molecules belong to one of two symmetry types. Examples of linear molecules are hydrogen (H_2), oxygen (O_2), hydrogen cyanide (HCN), hydrogen chloride (HCl), carbon monoxide (CO), carbon dioxide (CO_2), acetylene (C_2H_2), and beryllium fluoride (BeF_2). There are hundreds of diatomic molecules that are linear and they can be divided into mononuclear diatomic molecules (the first two examples) and heteronuclear diatomic molecules. A linear molecule has all atoms along a line. Thus, any linear molecule has a symmetry axis that passes through all atoms. The rotation of any angle around this axis is a symmetry operation because of the molecular shape (Figure 1.22). Therefore, the order of rotation axis is ∞. There are also an infinite number of vertical mirror planes present. In each example of Figure 1.22, only one vertical plane was represented. The molecule *HCN* that was illustrated and *all heteronuclear diatomic molecules belong to the group called* $C_{\infty v}$. For the $C_{\infty v}$ group, symmetry elements are E, C_∞, and $\infty\Sigma_v$. In case the linear molecule consists of two equivalent halves as the carbon dioxide illustration of Figure 1.22b, there is a supplementary and unique plane of symmetry σ_h and an infinite number of twofold axes perpendicular to the molecular axis. We represented only one c_2 axis to illustrate their positions. Because of the presence of a c_2 axis and a horizontal plane, both perpendicular to the principal axis, the group is called $\boldsymbol{D}_{\infty h}$.

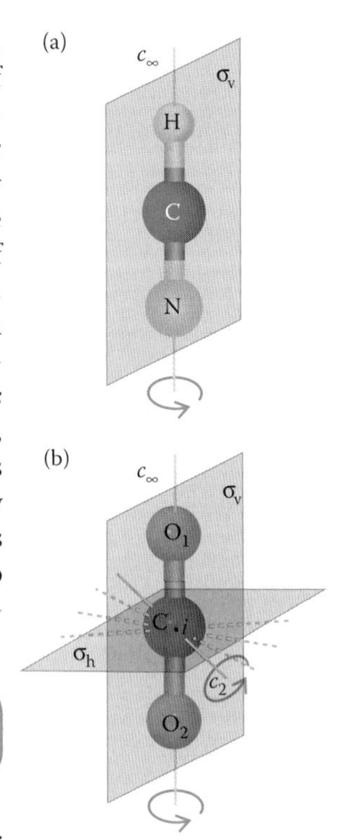

FIGURE 1.22 Symmetry elements of linear molecules: (a) hydrogen cyanide, which belongs to the $\boldsymbol{C}_{\infty v}$ group, and (b) carbon dioxide, which belongs to the $\boldsymbol{D}_{\infty h}$ group.

All linear molecules consisting of two equivalent halves belong to the $D_{\infty h}$ group.

For the $\boldsymbol{D}_{\infty h}$ group, symmetry operations are E, C_∞, $\infty\Sigma_v$, S_∞, I, and ∞C_2. Remember that the reflection through σ_h is included in S_∞ and it is not found as a distinct operation. Both $\boldsymbol{C}_{\infty v}$ and $\boldsymbol{D}_{\infty h}$ groups have a *low symmetry.*

1.3.4 TETRAHEDRAL MOLECULES

Multiple high-order axes molecules belong to the most interesting groups we can find in nature: $\boldsymbol{T}_d$ and $\boldsymbol{O}_h$. They belong to the same family of *cubic* groups. For this reason, we deal with the subject with careful consideration.

Let us look at the molecules illustrated in Figure 1.23, where four different molecules are represented: methane (CH_4), carbon tetrachloride (CCl_4), silane (SiH_4), and sulfate ion (SO_4^{2-}). All have the same spatial configuration and we can deduce from a simple visual inspection that they have the same symmetry. We note that in sulfate ion, you will see two double bonds and two single bonds S–O, when you draw this molecule

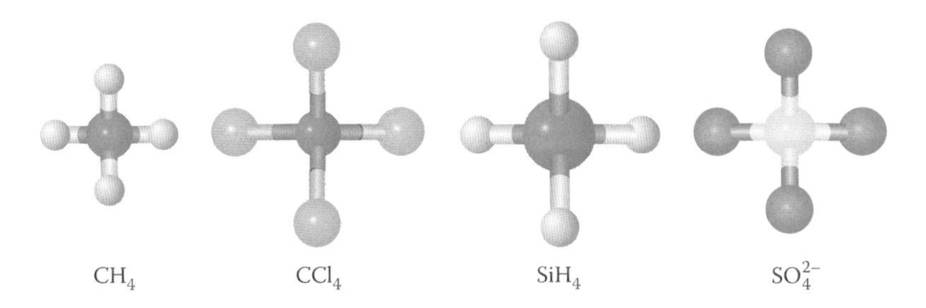

FIGURE 1.23 Four different molecules have the same configuration and, thus, the same symmetry.

using the Jmol viewer. But finally, the four oxygen atoms are equivalents and the molecule has the same symmetry as methane.

If we now look to the molecules represented in Figure 1.24a and b, can we say that they have the same symmetry? If we rotate in some way the molecules of methane and adamantane ($C_{10}H_{16}$), we can see that their projections on a plane are squares (Figure 1.24c and d). It is only one step to understand that both can be included in a cube (Figure 1.24e and f). Actually, they belong to the same symmetry group $\boldsymbol{T_d}$ (Nordman and Schmitkons 1965).

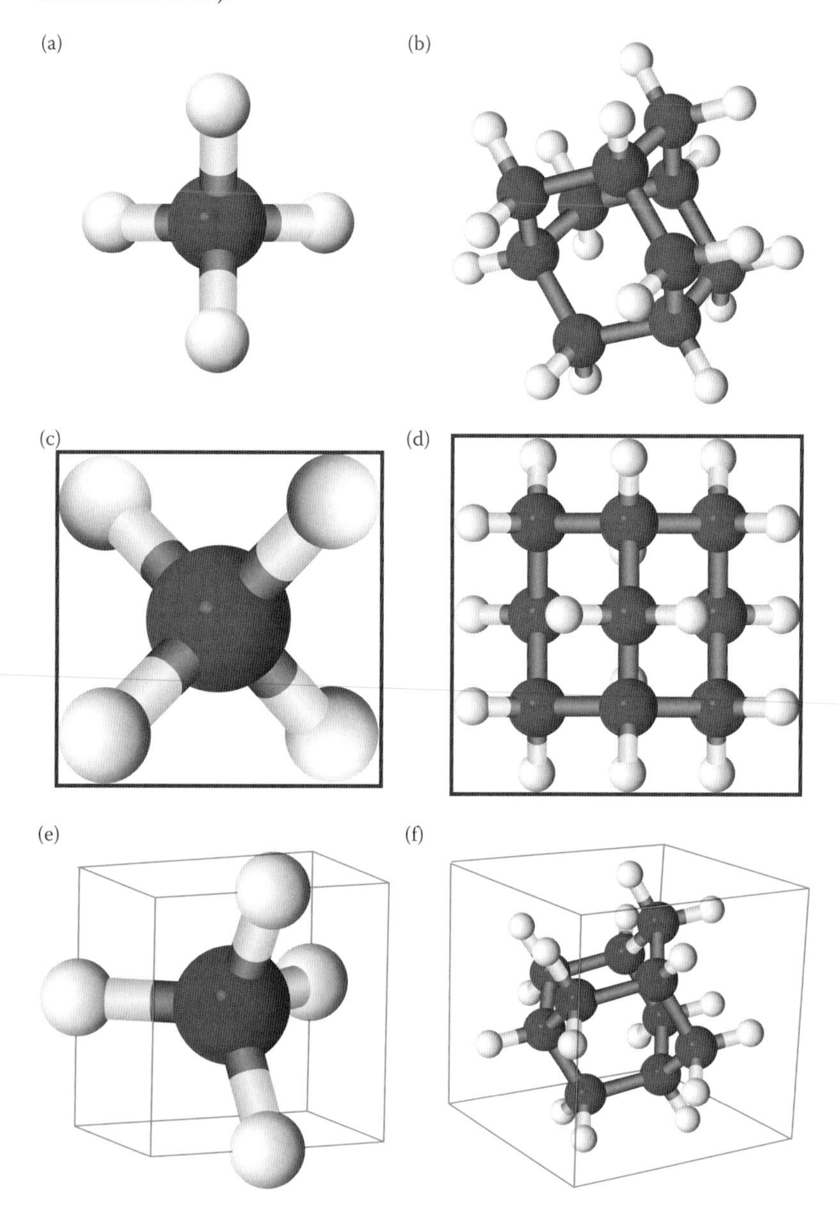

FIGURE 1.24 Methane (a) and adamantane (b) molecules have different structures, but both can be projected on a plane as a square (c) and (d). This suggests that both molecules can also be included in a cube (e) and (f).

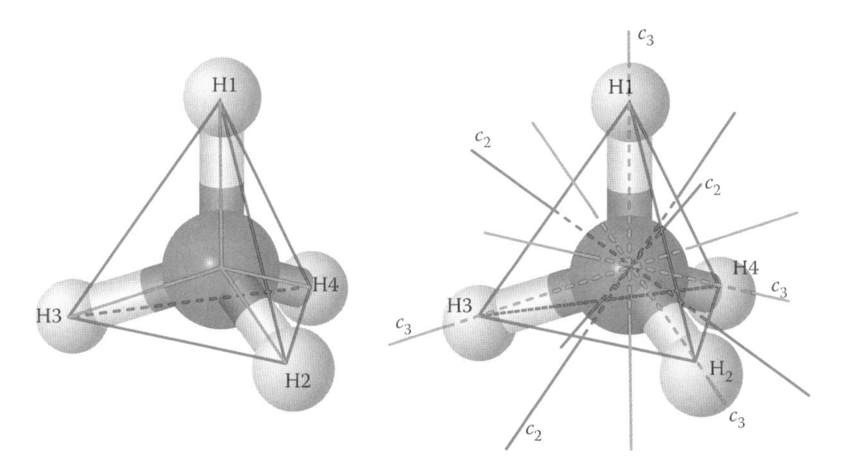

FIGURE 1.25 The structure of a methane molecule is a tetrahedron with seven proper rotation axes: four axes c_3 and three axes c_2.

We will consider further the simpler structure of the methane molecule. We rotate the molecule to find a known geometrical shape. We should get a line connecting every two hydrogen atoms in such a way that we find the structure of the methane molecule is a *tetrahedron* (Figure 1.25). A tetrahedron is a platonic solid with four equivalent equilateral triangular faces. It is only one convex polyhedron having four faces.

The tetrahedron has seven axes of symmetry:

- $4c_3$ (axes connecting vertices with the centers of opposite faces). Each of these axes generates two threefold rotations (C_3 and C_3^2). Therefore, *eight proper rotations* in all are generated.
- $3c_2$ (the axes connecting the midpoints of opposite edges). It is not obvious but they are perpendicular to each other. They also pass through the center of the tetrahedron (carbon atom). This means that they can be selected as coordinate axes x, y, and z. No one passes through a hydrogen atom. They generate *three proper rotations* C_2.

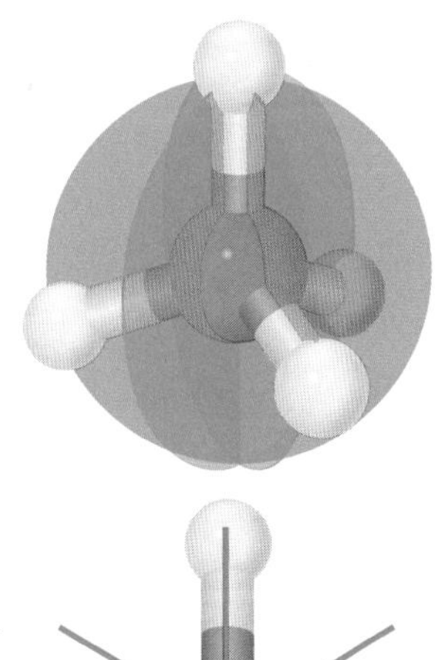

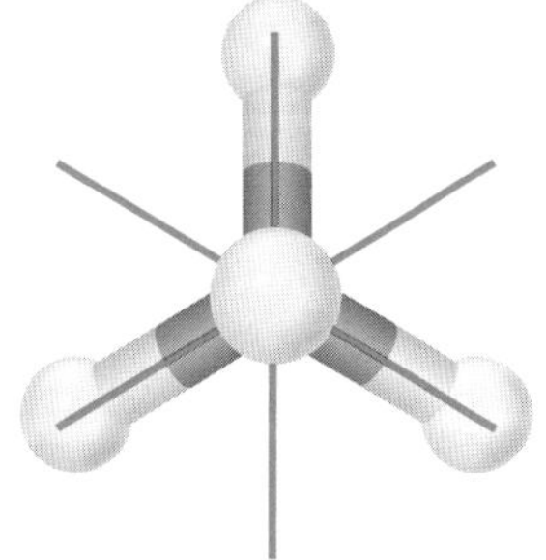

FIGURE 1.26 Reflection planes that pass through the H1 atom: perspective view (top) and top view (bottom).

It can be seen by simple inspection that the tetrahedron has six reflection planes:

- Three planes that pass through H1CH2, H1CH3, and H1CH4, respectively (Figure 1.26)
- Two planes that pass through H2CH3 and H2CH4, respectively
- One plane that passes through H3CH4

All the reflection planes are equivalent because the tetrahedron has not a principal axis to be considered as vertical. All the planes are named dihedral and they generate *six equivalent reflections*.

We can also see from Figure 1.27 the existence of three improper rotation axes s_4. They coincide with c_2 axes. Each of them generates two

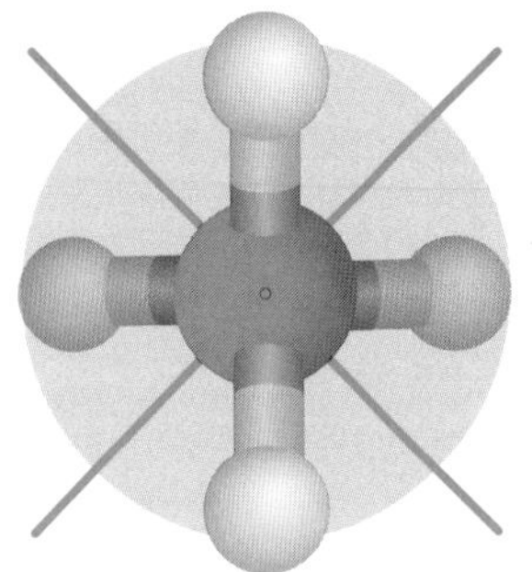

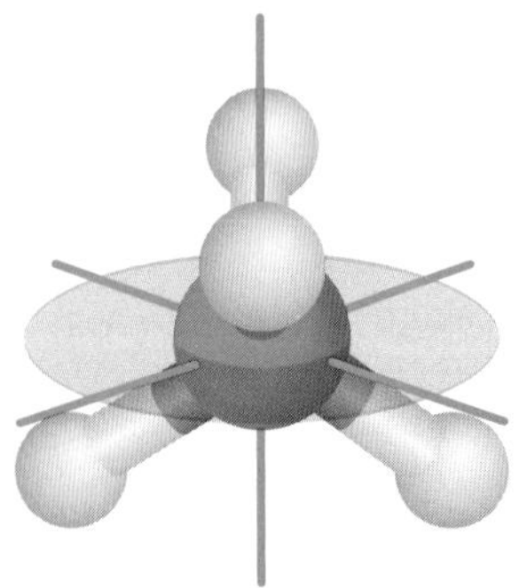

FIGURE 1.27 Improper rotation axes s_4 and one reflection plane used to perform improper rotation of tetrahedron represented here for the CH_4 molecule.

improper rotations, S_4^1 and S_4^3. We should remember the property summarized by Equation 1.3. $S_n^m = C_n^m$, when m is even and n is also even. Therefore, $S_4^2 = C_4^2 = C_2^1$. This proper rotation was previously taken into account. These axes generate *six equivalent improper rotations.*

The set of symmetry operations of a methane molecule, which is a tetrahedron, named $\boldsymbol{T}_d$, consists of 24 elements:

$$\boldsymbol{T}_d = \{E,\ 8C_3,\ 3C_2,\ 6\Sigma_d,\ 6S_4\}$$

1.3.5 OCTAHEDRAL MOLECULES

We now discuss the important symmetry of an *octahedron*. The geometrical body named octahedron is a platonic solid too, with eight equivalent equilateral triangular faces. It looks like it has been assembled by two equal pyramids (Figure 1.28). We take here as an example the $CoCl_6^{3-}$ octahedral complex. Other examples of molecules with the same configuration are sulfur hexafluoride (SF_6) and uranium hexafluoride (UF_6). This symmetry is often found in crystals or, in case of many ions, in solutions.

From the top view of the octahedron as shown in Figure 1.29a, we can easily discover the following:

- There is a c_2 rotation axis that generates a twofold proper rotation C_2. We selected the line that passes through atoms Cl6Co1Cl7 as the principal axis. There are also two other equivalent lines that pass through atoms Cl2CoCl3 and Cl4CoCl5, respectively, as shown in Figure 1.29b. Therefore, *three proper rotations* C_2 in all are generated.
- The principal axis and the other two equivalent axes are in fact c_4 rotation axes. Each of these generates two new fourfold proper rotations C_4^1 and C_4^3. The rotation $C_4^2 = C_2$ was previously discussed. Therefore, *six proper rotations* C_4 in all are generated.

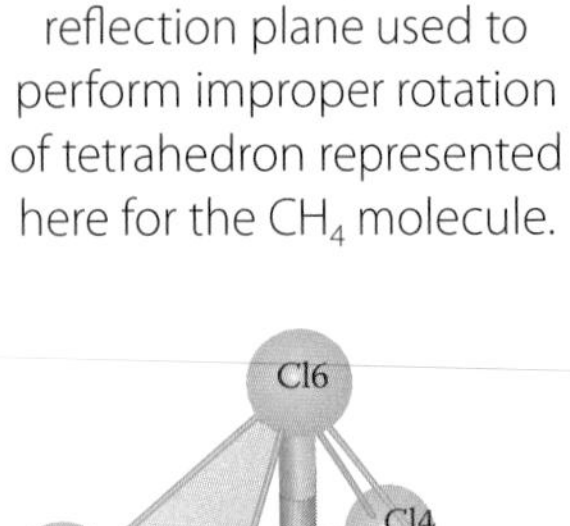

FIGURE 1.28 Octahedron solid has eight faces equilateral triangle (left-top face was filled to be easy seen). Here, it represents the molecular complex $[CoCl_6]^{3-}$.

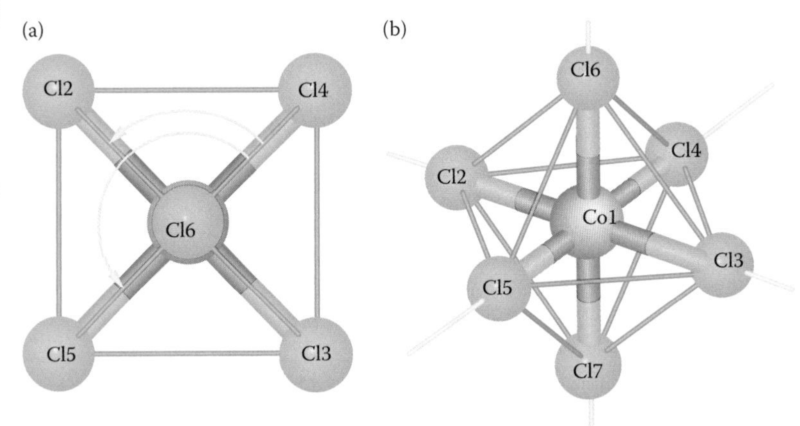

FIGURE 1.29 (a) Top view of the octahedron and (b) the three equivalent rotation axes, which are perpendicular to each other.

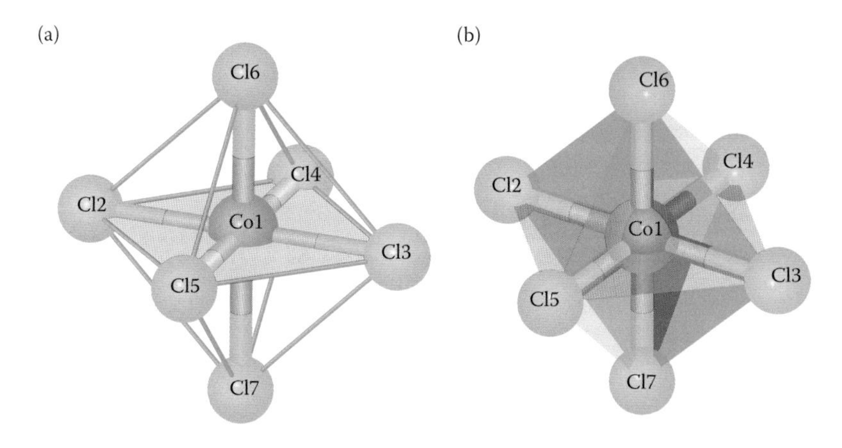

FIGURE 1.30 (a) Horizontal reflection plane and (b) three equivalent horizontal planes of octahedron.

- There is the horizontal plane σ_h illustrated in Figure 1.30a. The plane passes through atoms Cl2Cl3Cl4Cl5 and the central atom Co1. There are also two other equivalent planes as already discussed (Figure 1.30b). Therefore, *three reflections* Σ_h are generated.
- There is the inversion center i located on the central atom Co1. Thus, the *inversion operation I* is specific to the octahedron.
- There are three improper axes s_4, which are collinear with c_4 axes. Each of them generates two new improper rotations: S_4^1 and S_4^3. The result of the third improper rotation is not new because of the property already discussed; tetrahedron $S_4^2 = C_4^2 = C_2$. These axes generate *six equivalent improper rotations* S_4.

Now, we have to discover other symmetry operations that are not so visible. We can use the direct product of the already-found operations or carefully inspect the octahedron. The reader is invited to exercise the first method. We continue to illustrate the new operation looking at the octahedron from other points of view.

From an inspection of Figure 1.29a, we can see two other reflection planes. The first one passes through atoms Cl6Co1Cl7 and bisects the angle formed by Cl5Co1Cl3. The second plane passes through atoms Cl6Co1Cl7 and bisects the angle formed by Cl3Co1Cl4. The two dihedral planes are illustrated in Figure 1.31a. We have already shown that the octahedron has three equivalent axes. This implies that there are still four dihedral reflection planes, two planes for each axis: Cl2Co1Cl3 and Cl4Co1Cl5. Therefore, *six dihedral reflections* Σ_d in all are generated.

In Figure 1.31b, we can see two twofold rotation axes c_2' (the axes connecting the midpoints of opposite edges) perpendicular to the principal axis Cl6Co1Cl7. Finally, we can count other *six proper rotations* C_2' in all, generated by the new axes.

Can you see a 120° rotation as a symmetry operation in this molecule that has only right angles? Maybe not, but we can rotate the molecule to obtain a view as in Figure 1.32a. The equilateral triangle plane determined

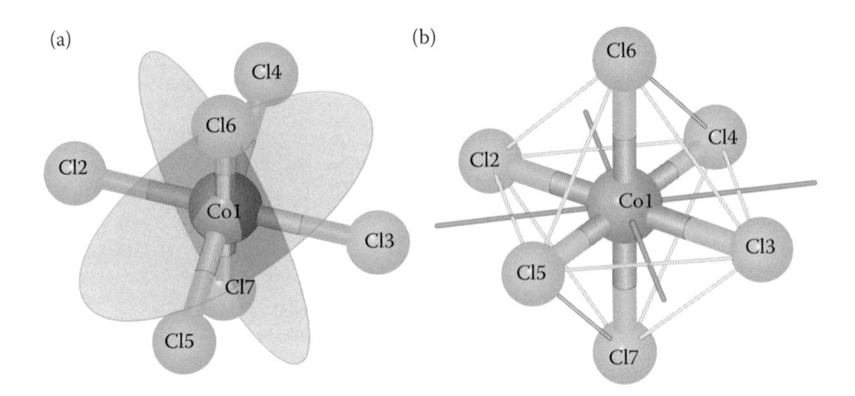

FIGURE 1.31 (a) Two of the dihedral reflection planes and (b) two twofold rotation axes that correspond to the principal axis Cl6Co1Cl7.

by atoms Cl6, Cl5, and Cl3 (better seen on the front view) and the other triangle plane determined by atoms Cl2, Cl7, and Cl4 (back side) suggests the existence of a 120° rotation around an axis perpendicular to the figure. This rotation will bring the octahedral molecule into a position equivalent to the initial one (Figure 1.32b). Because the octahedron is a platonic solid, with eight equivalent equilateral triangular faces, it has four proper rotation axes c_3 and, thus, we can count *eight proper rotations* C_3 in all generated by these new axes.

We can also see from Figure 1.33a that a 60° rotation around the inner axis will bring atom Cl3 in a position situated exactly above the initial position of atom Cl4 (Figure 1.33b). The new position of atom Cl4 is situated under the initial position of atom Cl6 and so on. A reflection through a plane situated midway between the two triangular planes (Figure 1.33c) will now bring atom Cl3 exactly in the initial position of atom Cl4, that is, into the plane situated behind. The reflection will bring atom Cl4 in the initial position of atom Cl6 and so on (Figure 1.33d). We know that

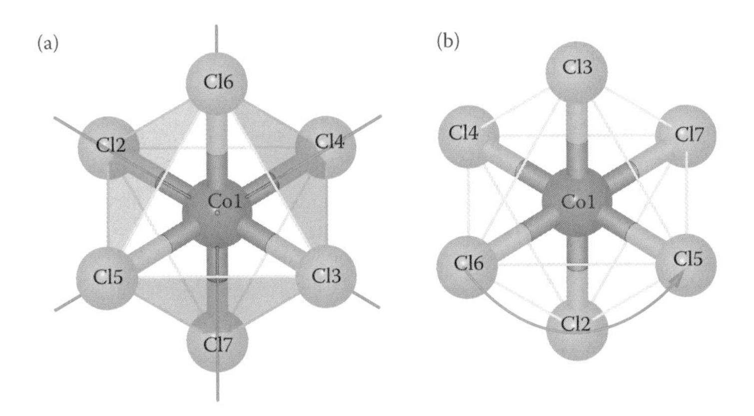

FIGURE 1.32 (a) Octahedron has four threefold axes, which pass only through the central atom and center of opposed faces. (b) Threefold rotation around the central axis of this image brings the molecule into an equivalent configuration.

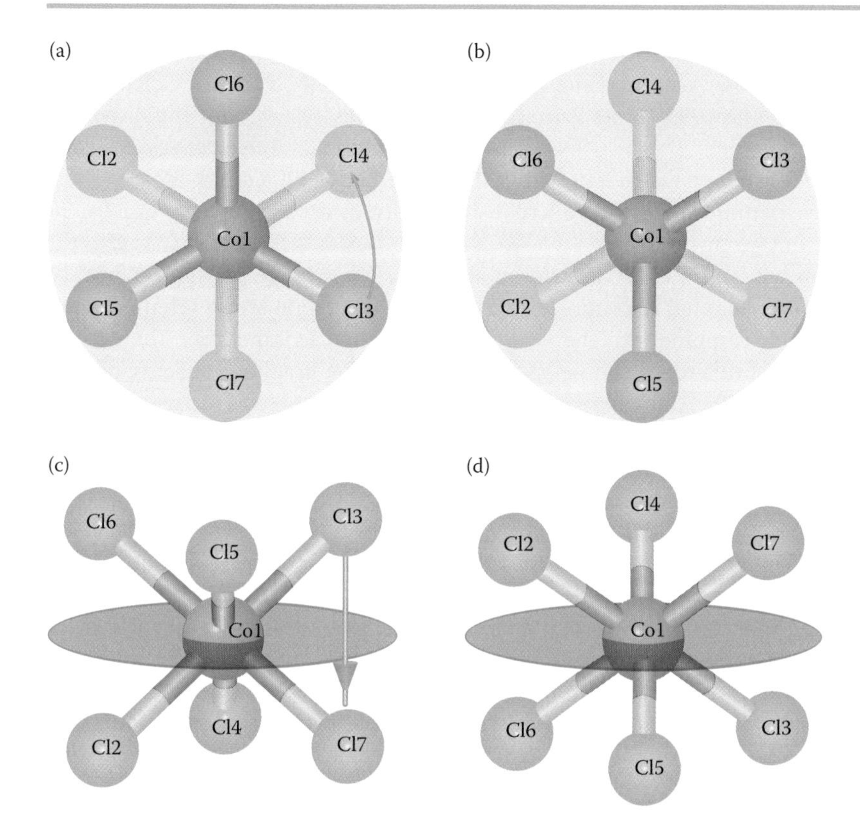

FIGURE 1.33 Effect of improper rotations S_6: (a and b) the first moment is represented by a 60° rotation and (c and d) the second moment is the reflection through a plane containing only the central atom of the molecule.

this operation, composed of a rotation followed by a reflection through a plane perpendicular to the rotation axis, is an improper rotation S_6^1. The midway "horizontal" plane passes only through central atom Co1. Therefore, the octahedron has four improper axes s_6 collinear with proper axes c_3. Each improper axis generates a set of symmetry operations: S_6^1, $S_6^2 = C_3$, $S_6^3 = C_2''$, $S_6^4 = C_3^2$, S_6^5, $S_6^6 = E$. The rotation C_2'' has the same effect as inversion I. Therefore, we can count *eight improper rotations* S_6 in all generated by these improper s_6 axes.

Now, we can make the final counting of 48 symmetry operations of octahedron just in the order we have discovered them:

$$\boldsymbol{O}_\mathbf{h} = \{E,\ 3C_2,\ 6C_4,\ 3\Sigma_h, I,\ 6S_4,\ 6\Sigma_d,\ 6C_2',\ 8C_3,\ 8S_6\}$$

1.3.6 ICOSAHEDRAL MOLECULE

The *icosahedron* has 20 equilateral triangle faces and 12 vertices. The icosahedral molecules belong to the $\boldsymbol{I}_h$ group. Some exotic chemical examples for icosahedral symmetry are the fullerene molecule C_{60} and the *closo*-dodecaborate anion ($[B_{12}H_{12}]^{2-}$).

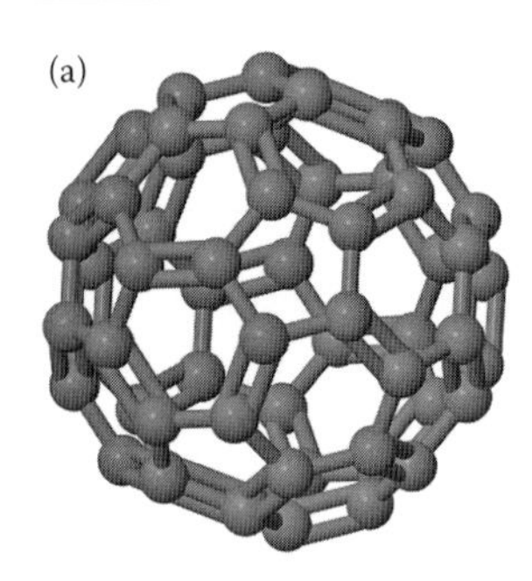

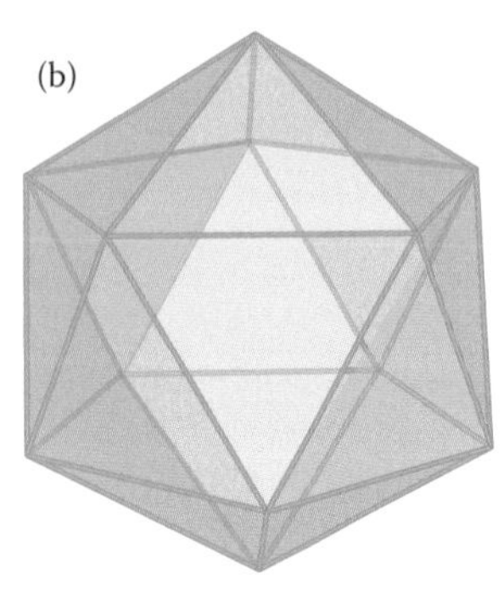

FIGURE 1.34 The fullerene molecule (a) can be inscribed into an icosahedron (b).

Carbon atoms can bond with other carbon atoms to form the well-known diamond and graphite, but, also, new structures called fullerenes. The 1996 Nobel Prize in Chemistry was awarded for the discovery of fullerenes (Nobel 1996). The shape of this big molecule is like a cage that can be constructed within the range of 20–140 carbon atoms. Thus, the symmetry of the fullerene polyhedron can vary within exactly 28 point groups: $\boldsymbol{I}_h$, $\boldsymbol{I}$, $\boldsymbol{T}_d$, $\boldsymbol{T}_h$, $\boldsymbol{T}$, $\boldsymbol{D}_{nh}$, $\boldsymbol{D}_{nd}$, $\boldsymbol{D}_n$ ($n = 2, 3, 5, 6$), $\boldsymbol{S}_{2m}$, $\boldsymbol{C}_{mh}$, $\boldsymbol{C}_{mv}$, $\boldsymbol{C}_m$ ($m = 2, 3$), $\boldsymbol{C}_s$, $\boldsymbol{C}_i$, $\boldsymbol{C}_1$ (Fowler et al. 1993). The strange polyhedral form of carbon, named buckminsterfullerene, has 60 carbon atoms (Figure 1.34a), which combine into the shape of a truncated icosahedron. The carbon cage of fullerenes can be generated from a flat graphene sheet (Chuvilin et al. 2010). This discovery was also awarded the Nobel Prize in Physics in 2010. A lot of research is conducted to discover extremely interesting applications of this new material. This sophisticated structure can be inscribed into an icosahedron, which is one of the perfect shapes of cubic group (Figure 1.34b).

Dodecaborate anion ($[B_{12}H_{12}]^{2-}$) with the complete nomenclature dodecahydro-*closo*-dodecaborate(2-) ion (IUPAC 1971), which belongs to the largest point group $\boldsymbol{I}_h$ (Figure 1.35), was also of great interest to the chemistry community in the last decade because of the application of agents derived from it in medical diagnostics (Sivaev et al. 2002).

A very nice animated illustration of the icosahedral symmetry can also be found on the Internet for 280-molecule icosahedral water clusters (Chaplin 2013).

Because they are still rare in nature and their symmetry is very difficult to illustrate, we only mention the set of 120 symmetry operations of this group:

$$\boldsymbol{I}_h = \{E,\ 12C_5,\ 12C_5^2,\ 20C_3,\ 15C_2, I,\ 12S_{10},\ 12S_{10}^3,\ 20S_6,\ 15\Sigma_d\}$$

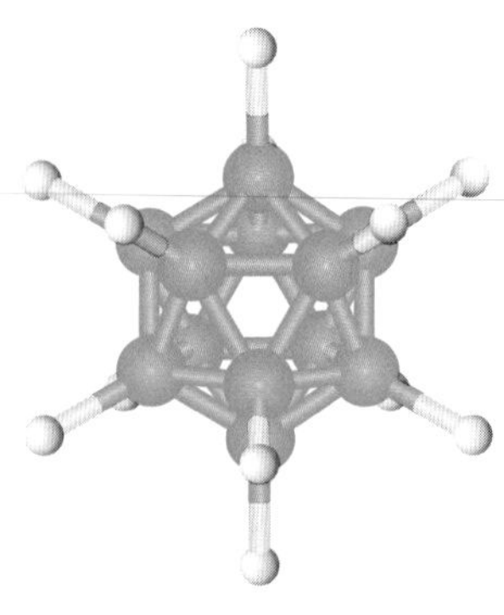

FIGURE 1.35 The molecule of dodecaborate belongs to the $\boldsymbol{I}_h$ point group.

This is the last regular polyhedron we are interested in, so it is time to discuss the five *platonic solids*. They can only exist as regular polyhedra. They are characterized by having regular polygons as faces, equivalent vertices, and equivalent edges, and only five of them are possible. These properties are strong reasons to believe that these bodies have a high symmetry. The platonic solids are tetrahedron, cube, octahedron, dodecahedron, and icosahedron. All studied polyhedra can be inscribed within a cube (Figure 1.36). We underline that the octahedron has the same symmetry as that of the cube.

1.3.7 *C* OR *D* GROUP?

Let us go back to the flowchart of molecular systematic classification (Figure 1.21). If the answer was *no* to the first two questions, it means the molecule does not have an evident symmetry. We focus now to the second question: Does the molecule have *multiple high-order axes*? The problem is are there several axes or only one axis? If there are several axes, the molecule belongs to one previous example: $\boldsymbol{T}_d$ has three axes c_3, $\boldsymbol{O}_h$ has three axes c_4, and three axes c_3, and $\boldsymbol{I}_h$ has six axes c_5. If the

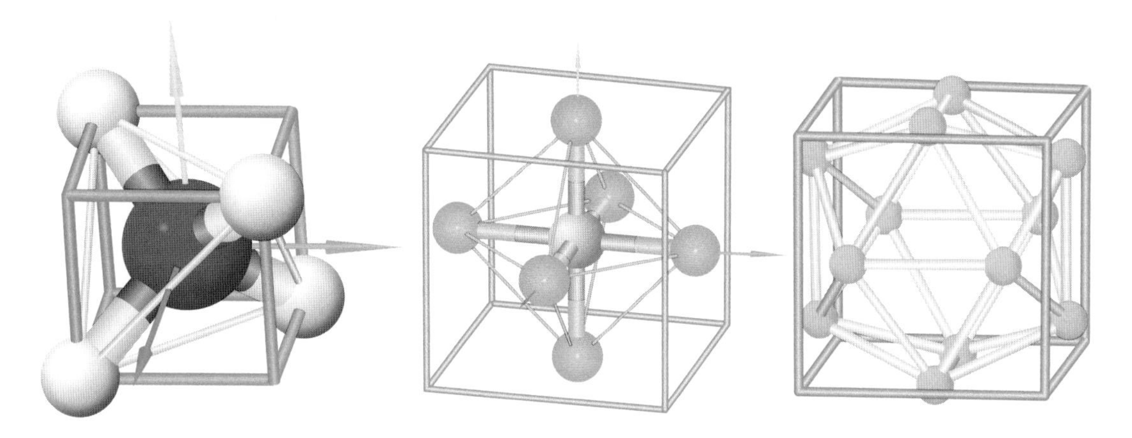

FIGURE 1.36 Tetrahedron, octahedron, and icosahedron within a cube.

molecule has only one high-order axis, we take this as the principal axis c_n and move on to the next question: *Is there a c_2 axis perpendicular to the principal axis? Yes* or *no* will make two main branches of the flowchart. A positive answer will send us to the ***D***-type groups. A negative answer will send us to the ***C*** (or ***S***)-type groups.

We note that going down on the flowchart the symmetry becomes lower and the number of group elements decreases.

1.3.7.1 $\boldsymbol{D}_{3h}$

We have already worked with the molecule boron trifluoride (BF_3).

1. It is not a linear molecule.
2. It is not a cubic molecule.
3. It has as principal axis a c_3 proper rotation axis (Figure 1.9).
4. It has three equivalent c_2 axes, each of them perpendicular to the principal axis (Figure 1.10). So, we are going on ***D*** branch.
5. It is a planar molecule, so it has a horizontal plane σ_h.

Therefore, we assign as $\boldsymbol{D}_{3h}$ the point group of molecule BF_3. From simple inspection, we can see that it also has three equivalent vertical planes. Each plane passes through the central atom of boron and through one outer atom of fluoride (Figure 1.37a). The last symmetry element is a threefold improper axis s_3 collinear with principal axis c_3. Therefore, BF_3 molecule belongs to the $\boldsymbol{D}_{3h}$ group that contains 12 symmetry operations (Figure 1.37b):

$$\boldsymbol{D}_{3h} = \{E,\ 2C_3,\ 3C_2,\ 2S_3,\ \Sigma_h,\ 3\Sigma_v\}$$

1.3.7.2 $\boldsymbol{D}_{5d}$

Ferrocene (stag) is the new molecule to be investigated. It is composed of 10 carbon atoms, 10 hydrogen atoms, which form two planar pentagons parallel to each other, and one Fe atom placed midway between

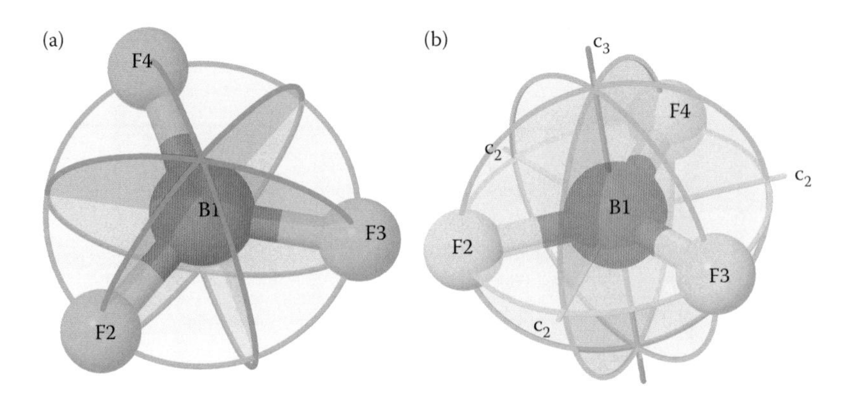

FIGURE 1.37 BF_3 molecule has three vertical planes (a) and belongs to the $\boldsymbol{D}_{3h}$ group (b).

the two pentagons (Figure 1.38a). During the last 60 years, *ferrocene* was one of the most studied metallocenes. First of all, we must rotate the molecule to find the view that offers us the maximum information. From Figure 1.38b, the existence of the proper rotation fivefold axis c_5 is obvious. This appears to be the principal axis and it generates a set of proper rotations: $C_5^1, C_5^2, C_5^3, C_5^4$, and $C_5^5 = E$. Each pentagonal molecule has five c_2 axes but not one is good for ferrocene. If we take a twofold axis that passes through the central atom and is collinear with the y axis, the 180° rotation around this axis brings the molecule into an equivalent configuration (Figure 1.39). The difference is that the planes C12C14C16C18C20 and C2C4C6C8C10 interchange their positions and also related hydrogen atoms. Because the principal axis is a fivefold axis, we know that there are five twofold axes perpendicular to it. Starting from illustrations of Figure 1.38b, we can also imagine the existence of

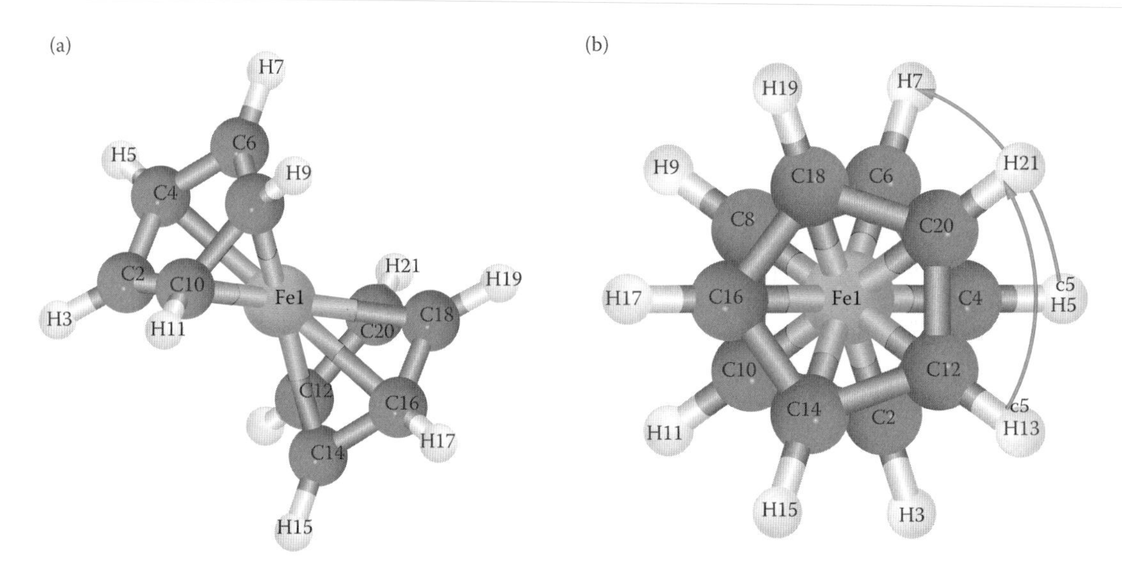

FIGURE 1.38 The ferrocene molecule has a complicated spatial structure (a), but it is easy to see the existence of the proper rotation axis c_5 as principal axis (b).

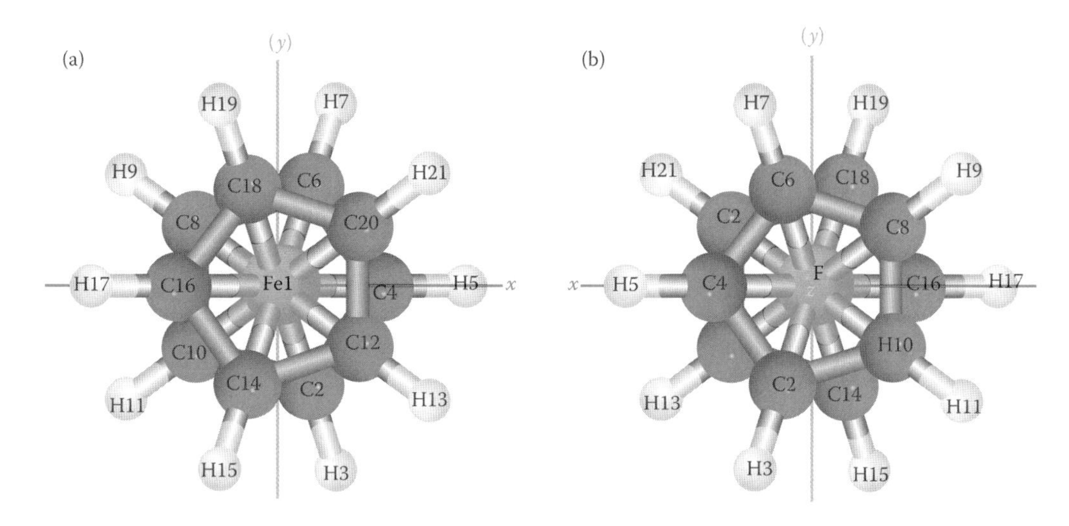

FIGURE 1.39 180° rotation around the *y* axis, which passes through the central atom of Fe, brings the molecule into an equivalent configuration: (a) before rotation and (b) after rotation.

a reflection plane that passes through the central atom and outer atoms H5C4C16H17. The ferrocene molecule has five equivalent reflection planes as you can see in Figure 1.40a.

We now have sufficient data to answer the flowchart questions:

1. It is not a linear molecule.
2. It is not a cubic molecule.
3. It has as principal axis a c_5 proper rotation axis (Figure 1.38b).
4. It has five equivalent c_2 axes, each of them perpendicular to the principal axis (Figure 1.39). So, we are going on the ***D*** branch.
5. It does not have a horizontal plane.
6. It has five dihedral planes σ_d.

Therefore, we assign as $\boldsymbol{D}_{5d}$ the symmetry of the ferrocene molecule. We can also observe the improper rotations from Figure 1.38b. A 36°

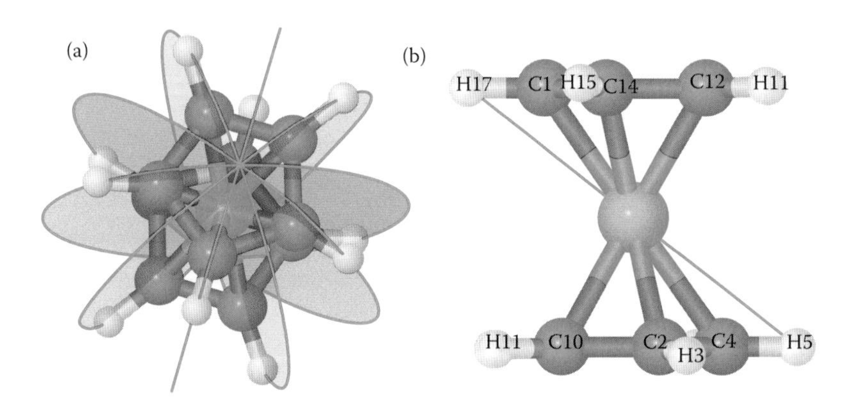

FIGURE 1.40 Reflection planes of ferrocene molecule (a) and the inversion operation (b).

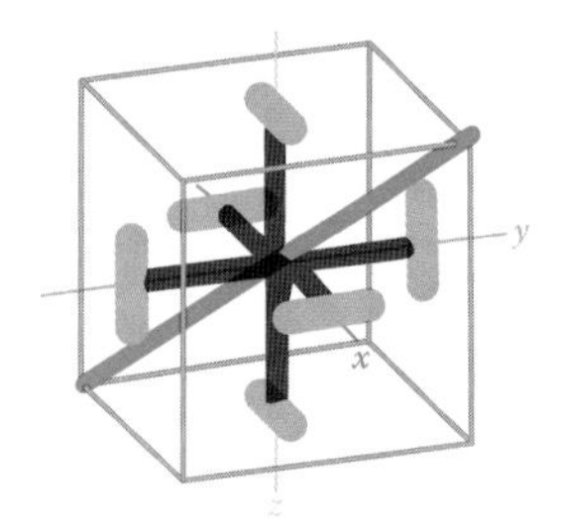

rotation brings H13 in a position from where a reflection through a horizontal plane will bring it to the initial position of H5. So, the molecule has an improper rotation axis s_{10} collinear with the principal axis. Figure 1.40b shows that the connection line between H17 and H5 atoms passes through the center of the molecule. This suggests the existence of the inversion operation.

We conclude that the ferrocene molecule belongs to the $\boldsymbol{D}_{5d}$ group formed by 21 symmetry operations:

$$\boldsymbol{D}_{5d} = \{E,\ 2C_5,\ 2C_5^2,\ 5C_2, I,\ 3S_{10}^3,\ 2S_{10},\ 5\Sigma_d\}$$

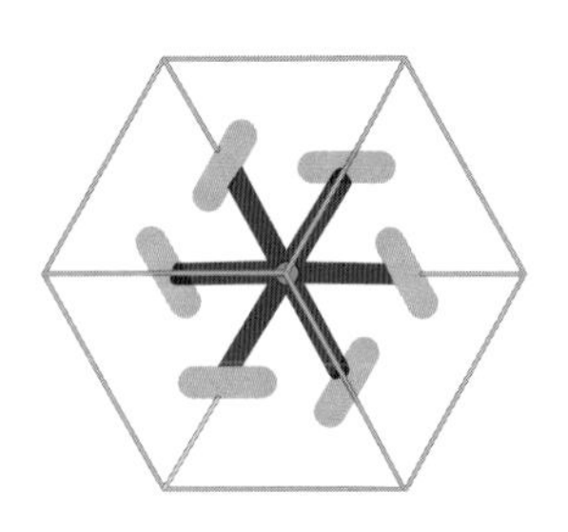

FIGURE 1.41 Sketch of an example of $\boldsymbol{D}_3$ symmetry.

1.3.7.3 $\boldsymbol{D}_3$

A simple $\boldsymbol{D}_3$ molecule exists, but it is rare. We sketched a body that belongs to this symmetry in Figure 1.41.

The $\boldsymbol{D}_3$ group is formed by six symmetry operations:

$$\boldsymbol{D}_3 = \{E,\ 2C_3,\ 3C_2\}$$

1.3.7.4 $\boldsymbol{C}_{3h}$

We have finished the flow chart branch determined by the existence of a twofold rotation axis perpendicular to the principal axis. We now take the molecule of boric acid (Figure 1.42). The molecule is planar and has a proper threefold rotation axis. This is the principal axis. This structure does not have a twofold rotation axis.

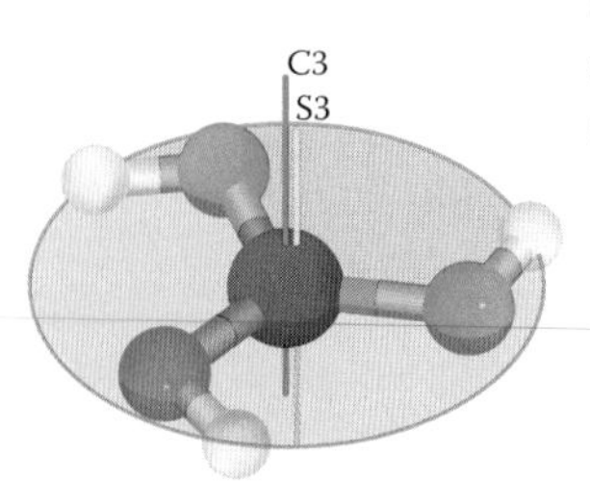

FIGURE 1.42 The molecule of boric acid is planar and has proper and improper rotation axes.

Following the systematic classification, we answer that

1. It is not a linear molecule.
2. It is not a cubic molecule.
3. It has as the principal axis a c_3 proper rotation axis.
4. It does not have c_2 axes perpendicular to the principal axis. So, we are going on the *C* branch.
5. It has a horizontal plane.

Therefore, the group must be $\boldsymbol{C}_{3h}$. Starting from c_3 and σ_h elements, we can imagine the existence of an improper threefold rotation axis s_3. Finally, the set of symmetry operations is

$$\boldsymbol{C}_{3h} = \{E, C_3, C_3^2,\ \Sigma_h, S_3, S_3^5\}$$

1.3.7.5 $\boldsymbol{C}_{3v}$

The ammonia molecule (NH_3) is the new example. It is a well-known molecule because of its much disputed utilization in agriculture and food industry. Do not be confused between ammonia (NH_3) and nitrate ion $[NO_3]^-$, which are equally disputed subjects over their use. However, the high-resolution study of ammonia spectra conducted to the brilliant idea to obtain practically the inversion of population. Thus, the first coherent radiation source (in microwave range) was realized by Nobel Prize laureates Basov, Prokhorov, and Townsend. This small molecule has two energetic states relatively easy to be used to obtain the inversion of population.

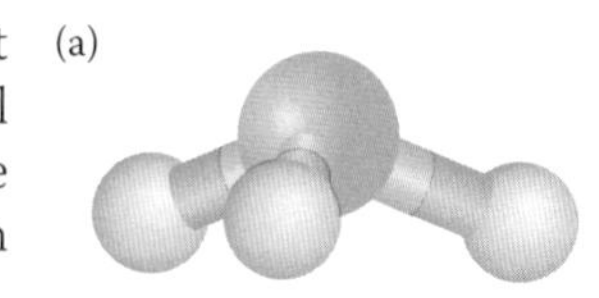

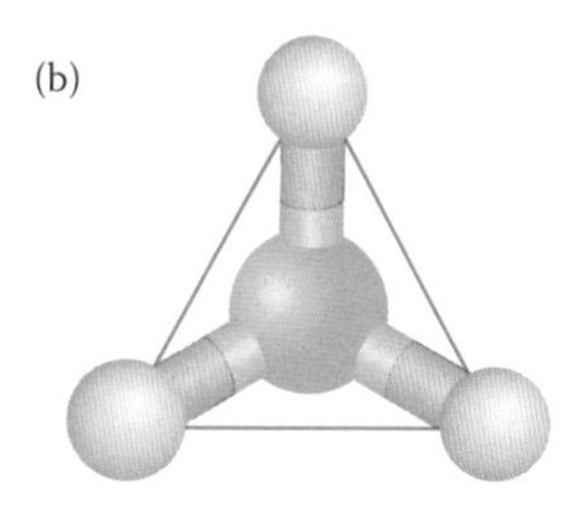

FIGURE 1.43 Ammonia molecule: spatial structure (a) and top view (b).

Nitrogen atom has three identical "legs" of hydrogen (Figure 1.43a). Viewing the molecule from the top, it appears to be inscribed in an equilateral triangle as shown in Figure 1.43b. As a consequence, the molecule has both a threefold rotation axis c_3 as the principal axis and three reflection planes, each of them passing through the nitrogen atom and one hydrogen atom. From the left image, we can see that the molecule has neither a twofold axis nor a horizontal reflection plane. It also does not have an improper axis.

We are now able to follow the way of systematic classification:

1. It is not a linear molecule.
2. It is not a cubic molecule.
3. It has as principal axis a c_3 proper rotation axis (Figure 1.43b).
4. It does not have c_2 axes perpendicular to the principal axis. So, we are going on the *C* branch.
5. It does not have a horizontal plane.
6. It has vertical planes.

Therefore, the ammonia molecule belongs to the $\boldsymbol{C}_{3v}$ group, which has only six symmetry operations, easy to be presented all in one illustration (Figure 1.44):

$$\boldsymbol{C}_{3v} = \{E, 2C_3, 3\Sigma_v\}$$

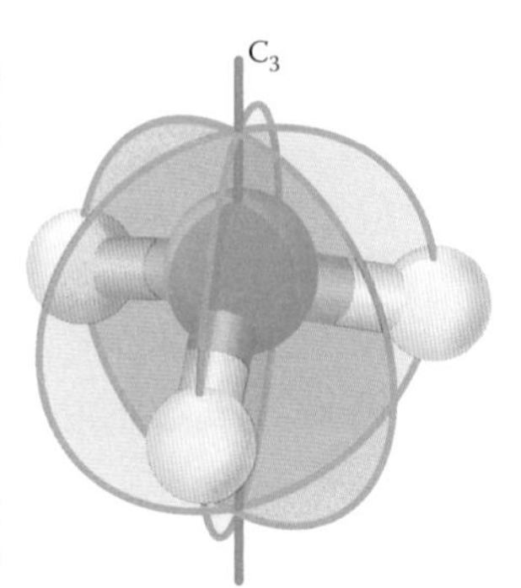

FIGURE 1.44 Complete set of symmetry elements of the ammonia molecule.

1.3.7.6 C_{2v}

We started the introduction in group theory with the water molecule. It is time to classify this molecule, which is very simple but it is extremely important for Earth's life. We mentioned that it has a proper c_2 axis (Figure 1.8) and two vertical planes (Figure 1.3). The vertical planes are not equivalent: σ_v passes through all the atoms and σ'_v passes only through the oxygen atom. So, our life depends on this molecule with very low symmetry, which belongs to the $\boldsymbol{C}_{2v}$ group (Figure 1.45):

$$\boldsymbol{C}_{2v} = \{E, C_2, \Sigma_v, \Sigma'_v\}$$

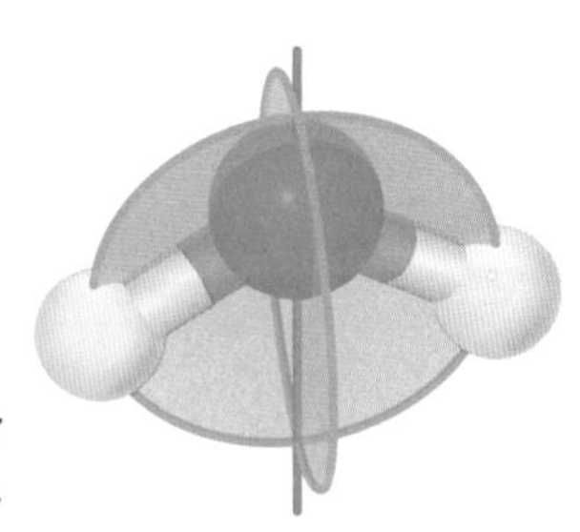

FIGURE 1.45 All symmetry elements of a water molecule.

1.3.7.7 S_{2n}

If the molecule does not have vertical planes, then we should look for improper rotations S_{2n}. There are two possible configurations: S_4 and S_6. The molecules of this type are rare and extremely complicated. 1,3,5,7-tetrafluoracyclooctatetrane is a common example of the S_4 symmetry group in certain sites of chemistry. We prefer to illustrate this group with a hypothetical molecular complex as shown in Figure 1.46a. It is simpler

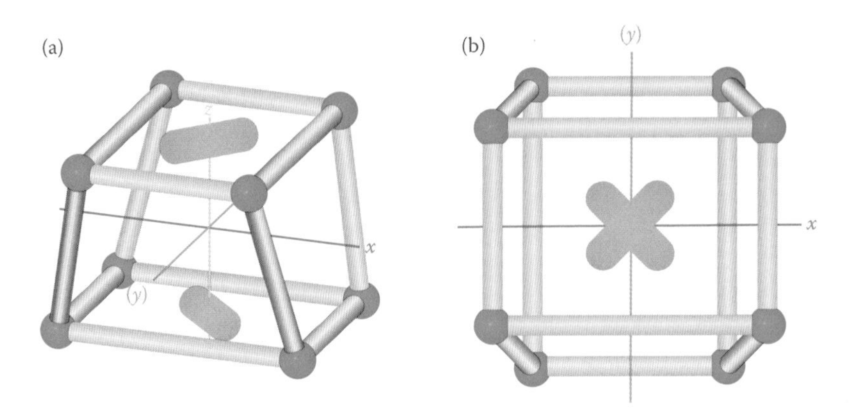

FIGURE 1.46 Hypothetical molecular complex illustrates S_4 group: (a) perspective view; (b) top view.

to discover the symmetry elements looking down from top to the molecule (Figure 1.46b): one vertical proper twofold c_2 axis and one improper rotation $s_4 = \sigma_h \cdot c_4$ axis.

The $\boldsymbol{S}_4$ group contains only six symmetry operations:

$$S_4 = \{E, C_2, 2S_4\}$$

1.3.7.8 C_2

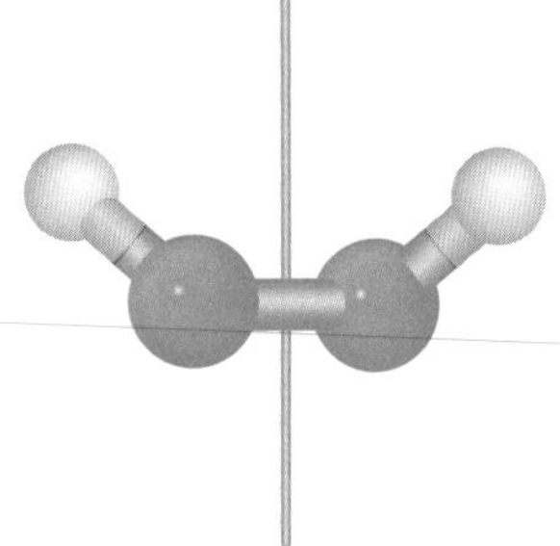

FIGURE 1.47 Nonplanar H_2O_2 molecule has only one twofold rotation axis.

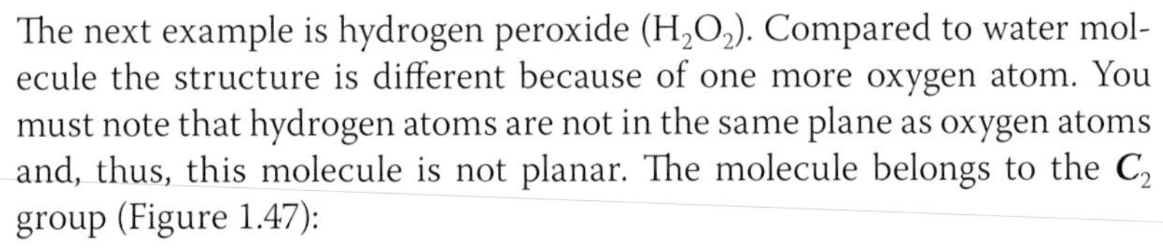

The next example is hydrogen peroxide (H_2O_2). Compared to water molecule the structure is different because of one more oxygen atom. You must note that hydrogen atoms are not in the same plane as oxygen atoms and, thus, this molecule is not planar. The molecule belongs to the $\boldsymbol{C}_2$ group (Figure 1.47):

$$\boldsymbol{C}_2 = \{E, C_2\}$$

There are two other configurations of hydrogen peroxide that are discussed in the literature and they belong to different point groups.

Does molecule exist without symmetry? Yes, it does. For example, the FClSO (see Figure 1.48) or CFClBrH molecules, which are composed of different atoms. They can only have identity operation as the symmetry element and belong to the C_1 group.

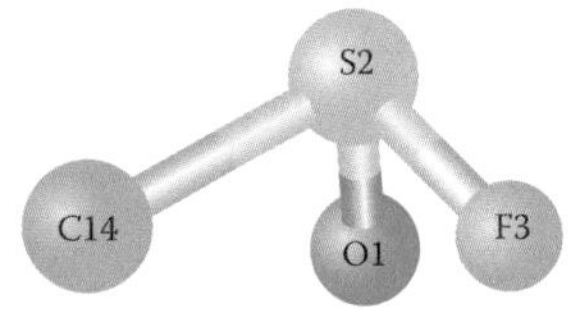

FIGURE 1.48 FClSO molecule belongs to the C_1 group.

In Section 1.2.3, we established the complete set of symmetry operations of the allene molecule: $E, C_2^1, \Sigma_d', \Sigma_d'', C_2^1, \Sigma_d', \Sigma_d'', S_4^1, S_4^3, C_2', C_2''$. What is the point group of allene?

1. It is not a linear molecule.
2. It is not a cubic molecule.
3. It has as principal axis a c_2 proper rotation axis (Figure 1.20).
4. It has two equivalent c_2 axes, each of them perpendicular to principal axis (Figure 1.20). So, we are going on the D branch.

5. It does not have a horizontal plane.
6. It has two dihedral planes σ_d.

Therefore, we assign as $\boldsymbol{D}_{2d}$ the point group of the allene molecule that is formed by eight symmetry operations:

$$\boldsymbol{D}_{2d} = \{E, C_2, 2C_2', 2S_4, 2\Sigma_d\}$$

1.4 MATRIX REPRESENTATION OF SYMMETRY TRANSFORMATION

In Section 1.2.2, we started to simplify the search for the effect of symmetry operations from more or less complicated illustrations into a simpler analytical way. We should know how to write the effect of rotation C_2 or reflection Σ on the coordinates of a point. We can also write the effect of rotation C_4 but we cannot do that for a rotation C_3 or of higher order. The matrix representation is an algebraic way to describe any symmetry operation.

We will study the effect of a *rotation* by angle θ around the z axis. Let us take a point P in the plane xOy. The initial position of the point P can be defined by the vector $\vec{r}_1$ (Figure 1.49) with components x_1 and y_1. Any rotation around the z axis does not change the coordinate z. The final position vector $\vec{r}_2$ has the components x_2 and y_2 and we try to relate them to the initial coordinates as follows:

$$(1.22) \quad \begin{cases} x_2 = r_2 \cos\alpha_2 = r_2 \cos(\alpha_1 + \theta) = r_2(\cos\alpha_1 \cos\theta - \sin\alpha_1 \sin\theta) \\ y_2 = r_2 \sin\alpha_2 = r_2 \sin(\alpha_1 + \theta) = r_2(\sin\alpha_1 \cos\theta + \cos\alpha_1 \sin\theta) \end{cases}$$

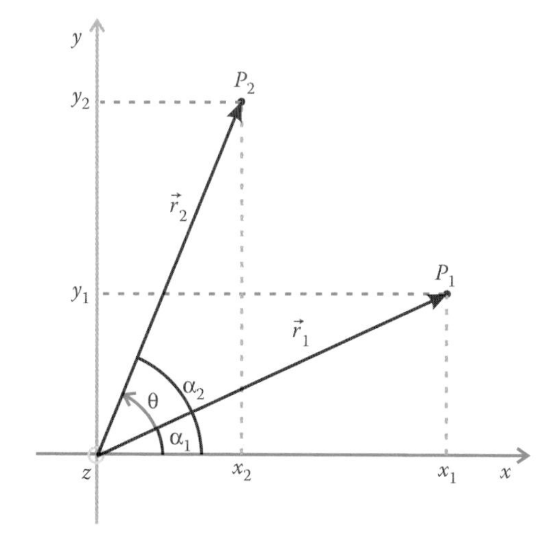

FIGURE 1.49 Rotation of the position vector through an angle θ around the z axis.

But $r_2 = r_1$, then

$$\begin{cases} x_2 = r_1 \cos\alpha_1 \cos\theta - r_1 \sin\alpha_1 \sin\theta = x_1 \cos\theta - y_1 \sin\theta \\ y_2 = r_1 \sin\alpha_1 \cos\theta + r_1 \cos\alpha_1 \sin\theta = x_1 \sin\theta + y_1 \cos\theta \end{cases} \tag{1.23}$$

The relationship in 1.23 between the initial and final coordinates can be written using matrices as

$$\begin{bmatrix} x_2 \\ y_2 \end{bmatrix} = \begin{bmatrix} \cos\theta & -\sin\theta \\ \sin\theta & \cos\theta \end{bmatrix} \begin{bmatrix} x_1 \\ y_1 \end{bmatrix} \tag{1.24}$$

Therefore, the matrix expression of rotation C_θ for tridimensional space is

$$\begin{bmatrix} x_2 \\ y_2 \\ z_2 \end{bmatrix} = \begin{bmatrix} \cos\theta & -\sin\theta & 0 \\ \sin\theta & \cos\theta & 0 \\ 0 & 0 & 1 \end{bmatrix} \begin{bmatrix} x_1 \\ y_1 \\ z_1 \end{bmatrix} \Rightarrow C_\theta = \begin{bmatrix} \cos\theta & -\sin\theta & 0 \\ \sin\theta & \cos\theta & 0 \\ 0 & 0 & 1 \end{bmatrix} \tag{1.25}$$

We will verify the correctness of the result for a particular angle, 180°. In Section 1.2.2, we have written based on illustration:

$$C_2(z)[x_1, y_1, z_1] = [-x_1, -y_1, +z_1] \tag{1.26}$$

Using matrix, we can now obtain the same result:

$$\begin{bmatrix} \cos 180^\circ & -\sin 180^\circ & 0 \\ \sin 180^\circ & \cos 180^\circ & 0 \\ 0 & 0 & 1 \end{bmatrix} \begin{bmatrix} x_1 \\ y_1 \\ z_1 \end{bmatrix} = \begin{bmatrix} -1 & 0 & 0 \\ 0 & -1 & 0 \\ 0 & 0 & 1 \end{bmatrix} \begin{bmatrix} x_1 \\ y_1 \\ z_1 \end{bmatrix} = \begin{bmatrix} -x_1 \\ -y_1 \\ z_1 \end{bmatrix} \tag{1.27}$$

The matrix method seems to be more complicated but it *works for any value* of the rotation angle.

We note that the matrix representation found *is valuable for counter-clockwise direction (trigonometric direction)*, where the angle has positive values. For clockwise direction, where the angle has negative values, the matrix of $C(z)$ rotation will be

$$\begin{bmatrix} \cos\theta & \sin\theta & 0 \\ -\sin\theta & \cos\theta & 0 \\ 0 & 0 & 1 \end{bmatrix} \tag{1.28}$$

What is the matrix for the *reflection* operation through the plane yOz? This reflection will change only the sign of the x coordinate

(Figure 1.50). Thus, we can write the matrix equation for the particular reflection $\Sigma(y,z)$:

$$\underbrace{\begin{bmatrix} -1 & 0 & 0 \\ 0 & 1 & 0 \\ 0 & 0 & 1 \end{bmatrix}}_{\Sigma(y,z)} \begin{bmatrix} x_1 \\ y_1 \\ z_1 \end{bmatrix} = \begin{bmatrix} -x_1 \\ y_1 \\ z_1 \end{bmatrix} \tag{1.29}$$

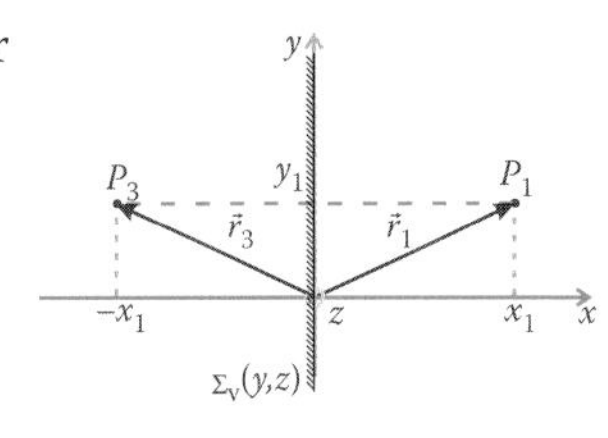

FIGURE 1.50 Reflection through the *yOz* plane changes the sign of the *x* coordinate.

It is obvious that the matrix of the reflection through the xOz plane and through the xOy plane will have similar expressions:

$$\Sigma(x,z): \begin{bmatrix} 1 & 0 & 0 \\ 0 & -1 & 0 \\ 0 & 0 & 1 \end{bmatrix} \tag{1.30}$$

$$\Sigma(x,y): \begin{bmatrix} 1 & 0 & 0 \\ 0 & 1 & 0 \\ 0 & 0 & -1 \end{bmatrix} \tag{1.31}$$

The matrix of the reflection through the dihedral plane depends on the angle between the plane and the coordinate axes. For the reflection plane, which contains the z axis and the bisector of the xOy plane, the effect is to interchange the coordinates x and y (Figure 1.51).

Therefore, the matrix equation is

$$\begin{bmatrix} 0 & 1 & 0 \\ 1 & 0 & 0 \\ 0 & 0 & 1 \end{bmatrix} \begin{bmatrix} x_1 \\ y_1 \\ z_1 \end{bmatrix} = \begin{bmatrix} y_1 \\ x_1 \\ z_1 \end{bmatrix} \tag{1.32}$$

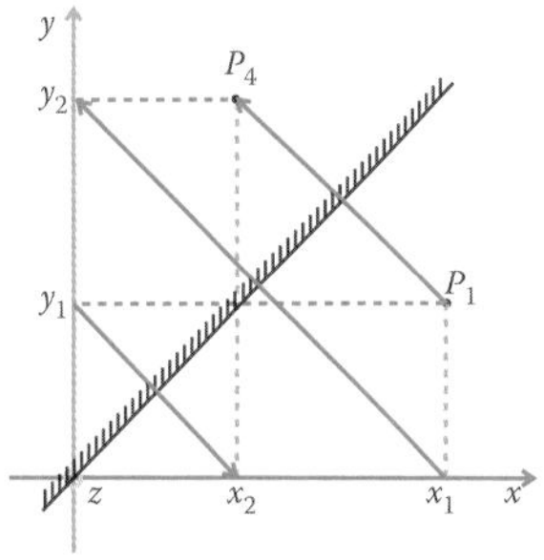

FIGURE 1.51 Effect of reflection through a plane, which contains the *z* axis and the bisector of the *xOy* plane, is to interchange the coordinates *x* and *y*.

The dihedral plane that contains the z axis and is oriented at 135° to the x axis has the matrix

$$\begin{bmatrix} 0 & -1 & 0 \\ -1 & 0 & 0 \\ 0 & 0 & 1 \end{bmatrix} \tag{1.33}$$

Let us now find the matrix of a *general reflection* through the plane that contains the z axis and makes an arbitrary angle θ with the x axis (Figure 1.52). We have to perform three consecutive transformations of initial coordinates for which we know the matrices:

- The first moment is to find the point coordinates into a system $x'Oy'$ rotated counterclockwise by θ. The matrix of such

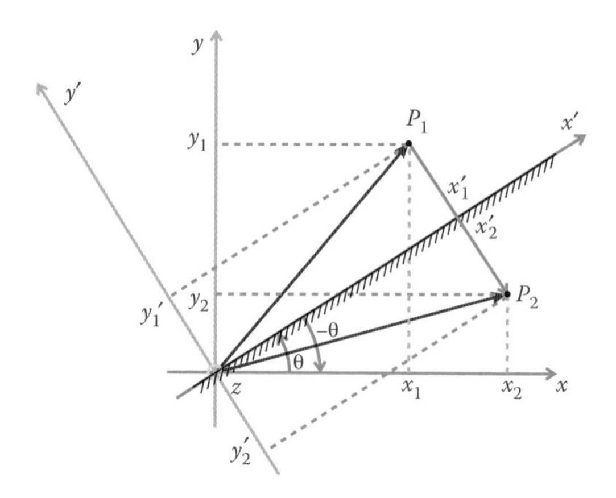

FIGURE 1.52 Reflection through the plane containing the *z* axis and an arbitrary axis in the *xy* plane can be performed by three consecutive operations with known matrices.

transformation is similar to a fixed system, and performing a clockwise rotation by θ (Equation 1.28):

$$C_\theta \begin{bmatrix} x_1 \\ y_1 \\ z_1 \end{bmatrix} = \begin{bmatrix} \cos\theta & \sin\theta & 0 \\ -\sin\theta & \cos\theta & 0 \\ 0 & 0 & 1 \end{bmatrix} \begin{bmatrix} x_1 \\ y_1 \\ z_1 \end{bmatrix} = \begin{bmatrix} x'_1 \\ y'_1 \\ z'_1 \end{bmatrix} \tag{1.34}$$

- The second moment is to find the coordinates of the image point P_2 as a result of the reflection of the point P_1 through the mirror plane, which coincides with the $x'Oz$ plane. This means to apply the reflection matrix from Equation 1.30:

$$\Sigma_{x'z} \begin{bmatrix} x'_1 \\ y'_1 \\ z'_1 \end{bmatrix} = \begin{bmatrix} 1 & 0 & 0 \\ 0 & -1 & 0 \\ 0 & 0 & 1 \end{bmatrix} \begin{bmatrix} x'_1 \\ y'_1 \\ z'_1 \end{bmatrix} = \begin{bmatrix} x'_2 \\ y'_2 \\ z'_2 \end{bmatrix} \tag{1.35}$$

- The third moment is to find the coordinates of the image point P_2 in the original system xOy. It means to rotate back (clockwise) the coordinates systems to the initial position. The matrix is the same as the counterclockwise rotation of the point (Equation 1.25):

$$C_{-\theta} \begin{bmatrix} x'_2 \\ y'_2 \\ z'_2 \end{bmatrix} = \begin{bmatrix} \cos\theta & -\sin\theta & 0 \\ \sin\theta & \cos\theta & 0 \\ 0 & 0 & 1 \end{bmatrix} \begin{bmatrix} x'_2 \\ y'_2 \\ z'_2 \end{bmatrix} = \begin{bmatrix} x_2 \\ y_2 \\ z_2 \end{bmatrix} \tag{1.36}$$

Finally, we have to make the matrix product of involved transformations in order to find the matrix of a general reflection. Therefore, we should look for the matrix that is the product:

(1.37)

$$\Sigma_\theta = C_{-\theta}\Sigma_{x'z}C_\theta = \begin{bmatrix} \cos\theta & -\sin\theta & 0 \\ \sin\theta & \cos\theta & 0 \\ 0 & 0 & 1 \end{bmatrix}\begin{bmatrix} 1 & 0 & 0 \\ 0 & -1 & 0 \\ 0 & 0 & 1 \end{bmatrix}\begin{bmatrix} \cos\theta & \sin\theta & 0 \\ -\sin\theta & \cos\theta & 0 \\ 0 & 0 & 1 \end{bmatrix}$$

$$= \begin{bmatrix} \cos^2\theta - \sin^2\theta & 2\sin\theta\cos\theta & 0 \\ 2\sin\theta\cos\theta & \sin^2\theta - \cos^2\theta & 0 \\ 0 & 0 & 1 \end{bmatrix}$$

Using trigonometric relationships

(1.38)
$$\begin{cases} \sin 2\theta = 2\sin\theta\cos\theta \\ \cos 2\theta = \cos^2\theta - \sin^2\theta \end{cases}$$

the general matrix of a reflection through a plane containing the z axis and making angle θ with the x axis in the counterclockwise direction can be written as

(1.39)
$$\Sigma_\theta = \begin{bmatrix} \cos 2\theta & \sin 2\theta & 0 \\ \sin 2\theta & -\cos 2\theta & 0 \\ 0 & 0 & 1 \end{bmatrix}$$

We can test this formula for already-known reflections:

- $\Sigma(x,z)$ means $\theta = 0°$, so

$$\Sigma_{x,z} = \begin{bmatrix} 1 & 0 & 0 \\ 0 & -1 & 0 \\ 0 & 0 & 1 \end{bmatrix}$$

  is exactly the matrix given by Equation 1.30.
- $\Sigma_{d45°}$ means $\theta = 45°$, so

$$\Sigma_{d,45°} = \begin{bmatrix} 0 & 1 & 0 \\ 1 & 0 & 0 \\ 0 & 0 & 1 \end{bmatrix}$$

  which is the matrix given by Equation 1.32.

We underline that the formula given by Equation 1.39 is valuable only for a plane containing the z axis that is perpendicular to the xy plane (i.e., vertical or dihedral), such that the matrix element a_{33} is always equal to 1, which means the z coordinate is left unchanged. But this formula can be easily adapted for any plane containing the x axis and perpendicular to the yz plane or containing the y axis and perpendicular to the xz plane.

The *inversion* has a simple matrix that must change the sign of every coordinate:

(1.40)
$$I = \begin{bmatrix} -1 & 0 & 0 \\ 0 & -1 & 0 \\ 0 & 0 & -1 \end{bmatrix}$$

The *identity* is simply described by the unit matrix, which has no effect to any coordinate:

(1.41)
$$E = \begin{bmatrix} 1 & 0 & 0 \\ 0 & 1 & 0 \\ 0 & 0 & 1 \end{bmatrix}$$

To find the matrix of *improper rotation*, we proceed as in the previous case of the reflection through a plane at any angle. The effect of an improper rotation through the angle θ around the z axis is similar to the effect of a proper rotation through the same angle θ around the z axis followed by a reflection through the xy plane, which changes the sign of the z coordinate. Thus, keeping in mind the rule to order the matrices from the right to the left, we can write for an *improper rotation around the z axis in the counterclockwise direction*:

(1.42)
$$S_\theta = \Sigma_{xy} C_\theta = \begin{bmatrix} 1 & 0 & 0 \\ 0 & 1 & 0 \\ 0 & 0 & -1 \end{bmatrix} \begin{bmatrix} \cos\theta & -\sin\theta & 0 \\ \sin\theta & \cos\theta & 0 \\ 0 & 0 & 1 \end{bmatrix} = \begin{bmatrix} \cos\theta & -\sin\theta & 0 \\ \sin\theta & \cos\theta & 0 \\ 0 & 0 & -1 \end{bmatrix}$$

In Section 1.2.3, we worked to find the complete set of symmetry elements and to establish the multiplication table of the point group. The same work can be done using matrices. But to exercise matrices on the allene molecule is not very useful because the symmetry elements are rotation axes c_2 collinear with coordinate axes and reflection planes Σ_d containing coordinate axes, such that we only have to apply directly the matrices that we just established.

Examples of not-so-trivial molecules are boron trifluoride (BF_3) and methane (CH_4).

1. We take as the first example the BF_3 molecule from Figure 1.53. We have already found that there are three threefold rotations C_3^1, C_3^2, and C_3^3. We have also seen other three twofold rotations C_2', C_2'', and C_2'''.

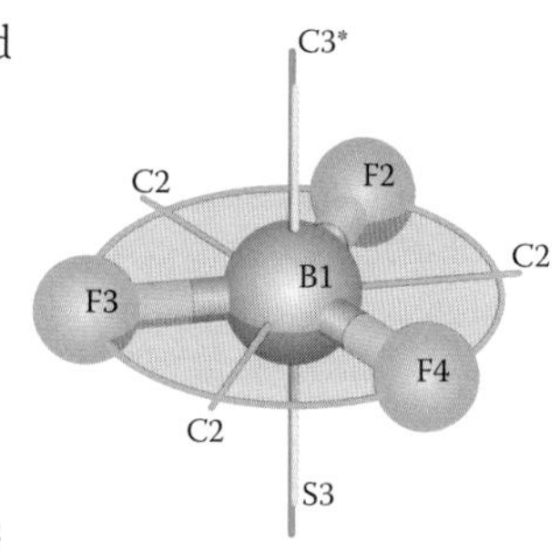

FIGURE 1.53
Configuration of the BF_3 molecule and rotation axes C_3, S_3, and C_2.

The corresponding matrices of counterclockwise rotations C_3^1 and C_3^2 around the z axis are

$$C_3^1 = \begin{bmatrix} \cos 120^\circ & -\sin 120^\circ & 0 \\ \sin 120^\circ & \cos 120^\circ & 0 \\ 0 & 0 & 1 \end{bmatrix} = \begin{bmatrix} -\frac{1}{2} & -\frac{\sqrt{3}}{2} & 0 \\ \frac{\sqrt{3}}{2} & -\frac{1}{2} & 0 \\ 0 & 0 & 1 \end{bmatrix} \tag{1.43}$$

$$C_3^2 = \begin{bmatrix} \cos 240^\circ & -\sin 240^\circ & 0 \\ \sin 240^\circ & \cos 240^\circ & 0 \\ 0 & 0 & 1 \end{bmatrix} = \begin{bmatrix} -\frac{1}{2} & \frac{\sqrt{3}}{2} & 0 \\ -\frac{\sqrt{3}}{2} & -\frac{1}{2} & 0 \\ 0 & 0 & 1 \end{bmatrix} \tag{1.44}$$

Is matrix C_3^2 the inverse of matrix C_3^1? We write the inverse of matrix C_3^1 by transposing each row of the matrix as a new column; therefore

$$(C_3^1)^{-1} = \begin{bmatrix} -\frac{1}{2} & -\frac{\sqrt{3}}{2} & 0 \\ \frac{\sqrt{3}}{2} & -\frac{1}{2} & 0 \\ 0 & 0 & 1 \end{bmatrix}^{-1} = \begin{bmatrix} -\frac{1}{2} & \frac{\sqrt{3}}{2} & 0 \\ -\frac{\sqrt{3}}{2} & -\frac{1}{2} & 0 \\ 0 & 0 & 1 \end{bmatrix} = C_3^2 \tag{1.45}$$

So, they form a class that will be written as $2C_3^1$.

If the axis of rotation C_2' is collinear with the x axis, its matrix can be easily written by knowing the effect of unchanging the coordinate x and changing the sign of coordinates y and z:

(1.46)
$$C_2' = \begin{bmatrix} 1 & 0 & 0 \\ 0 & -1 & 0 \\ 0 & 0 & -1 \end{bmatrix}$$

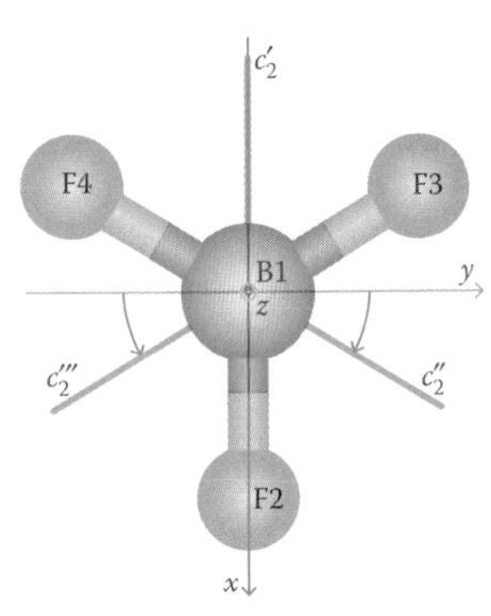

FIGURE 1.54
Orientation of rotation axes c_2'' and c_2''' reported to coordinate axes.

We can also apply Equation 1.25 but taking into account that the rotation axis is now x axis:

(1.47)
$$C_2' = \begin{bmatrix} 1 & 0 & 0 \\ 0 & \cos\theta & -\sin\theta \\ 0 & \sin\theta & \cos\theta \end{bmatrix} = \begin{bmatrix} 1 & 0 & 0 \\ 0 & -1 & 0 \\ 0 & 0 & -1 \end{bmatrix}$$

The matrices of C_2'' and C_2''' rotations are not evident because the rotation axes are not collinear with any coordinate axis (Figure 1.54).

The rotation axis c_2'' is at an angle of 30° to the y axis. We suggest applying the method previously used to establish the matrix of a general rotation given by Equation 1.39. We will rotate the coordinates system by 30° around the z axis, such that the y axis becomes collinear to the c_2'' axis named y'. Then, we will perform a C_2 rotation around the new y' axis and will return to the initial position of the y axis through a rotation by −30° around the z axis. The 30° clockwise rotation of the coordinate system is similar to the counterclockwise rotation C_{12}. The last rotation will be similar to C_{-12}. Therefore, the matrix of rotation C_2'' around the c_2'' axis will be given by the product:

(1.48)

$$C_2'' = C_{-12}C_{2y'}C_{12}$$

$$= \begin{bmatrix} \cos 30^\circ & \sin 30^\circ & 0 \\ -\sin 30^\circ & \cos 30^\circ & 0 \\ 0 & 0 & 1 \end{bmatrix} \begin{bmatrix} -1 & 0 & 0 \\ 0 & 1 & 0 \\ 0 & 0 & -1 \end{bmatrix} \begin{bmatrix} \cos 30^\circ & -\sin 30^\circ & 0 \\ \sin 30^\circ & \cos 30^\circ & 0 \\ 0 & 0 & 1 \end{bmatrix}$$

$$= \begin{bmatrix} \cos 30^\circ & \sin 30^\circ & 0 \\ -\sin 30^\circ & \cos 30^\circ & 0 \\ 0 & 0 & 1 \end{bmatrix} \begin{bmatrix} -\cos 30^\circ & \sin 30^\circ & 0 \\ \sin 30^\circ & \cos 30^\circ & 0 \\ 0 & 0 & -1 \end{bmatrix}$$

$$= \begin{bmatrix} -\frac{1}{2} & \frac{\sqrt{3}}{2} & 0 \\ \frac{\sqrt{3}}{2} & \frac{1}{2} & 0 \\ 0 & 0 & -1 \end{bmatrix}$$

We will apply the same method to find the matrix of rotation C_2''' around c_2''' axis that will be given by the product:

(1.49)

$$C_2''' = C_{12}C_{2y'}C_{-12}$$

$$= \begin{bmatrix} \cos 30^\circ & -\sin 30^\circ & 0 \\ \sin 30^\circ & \cos 30^\circ & 0 \\ 0 & 0 & 1 \end{bmatrix} \begin{bmatrix} -1 & 0 & 0 \\ 0 & 1 & 0 \\ 0 & 0 & -1 \end{bmatrix} \begin{bmatrix} \cos 30^\circ & \sin 30^\circ & 0 \\ -\sin 30^\circ & \cos 30^\circ & 0 \\ 0 & 0 & 1 \end{bmatrix}$$

$$= \begin{bmatrix} -\frac{1}{2} & -\frac{\sqrt{3}}{2} & 0 \\ -\frac{\sqrt{3}}{2} & \frac{1}{2} & 0 \\ 0 & 0 & -1 \end{bmatrix}$$

Based on these rotations, we start to complete Table 1.8. The part containing known operations is white while the gray-colored cells are to be discovered. The existence of vertical planes σ_v',σ_v'', and σ_v''' is obvious by a simple inspection of Figures 1.53 and 1.54. σ_h is the molecular plane containing all atoms. The improper rotation axis S_3 is superimposed on the c_3 axis. The improper rotation S_3^1 is a combination of rotation C_3^1 followed by reflection Σ_h.

TABLE 1.8 Symmetry Operations of the BF_3 Molecule

Symmetry Elements	Symmetry Operations	Irreducible Operations	Group Elements
C_3	C_3^1	C_3^1	C_3^1
	C_3^2	C_3^2	C_3^2
	C_3^3	E	C_2'
C_2'	C_2'	C_2'	C_2''
	$C_2'^2$	E	C_2'''
C_2''	C_2''	C_2''	Σ_v'
	$C_2'^2$	E	Σ_v''
C_2'''	C_2'''	C_2'''	Σ_v'''
	$C_2''^2$	E	Σ_h
σ_v'	Σ_v'	Σ_v'	S_3^1
	$\Sigma_v'^2$	E	S_3^2
σ_v''	Σ_v''	Σ_v''	E
	$\Sigma_v''^2$	E	
σ_v'''	Σ_v'''	Σ_v'''	
	$\Sigma_v'''^2$	E	
σ_h	Σ_h	Σ_h	
	Σ_h^2	E	
S_3	S_3^1	S_3^1	
	S_3^2	S_3^2	
	S_3^3	E	

What are the matrices of reflections through the planes σ'_v, σ''_v, and σ'''_v? The plane σ'_v is the plane xz and the matrix is (according to Equation 1.30)

$$\Sigma'_v = \Sigma_{xz} = \begin{bmatrix} 1 & 0 & 0 \\ 0 & -1 & 0 \\ 0 & 0 & 1 \end{bmatrix} \tag{1.50}$$

The plane σ''_v contains axes z and c''_2. Because we know that the value of the angle between the plane σ''_v and axis x is 60°, we can use the relationship given by Equation 1.39:

$$\Sigma''_v = \begin{bmatrix} \cos 120^\circ & \sin 120^\circ & 0 \\ \sin 120^\circ & -\cos 120^\circ & 0 \\ 0 & 0 & 1 \end{bmatrix} = \begin{bmatrix} -\frac{1}{2} & \frac{\sqrt{3}}{2} & 0 \\ \frac{\sqrt{3}}{2} & \frac{1}{2} & 0 \\ 0 & 0 & 1 \end{bmatrix} \tag{1.51}$$

In a similar way, the matrix of the plane σ'''_v, which contains axes z and c'''_2, and making the angle of 60° with the x axis in the clockwise direction, is

$$\Sigma'''_v = \begin{bmatrix} \cos 120^\circ & -\sin 120^\circ & 0 \\ -\sin 120^\circ & -\cos 120^\circ & 0 \\ 0 & 0 & 1 \end{bmatrix} = \begin{bmatrix} -\frac{1}{2} & -\frac{\sqrt{3}}{2} & 0 \\ -\frac{\sqrt{3}}{2} & \frac{1}{2} & 0 \\ 0 & 0 & 1 \end{bmatrix} \tag{1.52}$$

The matrix of the horizontal plane xy is

$$\Sigma_h = \Sigma_{xy} = \begin{bmatrix} 1 & 0 & 0 \\ 0 & 1 & 0 \\ 0 & 0 & -1 \end{bmatrix} \tag{1.53}$$

We have still to find two improper rotations S_3^1 and S_3^2 around the same axis as C_3 rotation and we can apply the relationship given by Equation 1.42:

$$S_3^1 = \begin{bmatrix} \cos 120^\circ & -\sin 120^\circ & 0 \\ \sin 120^\circ & \cos 120^\circ & 0 \\ 0 & 0 & -1 \end{bmatrix} = \begin{bmatrix} -\frac{1}{2} & -\frac{\sqrt{3}}{2} & 0 \\ \frac{\sqrt{3}}{2} & -\frac{1}{2} & 0 \\ 0 & 0 & -1 \end{bmatrix} \tag{1.54}$$

$$(1.55)\qquad S_3^2 = \begin{bmatrix} \cos 240^\circ & -\sin 240^\circ & 0 \\ \sin 240^\circ & \cos 240^\circ & 0 \\ 0 & 0 & -1 \end{bmatrix} = \begin{bmatrix} -\frac{1}{2} & \frac{\sqrt{3}}{2} & 0 \\ -\frac{\sqrt{3}}{2} & -\frac{1}{2} & 0 \\ 0 & 0 & -1 \end{bmatrix}$$

We can check whether the matrix S_3^2 is the inverse of the matrix S_3^1 by simple transposing of the first one:

$$(1.56)\qquad (S_3^1)^{-1} = \begin{bmatrix} -\frac{1}{2} & -\frac{\sqrt{3}}{2} & 0 \\ \frac{\sqrt{3}}{2} & -\frac{1}{2} & 0 \\ 0 & 0 & -1 \end{bmatrix}^{-1} = \begin{bmatrix} -\frac{1}{2} & \frac{\sqrt{3}}{2} & 0 \\ -\frac{\sqrt{3}}{2} & -\frac{1}{2} & 0 \\ 0 & 0 & -1 \end{bmatrix} = S_3^2$$

and comparing to Equation 1.55 we find that they are inverse to each other.

So, we have a set of symmetry operations:

$$E, C_3^1, C_3^2, C_2', C_2'', C_2''', \Sigma_v', \Sigma_v'', \Sigma_v''', \Sigma_h, S_3^1, S_3^2$$

and we have to see how these operations are collected into classes. Because this molecule has a relatively simple geometry, we can collect them in six classes by simple inspection (this time geometry is faster than algebra):

$$1 = E;\ 2 = C_3^1, C_3^2;\ 3 = C_2', C_2'', C_2''';\ 4 = \Sigma_v', \Sigma_v'', \Sigma_v''';\ 5 = \Sigma_\mathrm{h};\ 6 = S_3^1, S_3^2.$$

We will now show how to use the general method to collect operations in classes. The first step is to make the *similarity transformation* of every operation A of the group $X^{-1}AX$, using as X all the operations in the group, including itself. The result of similarity transformation $X^{-1}AX$ is B, the other member of group, named *conjugate* of A.

The class is the complete set of elements that are conjugated to each other.

Let us start with E:

$$E^{-1}EE = E$$

$$(C_3^1)^{-1}EC_3^1 = \left(C_3^1\right)^{-1}C_3^1 = E$$

....

The conclusion is that E forms a class by itself.
We continue with C_3^1:

$$E^{-1}C_3^1E = C_3^1$$

$$(C_3^1)^{-1}C_3^1C_3^1 = C_3^1$$

$$(C_3^2)^{-1}C_3^1C_3^2 = (C_3^2)^{-1}C_3^3 = (C_3^2)^{-1} = C_3^1$$

To go forward, we need to introduce the matrices

$$(C_2')^{-1}C_3^1C_2' = \begin{bmatrix} 1 & 0 & 0 \\ 0 & -1 & 0 \\ 0 & 0 & -1 \end{bmatrix} \begin{bmatrix} -\frac{1}{2} & -\frac{\sqrt{3}}{2} & 0 \\ \frac{\sqrt{3}}{2} & -\frac{1}{2} & 0 \\ 0 & 0 & 1 \end{bmatrix} \begin{bmatrix} 1 & 0 & 0 \\ 0 & -1 & 0 \\ 0 & 0 & -1 \end{bmatrix}$$

$$= \begin{bmatrix} -\frac{1}{2} & \frac{\sqrt{3}}{2} & 0 \\ -\frac{\sqrt{3}}{2} & -\frac{1}{2} & 0 \\ 0 & 0 & 1 \end{bmatrix} = C_3^2$$

$$(C'')^{-1}C_3^1C_2'' = \begin{bmatrix} -\frac{1}{2} & \frac{\sqrt{3}}{2} & 0 \\ \frac{\sqrt{3}}{2} & \frac{1}{2} & 0 \\ 0 & 0 & -1 \end{bmatrix} \begin{bmatrix} -\frac{1}{2} & -\frac{\sqrt{3}}{2} & 0 \\ \frac{\sqrt{3}}{2} & -\frac{1}{2} & 0 \\ 0 & 0 & 1 \end{bmatrix} \begin{bmatrix} -\frac{1}{2} & \frac{\sqrt{3}}{2} & 0 \\ \frac{\sqrt{3}}{2} & \frac{1}{2} & 0 \\ 0 & 0 & -1 \end{bmatrix}$$

$$= \begin{bmatrix} -\frac{1}{2} & \frac{\sqrt{3}}{2} & 0 \\ -\frac{\sqrt{3}}{2} & -\frac{1}{2} & 0 \\ 0 & 0 & 1 \end{bmatrix} = C_3^2$$

$$(C_2''')^{-1}\,C_3^1C_2'''=\begin{bmatrix}-\frac{1}{2} & -\frac{\sqrt{3}}{2} & 0\\ -\frac{\sqrt{3}}{2} & \frac{1}{2} & 0\\ 0 & 0 & -1\end{bmatrix}\begin{bmatrix}-\frac{1}{2} & -\frac{\sqrt{3}}{2} & 0\\ \frac{\sqrt{3}}{2} & -\frac{1}{2} & 0\\ 0 & 0 & 1\end{bmatrix}\begin{bmatrix}-\frac{1}{2} & -\frac{\sqrt{3}}{2} & 0\\ -\frac{\sqrt{3}}{2} & \frac{1}{2} & 0\\ 0 & 0 & -1\end{bmatrix}$$

$$=\begin{bmatrix}-\frac{1}{2} & \frac{\sqrt{3}}{2} & 0\\ -\frac{\sqrt{3}}{2} & -\frac{1}{2} & 0\\ 0 & 0 & 1\end{bmatrix}=C_3^2$$

$$(\Sigma_v')^{-1}C_3^1\Sigma_v' =\begin{bmatrix}1 & 0 & 0\\ 0 & -1 & 0\\ 0 & 0 & 1\end{bmatrix}\begin{bmatrix}-\frac{1}{2} & -\frac{\sqrt{3}}{2} & 0\\ \frac{\sqrt{3}}{2} & -\frac{1}{2} & 0\\ 0 & 0 & 1\end{bmatrix}\begin{bmatrix}1 & 0 & 0\\ 0 & -1 & 0\\ 0 & 0 & 1\end{bmatrix}$$

$$=\begin{bmatrix}-\frac{1}{2} & \frac{\sqrt{3}}{2} & 0\\ -\frac{\sqrt{3}}{2} & -\frac{1}{2} & 0\\ 0 & 0 & 1\end{bmatrix}=C_3^2$$

$$(\Sigma_v''')^{-1}C_3^1\Sigma_v''' =\begin{bmatrix}-\frac{1}{2} & -\frac{\sqrt{3}}{2} & 0\\ -\frac{\sqrt{3}}{2} & \frac{1}{2} & 0\\ 0 & 0 & 1\end{bmatrix}\begin{bmatrix}-\frac{1}{2} & -\frac{\sqrt{3}}{2} & 0\\ \frac{\sqrt{3}}{2} & -\frac{1}{2} & 0\\ 0 & 0 & 1\end{bmatrix}\begin{bmatrix}-\frac{1}{2} & -\frac{\sqrt{3}}{2} & 0\\ -\frac{\sqrt{3}}{2} & \frac{1}{2} & 0\\ 0 & 0 & 1\end{bmatrix}$$

$$=\begin{bmatrix}-\frac{1}{2} & -\frac{\sqrt{3}}{2} & 0\\ \frac{\sqrt{3}}{2} & -\frac{1}{2} & 0\\ 0 & 0 & 1\end{bmatrix}=C_3^1$$

$$\left(\Sigma_{\mathrm{h}}\right)^{-1} C_3^1 \Sigma_{\mathrm{h}} = \begin{bmatrix} 1 & 0 & 0 \\ 0 & 1 & 0 \\ 0 & 0 & -1 \end{bmatrix} \begin{bmatrix} -\frac{1}{2} & -\frac{\sqrt{3}}{2} & 0 \\ \frac{\sqrt{3}}{2} & -\frac{1}{2} & 0 \\ 0 & 0 & 1 \end{bmatrix} \begin{bmatrix} 1 & 0 & 0 \\ 0 & 1 & 0 \\ 0 & 0 & -1 \end{bmatrix}$$

$$= \begin{bmatrix} -\frac{1}{2} & -\frac{\sqrt{3}}{2} & 0 \\ \frac{\sqrt{3}}{2} & -\frac{1}{2} & 0 \\ 0 & 0 & 1 \end{bmatrix} = C_3^1$$

$$\left(S_3^1\right)^{-1} C_3^1 S_3^1 = \begin{bmatrix} -\frac{1}{2} & \frac{\sqrt{3}}{2} & 0 \\ -\frac{\sqrt{3}}{2} & -\frac{1}{2} & 0 \\ 0 & 0 & -1 \end{bmatrix} \begin{bmatrix} -\frac{1}{2} & -\frac{\sqrt{3}}{2} & 0 \\ \frac{\sqrt{3}}{2} & -\frac{1}{2} & 0 \\ 0 & 0 & 1 \end{bmatrix} \begin{bmatrix} -\frac{1}{2} & -\frac{\sqrt{3}}{2} & 0 \\ \frac{\sqrt{3}}{2} & -\frac{1}{2} & 0 \\ 0 & 0 & -1 \end{bmatrix}$$

$$= \begin{bmatrix} -\frac{1}{2} & -\frac{\sqrt{3}}{2} & 0 \\ \frac{\sqrt{3}}{2} & -\frac{1}{2} & 0 \\ 0 & 0 & 1 \end{bmatrix} = C_3^1$$

$$\left(S_3^2\right)^{-1} C_3^1 S_3^2 = \begin{bmatrix} -\frac{1}{2} & -\frac{\sqrt{3}}{2} & 0 \\ \frac{\sqrt{3}}{2} & -\frac{1}{2} & 0 \\ 0 & 0 & -1 \end{bmatrix} \begin{bmatrix} -\frac{1}{2} & -\frac{\sqrt{3}}{2} & 0 \\ \frac{\sqrt{3}}{2} & -\frac{1}{2} & 0 \\ 0 & 0 & 1 \end{bmatrix} \begin{bmatrix} -\frac{1}{2} & \frac{\sqrt{3}}{2} & 0 \\ -\frac{\sqrt{3}}{2} & -\frac{1}{2} & 0 \\ 0 & 0 & -1 \end{bmatrix}$$

$$= \begin{bmatrix} -\frac{1}{2} & -\frac{\sqrt{3}}{2} & 0 \\ \frac{\sqrt{3}}{2} & -\frac{1}{2} & 0 \\ 0 & 0 & 1 \end{bmatrix} = C_3^1$$

The conclusion of the similarity transformations applied to C_3^1 and, then, to C_3^2 (the reader should perform this exercise) is that they are conjugated to each other and together form a class of group. The exercise can be repeated for all other elements of the group.

TABLE 1.9 Multiplication Table of the D_{3h} Group

D_{3h}	E	C_3^1	C_3^2	C_2'	C_2''	C_2'''	Σ_v'	Σ_v''	Σ_v'''	Σ_h	S_3^1	S_3^2
E	E	C_3^1	C_3^2	C_2'	C_2''	C_2'''	Σ_v'	Σ_v''	Σ_v'''	Σ_h	S_3^1	S_3^2
C_3^1	C_3^1	C_3^2	E	C_2''	C_2'''	C_2'	Σ_v''	Σ_v'''	Σ_v'	S_3^1	S_3^2	Σ_h
C_3^2	C_3^2	E	C_3^1									
C_2'	C_2'			E								
C_2''	C_2''				E							
C_2'''	C_2'''					E						
Σ_v'	Σ_v'						E					
Σ_v''	Σ_v''							E				
Σ_v'''	Σ_v'''								E			
Σ_h	Σ_h									E		
S_3^1	S_3^1										S_3^2	E
S_3^2	S_3^2										E	S_3^1

Finally, we can write the multiplication table of the D_{3h} group in Table 1.9.

2. We take as second example of the molecule of CH_4, which was studied in detail in Section 1.3.4. We are now interested only in the matrices generated by the four c_3 axes, which are not collinear with any coordinate axis. The rotation axes are oriented from the central carbon atom to each hydrogen atom, which are placed in the vertices of the cube (Figure 1.55).

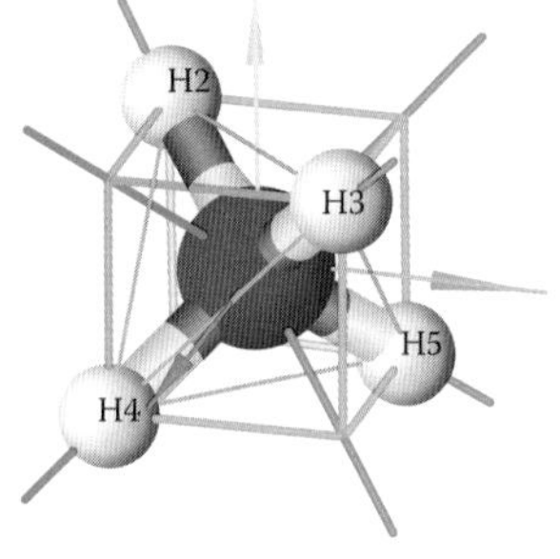

FIGURE 1.55 Methane molecule has four proper rotation axes c_3 oriented to the vertices of the cube in which it can be inscribed.

Let us take the rotation around the axis oriented to the vertex H3 of coordinates (1, 1, 1). The counterclockwise C_3 rotation will bring the x axis into the y axis, the y axis into the z axis, and the z axis into the x axis without changing the sign of the coordinate. Thus, we can discover the matrix as a result of the transformation that it can make:

$$\begin{bmatrix} z \\ x \\ y \end{bmatrix} = C_3(1,1,1)\begin{bmatrix} x \\ y \\ z \end{bmatrix} = \begin{bmatrix} a_{11} & a_{12} & a_{13} \\ a_{21} & a_{22} & a_{23} \\ a_{31} & a_{32} & a_{33} \end{bmatrix}\begin{bmatrix} x \\ y \\ z \end{bmatrix} = \begin{bmatrix} 0 & 0 & 1 \\ 1 & 0 & 0 \\ 0 & 1 & 0 \end{bmatrix}\begin{bmatrix} x \\ y \\ z \end{bmatrix}$$

Thus, the matrix of counterclockwise rotation C_3 is

$$C_3(1,1,1) = \begin{bmatrix} 0 & 0 & 1 \\ 1 & 0 & 0 \\ 0 & 1 & 0 \end{bmatrix} \tag{1.57}$$

The counterclockwise rotation C_3^2 around the same axis, which brings the x axis into the z axis, the y axis into the x axis, and the z

axis into the y axis without changing the sign of the coordinate, has the matrix

(1.58)
$$C_3^2(1,1,1) = \begin{bmatrix} 0 & 1 & 0 \\ 0 & 0 & 1 \\ 1 & 0 & 0 \end{bmatrix}$$

and this is the transpose of the matrix given by Equation 1.57. We expected the rotations C_3^1 and C_3^2 to be inverse to each other.

In an analog manner, we can write the matrices of the C_3^1 rotation around the other axes:

(1.59)
$$\text{H2:} \quad C_3(-1,-1,1) = \begin{bmatrix} 0 & 1 & 0 \\ 0 & 0 & -1 \\ -1 & 0 & 0 \end{bmatrix}$$

(1.60)
$$\text{H4:} \quad C_3(1,-1,-1) = \begin{bmatrix} 0 & 0 & -1 \\ -1 & 0 & 0 \\ 0 & 1 & 0 \end{bmatrix}$$

(1.61)
$$\text{H5:} \quad C_3(-1,1,-1) = \begin{bmatrix} 0 & 0 & 1 \\ -1 & 0 & 0 \\ 0 & -1 & 0 \end{bmatrix}$$

The reader should continue to exercise on all matrices of the $\boldsymbol{T}_d$ group. The knowledge or intuition to discover matrices of the symmetry operations and to work with them will be very useful later in this book.

1.5 GROUP REPRESENTATIONS

We have seen until now the point group that describes the symmetry of the molecule, which is a set of symmetry operations that we can apply to the molecule without changing its size or shape. We can complete the multiplication table with these operations using the property that the product of two operations of the group will always result in another operation of the group. This means the operations can generate each other and the group is complete. So, the group can be represented by these symmetry operations.

We have also found that each operation applied to a point determined by three coordinates can generate a 3×3 matrix. There is a distinct matrix for each operation of the group. Thus, we can generate a complete set of matrices to describe the symmetry of the molecule. Therefore,

> We can generate a 3×3 matrix representation of the group starting from a set of three coordinates named *base*.

Thus we can re-write the multiplication table of the $\boldsymbol{D}_{3h}$ group as shown in the Table 1.10 by using matrix representations (only for few elements because of the limited page size).

As you can see, the matrix representation is rigorous, complete, but extremely dense. It has the advantage to be extremely efficient from the computational point of view. Can we find other representations of the group? To answer this question, we must remember how we generated the 3×3 matrices. We started to discover the transformations of a point

TABLE 1.10 Multiplication Table of the D_{3h} Group in Matrix Format (Partly Shown Here)

D_{3h}	E	C_3^1	...	Σ_v'	...	Σ_h	...
E	$\begin{bmatrix} 1 & 0 & 0 \\ 0 & 1 & 0 \\ 0 & 0 & 1 \end{bmatrix}$	$\begin{bmatrix} -\frac{1}{2} & -\frac{\sqrt{3}}{2} & 0 \\ \frac{\sqrt{3}}{2} & -\frac{1}{2} & 0 \\ 0 & 0 & 1 \end{bmatrix}$	...	$\begin{bmatrix} 1 & 0 & 0 \\ 0 & -1 & 0 \\ 0 & 0 & 1 \end{bmatrix}$	...	$\begin{bmatrix} 1 & 0 & 0 \\ 0 & 1 & 0 \\ 0 & 0 & -1 \end{bmatrix}$	...
C_3^1	$\begin{bmatrix} -\frac{1}{2} & -\frac{\sqrt{3}}{2} & 0 \\ \frac{\sqrt{3}}{2} & -\frac{1}{2} & 0 \\ 0 & 0 & 1 \end{bmatrix}$	$\begin{bmatrix} -\frac{1}{2} & \frac{\sqrt{3}}{2} & 0 \\ -\frac{\sqrt{3}}{2} & -\frac{1}{2} & 0 \\ 0 & 0 & 1 \end{bmatrix}$	...	$\begin{bmatrix} -\frac{1}{2} & \frac{\sqrt{3}}{2} & 0 \\ \frac{\sqrt{3}}{2} & \frac{1}{2} & 0 \\ 0 & 0 & 1 \end{bmatrix}$	...	$\begin{bmatrix} -\frac{1}{2} & -\frac{\sqrt{3}}{2} & 0 \\ \frac{\sqrt{3}}{2} & -\frac{1}{2} & 0 \\ 0 & 0 & -1 \end{bmatrix}$	...
...	...	...	...	...	...	...	...
Σ_v'	$\begin{bmatrix} 1 & 0 & 0 \\ 0 & -1 & 0 \\ 0 & 0 & 1 \end{bmatrix}$	$\begin{bmatrix} -\frac{1}{2} & -\frac{\sqrt{3}}{2} & 0 \\ -\frac{\sqrt{3}}{2} & \frac{1}{2} & 0 \\ 0 & 0 & 1 \end{bmatrix}$	...	$\begin{bmatrix} 1 & 0 & 0 \\ 0 & 1 & 0 \\ 0 & 0 & 1 \end{bmatrix}$	...	$\begin{bmatrix} 1 & 0 & 0 \\ 0 & -1 & 0 \\ 0 & 0 & -1 \end{bmatrix}$	...
...	...	...	...	...	...	...	...
Σ_h	$\begin{bmatrix} 1 & 0 & 0 \\ 0 & 1 & 0 \\ 0 & 0 & -1 \end{bmatrix}$	$\begin{bmatrix} -\frac{1}{2} & -\frac{\sqrt{3}}{2} & 0 \\ \frac{\sqrt{3}}{2} & -\frac{1}{2} & 0 \\ 0 & 0 & -1 \end{bmatrix}$	...	$\begin{bmatrix} 1 & 0 & 0 \\ 0 & -1 & 0 \\ 0 & 0 & -1 \end{bmatrix}$	...	$\begin{bmatrix} 1 & 0 & 0 \\ 0 & 1 & 0 \\ 0 & 0 & 1 \end{bmatrix}$	...
...	...	...	...	...	...	...	...

determined by three coordinates. If we are interested in the transformation of a point determined by two coordinates, the matrices size will be 2×2. Thus, the new set of 2×2 matrices will give us a new representation of the same group $\boldsymbol{D}_{3h}$. We can even have a one-dimensional representation if we are only interested in one coordinate. Therefore, *the number and size of group representations is not limited and depends on the base* that we chose.

In the previous example, we studied the transformation of one fixed point of the molecule. But this point could be any atom of the molecule. If the molecule is rigid and the interatomic distances are constant, we obtain a correct and sufficient representation. Let us imagine that the molecule vibrates and all the atoms move. Then, we need to study an ensemble of n atoms moving without any correlation and, as a consequence, we must describe the position of each atom by a tridimensional vector. Therefore, each symmetry element can be expressed by a $(3n) \times (3n)$ matrix (!). In the case of the BF_3 molecule, we need to write 12×12 matrices.

So, the question is how can we reduce the size of the representation such as to keep enough information about the symmetry of the molecule?

Let us suppose that we have a matrix of high order, such as

$$A = \begin{bmatrix} a_{11} & a_{12} & a_{13} & a_{14} & a_{15} & a_{16} \\ a_{21} & a_{22} & a_{23} & a_{24} & a_{25} & a_{26} \\ a_{31} & a_{32} & a_{33} & a_{34} & a_{35} & a_{36} \\ a_{41} & a_{42} & a_{43} & a_{44} & a_{45} & a_{46} \\ a_{51} & a_{52} & a_{53} & a_{54} & a_{55} & a_{56} \\ a_{61} & a_{62} & a_{63} & a_{64} & a_{65} & a_{66} \end{bmatrix} \tag{1.62}$$

If we can now transform the matrix into a form containing all nonzero elements in square blocks along the diagonal, we find a block-factored matrix such as, for example

$$A' = \begin{bmatrix} a'_{11} & a'_{12} & 0 & 0 & 0 & 0 \\ a'_{21} & a'_{22} & 0 & 0 & 0 & 0 \\ 0 & 0 & a'_{33} & 0 & 0 & 0 \\ 0 & 0 & 0 & a'_{44} & a'_{45} & a'_{46} \\ 0 & 0 & 0 & a'_{54} & a'_{55} & a'_{56} \\ 0 & 0 & 0 & a'_{64} & a'_{65} & a'_{66} \end{bmatrix} \tag{1.63}$$

The blocks of the transformed matrix are lower-order (2, 1, and 3) matrices. This suggests that the transformed matrix can be reduced to simpler matrices ($A'_2, A'_1,$ and A'_3).

If we now have a group, that is a set of several matrices A, B, C, ..., which form a representation of the group, we search to perform the *same similarity transformation on each matrix*, that is

$$\begin{cases} X^{-1}AX = A' \\ X^{-1}BX = B' \\ X^{-1}CX = C' \\ \vdots \end{cases} \tag{1.64}$$

The new set of matrices A', B', C',... also forms a representation of the group. If all matrices were transformed by the similarity transformation to be *blocked out in the same way* (as in our example 2, 1, and 3), then *the initial representation can be reduced to lower-dimension representations* of the group.

In our example, the six-dimensional initial representation will be reduced to three simpler representations:

Two dimension

$$\Gamma_2 = \{A_2', B_2', C_2', ...\}$$

One dimension

$$\Gamma_1 = \{A_1', B_1', C_1', ...\}$$

Three dimension

$$\Gamma_3 = \{A_3', B_3', C_3', ...\}$$

Therefore, the initial representation formed by the matrices A, B, C,... is a *reducible* representation. If it is not possible to find a similarity transformation to reduce all initial matrices in the same way, the representation is named *irreducible.*

1.6 PROPERTIES OF IRREDUCIBLE REPRESENTATIONS

1.6.1 GREAT ORTHOGONALITY THEOREM

The *great orthogonality theorem* is a relationship between the matrix elements of the irreducible representations of a group $\boldsymbol{G}$ and may be stated as

$$\boxed{\sum_R (\Gamma_i(R)_{mn}) \cdot (\Gamma_j(R)_{m'n'})^* = \frac{h}{\sqrt{l_i l_j}} \delta_{ij}\delta_{mm'}\delta_{nn'}} \tag{1.65}$$

with

h: the order of the group
R: a symmetry operation of the group
Γ_i: an irreducible representation of type i
$\Gamma_i(R)_{mn}$: the element from m row and n column of the matrix R of irreducible representation Γ_i
l_i: the dimension of the irreducible representation type i, that is, the order of etach matrix of representation

Note: In case complex numbers are involved, we must take the complex conjugated number $((\Gamma_i(R)_{mn})^*)$.

The great orthogonality theorem is an expression of the fact that the vectors consisting of corresponding elements of the representation matrices are orthogonal and normalized. It is another way to say that if two vectors are orthogonal to each other, their scalar product is zero:

$$\begin{cases} \sum_R (\Gamma_i(R)_{mn}) \cdot (\Gamma_j(R)_{mn}) = 0 \quad \text{if } i \neq j \\ \sum_R (\Gamma_i(R)_{mn}) \cdot (\Gamma_i(R)_{m'n'}) = 0 \quad \text{if } m \neq m' \text{ and } n \neq n' \end{cases} \tag{1.66}$$

And the square of the length of a vector equals h/l_i:

$$\sum_R (\Gamma_i(R)_{mn}) \cdot (\Gamma_i(R)_{mn}) = \frac{h}{l_i} \tag{1.67}$$

1.6.2 CONSEQUENCES OF GREAT ORTHOGONALITY THEOREM

We will show how this theorem can be manipulated into several expressions solely in terms of the traces of the matrices of the irreducible representations.

The matrix trace corresponding to all the elements of a class of an irreducible representation has the same value. This is called the character of the class. So, we define the *character* "χ" of a class as the sum of the diagonal elements of the group square matrix:

$$\chi = \sum_i a_{ii} \tag{1.68}$$

The character of a matrix has an important property: it is *invariant upon similarity transformation*. The matrix character is a just number and it is quite practical to work with characters instead of the full representation matrices for a specific base. Further, the conjugated matrices have the same character (see Cotton 1990, p. 84), and thus all symmetry operations of a class are represented only by one number.

The great theorem of orthogonality has five important practical consequences in terms of characters:

Rule 1

$$\sum_i l_i^2 = \sum_i (\chi_i(E))^2 = h \tag{1.69}$$

The sum of the dimension squares of the irreducible representations (sum of the character squares of the identity element) *is equal to the order of the group.*

$\chi_i(E)$ is the character of the identity operation E from irreducible representation Γ_i.

Rule 2

$$\sum_R (\chi_i(R))^2 = h \tag{1.70}$$

The sum of character squares over all symmetry operation for a given irreducible representation is equal to the order of the group.

$\chi_i(R)$ is the character of symmetry operation R from irreducible representation Γ_i.

Rule 3

$$\sum_R \chi_i(R)\chi_j(R) = 0 \quad \text{if } i \neq j \tag{1.71}$$

The vectors whose components are characters of different irreducible representations are orthogonal.

Rule 4

The characters of all matrices of operations belonging to the same class are identical for a given type of irreducible or reducible representation.

Rule 5

The number of classes is equal to the number of irreducible representations of the group.

To illustrate the practical importance of the five rules, we will take as an example the symmetry group of water molecule $\boldsymbol{C}_{2v}$ (Section 1.3.7).

The group is composed of four symmetry operations, each in a separate class.

$\mathbf{C}_{2v}$	E	C_2	Σ_V	Σ_V'

This means that we have two important pieces of information:

- According to Rule 5, the number of irreducible representation is equal to the number of classes; thus, there are four representations.
- The group order $h=4$. According to Rule 1, the sum of the dimension squares of the irreducible representations is equal to the order of the group. Thus

$$l_1^2 + l_2^2 + l_3^2 + l_4^2 = 4$$

It is obvious that the single solution is $l_1^2 = l_2^2 = l_3^2 = l_4^2 = 1$. Since the dimension of the representation must be a positive number, results in the group have four one-dimensional irreducible representations.

One representation is always the totally symmetric representation, which has all characters equal to 1. This obeys Rule 2; the sum of character squares for a given irreducible representation is equal to the order of the group; therefore

$$\sum_R (\chi_i(R))^2 = 1^2 + 1^2 + 1^2 + 1^2 = 4 \tag{1.72}$$

Hence, we already have the first irreducible representation:

$\mathbf{C}_{2v}$	$\mathbf{E}$	C_2	Σ_V	Σ_V'
Γ_1	1	1	1	1

We know the character of the identity matrix is always a positive number (1, 2, 3, ...). Hence, the character table is as follows:

$\mathbf{C}_{2v}$	E	C_2	Σ_V	Σ_V'
Γ_1	1	1	1	1
Γ_2	1			
Γ_3	1			
Γ_4	1			

We apply again Rule 2 to find characters of other representations. From Equation 1.72, we know the characters must be combinations of ±1. Thus, we can complete the second row with ±1 but taking into account Rule 3, any two different irreducible representations are orthogonal:

$$\Gamma_1 \cdot \Gamma_2 = (1)(1) + (1)(1) + (1)(-1) + (1)(-1) = 0$$

$$\Gamma_1 \cdot \Gamma_3 = (1)(1) + (1)(-1) + (1)(-1) + (1)(1) = 0$$

$$\Gamma_1 \cdot \Gamma_4 = (1)(1) + (1)(-1) + (1)(1) + (1)(-1) = 0$$

TABLE 1.11 Table of Characters of Irreducible Representations Deduced for the Symmetry Group C_{2v}

C_{2v}	E	C_2	Σ_v	Σ'_v
Γ_1	1	1	1	1
Γ_2	1	1	−1	−1
Γ_3	1	−1	−1	1
Γ_4	1	−1	1	−1

The reader can check whether the representations Γ_2, Γ_3, and Γ_4 are orthogonal to each other. Therefore, we just discovered all irreducible representations of the symmetry group C_{2v} as shown in Table 1.11.

The previous example was more an exercise on how to apply the consequences of the great theorem of orthogonality than a method to discover the irreducible representations of groups, because most of them are more complicated (higher group order) and any textbook of group theory contains as attachment all this information in the character tables.

1.6.3 DECOMPOSITION THEOREM OF REDUCIBLE REPRESENTATIONS

The following consequence is a key point for a large number of applications of the group theory to molecular problems. The purpose is to find the relationship between any reducible representation and the irreducible representations of the group. The procedure is analogous to vector decomposition onto a set of complete orthogonal basis vectors. We already know that for any reducible representation, all matrices can be blocked along the diagonal by a proper similarity transformation. Then, each block will belong to an irreducible representation of the group.

We assume that $\Gamma(R)$ is a reducible representation of the symmetry group with the corresponding characters $\chi(R)$. We would like to know how many irreducible representations of symmetry Γ_i are contained in $\Gamma(R)$. We assume that we have transformed $\Gamma(R)$ to its blocked form $\Gamma'(R)$. As the characters are invariant with respect to this transformation, we obtain

$$\chi(R) = \chi'(R) = \sum_j a_j \chi_j(R) \tag{1.73}$$

where a_j is the number of times that irreducible representation Γ_j is contained in $\Gamma(R)$; therefore, it is the number of times the block part of irreducible representation appears along the diagonal of the block-diagonalized matrix of reducible representation. χ_j is the character of Γ_j irreducible representation of group.

We multiply each side of Equation 1.73 with $\chi_i(R)$ and then we sum over all operations of the group:

$$\sum_R \chi(R)\chi_i(R) = \sum_R \sum_j a_j \chi_j(R)\chi_i(R)$$

$$= \sum_j \sum_R a_j \chi_j(R)\chi_i(R) = \sum_j a_j \sum_R \chi_j(R)\chi_i(R) \quad (1.74)$$

We will use a combination of Rules 2 and 3 (Equations 1.70 and 1.71) in one relationship:

$$\sum_R \chi_i(R)\chi_j(R) = h\delta_{ij} \quad (1.75)$$

And we obtain from Equation 1.74

$$\sum_R \chi(R)\chi_i(R) = \sum_j a_j h\delta_{ij} = ha_i \quad (1.76)$$

Therefore, the number of irreducible representation Γ_i, which appears in a reducible representation Γ, is given by

$$\boxed{a_i = \frac{1}{h}\sum_R \chi(R)\chi_i(R)} \quad (1.77)$$

So, when we have the characters of a reducible representation, we need as input data only the characters of the irreducible representations of the group. These are listed in the tables of characters.

We take as an example the C_{4v} group, which has the character table for irreducible representations as shown in Table 1.12.

Let us suppose we found an eight-dimensional representation Γ, which can be decomposed in irreducible representations using Equation 1.77:

$$a_1 = \frac{1}{8}[1 \cdot (1)(8) + 2 \cdot (1)(0) + 1 \cdot (1)(0) + 2 \cdot (1)(2) + 2 \cdot (1)(2)] = 2$$

TABLE 1.12 Table of Characters of Irreducible Representations Deduced for the Symmetry Group C_{4v}

C_{4v}	E	$2C_4$	C_2	$2\Sigma_v$	$2\Sigma_d$
Γ_1	1	1	1	1	1
Γ_2	1	1	1	−1	−1
Γ_3	1	−1	1	1	−1
Γ_4	1	−1	1	−1	1
Γ_5	2	0	−2	0	0

The numbers out of the brackets are the numbers of elements of each class.

$$a_2 = \frac{1}{8}[1(1)(8) + 2(1)(0) + 1(1)(0) + 2(-1)(2) + 2(-1)(2)] = 0$$

$$a_3 = \frac{1}{8}[1(1)(8) + 2(-1)(0) + 1(1)(0) + 2(1)(2) + 2(-1)(2)] = 1$$

$$a_4 = \frac{1}{8}[1 \cdot (1)(8) + 2 \cdot (-1)(0) + 1 \cdot (1)(0) + 2 \cdot (-1)(2) + 2 \cdot (1)(2)] = 1$$

$$a_5 = \frac{1}{8}[1 \cdot (2)(8) + 2 \cdot (0)(0) + 1 \cdot (-2)(0) + 2 \cdot (0)(2) + 2 \cdot (0)(2)] = 2$$

Therefore, the representation Γ can be decomposed into irreducible representations as follows:

$$\Gamma = 2\Gamma_1 + \Gamma_3 + \Gamma_4 + 2\Gamma_5$$

1.7 TABLES OF CHARACTERS

1.7.1 DESCRIPTION OF TABLE OF CHARACTERS

The character tables are used throughout all applications of the group theory to physical problems, especially those involving the decomposition of reducible representations into their irreducible components. A selection of character tables often found is given in the Appendix to this book. We will discuss the character table of the $\boldsymbol{C}_{4v}$ group, as it is presented in most books (Figure 1.56). By convention, character tables are displayed with the columns labeled by the classes and the rows by the irreducible representations.

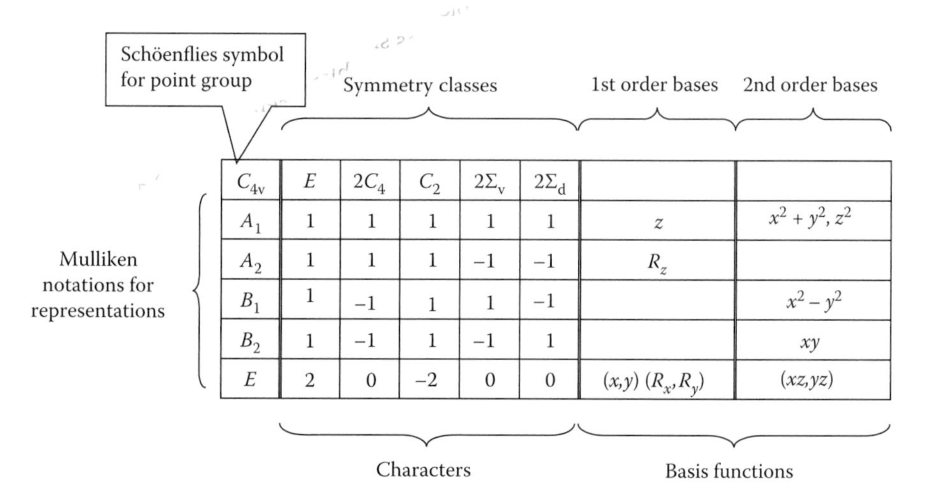

C_{4v}	E	$2C_4$	C_2	$2\Sigma_v$	$2\Sigma_d$		
A_1	1	1	1	1	1	z	$x^2 + y^2, z^2$
A_2	1	1	1	−1	−1	R_z	
B_1	1	−1	1	1	−1		$x^2 - y^2$
B_2	1	−1	1	−1	1		xy
E	2	0	−2	0	0	(x,y) (R_x, R_y)	(xz, yz)

FIGURE 1.56 Complete format of the character table of the $\boldsymbol{C}_{4v}$ group. First- and second-order bases are often named linear and quadratic bases, respectively.

TABLE 1.13 Significance of Mulliken Symbols Used to Describe the Irreducible Representations

1. ***One-dimensional*** representations are designated by symbols *A* ***and*** *B*.
   Two-dimensional representations are designated by symbol *E*.
   Three-dimensional representations are designated by symbol *T*.
   In the particular case of icosahedral group, the symbols *G* and *H* are used to designate four- and five-dimensional representations, respectively.
2. Any one-dimensional representation that is ***symmetric*** with respect to the principal rotation ($\chi(C_4) = 1$ in our example) is designated by *A*.
   Any one-dimensional representation that is ***antisymmetric*** with respect to the principal rotation ($\chi(C_4) = -1$ in our example) is designated by *B*.
3. The ***subscripts 1 and 2*** are used to designate representations, which are ***symmetric and antisymmetric with respect to*** C_2 rotation perpendicular to a principal axis or to a vertical plane.
4. ***Prime and double prime*** are attached to the letter if the representation is ***symmetric or antisymmetric*** with respect to a ***horizontal plane***.
5. The presence of inversion is marked by ***subscript*** *g* (from German *gerade*, which means even) ***or*** *u* (from German *ungerade*, which means odd) if the representation is ***symmetric or antisymmetric*** with respect to ***inversion***.

The first column contains all irreducible representations of the group in Mulliken notation. The *significance of Mulliken symbols* is shown in Table 1.13.

The next columns of the table under the classes are dedicated to display the characters of all irreducible representations of the group.

The last two columns are dedicated to show the basis of first order and second order. The Cartesian coordinates x, y, and z can serve as the basis for a representation together or separately. In table of Figure 1.56, coordinate z is a basis for totally symmetric one-dimensional representation A_1 and the pair (x, y) can generate the two-dimensional representation E. It means the coordinates x and y transform to each other only together according to the representation E and never separately. On the contrary, all the representations never mix the z coordinate with x and y (in our example!). The reason is that the z axis is collinear to all rotation axes and, on the contrary all symmetry planes contain the z axis. Thus, it is obvious that the z coordinate will not be affected by any symmetry operation of the group. If we can find an operation that mixes all the coordinates, as in case of cubic groups, then the set (x, y, z) will generate a three-dimensional representation. It happens because there are axes that are not collinear to any coordinate axis.

In the last column, we can see that squares and binary product of coordinates can serve as the basis for several one-dimensional representations. Pairs of binary product can generate only two-dimensional representation E. We will discuss later the physical significance of these types of basis.

Instead of the vector, which is a right arrow and can be designated by its x, y, z coordinates, the representations can also be generated by a curved arrow around a coordinate axis, designated by the symbol R_x, R_y, or R_z, which represents a rotation around the axis specified by the subscript. Figure 1.57 shows the effect of rotation around an axis collinear to the

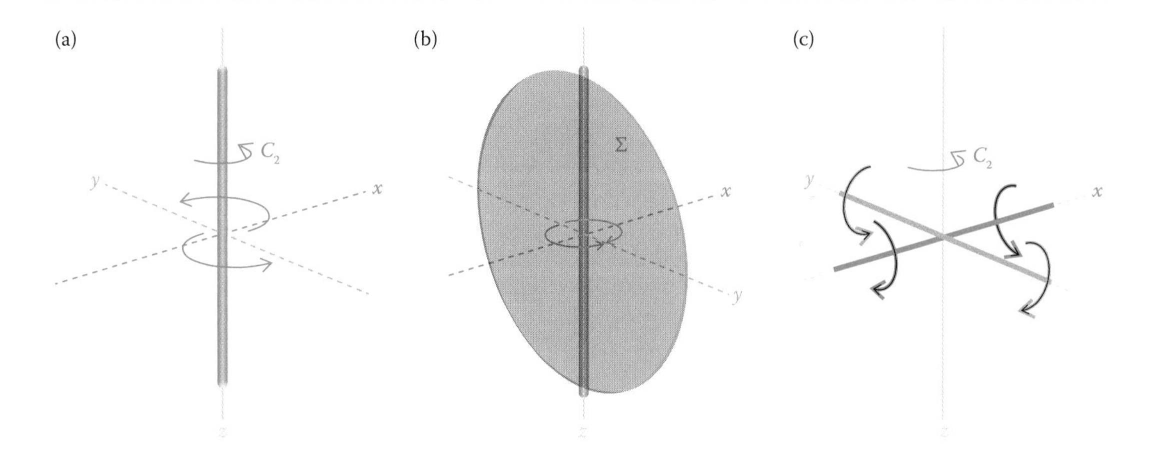

FIGURE 1.57 (a) Rotation $C_2(z)$ does not change the direction of rotation R_z. (b) The reflection through a vertical plane changes the direction of rotation R_z. (c) The rotations $C_2(z)$ reverses the direction of both rotation R_x and R_y.

z axis and, respectively, of reflection through a vertical plane on the R_z rotation. Any rotation (in our example C_2) around an axis collinear to the z axis does not change the direction of R_z rotation (Figure 1.57a). Thus, the character of C_4 and C_2 rotations generated by the basis R_z is obvious, +1. Any reflection through a vertical or dihedral plane containing the z axis changes the direction of rotation (Figure 1.57b) and the character is −1. However, the rotation $C_2(z)$ changes the direction of both R_x rotation and R_y rotation because it is perpendicular to both axes and the two-dimensional matrix will have the trace −2. The rotation $C_4(z)$ interchanges R_x and R_y rotations and the two-dimensional matrix will have the trace 0. Pairs of rotations can generate two- or three-dimensional (for cubic groups) representations. They never appear in the column of second-order bases.

1.7.2 EXAMPLE OF CHARACTER TABLE CONSTRUCTION

Let us take a pyramid with regular pentagon as base (Figure 1.58a) or a molecular complex (Figure 1.58b) with the same symmetry. The $\boldsymbol{C}_{5v}$ group is a set of 10 symmetry operations:

- Fivefold rotations around the z axis grouped in three classes C_5^1 and C_5^4, C_5^2 and C_5^3, and $C_5^5 = E$
- Five vertical reflections grouped in one class (they can be transformed from one to another by a C_5^1 rotation): Σ_V

The order of the group is $h = 10$. The number of irreducible representations equals the number of classes; hence, there are four representations. We reduce the complexity of the problem starting from the observation that in every group, there must be a totally symmetric representation, which left the molecule unchanged. All characters of this irreducible representation are equal to 1. Thus, we can start to complete the character table as in Table 1.14.

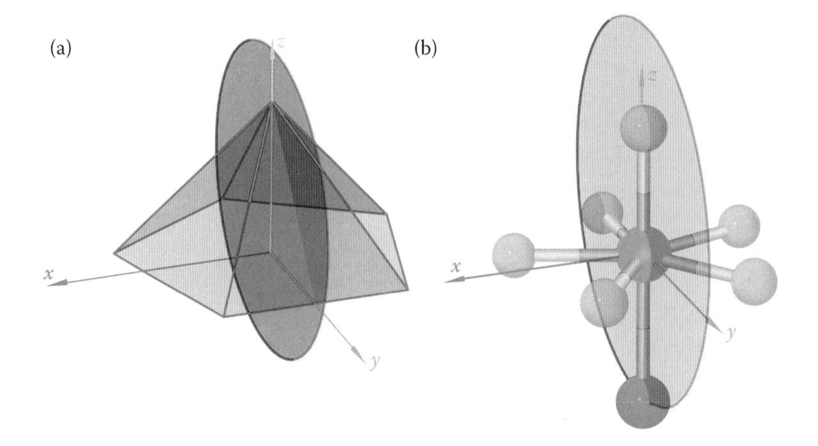

FIGURE 1.58 Pentagonal pyramid (a) and a molecular complex (b). Both have the same C_{5v} symmetry. One vertical reflection plane is also represented.

According to Rule 1, the sum of the dimension squares of the irreducible representations is equal to the order of the group. Thus,

$$1^2 + l_2^2 + l_3^2 + l_4^2 = 10$$

It is obvious that the only solution of this equation is: $l_2^2 = 1$; $l_3^2 = l_4^2 = 4$. Since the dimension of the representation must be a positive integer, the result in the group has two one-dimensional and two two-dimensional irreducible representations. We can complete the first column of the table as shown in Table 1.15.

The characters of the remaining representations must obey the rule of orthogonality. We write the second representation orthogonal to the first one:

$$A_1 \cdot A_2 = 1 \cdot (1) \cdot (1) + 2 \cdot (1) \cdot (\chi_{A2}(C_5)) + 2 \cdot (1) \cdot \left(\chi_{A2}\left(C_5^2\right)\right) + 5 \cdot (1) \cdot (\chi_{A2}(\Sigma_v)) = 0$$

$$1 + 2 \cdot (\chi_{A2}(C_5)) + 2 \cdot \left(\chi_{A2}\left(C_5^2\right)\right) = -5 \cdot (\chi_{A2}(\Sigma_v))$$

The only solution is to assign the character +1 to C_5 and C_5^2, and the character −1 to Σ_V. The one-dimensional representation that is

TABLE 1.14 Totally Symmetric Representation, Where All Characters Are Equal to 1, Is Always Present in the First Row of Every Character Table

C_{5v}	E	$2C_5$	$2C_5^2$	$5\Sigma_v$
A_1	1	1	1	1
Γ_2				
Γ_3				
Γ_4				

TABLE 1.15 Characters in the First Column Correspond to the Identity Operation and Show the Dimension of Each Representation of the Group

C_{5v}	E	$2C_5$	$2C_5^2$	$5\Sigma_v$
A_1	1	1	1	1
$A_2(B?)$	1			
E_1	2			
E_2	2			

antisymmetric with respect to a vertical plane is denoted by Mulliken (Rule 3) by subscript 2, that is, A_2.

To find the characters of the next representation (E_1), we apply the orthonormalization rule: it is orthogonal to the first two representations and normalized to itself. For simplicity, we denote by x, y, and z the unknown characters (do not confuse with coordinates):

$$\begin{cases} 2 + 2x + 2y + 5z = 0 \\ 2 + 2x + 2y - 5z = 0 \\ 4 + 2x^2 + 2y^2 + 5z^2 = 10 \end{cases}$$

The difference of the first two equations implies $z = 0$ and the system will be reduced to

$$\begin{cases} 2 + 2x + 2y = 0 \\ 4 + 2x^2 + 2y^2 = 10 \end{cases} \Leftrightarrow \begin{cases} x + y = -1 \\ x^2 + y^2 = 3 \end{cases}$$

After the substitution of y from the first equation into the second equation, we must find the solutions of the equation:

$$x^2 + x - 1 = 0$$

Therefore,

$$x_{1,2} = \frac{-1 \pm \sqrt{5}}{2} = \frac{-1 \pm 2.236}{2} = \begin{cases} = \dfrac{1.236}{2} = 0.618 \\ = \dfrac{-3.236}{2} = -1.618 \end{cases}$$

$$y_{1,2} = -1 - x = \begin{cases} = -1.618 \\ = 0.618 \end{cases}$$

Both solutions are pairs of irrational numbers. One pair of numbers (e.g., $x = 0.618$ and $y = -1.618$) is composed of the characters of the E_1 representation and the other pair (e.g., -1.618 and 0.618) is composed of the characters of the E_2 representation. The reader can check that the

TABLE 1.16 Characters Can Often Be Obtained Using the Properties of Irreducible Representations

C_{5v}	E	$2C_5$	$2C_5^2$	$5\Sigma_v$
A_1	1	1	1	1
A_2	1	1	1	−1
E_1	2	0.618	−1.618	0
E_2	2	−1.618	0.618	0

two representations are also orthogonal to each other by simple computation. For an exact computation, the expressions containing $\sqrt{5}$ must be taken.

Thus, we have obtained all characters of the group as shown in Table 1.16.

The last-obtained values are correct but, apparently, without any significance. To find the physical significance of these characters, we must examine the matrices generated when each operation is applied to the (x,y) coordinates. In the geometry of the illustrations of Figure 1.58, the z coordinate is not changed by any symmetry operation. Thus, the matrices will be only of 2×2 dimensions.

The matrices of symmetry operations and their characters are

- Identity E has the matrix

$$\begin{bmatrix} 1 & 0 \\ 0 & 1 \end{bmatrix}$$

  and the character is $\chi(E) = 2$.
- Rotation C_5 has the matrix

$$\begin{bmatrix} \cos\frac{2\pi}{5} & -\sin\frac{2\pi}{5} \\ \sin\frac{2\pi}{5} & \cos\frac{2\pi}{5} \end{bmatrix}$$

  and the character is $\chi(C_5) = 2\cos(2\pi/5)$. In other words, the character is 2 cos 72°, that is, 0.618.
- Rotation C_5^2 has the matrix

$$\begin{bmatrix} \cos 2\frac{2\pi}{5} & -\sin 2\frac{2\pi}{5} \\ \sin 2\frac{2\pi}{5} & \cos 2\frac{2\pi}{5} \end{bmatrix}$$

  and the character is $\chi(C_5^2) = 2\cos 2(2\pi/5)$. In other words, the character is 2 cos 144°, that is, −1.618.
- Reflection Σ_V illustrated in Figure 1.58 has the matrix

$$\begin{bmatrix} -1 & 0 \\ 0 & 1 \end{bmatrix}$$

  and the character is $\chi(\Sigma_V) = 0$.

Hence, we can see that the irrational number $-1 + \sqrt{5}/2$, obtained as character using only the orthonormalization relation, is the value of the expression strongly related to the rotation angle (72°) of the C_5 operation. Similar to the irrational number $-1 - \sqrt{5}/2$ is the value of the expression related to the rotation angle (144°) of the C_5^2 operation.

We now have to complete the last two columns, that is, to find the first- and second-order bases. The bases can be polar vectors (e.g., Cartesian coordinates, translations), axial vectors (e.g., rotations), and algebraic functions of Cartesian coordinates (tensors).

We have already seen that the z coordinate belongs to the totally symmetric representation A_1 and the x and y coordinates jointly belong to E_1 representation. The direction of rotation R_z around the z axis is left unchanged by any rotation around the z axis (+1 character) and is inversed by each vertical reflection (character −1). Hence, the rotation R_z belongs to the A_2 representation. The R_x and R_y jointly belong to a two-dimensional representation in each group where the principal axis is different than C_2. They are mixed by any $C_n(z)$ rotation (except C_2). The reflection illustrated in Figure 1.58 (through the plane yOz) left unchanged the sign of R_x and changed the sign of R_y.

We will see later that atomic orbitals can be expressed by algebraic functions of the Cartesian coordinates. The s and d orbital wavefunctions are associated with $x^2 + y^2 + z^2$, $2z^2–(x^2 + y^2)$, $x^2 - y^2$, xy, xz, and yz, which appear only as second-order bases individual (for one-dimensional representation) or in pairs (for two-dimensional representations). They can also be seen separately as in most cases $x^2 + y^2$ and z^2. The last two functions generate the representation totally symmetric, with the exception of cubic groups.

We take the pair of functions (xz, yz) to calculate the matrices generated by it under the action of symmetry operations:

$$C_5^1(z)\begin{cases} x \to x\cos\frac{2\pi}{5} - y\sin\frac{2\pi}{5} \\ y \to x\sin\frac{2\pi}{5} + y\cos\frac{2\pi}{5} \\ z \to z \end{cases} \Rightarrow \begin{cases} xz \to xz\cos\frac{2\pi}{5} - yz\sin\frac{2\pi}{5} \\ yz \to xz\sin\frac{2\pi}{5} + yz\cos\frac{2\pi}{5} \end{cases}$$

$$\Rightarrow \text{trace}\begin{bmatrix} \cos\frac{2\pi}{5} & -\sin\frac{2\pi}{5} \\ \sin\frac{2\pi}{5} & \cos\frac{2\pi}{5} \end{bmatrix} = 2\cos\frac{2\pi}{5}$$

$$C_5^2(z)\begin{cases} x \to x\cos2\frac{2\pi}{5} - y\sin2\frac{2\pi}{5} \\ y \to x\sin2\frac{2\pi}{5} + y\cos2\frac{2\pi}{5} \\ z \to z \end{cases} \Rightarrow \begin{cases} xz \to xz\cos2\frac{2\pi}{5} - yz\sin2\frac{2\pi}{5} \\ yz \to xz\sin2\frac{2\pi}{5} + yz\cos2\frac{2\pi}{5} \end{cases}$$

$$\Rightarrow \text{trace}\begin{bmatrix} \cos\frac{4\pi}{5} & -\sin\frac{4\pi}{5} \\ \sin\frac{4\pi}{5} & \cos\frac{4\pi}{5} \end{bmatrix} = 2\cos\frac{4\pi}{5}$$

TABLE 1.17 Character Table of the C_{5v} Group Contains the Characters and Bases of All Irreducible Representations

C_{5v}	E	$2C_5$	$2C_5^2$	$5\Sigma_v$	First-Order Bases (Linear)	Second-Order Bases (Quadratic)
A_1	1	1	1	1	z	x^2+y^2, z^2
A_2	1	1	1	−1	R_z	
E_1	2	2 cos 72°	2 cos 144°	0	(x,y) (R_x,R_y)	(xz,yz)
E_2	2	2 cos 144°	2 cos 72°	0		$(x^2-y^2, 2xy)$

$$\Sigma_v(yz)\begin{cases} x \to -x \\ y \to y \\ z \to z \end{cases} \Rightarrow \begin{cases} xz \to -xz \\ yz \to yz \end{cases} \Rightarrow \text{trace}\begin{bmatrix} -1 & 0 \\ 0 & 1 \end{bmatrix} = 0$$

Thus, the function (xz, yz) has the following characters under the symmetry elements of the group: 2, 2 cos 72°, 2 cos 144°, 0. It is therefore assigned to the E_1 representation.

You should exercise to assign the pairs $(x^2-y^2,\ 2xy)$ to the E_2 representation.

The final and complete form of the character table of the C_{5v} group is shown in Table 1.17.

1.7.3 WAVEFUNCTIONS AS BASES FOR IRREDUCIBLE REPRESENTATIONS

We have just introduced in a particular case the orbital wavefunctions as bases for irreducible representations. We will discuss in detail this subject, which is the fundamental idea of this book. Optical spectroscopy can be a powerful method of investigation of matter when the matter interacts with radiation. The interaction can result in change of state of matter, that is, of state function.

We worked until now with molecules that are composed of many atoms and, thus, the state function of the molecule is a combination of state functions of all atoms. A symmetry operation applied to the molecule changes the position of atoms, that is, one atom including its orbitals takes the place of other atom in such a way that the total energy of the molecule must be unchanged.

The atomic state is described by the state function assigned by the symbol Ψ_i. The value of the energy of the atomic system in this state is E_i. The relationship between the state function Ψ_i and its corresponding energy E_i is given by the stationary Schrödinger equation

$$\hat{H}\Psi_i = E_i\Psi_i \tag{1.78}$$

where $\hat{H}$ is the Hamiltonian operator. The function Ψ_i is named in quantum mechanics as the *eigenfunction* of $\hat{H}$ and the energy E_i is named the *eigenvalue* of $\hat{H}$. The Hamiltonian eigenfunctions are orthogonal and normalized, that is, they obey the *orthonormality* rule:

$$\int \Psi_i^* \Psi_j \, d\tau = \delta_{ij} \quad (1.79)$$

We now turn to the symmetry problem. A symmetry operation can interchange the atoms of the molecule, whether the Hamiltonian must remain unchanged or not; meaning the system energy must be the same. It means any symmetry operation $\hat{R}$ must commute with the Hamiltonian operator:

$$\hat{R}\hat{H} = \hat{H}\hat{R} \quad (1.80)$$

We are ready to show that the eigenfunctions for a molecule are the bases for irreducible representations of the symmetry group to which the molecule belongs.

1.7.3.1 NONDEGENERATE CASE

Let us first take the case of nondegenerate eigenvalues. This means for each value of energy E_i only one function Ψ_i corresponds. We apply the symmetry operation $\hat{R}$ to each side of Equation 1.78.

$$\hat{R}\hat{H}\Psi_i = \hat{R}E_i\Psi_i \quad (1.81)$$

Taking into account the commutativity relation 1.80 and the fact that E_i is a constant, we can write

$$\hat{H}(\hat{R}\Psi_i) = E_i(\hat{R}\Psi_i)$$

Hence, the new function $\hat{R}\Psi_i$ is a Hamiltonian eigenfunction too. As an eigenfunction, it must be normalized:

$$\int (\hat{R}\Psi_i)^* (\hat{R}\Psi_i) d\tau = 1 \Rightarrow (\hat{R})^2 \int (\Psi_i)^* (\Psi_i) d\tau = 1 \Rightarrow \hat{R} = \pm 1$$

So, applying all group operations to an eigenfunction, we generated a one-dimensional irreducible representation, whose characters are ±1.

1.7.3.2 DEGENERATE CASE

It is possible for the atomic system to have several states, that is, several eigenfunctions Ψ_{ij}, for one value of energy, that is, eigenvalue E_i. It means this value of energy is degenerated. We write Equation 1.78 as follows:

$$\hat{H}\Psi_{ij} = E_i\Psi_{ij} \quad \text{where } j = 1,2,\ldots,k \quad (1.82)$$

These eigenfunctions obey the orthonormality condition too:

$$\int \Psi_{il}^{*}\Psi_{jm}d\tau = \delta_{ij}\delta_{lm} \tag{1.83}$$

In the degenerate case, any linear combination of the eigenfunctions can also be a solution of Equation 1.82 for the same energy:

$$\hat{H}\sum_{j=1}^{k} a_{ij}\Psi_{ij} = \sum_{j=1}^{k} a_{ij}\hat{H}\Psi_{ij} = \sum_{j=1}^{k} a_{ij}E_i\Psi_{ij} = E_i\sum_{j=1}^{k} a_{ij}\Psi_{ij}$$

To be true eigenfunctions, these linear combination functions must be normalized:

(1.84)

$$\int\left(\sum_{p=1}^{k} a_{ip}\Psi_{ip}\right)^{*}\left(\sum_{q=1}^{k} a_{jq}\Psi_{jq}\right)d\tau = \int\left(\sum_{p=1}^{k} a_{ip}\Psi_{ip}^{*}\right)\left(\sum_{q=1}^{k} a_{jq}\Psi_{jq}\right)d\tau = 1$$

All the products vanish if $i \neq j$ and $p \neq q$. So, the relation 1.84 reduces to

$$\int\left(\sum_{p=1}^{k} a_{ip}\Psi_{ip}^{*}a_{ip}\Psi_{ip}\right)d\tau = \sum_{p=1}^{k} a_{ip}a_{ip}\int \Psi_{ip}^{*}\Psi_{ip}d\tau = \sum_{p=1}^{k} a_{ip}^{2} = 1 \tag{1.85}$$

We apply two different symmetry operations $\hat{R}$ and $\hat{S}$ of the group to each side of Equation 1.82:

Let there be two eigenfunctions Ψ_{il} and Ψ_{im}	$\hat{H}\Psi_{il} = E_i\Psi_{il}$	$\hat{H}\Psi_{im} = E_i\Psi_{im}$
Let us apply the symmetry operations $\hat{R}$ and $\hat{S}$, respectively	$\hat{R}\hat{H}\Psi_{il} = \hat{R}E_i\Psi_{il}$	$\hat{S}\hat{H}\Psi_{im} = \hat{S}E_i\Psi_{im}$
Commutativity	$\hat{H}(\hat{R}\Psi_{il}) = E_i(\hat{R}\Psi_{il})$	$\hat{H}(\hat{S}\Psi_{im}) = E_i(\hat{S}\Psi_{im})$
Hence, the new functions are Hamiltonian eigenfunctions too	$\hat{R}\Psi_{il}$	$\hat{S}\Psi_{im}$
And can be written as a linear combination of initial eigenfunctions	$\hat{R}\Psi_{il} = \sum_{m=1}^{k} r_{ml}\Psi_{im}$	$\hat{S}\Psi_{im} = \sum_{p=1}^{k} s_{pm}\Psi_{ip}$

Because $\hat{R}$ and $\hat{S}$ are members of the group, their product $\hat{T} = \hat{S}\hat{R}$ must be a member of the group too and the new eigenfunctions generated by operation $\hat{T}$ can be written as a linear combination of initial eigenfunctions:

(1.86)
$$\hat{T}\Psi_{il} = \sum_{p=1}^{k} t_{pl}\Psi_{ip}$$

At the same time, $\hat{T}$ can be written as the result of successive application of $\hat{R}$ and $\hat{S}$ operations:

(1.87)
$$\hat{T}\Psi_{il} = \hat{S}\hat{R}\Psi_{il} = \hat{S}\sum_{m=1}^{k} r_{ml}\Psi_{im} = \sum_{m=1}^{k} r_{ml}\left(\hat{S}\Psi_{im}\right) = \sum_{m=1}^{k} r_{ml}\left(\sum_{p=1}^{k} s_{pm}\Psi_{ip}\right)$$
$$= \sum_{p=1}^{k}\sum_{m=1}^{k} r_{ml}s_{pm}\Psi_{ip}$$

Comparing Equations 1.86 and 1.87, we obtain

(1.88)
$$t_{pl} = \sum_{m=1}^{k} r_{ml}s_{pm}$$

that is, the relation between the matrix elements of $\hat{T}$, $\hat{R}$, and $\hat{S}$. We can conclude that a set of symmetry operations applied to k eigenfunctions $\Psi_{i1}, \Psi_{i2},\ldots, \Psi_{ik}$ generate a set of matrices of $k \times k$ dimension, which is a k-dimensional representation of the group.

The next question follows: is this an irreducible representation? We start to suppose it is not irreducible. So, it can be decomposed in irreducible representations, which means each $k \times k$ matrix can be blocked in lower-size matrices. Each set of lower-size matrices has a subset of initial eigenfunctions as the basis. Then, different subsets could have different values of energy that is contrary to the initial conditions: all eigenfunctions correspond to the same eigenvalue.

The final conclusion is that a set of k eigenfunctions $\Psi_{i1}, \Psi_{i2},\ldots, \Psi_{ik}$ of the same eigenvalue is a *basis of a k-dimensional irreducible representation of the symmetry group.*

1.7.4 SYMMETRY OF ATOMIC ORBITALS

In Section 1.7.2, we showed that a pair of orbitals d, which can be expressed by algebraic functions (xz, yz), can be the basis of a two-dimensional irreducible representation of the $\boldsymbol{C}_{5v}$ group. We applied the symmetry operations about the Cartesian coordinates. We now start from the expressions of p orbital functions as given by quantum mechanics using spherical coordinates (Figure 1.59):

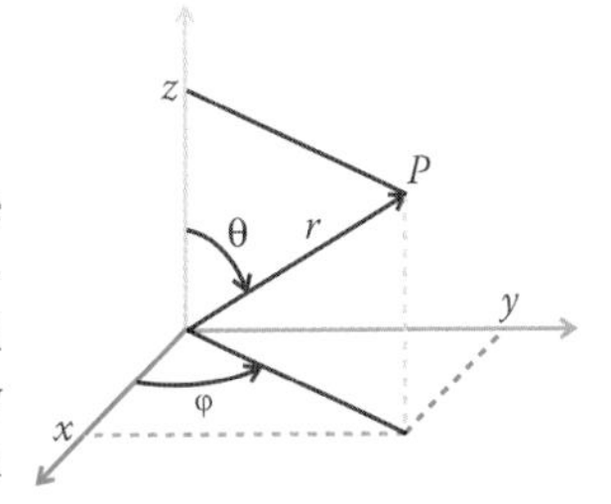

FIGURE 1.59 Spherical coordinates r, θ, and φ related to the Cartesian coordinates of point P.

(1.89)
$$\begin{cases} p_x = r\sin\theta\cos\varphi \\ p_y = r\sin\theta\sin\varphi \end{cases}$$

We will generate the matrices that represent the transformations of both functions by each symmetry operation: E, $C_5(z)$, $C_5^2(z)$, $\Sigma_v(yz)$.

First of all, we see that angle θ is left unchanged by each operation. Each operation affects only angle φ. What will be the effect after the rotation of angle α around the z axis? The final state will be given by the trigonometric relations

$$\begin{cases} \cos(\varphi + \alpha) = \cos\varphi\cos\alpha - \sin\varphi\sin\alpha \\ \sin(\varphi + \alpha) = \sin\varphi\cos\alpha + \cos\varphi\sin\alpha \end{cases} \tag{1.90}$$

The reflection through the vertical plane yz will change the sign of angle φ, so the final state after reflection will also be given by trigonometric relations:

$$\begin{cases} \cos(-\varphi) = \cos\varphi \\ \sin(-\varphi) = -\sin\varphi \end{cases} \tag{1.91}$$

We can now write the transformations of orbital functions produced by each symmetry operation using relations 1.90 and 1.91 as shown in Table 1.18.

We obtained the characters equal to 2, 2 cos 72°, 2 cos 144°, and 0 as can be seen from Table 1.17 (see Table 1.19).

We will shortly discuss the symmetry of all orbitals s, p, d, and f, which will appear in this book.

TABLE 1.18 Transformation of *p* Orbital Functions Produced by Each Symmetry Operation of C_{5v} Group Gives the Matrix Corresponding to the Operation

	Initial State	Final State	Matrix
E	$\begin{bmatrix} p_x \\ p_y \end{bmatrix} = \begin{bmatrix} r\sin\theta\cos\varphi \\ r\sin\theta\sin\varphi \end{bmatrix}$	$\begin{bmatrix} r\sin\theta\cos\varphi \\ r\sin\theta\sin\varphi \end{bmatrix} = \begin{bmatrix} p_x \\ p_y \end{bmatrix}$	$\begin{bmatrix} 1 & 0 \\ 0 & 1 \end{bmatrix}$
C_5	$\begin{bmatrix} p_x \\ p_y \end{bmatrix} = \begin{bmatrix} r\sin\theta\cos\varphi \\ r\sin\theta\sin\varphi \end{bmatrix}$	$\begin{bmatrix} r\sin\theta(\cos\varphi\cos 72° - \sin\varphi\sin 72°) \\ r\sin\theta(\sin\varphi\cos 72° + \cos\varphi\sin 72°) \end{bmatrix} = \begin{bmatrix} p_x\cos 72° - p_y\sin 72° \\ p_x\sin 72° + p_y\cos 72° \end{bmatrix}$	$\begin{bmatrix} \cos 72° & -\sin 72° \\ \sin 72° & \cos 72° \end{bmatrix}$
C_5^2	$\begin{bmatrix} p_x \\ p_y \end{bmatrix} = \begin{bmatrix} r\sin\theta\cos\varphi \\ r\sin\theta\sin\varphi \end{bmatrix}$	$\begin{bmatrix} r\sin\theta(\cos\varphi\cos 144° - \sin\varphi\sin 144°) \\ r\sin\theta(\sin\varphi\cos 144° + \cos\varphi\sin 144°) \end{bmatrix} = \begin{bmatrix} p_x\cos 144° - p_y\sin 144° \\ p_x\sin 144° + p_y\cos 144° \end{bmatrix}$	$\begin{bmatrix} \cos 144° & -\sin 144° \\ \sin 144° & \cos 144° \end{bmatrix}$
Σ_V	$\begin{bmatrix} p_x \\ p_y \end{bmatrix} = \begin{bmatrix} r\sin\theta\cos\varphi \\ r\sin\theta\sin\varphi \end{bmatrix}$	$\begin{bmatrix} r\sin\theta\cos\varphi \\ -r\sin\theta\sin\varphi \end{bmatrix} = \begin{bmatrix} p_x \\ -p_y \end{bmatrix}$	$\begin{bmatrix} 1 & 0 \\ 0 & -1 \end{bmatrix}$

TABLE 1.19 Matrix Traces Obtained in Table 1.18 Correspond to the Characters of Two-Dimensional Representation E_1 of C_{5v} Group Generated by (*x*,*y*) Basis, Which Corresponds to *p*-Type Orbitals Pair p_x and p_y

C_{5v}	E	$2C_5$	$2C_5^2$	$5\Sigma_v$		
A_1	1	1	1	1	Z	x^2+y^2, z^2
A_2	1	1	1	−1	R_z	
E_1	2	2 cos 72°	2 cos 144°	0	(x,y) (R_x,R_y)	(xz,yz)
E_2	2	2 cos 144°	2 cos 72°	0		$(x^2-y^2, 2xy)$

We begin with the eigenfunctions of the hydrogen atom. They are the product between a radial function and an angular function:

$$\Psi = R_n(r) \cdot Y_{lm}(\theta,\varphi) \tag{1.92}$$

This property is important because there is no symmetry operation that can change the radial function $R(r)$ and we are not interested in it. The symmetry operations can affect only angular functions, named spherical harmonics. The squared eigenfunctions, that is, $|\Psi^*\Psi|^2$, gives a measure of the probability of finding the electron at a given position in space around the nucleus. The place, that is, a small volume where the probability is higher, is represented by a high density of points. The radial function gives the distance where the density is higher. The angular function gives the three-dimensional shape of the distribution of points that is named orbital. The illustration of the orbital shape is intended to describe the regions in space where the electrons occupying the orbital are likely to be found. The orbital is like a "home" for an electron, which prefers to "stay" longer in certain "rooms." The illustration of an orbital is a representation of contour surface where the probability density is constant and there is a high probability (e.g., 90%) of finding the electron within the volume bounded by this surface. We underline that the angular function does not depend in first approximation on the principal quantum number n. Thus, all orbitals of the same type have the same angular function. We present in Figure 1.60 the expression and shape of spherical harmonics of orbitals *s*, *p*, *d*, and *f*, as you can find in most of the atomic physics textbooks. These four labels of orbitals were due to Friedrich Hund (Jensen 2007), who replaced the values of orbital quantum number with the spectral series notations: *s* from *sharp*, *p* from *principal*, *d* from *diffuse*, and *f* from *fundamental*.

In several textbooks of quantum mechanics or atomic physics, the eigenfunctions Y_{11} and Y_{1-1} are given together in one complex valued expression:

$$Y_{1,\pm1} = \sqrt{\frac{3}{8\pi}} \sin\theta e^{\pm i\varphi}$$

(a)

Function	Orbital	Algebraic expression	Shape
Y_{00}	s	$\sqrt{\frac{1}{4\pi}}$	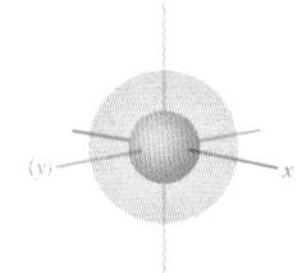
Y_{10}	p_z	$\sqrt{\frac{3}{4\pi}}\cos\theta$	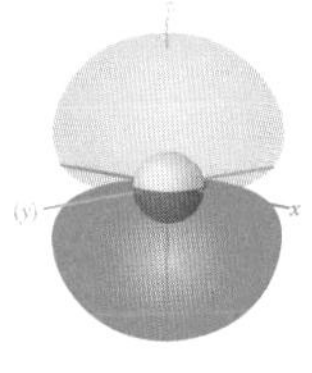
Y_{11}	p_x	$\sqrt{\frac{3}{8\pi}}\sin\theta\cos\varphi$	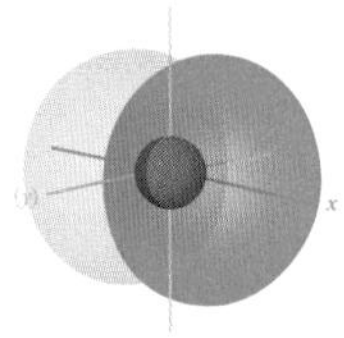
Y_{1-1}	p_y	$\sqrt{\frac{3}{8\pi}}\sin\theta\sin\varphi$	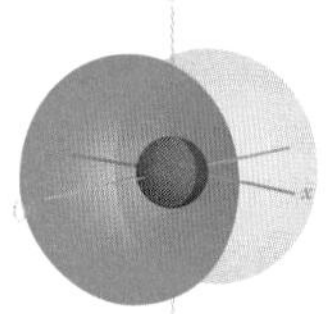

FIGURE 1.60 Real valued spherical harmonics of orbitals and their shapes: (a) *s* and *p* type, (b) *d* type.

We wrote the two functions separately because we are interested in the shape and symmetry of each orbital p_x and p_y.

The pairs of eigenfunctions Y_{21} and Y_{2-1} and Y_{22} and Y_{2-2} are also given by the complex functions, respectively:

$$Y_{2,\pm 1} = \sqrt{\frac{15}{8\pi}}\sin\theta\cos\theta e^{\pm i\varphi}$$

$$Y_{2,\pm 2} = \sqrt{\frac{15}{32\pi}}\sin^2\theta e^{\pm i2\varphi}$$

In the above illustrations, the nucleus size is exaggerated. Its real size is approximately 10^5 times smaller than the orbital size, impossible to be illustrated. You can also generate very nice illustrations of orbitals using the free software *Orbital Viewer* (Manthey 2013).

(b)

Function	Orbital	Algebraic expression	Shape
Y_{20}	$d_{z^2} = d_{2z^2-x^2-y^2}$	$\sqrt{\frac{5}{16\pi}}(3\cos^2\theta - 1)$	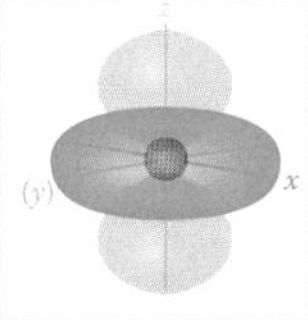 (y) x
Y_{21}	d_{xz}	$\sqrt{\frac{15}{8\pi}}\sin\theta\cos\theta\cos\varphi$	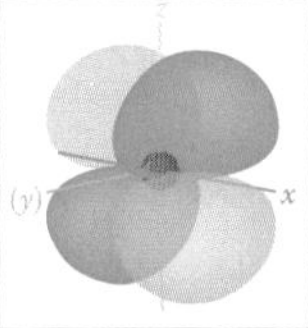 (y) x
Y_{2-1}	d_{yz}	$\sqrt{\frac{15}{8\pi}}\sin\theta\cos\theta\sin\varphi$	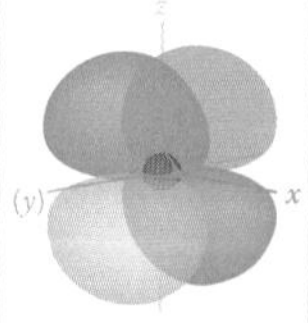 (y) x
Y_{22}	$d_{x^2-y^2}$	$\sqrt{\frac{15}{32\pi}}\sin^2\theta\cos 2\varphi$	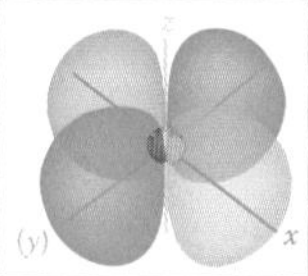 (y) x
Y_{2-2}	d_{xy}	$\sqrt{\frac{15}{32\pi}}\sin^2\theta\sin 2\varphi$	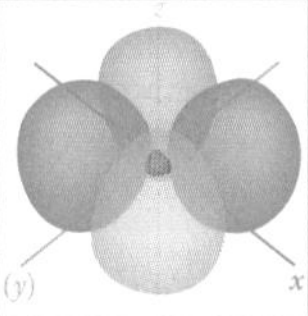 (y) x

FIGURE 1.60 (continued) Real valued spherical harmonics of orbitals and their shapes: (a) *s* and *p* type, (b) *d* type.

We note that the "*s*" wavefunction has no angular dependence (i.e., θ and φ do not appear). The probability density is spherical and the symmetry of the *s* orbital is totally symmetric. Hence, *it always generates the totally symmetric irreducible representation* and we can write as basis an algebraic function of coordinates:

- In case of high-symmetry groups, for example, cubic and icosahedral, which contain rotations that mix all coordinates, the basis $x^2 + y^2 + z^2$ (i.e., the equation of sphere) represents the orbital *s*.
- In case of most symmetry groups, which have operations that mix just the *x* and *y* coordinates, the bases $x^2 + y^2$ and z^2, respectively (i.e., the equation of a sphere projected on the *xy* plane) represent the orbital *s*.
- In case of low-symmetry groups, which have operations that do not mix the coordinates, the bases x^2, y^2, and z^2, respectively, represent the orbital *s*.

Concerning "p" wavefunction, we discussed earlier that the p orbitals transform as the Cartesian coordinates:

$$\begin{cases} p_x = r\sin\theta\cos\varphi = x \\ p_y = r\sin\theta\sin\varphi = y \\ p_z = r\cos\theta = z \end{cases} \tag{1.93}$$

The orbitals p are always represented by x, y, z or (x,y) and z or (x, y, z). They are *antisymmetric* to inversion.

The "d" wavefunctions give five orbitals that are related to the Cartesian coordinates as follows:

1. Y_{20}

$$3\cos^2\theta - 1 = 3\cos^2\theta - (\cos^2\theta + \sin^2\theta) = 2\cos^2\theta - \sin^2\theta$$

$$= 2\cos^2\theta - (\sin^2\theta)(\cos^2\varphi + \sin^2\varphi)$$

$$= 2\cos^2\theta - (\sin^2\theta\cos^2\varphi + \sin^2\theta\sin^2\varphi)$$

$$= \frac{2z^2}{r^2} - \frac{x^2 + y^2}{r^2} = \frac{2z^2 - (x^2 + y^2)}{r^2}$$

   Therefore, the symmetry of this orbital is given by the function $2z^2 - (x^2 + y^2)$ in cubic and icosahedral groups or simpler by z^2 in all other groups.

2. Y_{22}

$$\sin^2\theta\cos 2\varphi = \sin^2\theta(\cos^2\varphi - \sin^2\varphi) = \sin^2\theta\cos^2\varphi - \sin^2\theta\sin^2\varphi$$

$$= \frac{x^2 - y^2}{r^2}$$

   Therefore, the symmetry of this orbital is given by the function $x^2 - y^2$. In cubic and icosahedral groups, $d_{x^2-y^2}$ and d_{z^2} form a basis for a two-dimensional irreducible representation.

3. Y_{21}

$$\sin\theta\cos\theta\cos\varphi = (\sin\theta\cos\varphi)\cos\theta = \frac{xz}{r^2}$$

4. Y_{2-1}

$$\sin\theta\cos\theta\sin\varphi = (\sin\theta\sin\varphi)\cos\theta = \frac{yz}{r^2}$$

   In most of symmetry groups, the orbitals d_{xz} and d_{yz} together form a basis for two-dimensional representation.

5. Y_{2-2}

$$\sin^2\theta\sin 2\varphi = (\sin^2\theta)2\sin\varphi\cos\varphi = 2(\sin\theta\cos\varphi)(\sin\theta\sin\varphi) = \frac{xy}{r^2}$$

which is named d_{xy}.

We must keep in mind that *all orbitals "d" are symmetric to the inversion* and this important property can be observed by simple visual inspection of the shape and sign (or color).

Concerning the *orbitals "f"* type, it is important to note their *antisymmetric character to inversion.*

1.7.5 CHARACTER TABLES OF CYCLIC GROUPS

We will now discuss the symbol ε, which appears in the character tables of cyclic groups. The *cyclic group* of order h is a collection of symmetry operations composed of operation A and all of its powers up to A^h (=E). Any cyclic group has an important property of commutativity. It means all multiplications inside the group are commutative $A \cdot B = B \cdot A$. Another property is that each element is in a separate class; hence, the number of classes is equal to h.

Let us take as an example the $\mathbf{C}_4$ group. The group consists of four rotations: C_4^1, C_4^2, C_4^3, and $C_4^4 = E$. So, the group order is $h = 4$. According to Rule 5, the number of irreducible representations of the group is equal to 4 (number of classes): Γ_1, Γ_2, Γ_3, Γ_4. According to Rule 1

$$l_1^2 + l_2^2 + l_3^2 + l_4^2 = 4$$

This equation has only one solution: $l_1^2 = l_2^2 = l_3^2 = l_4^2 = 1$. Hence, the group has four one-dimensional irreducible representations.

We ignore the fact that one of them is the totally symmetric representation, which is completely determined. All representations must obey the rule of orthonormalization (Rule 2 plus Rule 3):

$$\sum_{m=1}^{n}\left[\chi_p(C_n^m)\right]\cdot\left[\chi_q(C_n^m)\right]^* == h\delta_{pq} \tag{1.94}$$

We have used here subscripts p and q to avoid confusion with i and j, which are recognized symbols for imaginary number $\sqrt{-1}$. The effect of rotation is generally given by the trigonometric functions $\sin(2\pi/n)$ and $\cos(2\pi/n)$. Using the complex number, we can denote the first character in the table as

$$\chi^{\Gamma_p}(C_4) = \cos\left[\left(\frac{2\pi}{4}\right)p\right] + i\sin\left[\left(\frac{2\pi}{4}\right)p\right] = e^{i\left(\frac{2\pi}{4}\right)p} \equiv \varepsilon^p \tag{1.95}$$

TABLE 1.20 Raw Form of the Character Table of Cyclic Group C_4 as Resulted from the Properties of Irreducible Representations

C_4	C_4^1	C_4^2	C_4^3	C_4^4
Γ_1	ε^1	ε^2	ε^3	ε^4
Γ_2	ε^2	ε^4	ε^6	ε^8
Γ_3	ε^3	ε^6	ε^9	ε^{12}
Γ_4	ε^4	ε^8	ε^{12}	ε^{16}

According to this notation, we can complete the character table by ε and its powers (Table 1.20).

Now, we must check whether the founded representations satisfy the orthonormalization rule (Equation 1.94), taking into account that the second character is complex conjugated. Given two representations Γ_p and Γ_q, we can write their direct product as follows

$$\varepsilon^{p-q} + \varepsilon^{2(p-q)} + \varepsilon^{3(p-q)} + \varepsilon^{4(p-q)} \tag{1.96}$$

It is obvious that for $p = q$ ($\varepsilon^0 = 1$), the sum equals 4 (the order of the group). So, the *normalization rule is satisfied.*

For $p \neq q$, we denote $p - q = r$ and we write again the sum given by Equation 1.96 as

$$\varepsilon^{p-q} + \varepsilon^{2(p-q)} + \varepsilon^{3(p-q)} + \varepsilon^{4(p-q)} = \varepsilon^{p-q}(\varepsilon^1 + \varepsilon^2 + \varepsilon^3 + \varepsilon^4) \tag{1.97}$$

We must calculate the sum:

$$(\varepsilon^1 + \varepsilon^2 + \varepsilon^3 + \varepsilon^4) = \sum_{n=1}^{4} e^{i\left(\frac{2\pi}{4}\right)n} = \sum_{n=1}^{4} \cos\left(\frac{2\pi}{4}n\right) + i\sum_{n=1}^{4} \sin\left(\frac{2\pi}{4}n\right) \tag{1.98}$$

But

$$\sum_{n=1}^{4} \cos\left(\frac{2\pi}{4}n\right) = 0 \quad \text{and} \quad \sum_{n=1}^{4} \sin\left(\frac{2\pi}{4}n\right) = 0$$

So, the sum given by (1.96) is always equal to 0 and the irreducible representations Γ_p and Γ_q *are orthogonal.*

Before rewriting the character table (Table 1.21), we calculate all powers of ε starting with

$$\varepsilon = e^{i\frac{2\pi}{4}} = \cos\left(\frac{\pi}{2}\right) + i\sin\left(\frac{\pi}{2}\right) = 0 + i = i$$

TABLE 1.21 Character Table of Cyclic Group C_4 as Resulted from the Properties of Irreducible Representations

C_4	$E(=C_4^4)$	C_4^1	C_4^2	C_4^3
Γ_1	$\varepsilon^4 = 1$	$\varepsilon^1 = i$	$\varepsilon^2 = -1$	$\varepsilon^3 = -i$
Γ_2	$\varepsilon^8 = 1$	$\varepsilon^2 = -1$	$\varepsilon^4 = 1$	$\varepsilon^6 = -1$
Γ_3	$\varepsilon^{12} = 1$	$\varepsilon^3 = -i$	$\varepsilon^6 = -1$	$\varepsilon^9 = i$
Γ_4	$\varepsilon^{16} = 1$	$\varepsilon^4 = 1$	$\varepsilon^8 = 1$	$\varepsilon^{12} = 1$

TABLE 1.22 Final Form of Character Table of the Cyclic Group C_4

C_4	E	C_4^1	C_4^2	C_4^3
A	1	1	1	1
B	1	−1	1	−1
E	1	i	−1	$-i$
	1	$-i$	−1	i

The last representation Γ_4 is in fact the first totally symmetric representation A. The representation Γ_2 is the representation B. We must keep the one-dimensional representations Γ_1 and Γ_3 as required by *Rule* 5 and *Rule* 1, but we join them in a pair obtaining in fact a two-dimensional representation, which has real characters (2, 0, −2, 0). The resulting representation obtained by joining them is denoted by E. Thus, the final expression of the character table for the C_4 group is shown in Table 1.22.

We have used this method to find the irreducible representation due to cyclicity of the group. It means that all symmetry operations are obtained by multiplication of the first rotation. It is a common property of the groups that are obviously cyclic: $\boldsymbol{C}_3$, $\boldsymbol{C}_4$, $\boldsymbol{C}_5$, $\boldsymbol{C}_6$, $\boldsymbol{C}_7$, $\boldsymbol{C}_8$, $\boldsymbol{C}_{3h}$, $\boldsymbol{C}_{4h}$, $\boldsymbol{C}_{5h}$, $\boldsymbol{C}_{6h}$, $\boldsymbol{S}_4$, $\boldsymbol{S}_6$, $\boldsymbol{S}_8$, and to $\boldsymbol{T}$ and $\boldsymbol{T}_h$, which are not obviously cyclic.

1.8 SYMMETRY OF CRYSTALS AND SPACE GROUPS

1.8.1 INTRODUCTION

We discussed earlier about molecular symmetry, which is related to the existence of one point where all symmetry elements intersect. The symmetry operations have a common property that at least one point of the object is not moved by any operation. Thus, the symmetry groups generated in such a way are called *point groups*. We remember that the symmetry elements are rotation axis (proper and improper), reflection plane, and inversion center. The symmetry operations apply to a molecule that is a distinct spatial structure composed of atoms. The number and type of atoms can vary from several to tens or hundreds in biomolecules but is *finite*. The molecules are very small and we cannot see them directly. We can only imagine how they are.

Now, we discuss the *infinite* molecules and the symmetry elements that would lead to them. It is the case of a *crystal* that can be regarded as an infinite molecule.

We started talking about symmetry taking as an example the water molecule. We imagine that three small atoms bounded together form a spatial structure with three symmetry elements: one rotation axis and two reflection planes (see Section 1.3.7). We ignore the identity. We have never seen a water molecule but we can see billions of water molecules everywhere around us. It is liquid water or ice. The structure of liquid water (this common substance) is still a subject of research. Its structure could explain their unique properties. Liquid water is not our subject, but you can find on the Internet a lot of interesting information and very well illustrated (e.g., Chaplin 2013). At lower temperature, the liquid water becomes solid (that is ice). If we hit an ice block with hammer, we can see that each piece of ice presents several plane faces. Also look at a glacier calving (preferably on TV). What do these plane faces and right angle between them mean? They suggest the way of arrangement of water molecules forming the ice. There is a planar (*periodic*) arrangement of billions of molecules inside the ice. A solid where the atoms (or molecules) form a periodic arrangement is called a *crystal* and the science, which studies the structure of crystals, is called *crystallography*. The principal method of structure investigation is X-ray diffraction. We will show later that the optical spectroscopy could also be a tool to investigate the crystal structure. Crystallography evolved for a long time separately from molecular science and, thus, there are differences from the theory of point groups.

1.8.2 UNIT CELL

Let us look at a small salt grain. You should imagine that the ions of sodium and chloride are arranged as shown in Figure 1.61a. The structure of sodium chloride was first determined by W.H. Bragg and W.L. Bragg by X-ray diffraction. They were awarded the Nobel Prize in Physics

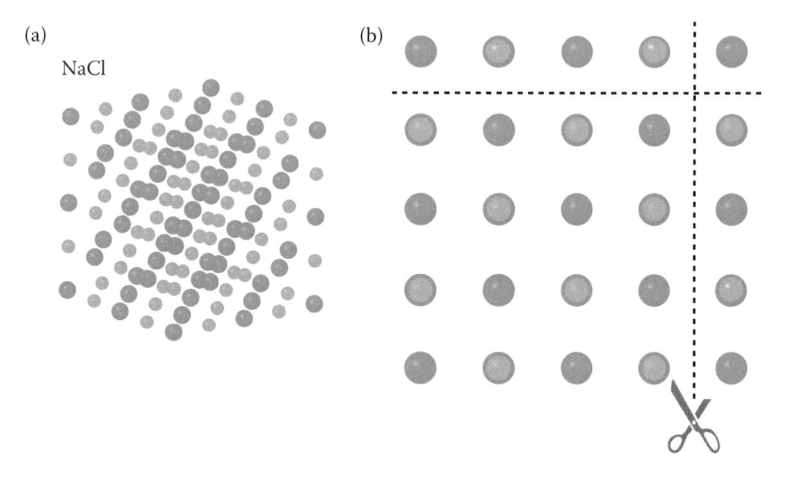

FIGURE 1.61 (a) Microscopic structure of a NaCl single crystal. (b) The ions are arranged in equidistant planes in orthogonal directions.

in 1915 (Nobel Prize 1915). From this representation, it is apparent that the crystal is not based on NaCl molecules but on sodium and chloride ions perfectly arranged in the sites of a *crystal lattice*. We represent here only 124 atoms but the small grain contains billions of billions of atoms. 1 mm^3 of solid NaCl contains roughly

$$N = N_A \cdot \frac{\rho V}{\mu} = 6 \cdot 10^{23}\,\frac{\text{molecules}}{\text{mol}} \cdot \frac{2.1\dfrac{\text{g}}{\text{cm}^3} \cdot 10^{-3}\text{cm}^3}{58.5\dfrac{\text{g}}{\text{mol}}}$$

$$= 21.5 \cdot 10^{18}\text{molecules!}$$

A single crystal of sodium chloride has the appearance of cubic crystal. If we have a perfect knife to remove the last plane of atoms (process named *cleavage*), we get a smaller crystal but with the same plane faces and right angles; therefore, the same symmetry as the first one as shown in Figure 1.61b. We can cut another plane and the result is the same. The next question arises: what is the smallest crystal that keeps the symmetry properties of an entire crystal? This is named the *unit cell* and the repetition in space of unit cell generates a crystal lattice.

Let us imagine now the reverse process: we can build the sodium chloride crystal atom by atom as a house is built brick by brick. It is only an exercise of imagination. Today, it is practically possible to manipulate atoms in the laboratory using the scanning tunneling microscope (STM), but we are far from the moment to move billions of atoms. Actually, the crystal growing (not building), which means to let million atoms on each second to be "naturally" arranged by the atomic forces in specific sites, is currently performed by different techniques in order to obtain crystals in laboratory or industry. We can pick up individual chloride anions and place them on a plane and then place sodium cations in the empty volume between chloride anions as shown in Figure 1.62a. The process described is named *coordination* in chemistry. Then, we complete the upper planes of atoms (Figure 1.62b and c). We note here that while atom size is known to decrease with Z number in a row of periodic table, for example, sodium atom size is larger than chlorine atom size, the corresponding ionic radii are inversely correlated, for example, Na^+ ion is smaller than Cl^- ion, as shown in Figure 1.62.

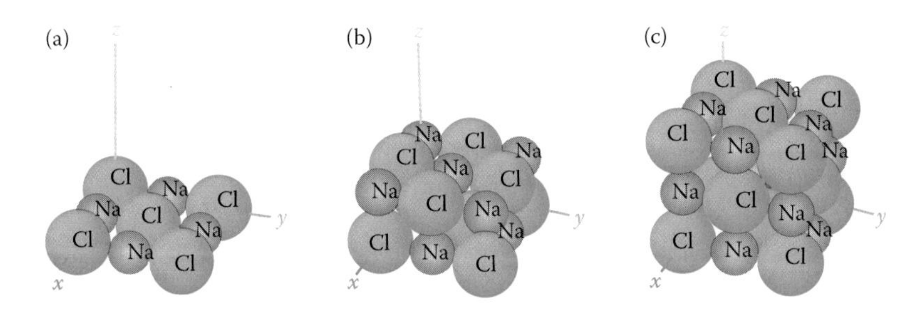

FIGURE 1.62 The building process, plane by plane of sodium chloride crystal.

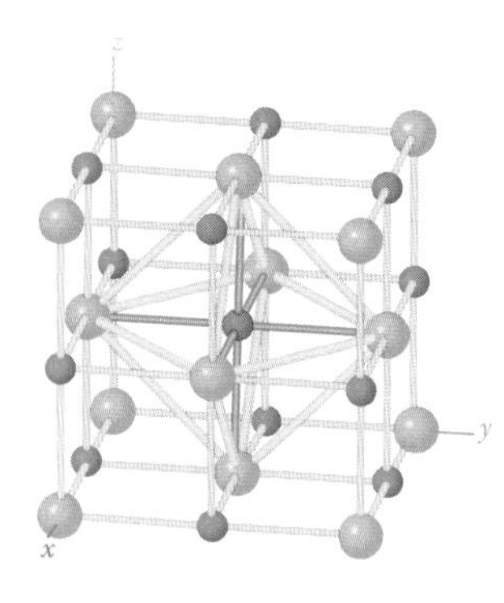

FIGURE 1.63 The negative charge of the first coordination sphere surrounding the sodium cation has octahedral symmetry (Na^+ is a small ball and Cl^- is a big ball).

What is the local symmetry "seen" by one sodium cation? We reduce the atom size and connect the atoms. Each cation of sodium is surrounded (coordinated) by six anions of chlorine, placed equidistant as nearest neighbors, and vice versa, as illustrated in Figure 1.63. It is easier to discover that the negative charge distribution around the sodium cation has *octahedral symmetry* O_h (see Section 1.3.5). The figure contains 12 lines, which are not atomic bonds, drawn just to suggest the octahedron shape. The symmetry is the same for the chlorine anion surrounded (coordinated) by six sodium cations. We note here that anywhere (only) *in the volume* of the sodium chloride crystal, the octahedral symmetry is valuable for every ion whatever its charge. We must remember later this observation, which is important to explain optical properties of doped crystals.

We return to the previous question: What is the smallest crystal unit that possesses the symmetry of the crystal? The atomic complex illustrated in Figures 1.62 and 1.63 is the *unit cell*. If we translate along the y axis, this complex, as illustrated in Figure 1.64, the structure and symmetry of the crystal are maintained. The crystal symmetry is maintained whatever translation direction we choose—x, y, or z.

Translation is a symmetry operation specific only to crystals not to molecules.

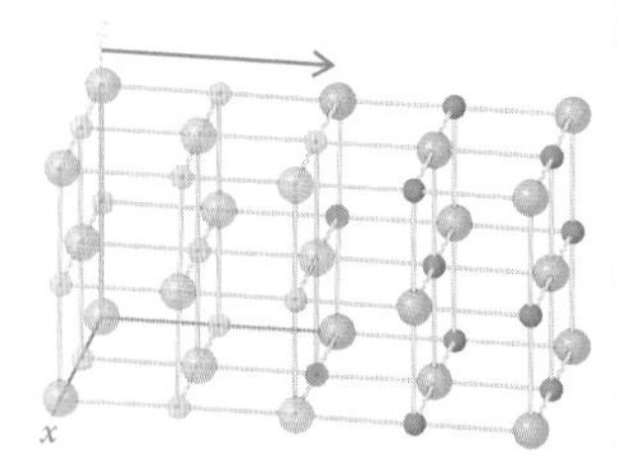

FIGURE 1.64 Unit cell is any cell that can be translated indefinitely in a crystal lattice, such that the overall crystal structure is maintained.

Thus, the unit cell of the NaCl crystal is more complicated than the small cube of eight atoms that we expect it to be.

The geometry of the unit cell can be used to classify crystals. To describe a spatial structure, we currently use the Cartesian coordinates: three orthogonal directions described by unitary vectors. In case of crystals, it is better to attach three (or four in some cases) vectors $\vec{a}, \vec{b}$, and $\vec{c}$ in each direction of the unit cell named *crystallographic axes*. Depending on the symmetry of the lattice, the directions may not be perpendicular to one another and we must also define three angles α, β, and γ between these vectors, as shown in Figure 1.65a. This means that we attach a particular coordinate system (not necessary Cartesian) for each type of crystal. These directions within the crystal become important when we study the optical properties of crystals that depend on the direction in the crystal (*optical anisotropy*). The lengths of a, b, and c are proportional to the unit cell edges. In case of the sodium chloride crystal, the unit cell is a perfect *cube* characterized by $a = b = c$ and $\alpha = \beta = \gamma = 90°$ (Figure 1.65b).

Sodium chloride is a member of alkali halides group. Are there the same symmetry and unit cell for all alkali halides crystals? Most of alkali halide crystals present the sodium chloride structure. Cesium combinations are exceptions. As we have previously seen, the arrangement of sodium cations depends on the space available between bigger chlorine anions. If we take cesium cations instead of sodium, the excess positive charge is the same but the radius of cesium is approximately 3 times bigger than the radius of sodium, that is, 1.69/0.60 (Evans 1964). This means the cesium cation cannot enter the same sites as sodium and the arrangement will be different (Figure 1.66a). Each cesium ion is surrounded by eight halogen ions placed in all the corners of the cube. However, each

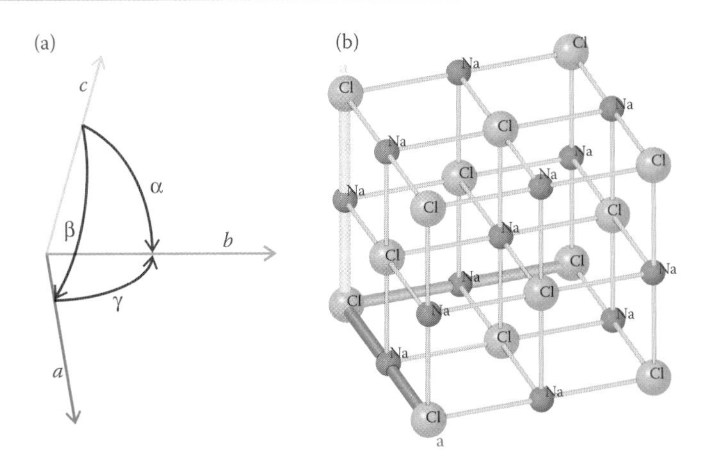

FIGURE 1.65 Coordinates system attached to the unit cell (a) and illustrated for the sodium chloride crystal (b).

halogen ion is surrounded by eight cesium ions too. Hence, the unit cell of both sodium chloride and cesium chloride has a cubic symmetry. The difference is given by the position of chlorine ions: the unit cell of cesium chloride is called *body-centered cubic (BCC)* shown in Figure 1.66a and the unit cell of sodium chloride is called *face-centered cubic (FCC)* shown in Figure 1.66b. We note here that there are two identical lattices FCC (one of sodium and other of chloride) interpenetrating in order to form together the structure of sodium chloride crystal.

> There are seven crystal systems in three dimensions: triclinic, monoclinic, orthorhombic, tetragonal, rhombohedral, hexagonal, and cubic (also called as isometric).

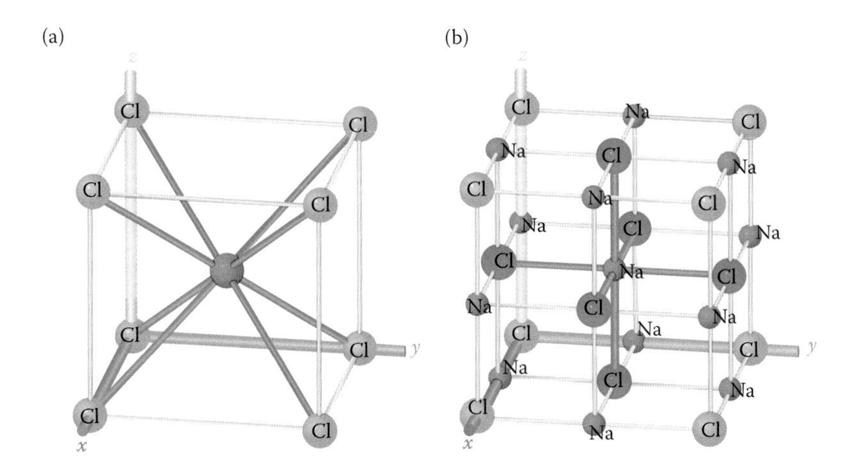

FIGURE 1.66 Unit cell of both cesium chloride and sodium chloride is cubic but the first one is body-centered (a) and the second one is face-centered (b).

This classification was done according to both the relation between length of the unit cell edges and the relation between angles. In several books dedicated to crystals or minerals, the rhombohedral and hexagonal systems are presented together as a unique system (hexagonal). They are illustrated in Figure 1.67. The 14 unit cells illustrated below represent the only possible way that space can be filled. Cubic cells have three orthogonal edges of equal length; BCC and FCC cells cannot be fully specified without also using translation operations in terms of half-cell distances. The tetragonal and orthorhombic cells also have edges that are orthogonal to each other, but either one edge differs in length from the other two

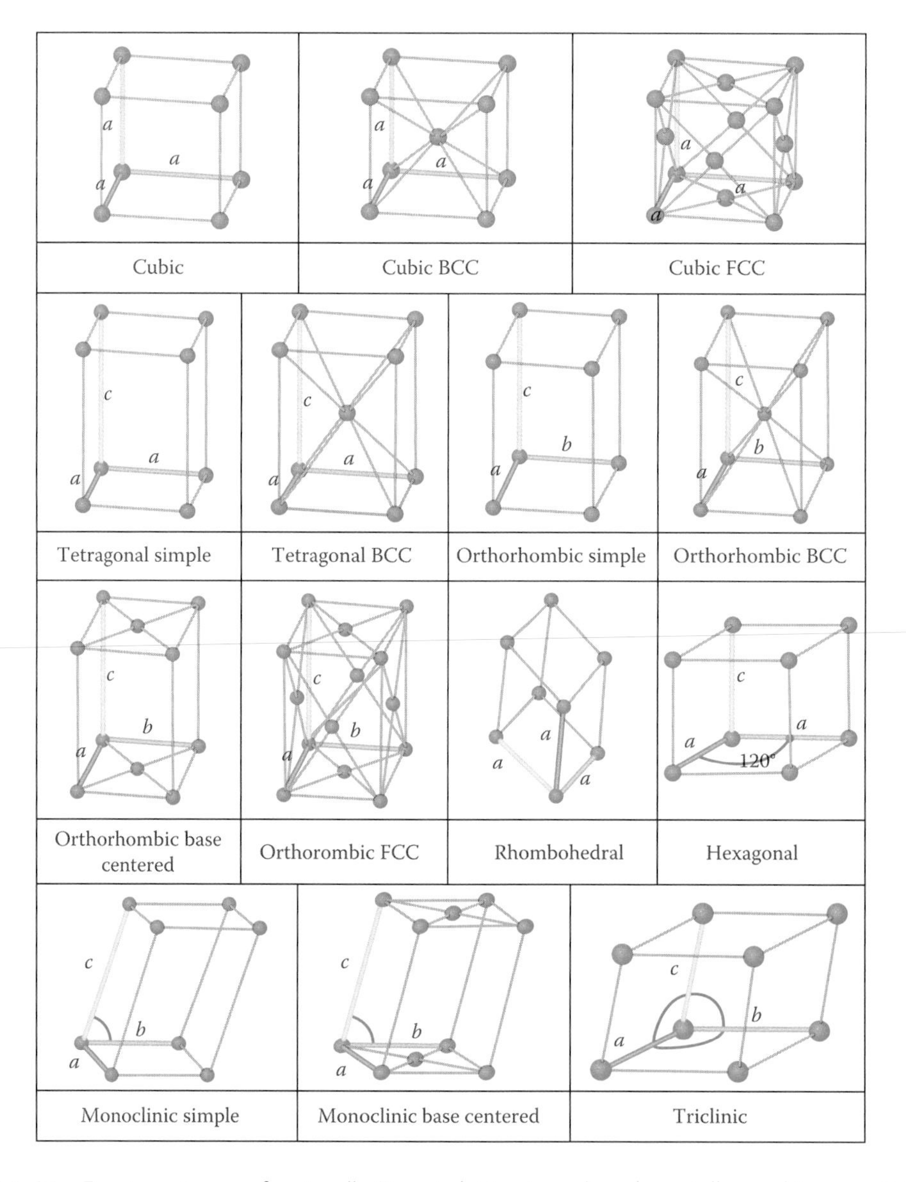

FIGURE 1.67 Fourteen types of unit cells (Bravais lattice types) and crystallographic axes distributed into seven crystal (lattice) systems: cubic, tetragonal, orthorhombic, rhombohedral, hexagonal, monoclinic, and triclinic.

TABLE 1.23 Summary of Relationships in the Crystal Systems

System	Relationships
Triclinic	$a \neq b \neq c$; $\alpha \neq \beta \neq \gamma \neq 90°$
Monoclinic	$a \neq b \neq c$; $\alpha = \gamma = 90°$, $\beta \neq 90°$
Hexagonal	$a = b \neq c$; $\alpha = \beta = 90°$, $\gamma = 120°$
Rhombohedral (trigonal)	$a = b = c$; $\alpha = \beta = \gamma \neq 90°$
Orthorhombic	$a \neq b \neq c$; $\alpha = \beta = \gamma = 90°$
Tetragonal	$a = b \neq c$; $\alpha = \beta = \gamma = 90°$
Cubic (isometric)	$a = b = c$; $\alpha = \beta = \gamma = 90°$

edges (tetragonal) or all three edges differ in length (orthorhombic). The monoclinic and triclinic cells have three lengths that are not equal. The difference is given by either one angle (monoclinic) or all three angles (triclinic) between the edges not equal to 90°. The rhombohedral cell can be understood as a cubic cell that has been stretched or squeezed along a diagonal. The three edges are equal and the three angles are equal too but are not 90°. The hexagonal cell has two angles of 90° and one of 120°. Only two edges are equal. We summarize the seven crystal systems from low to high symmetry in Table 1.23.

The symbols are also given to the lattices according to the centering as shown in Table 1.24. How can we calculate the number of atoms per unit cell? The atom placed in the center of the cell, as Cs atom in Figure 1.66a, is counted as 1 atom for that unit cell. The atom placed at the face center is counted as 1/2 atom for the unit cell, because the other half belongs to the neighboring cell. The atom placed at the cell corner is counted as 1/8 atom for the unit cell, because it is divided

TABLE 1.24 Symbols for Three-Dimensional Lattices According to Centering

Symbol	Name	Number of Atoms per Unit Cell	Position of Atoms
P	*Primitive*	1	Lattice points at corners only
R	*Rhombohedral*	1	Lattice points at corners only
A	*A-centered*	2	Lattice points at corners and centers of **b**, **c** faces
B	*B-centered*	2	Lattice points at corners and centers of **a**, **c** faces
C	*C-centered*	2	Lattice points at corners and centers of **a**, **b** faces
I	*Body-centered*	2	Lattice points at corners and body center
F	*Face-centered*	4	Lattice points at corners and all face centers

Source: Adapted from O'Keeffe, M. and B. G. Hyde. 1996. *Crystal Structures I: Patterns and Symmetry*. Mineralogical Society of America, Washington D.C. (available at http://www.public.asu.edu/~rosebudx/okeeffe.htm).

into eight cells that are in that corner. Therefore, the *I*-type cell (body-centered) has $(1 + 8 \cdot 1/8)$ atoms and the *F*-type cell (face-centered) has $(6 \cdot 1/2 + 8 \cdot 1/8)$ atoms.

1.8.3 HERMANN–MAUGUIN (INTERNATIONAL) SYMBOLS

The Hermann–Mauguin (or international) notation is the one most commonly used in crystallography. It is used to describe the crystal classes from the symmetry operations. It consists of a set of four symbols. The first symbol is the letter symbol of the cell according to the centering (*P, R, A, B, C, I, F*) as we previously presented in Table 1.24. The next three symbols describe the most important symmetry elements. For better understanding, we will start to analyze some examples.

We are using four elements to describe the molecular symmetry: proper and improper rotation axes, reflection planes, and inversion center. In crystallography, they are denoted as follows:

- The *n*-fold proper rotation axis is simply denoted by the number "*n*." Therefore, the twofold rotation is denoted by 2 and so on. In crystallography, only twofold, threefold, fourfold, and sixfold rotations are possible. Although the objects may appear to have fivefold, sevenfold, eightfold, or higher-fold rotation axes, these are not possible in crystals. The reason is that the external shape of a crystal is based on a geometric arrangement of atoms. Figure 1.68 illustrates how only the twofold, threefold, fourfold, and sixfold rotations can fill the space and why they are used in crystallography.

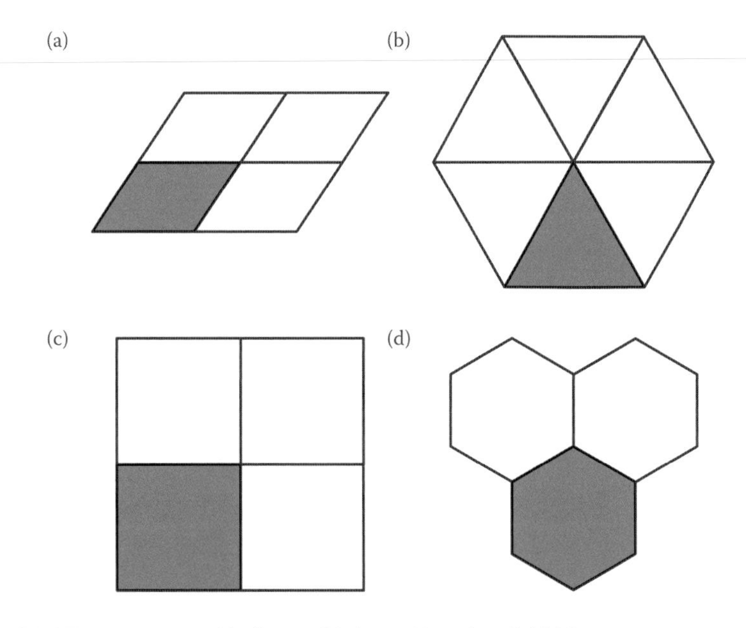

FIGURE 1.68 Two- (a), three- (b), four- (c) and sixfold (d) rotations can only completely fill the space.

- The reflection plane, called "mirror" by most authors of crystallographic books, is simply denoted by the letter "*m*."
- The inversion, seen as a reflection through a point, is denoted by "*i*."
- The improper rotation, which is a mixed *rotation–reflection* operation in the point groups, is replaced by a *different* operation called *rotation–inversion* and denoted by the symbol "–*n*" or "$\bar{n}$". In such a way, S_2 rotation becomes "–1" ("$\bar{1}$") rotoinversion! S_3 rotation becomes "–3" ("$\bar{3}$") rotoinversion, S_4 rotation becomes "–4" ("$\bar{4}$") rotoinversion, and S_6 rotation becomes "–3" "$\bar{3}$" rotoinversion.

Every unique axis, mirror, or rotoreflection will be denoted and counted. An element is unique if it exists by itself and is not produced by another symmetry operation.

EXAMPLE 1.1

The *orthorhombic* cell ($a \neq b \neq c$; $\alpha = \beta = \gamma = 90°$) of Figure 1.69 has three twofold rotation axes (C_2), three mirror planes (m), and a center of inversion (i). There is no operation to convert one rotation axis into the other; so each of three twofold axes is unique. Therefore, we can write a 2 (for twofold) for each axis. The symbol of the group will be 222. At the moment we ignore centering symbol (first symbol). The body has also three mirror planes that are also unique from the same reason as for rotations. So, we can write 2*m*2*m*2*m*. Each of the twofold axes is perpendicular to one mirror plane. In this case, we put a slash between the symbol of axis and the symbol of mirror plane, so our symbol finally becomes: 2/*m*2/*m*2/*m* or $\frac{2}{m}\frac{2}{m}\frac{2}{m}$. In the Schönflies notation, used extensively at molecules, the symmetry symbol of this body is $\boldsymbol{D}_{2h}$.

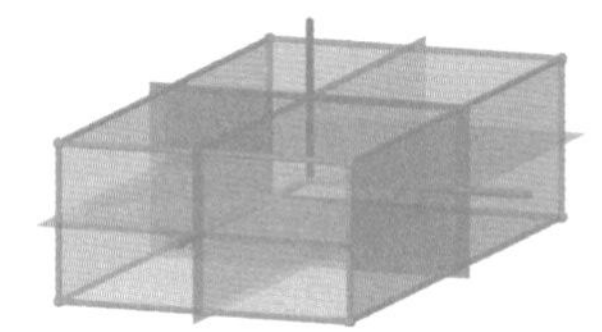

FIGURE 1.69 Orthorhombic cell has three twofold rotation axes (C_2), three mirror planes (m), and a center of inversion (i).

EXAMPLE 1.2

The *tetragonal* cell ($a = b \neq c$; $\alpha = \beta = \gamma = 90°$) of Figure 1.70 contains one fourfold axis, four twofold axes, five mirror planes, and a center of inversion. The fourfold rotation is unique. There are two twofold axes that are perpendicular to identical faces and we take only one as unique axis, because they are equivalent (members of the same classes). There are also two twofold axes that pass through the vertical edges of the crystal and they are equivalent. Thus, we take only one as unique axis. The fourfold axis perpendicular to the top face shifts one twofold axis to another. So, we can write the symbol of the group as 422. Concerning the mirror planes, the horizontal plane is unique, but the vertical planes are related two by two by the fourfold rotation. It means the two planes perpendicular to the vertical faces give only one unique mirror and the two planes that pass through the vertical edges give another unique mirror. So we can write 4*m*2*m*2*m*. Finally, each mirror plane is perpendicular to the one unique axis. Our symbol finally becomes 4/*m*2/*m*2/*m* or $\frac{4}{m}\frac{2}{m}\frac{2}{m}$. In the Schönflies notation, the symmetry symbol of this body is $\boldsymbol{D}_{4h}$.

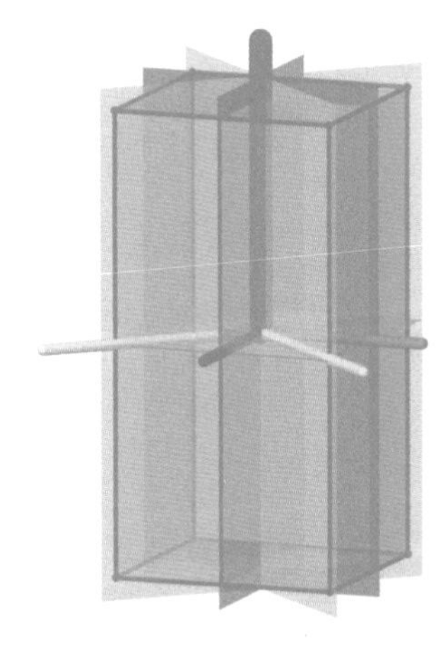

FIGURE 1.70 Tetragonal cell contains one fourfold axis, four twofold axes, five mirror planes, and a center of inversion.

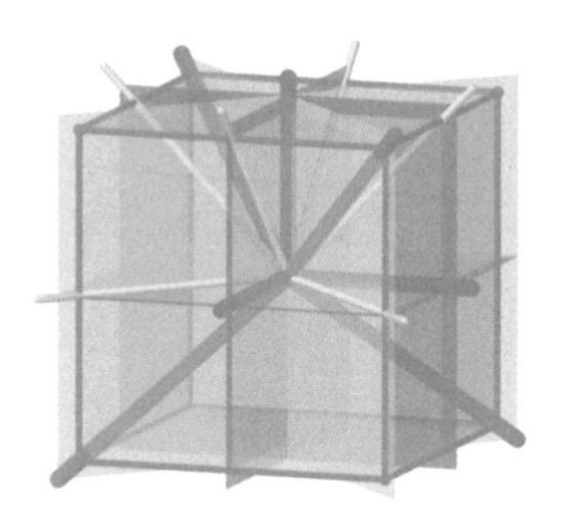

FIGURE 1.71 Cubic cell contains three fourfold axes, four threefold rotoinversions, six twofold axes through the middle of edges, nine mirror planes, and a center of inversion.

EXAMPLE 1.3

The *cubic* cell ($a = b = c$; $\alpha = \beta = \gamma = 90°$) of Figure 1.71 contains

- Three fourfold rotation axes; each of them is perpendicular to a square face
- Four threefold rotoinversion axes, each going out of the cube corners
- Six twofold rotation axes, going out of the cube edges
- Nine mirror planes
- A center of inversion

From the experience of the molecular cubic groups (see Section 1.3.5), the three fourfold rotations are members of the same class and we get 4 only once. The four threefold rotoinversions will be taken as $\bar{3}$ only once. There is also only one twofold axis unique and we get 2 only once. Each of the fourfold axis is perpendicular to a mirror plane and twofold axes too. Our symbol finally becomes $4/m$–$32/m$ or $\frac{4}{m}\bar{3}\frac{2}{m}$. In the Schönflies notation, the symmetry symbol of this body is $\boldsymbol{O}_h$.

We conclude that there are 32 possible combinations of symmetry operations (rotation, reflection, inversion, rotoinversion) that define the external symmetry of crystals. These combinations result in the 32 crystal classes known as the 32 point groups presented in Table 1.25. They are called crystal forms because this is what we can observe.

1.8.4 SYMMETRY OPERATIONS

The combination of all symmetry elements previously discussed (rotation, reflection, inversion, and rotoinversion) results in the 32 crystallographic point groups. Rotation and rotoinversion axes and mirror planes intersect in one fixed point of space and, as a consequence, such groups are called point groups. This description is good enough for the symmetry of unit cell, which is finite. The crystal, however, could be regarded as an infinite structure. Therefore, the crystal is obtained by multiplication of the unit cell throughout three-dimensional space. The lattice translation by a length of one unit cell in any crystallographic direction (which can be different from Cartesian directions) left the crystal in a final configuration indistinguishable from the initial one. Hence, the translation is a symmetry operation independent from the others mentioned above. The combination of rotations and reflections with translations results in two new classes of symmetry elements: screw axes and glide planes.

For better understanding of the operation generated by *screw axis*, you must look to your ballpoint pen. Most of them are composed of two pieces assembled together. If you rotate one piece while holding the other piece fixed, you will observe a translation of the moving piece during rotation. This operation is called screwing and the pen axis is the screw axis (Figure 1.72).

The operation that characterizes a screw axis is an n-fold rotation ($2\pi/n$) around axis followed by a translation of p/n in the direction of axis. It is denoted as n_p, where n is the rotation type and subscript p is the

TABLE 1.25 32 Point Groups (Crystal Classes) in Hermann–Maugain Notation and Their Equivalence in Schönflies Notation

Crystal System	Point Group (Hermann–Maugain)	Point Group (Schönflies)	Symmetry Elements
Triclinic	1	C_1	none
	$\bar{1}$	C_i	i
Monoclinic	2	C_2	$1C_2$
	m	C_s	$1m$
	$\frac{2}{m}$	C_{2h}	$i, 1C_2, 1m$
Orthorhombic	222	D_2	$3C_2$
	$2mm$	C_{2v}	$1C_2, 2m$
	$\frac{2}{m}\frac{2}{m}\frac{2}{m}$	D_{2h}	$i, 3C_2, 3m$
Tetragonal	4	C_4	$1C_4$
	$\bar{4}$	S_4	$1\bar{C}_4$
	$\frac{4}{m}$	C_{4h}	$i, 1C_4, 1m$
	422	D_4	$1C_4, 4C_2$
	$4mm$	C_{4v}	$1C_4, 4m$
	$\bar{4}2m$	D_{2d}	$1\bar{C}_4, 2C_2, 2m$
	$\frac{4}{m}\frac{2}{m}\frac{2}{m}$	D_{4h}	$i, 1C_4, 4C_2, 5m$
Rhombohedral (trigonal)	3	C_3	$1C_3$
	$\bar{3}$	S_6	$1\bar{C}_3$
	32	D_3	$1C_3, 3C_2$
	$3m$	C_{3v}	$1C_3, 3m$
	$\bar{3}\frac{2}{m}$	D_{3d}	$1\bar{C}_3, 3C_2, 3m$
Hexagonal	6	C_6	$1C_6$
	$\bar{6}$	C_{3h}	$1\bar{C}_6$
	$\frac{6}{m}$	C_{6h}	$i, 1C_6, 1m$
	622	D_6	$1C_6, 6C_2$
	$6mm$	C_{6v}	$1C_6, 6m$
	$\bar{6}m2$	D_{3h}	$1\bar{C}_6, 3C_2, 3m$
	$\frac{6}{m}\frac{2}{m}\frac{2}{m}$	D_{6h}	$i, 1C_6, 6C_2, 7m$
Cubic (isometric)	23	T	$3C_2, 4C_3$
	$\frac{2}{m}\bar{3}$	T_h	$3C_2, 3m, 4\bar{C}_3$
	432	O	$3C_4, 4C_3, 6C_2$
	$\bar{4}3m$	T_d	$3\bar{C}_4, 4C_3, 6m$
	$\frac{4}{m}\bar{3}\frac{2}{m}$	O_h	$3C_4, 4\bar{C}_3, 6C_2, 9m$

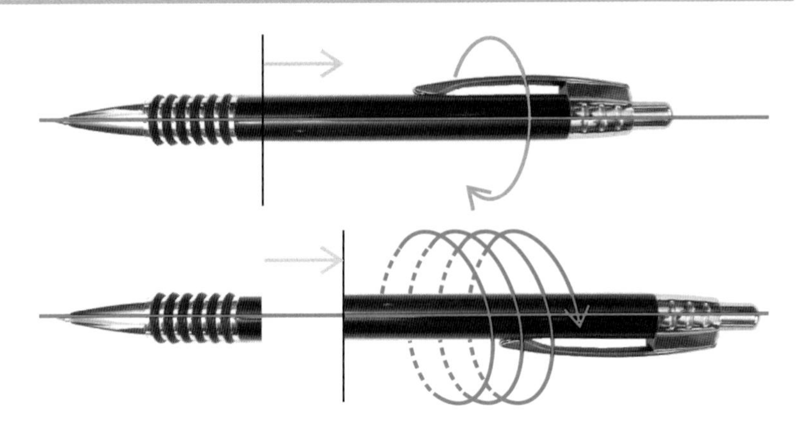

FIGURE 1.72 Combination of rotation and translation gives a screw movement.

translation length reported to the lattice vector parallel to the rotation axis. We give some examples of how to read this notation:

- The screw axis 2_1 (parallel to the *a* axis) in orthogonal case means a twofold rotation (180°) around the *a* axis followed by a 1/2 translation; therefore, one half of the unit vector in the direction of the *a* axis. Thus, the initial coordinates (x, y, z) of any point will change to $(x + 1/2, \bar{y}, \bar{z})$ as shown in Figure 1.73a.
- The screw axis 4_1 (parallel to the *b* axis) in orthogonal case means a fourfold rotation (90°) around the *b* axis followed by a 1/4 translation, therefore, one-fourth of the unit vector in the direction of the *b* axis. Thus, the initial coordinates (x, y, z) of any point will change to $(z, y + 1/4, \bar{x})$ as shown in Figure 1.73b.
- The screw axis 3_2 (parallel to the *c* axis) in hexagonal case means a threefold rotation (120°) around the axis followed by a 2/3 translation, therefore two-thirds of the unit vector in the direction of the *c* axis. Hence, the initial coordinates (x, y, z) of any point will change to $(\bar{y}, x - y, z + 2/3)$ as shown in Figure 1.73c.

Other possible operation results from the combination between reflection and translation and is called *glide plane*. It consists of reflection through a plane followed by translation parallel to that plane (Figure 1.74).

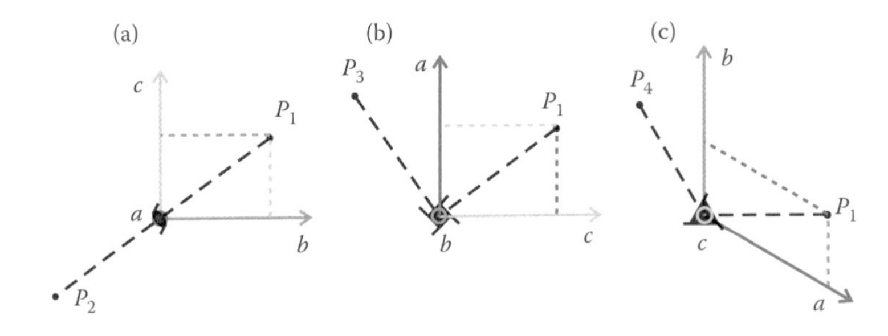

FIGURE 1.73 Examples of screw axes and their effect: (a) 2_1 parallel to *a* axis; (b) 4_1 parallel to *b* axis; (c) 3_2 parallel to *c* axis.

There are three types of glides:

- *Axial glide.* The glide parallel to the a axis and denoted as simply a consists of reflection through the plane and translation by $a/2$. The similar glides parallel to b and c axes are denoted by b and c.
- *Diagonal glide* denoted by n involves translation of $(a + b)/2$, $(b + c)/2$, or $(c + a)/2$; therefore, the direction of the glide is parallel to a diagonal face. It is possible to have a glide as $(a + b + c)/2$ in cubic, tetragonal, and hexagonal systems.
- *Diamond glide* denoted by d, where the translation is $(a \pm b)/4$, $(b \pm c)/4$, or $(c \pm a)/4$. In case of cubic and tetragonal systems, the translation $(a + b + c)/4$ is also possible.

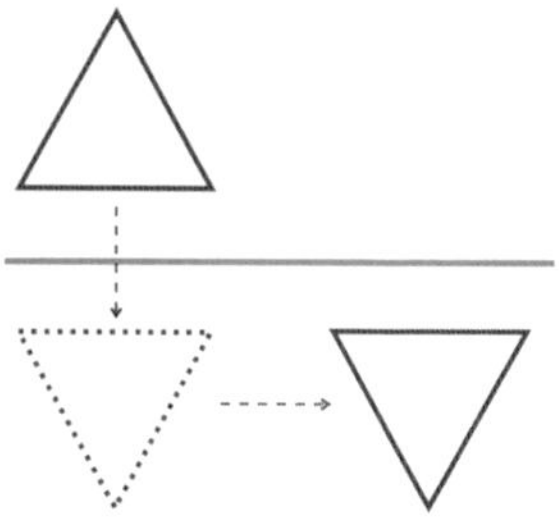

FIGURE 1.74 Glide operation is a combination between reflection through a plane followed by translation parallel to that plane.

To conclude, the symmetry operations of the crystals are identity, rotation, reflection, inversion, rotoinversion, screw, and glide.

1.8.5 SPACE GROUPS

A *space group* is a group that includes both the point group symmetry elements and the translation. The addition of translation greatly increases the possibilities of combination. The combinations of the 32 crystallographic point groups and the 14 Bravais lattices (which are again combinations of different crystal systems and noncentered or centered lattices) give rise to *230* different space groups instead of 32 point groups.

Hence, in order to determine the complete list of space groups, we must combine each point group with each Bravais lattice type. The international tables list all space groups by symbol and number, together with symmetry operators, origins, reflection conditions, and space group projection diagrams (see Rupp 2013). We do not intend to treat all cases in this book. We will try to provide a few examples.

We start with the point group $2/m$ (or $\mathbf{C}_{2h}$), which belongs to monoclinic system. In Section 1.8.2 (Figure 1.67), we have seen that the monoclinic-type cell can be only simple (P) or base-centered (C). Hence, we expect two types of space groups: $P2/m$ and $C2/m$. The symbol of space groups begins with a capital letter denoting the centering, in accordance with Table 1.24. Taking into account the new specific operations for space groups (screw and glade), we should replace the twofold rotation by the screw axis 2_1 and mirror plane m by the corresponding glade plane c. This gives new space groups as $P2_1/m$, $P2/c$, $P2_1/c$, and $C2/c$. From the search of all possible combinations, we eliminated those combinations that give equivalent result. For example, we eliminated $P2/a$ because it is the same as $P2/c$. The space group $C2_1/c$ is also equivalent to $C2/c$.

The second example is the point group $\boldsymbol{T}_d$ denoted as $\bar{4}3m$ in international notation. This group is part of cubic (isometric) systems. The cubic cell can be only simple (P) or body (volume)-centered (I) or face-centered (F). This means we have more possible combinations: $P\bar{4}3m$, $I\bar{4}3m$, $F\bar{4}3m$. We replace the mirror planes σ_d by the glade planes, which can be axial glade (c) in case of a face-centered cell, face glade (n) in case of a primitive cell, and diamond glade (d) in case of a body-centered cell; therefore, there are three more space groups: $P\bar{4}3n$, $I\bar{4}3d$, and $F\bar{4}3c$.

We note finally that the point group of a crystal may also be obtained from the space group by replacing each screw axis and glade plane by a proper rotation and a mirror plane, respectively.

1.9 ROTATION GROUPS AND OPERATORS

1.9.1 *SO*(3) GROUP

In quantum mechanics, Paul Dirac introduced a simplified notation: a physical state of the system is simply written as a *ket* vector denoted as $|\alpha\rangle$. For every ket $|\alpha\rangle$, there is a *bra* vector denoted as $\langle\alpha|$. Using this simplified notation, we are now able to write the inner product as $\langle\beta|\alpha\rangle$ (bra-ket). In terms of eigenfunctions, we can write

$$\int \Psi_i^* \Psi_j d\tau = \langle \Psi_i | \Psi_j \rangle = \delta_{ij}$$

This is the orthonormality condition.

A quantum state can be changed by the action of an operator:

$$\hat{F}\Psi = f\Psi \text{ or in Dirac notation } \hat{F}|\Psi\rangle = f\Psi$$

where the operator can be the Hamiltonian, potential, angular momentum, and so on. f is the eigenvalue of the operator and can have several values that form a set of eigenvalues. The most common example is the Schrödinger equation $\hat{H}|\Psi\rangle = E\Psi$.

The unitary operator leaves unchanged any vector (function): $\hat{1}|\Psi\rangle = \Psi$ and is generally denoted as $\hat{U}$. Let us consider the integrals $\int \Psi_i^* \Psi_j d\tau$. The value of the integral is unchanged by any rotation or reflection of the coordinate system. Thus, the symmetry operations do not destroy the orthonormality of the base functions. Therefore, the operators of symmetry operations are unitary (Landau and Lifshitz 1991). As a consequence, the rotation operator U(R) must be unitary too.

Any given rotation operator can be represented by a 3×3 orthogonal matrix. Let us suppose the vector $|\alpha\rangle$ is described by an orthonormate basis $\{|\Psi\rangle_{j=1,\ldots,n}\}$ and it can be written as a linear combination of the set functions:

$$|\alpha\rangle = \sum_j a_j |\Psi_j\rangle \quad (1.99)$$

where a_j are numbers. The $|\alpha\rangle$ vector is transformed by the $\hat{R}$ operator into the $|\beta\rangle$ vector as described by

$$|\beta\rangle = \hat{R}|\alpha\rangle = \hat{R}\sum_j a_j |\Psi_j\rangle = \sum_j a_j \hat{R} |\Psi_j\rangle \quad (1.100)$$

that can also be written as a linear combination of the original set $|\beta\rangle = \sum_k b_k |\Psi_k\rangle$. Thus, we can write

$$\sum_k b_k |\Psi_k\rangle = \sum_j a_j \hat{R} |\Psi_j\rangle$$

and then we can multiply the left side by $\langle\Psi_i|$ to obtain

$$b_i = \sum_j a_j \langle\Psi_i|\hat{R}|\Psi_j\rangle$$

also taking into account the orthonormality condition. So, the matrix components of the rotation operator will be given by

$$r_{ij} = \langle\Psi_i|\hat{R}\Psi_j\rangle \quad (1.101)$$

The rotations can constitute themselves into a group because

1. Two successive rotations form another rotation: $R(\theta_1)\cdot R(\theta_2) = R(\theta_1) + R(\theta_2)$.
2. The 0° rotation is identity element: $R(0°) = 1$.
3. The rotation product is associative: $R(\theta_1)\cdot[R(\theta_2)\cdot R(\theta_3)] = [R(\theta_1)\cdot R(\theta_2)]\cdot R(\theta_3)$.
4. There is an inverse rotation for each rotation: $R^{-1}(\theta) = R(-\theta)$.

Rotation preserves the origin, angles, distances, and orientations. Therefore, the rotation must preserve the origin, inner product, and vector product of any two vectors. Hence, we may see a rotation as a linear transformation $R{:}\mathfrak{R} \rightarrow \mathfrak{R}$. The group of proper rotations of $\mathfrak{R}^3$ (three-dimensional Euclidean space) around the origin is denoted by $SO(3)$ and called the orthogonal group, or the *rotation group* (Special Orthogonal group in 3 dimensions). It consists of *real 3 × 3 orthogonal matrices of determinant +1*. Any element R of $SO(3)$ may be expressed in terms of elementary rotations around x, y, and z axes, given by the following expressions:

$$R_x(\alpha) = \begin{bmatrix} 1 & 0 & 0 \\ 0 & \cos\alpha & -\sin\alpha \\ 0 & \sin\alpha & \cos\alpha \end{bmatrix}$$

$$R_y(\beta) = \begin{bmatrix} \cos\beta & 0 & \sin\beta \\ 0 & 1 & 0 \\ -\sin\beta & 0 & \cos\beta \end{bmatrix}$$

$$R_z(\gamma) = \begin{bmatrix} \cos\gamma & -\sin\gamma & 0 \\ \sin\gamma & \cos\gamma & 0 \\ 0 & 0 & 1 \end{bmatrix} \quad (1.102)$$

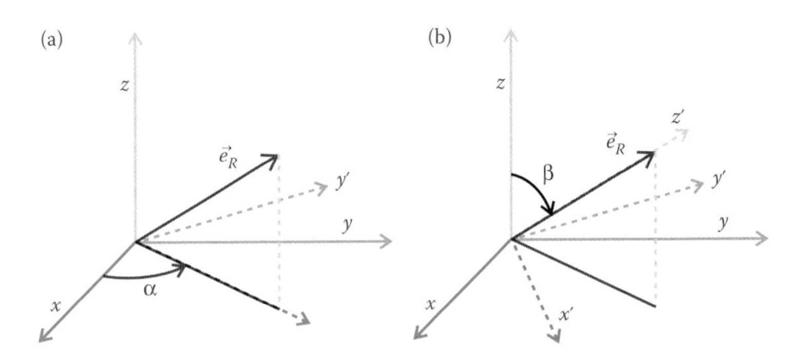

FIGURE 1.75 The rotation around $\vec{e}_R$ axis is characterized by angles α (a), β (b), and the rotation angle γ. The effect of rotation is obtained as a product of three rotational matrices over α, β, and γ.

Let us take an axis of rotation $\vec{e}_R$, which is a vector as shown in Figure 1.75. Now, we consider an arbitrary rotation with rotation axis determined by $\vec{e}_R$ over an angle γ and we denote the rotation by $R(\vec{e}_R, \gamma)$. Actually, the rotation can now be decomposed into three rotations. First, we rotate the system (x, y, z) around the z axis over the α angle to align the x axis to the projection of $\vec{e}_R$ on the (x,y) plane. Second, we rotate the new system around the new y axis position over the β angle until the new z' axis direction will be aligned to the $\vec{e}_R$ direction. Third, we rotate the new system (x', y', z') around the z' axis over the γ angle. Hence, we can write the effect of $R(\vec{e}_R, \gamma)$ as a product of three matrices:

(1.103)

$$R(\vec{e}_R,\gamma) = \begin{bmatrix} \cos\gamma & -\sin\gamma & 0 \\ \sin\gamma & \cos\gamma & 0 \\ 0 & 0 & 1 \end{bmatrix} \begin{bmatrix} \cos\beta & 0 & \sin\beta \\ 0 & 1 & 0 \\ -\sin\beta & 0 & \cos\beta \end{bmatrix} \begin{bmatrix} \cos\alpha & -\sin\alpha & 0 \\ \sin\alpha & \cos\alpha & 0 \\ 0 & 0 & 1 \end{bmatrix}$$

This is the way to obtain a system of coordinates on the manifold $SO(3)$. The three angles (α,β,γ) are called the *Euler angles*.

In classical mechanics, the angular momentum of a particle is a vector and it can vary continuously. For a system of many particles, the total angular momentum is the sum of all angular momenta. In quantum mechanics, the angular momentum is described by three operators $\hat{l}_x, \hat{l}_y$, and $\hat{l}_z$ that are the components of the $\hat{l}$ operator. Their detailed expressions can be found in textbooks of quantum mechanics (e.g., see page 83 of Landau and Lifshitz 1991 or page 162 of Sakurai and Napolitano 2011). Except the apparition of Planck constant and of the complex number i, the most important difference is that $U(R(\vec{e},\delta\varphi))$ is defined for a rotation angle $\delta\varphi$ *infinitesimal*; therefore, it must be infinitesimally close to the unit operator. For a many-particle system, the total angular momentum is defined by exactly the same relations. From now we start using the symbol J, measured in $\hbar$ units, instead of symbol L. The three J_k operators generate any rotation around the k axis:

(1.104) $$R = \exp(-i\delta\varphi e_k J_k)$$

where e_k are versors of the rotation axes and J_k are three operators, *generators* of the representations in $SO(3)$. These quantum generators of angular momentum obey the *fundamental commutation relations of angular momentum*:

(1.105) $$\left[\hat{J}_i, \hat{J}_j\right] = i\varepsilon_{ijk}\hat{J}_k$$

where ε_{ijk} is the completely antisymmetric tensor ($\varepsilon_{123} = +1$). The commutation relations show that the generators do not commute and, thus, the rotations around different axes do not commute. If J_k operators do not commute, it means they cannot have common eigenfunctions.

In the $SO(3)$ group, the generators can be expressed by 3×3 matrices as follows:

(1.106)

$$\hat{J}_x = -i\begin{pmatrix} 0 & 0 & 0 \\ 0 & 0 & 1 \\ 0 & -1 & 0 \end{pmatrix}, \quad \hat{J}_y = -i\begin{pmatrix} 0 & 0 & -1 \\ 0 & 0 & 0 \\ 1 & 0 & 0 \end{pmatrix}, \quad \hat{J}_z = -i\begin{pmatrix} 0 & 1 & 0 \\ -1 & 0 & 0 \\ 0 & 0 & 0 \end{pmatrix}$$

Any rotation given by Euler angles α, β, and γ will be expressed as

(1.107) $$R(\alpha,\beta,\gamma) = \exp(-i\alpha J_z) \cdot \exp(-i\beta J_y) \cdot \exp(-i\gamma J_z) \cdot$$

Remember, the first rotation is performed around the initial z axis, the second one is performed around the rotated y axis, and the last is again performed around the rotated z axis. These explain the use of J_z operator in both the first and the last terms.

Instead of the operators J_x and J_y, it is more convenient to use the complex combinations of them, therefore, the raising and lowering operators (see pages 85–86 of Landau and Lifshitz 1991):

$$\begin{cases} \hat{J}_+ = \hat{J}_x + i\hat{J}_y \\ \hat{J}_- = \hat{J}_x - i\hat{J}_y \end{cases}$$

They are not commutative. Why are they important? It can be shown that by taking an arbitrary eigenvalue m of the J_z operator we can obtain the upper eigenvalues (e.g., $m + 1$, $m + 2$, ...) using the raising operator or, on the contrary the lower eigenvalues (e.g., $m - 1$, $m - 2$, ...) using the lowering operator. Finally, the quantum number m can take all values in the range $-j, -j + 1, \ldots, j - 1, j$. The quantum number j can be integer or half-integer (0, 1/2, 1, 3/2, 2, 5/2, ...). We note that there are $2j + 1$ values of m quantum number.

It is a common convention to work with the square of angular momentum operator J^2 and J_z operator. They commute and thus, they have the same eigenfunctions. Their corresponding matrices are diagonal. The eigenvalues of J^2 are $j(j+1)$. The square of angular momentum is not changed by any rotation. We note that the energy level of the stationary state corresponding to the angular momentum J is $(2j+1)$-fold degenerate. It is called degeneracy in respect to the *direction* of the angular momentum. The state with $J=0$ is not degenerate and it has a spherical symmetry.

Using the above operators, it can be shown that the wavefunctions of the angular momentum are associate Legendre polynomials and hence the spherical harmonics, that depends only on l and m quantum numbers (for a single particle). Another consequence is the selection rule of transitions: $\Delta J = 0, \pm 1$ (except $J = 0 \to 0$) (see Landau and Lifshitz 1991).

In $SO(3)$, any j representation has an orthonormal basis that is a set of $|jm\rangle$ vectors (common eigenfunctions of J^2 and J_z operators). Let us see what the matrix components are of a unitary rotation $U(\alpha,\beta,\gamma)$.

$$U(\alpha,\beta,\gamma)\left|jm\right\rangle = e^{-i\alpha J_z}\cdot e^{-i\beta J_y}\cdot e^{-i\gamma J_z}\left|jm\right\rangle = e^{-i\alpha J_z}\cdot e^{-i\beta J_y}\left|jm\right\rangle\cdot e^{-i\gamma m}$$

We have used the property that the $|jm\rangle$ is an eigenfunction of the J_z operator and m is its eigenvalue. Then, the J_y operator will change m number to m' and the J_z operator leaves m' unchanged:

$$\begin{aligned}U(\alpha,\beta,\gamma)\left|jm\right\rangle &= e^{-i\alpha J_z}\left|jm'\right\rangle\cdot\left\langle jm'\right|e^{-i\beta J_y}\left|jm\right\rangle\cdot e^{-i\gamma m}\\ &= \left|jm'\right\rangle e^{-i\alpha m'}\cdot\left\langle jm'\right|e^{-i\beta J_y}\left|jm\right\rangle\cdot e^{-i\gamma m}\end{aligned}$$

So, the unitary operator U transformed the $|jm\rangle$ vector into $|jm'\rangle$ and the matrix elements are given by

$$D^{j}_{m'm}(\alpha,\beta,\gamma) = e^{-i(\alpha m'+\gamma m)}\cdot\left\langle jm'\right|e^{-i\beta J_y}\left|jm\right\rangle \tag{1.108}$$

The simplest example is for $j=1$. There are only three eigenfunctions. Readers can calculate the matrix elements of the unitary rotation operator using relations 3.5.39, 3.5.40, and 3.5.42 of Sakurai and Napolitano (2011). The matrix of the last term in Equation 1.108 will be reduced to

$$d^1(\beta,J_y) = \frac{1}{2}\begin{pmatrix}1+\cos\beta & -\sqrt{2}\sin\beta & 1-\cos\beta\\ \sqrt{2}\sin\beta & \cos\beta & -\sqrt{2}\sin\beta\\ 1-\cos\beta & \sqrt{2}\sin\beta & 1+\cos\beta\end{pmatrix}$$

It is the simplest case because all matrices have a 3×3 dimension. For a larger value of j, the matrix must be reduced to a block diagonal form, where each smaller square matrix cannot be broken into smaller blocks. The computation is time consuming.

As a practical conclusion the set of all rotations forms a group named $SO(3)$. The spherical harmonics $Y_l^m(\theta,\varphi)$ form a basis of a $(2\ell+1)$

dimensional irreducible representation. The characters of this irreducible representation are given by the following relationship:

$$\chi^{l}(\varphi) = \frac{\sin\left[(2l+1)\dfrac{\varphi}{2}\right]}{\sin\left(\dfrac{\varphi}{2}\right)} \tag{1.109}$$

In the general case, l is replaced by j.

1.9.2 *SU*(2) GROUP

In Section 1.9.1, a rotation was defined by $R(\vec{e}_R, \gamma)$, that is, rotation axis and rotation angle. The rotation axis can be expressed by a vector in three-dimensional spaces (Cartesian coordinates). The rotation axis can also be expressed by a complex number and this happens in the complex space. The group of rotations of the complex vector space $\boldsymbol{C}^2$ consisting of linear transformations that leave the orientation invariant is called the *S*pecial *U*nitary group and denoted as ***SU*(2)**.

We may write the unitary element $U \in SU(2)$ in terms of two complex numbers x and y as

$$U(x,y) = \begin{bmatrix} x & y \\ -y^* & x^* \end{bmatrix}, \quad \text{with } |x|^2 + |y|^2 = 1 \tag{1.110}$$

Hence, $\det U = 1$ (Thaler 2001).

We can readily see that the 2×2 matrix that characterizes a rotation of a spin 1/2 system can be written as $U(x,y)$.

Let us turn to the matrix calculation of Equation 1.108 for $j = 1/2$. The basis is formed by two spinor functions:

$$|+\rangle = \left|\frac{1}{2} \quad \frac{1}{2}\right\rangle = \begin{pmatrix} 1 \\ 0 \end{pmatrix} \quad \text{and} \quad |-\rangle = \left|\frac{1}{2} \quad -\frac{1}{2}\right\rangle = \begin{pmatrix} 0 \\ 1 \end{pmatrix}$$

We begin to calculate the matrix of the J_z operator because of its simplicity:

$$D^{\frac{1}{2}}(J_z) = \begin{pmatrix} \langle +|J_z|+\rangle & \langle +|J_z|-\rangle \\ \langle -|J_z|+\rangle & \langle -|J_z|-\rangle \end{pmatrix} = \begin{pmatrix} \frac{1}{2}\langle +|+\rangle & -\frac{1}{2}\langle +|-\rangle \\ \frac{1}{2}\langle -|+\rangle & -\frac{1}{2}\langle -|-\rangle \end{pmatrix}$$

$$= \begin{pmatrix} \frac{1}{2} & 0 \\ 0 & -\frac{1}{2} \end{pmatrix} = \frac{1}{2}\begin{pmatrix} 1 & 0 \\ 0 & -1 \end{pmatrix} = \frac{1}{2}\sigma_z$$

We have used that the J_z operator does not change the m number and the orthonormality property of the functions. The last matrix denoted by σ_z is the Pauli matrix. Then, we calculate the matrices for raising and lowering operators:

$$D^{\frac{1}{2}}(J_+) = \begin{pmatrix} \langle +|J_+|+\rangle & \langle +|J_+|-\rangle \\ \langle -|J_+|+\rangle & \langle -|J_+|-\rangle \end{pmatrix} = \begin{pmatrix} 0 & 1\langle +|+\rangle \\ 0 & 0 \end{pmatrix} = \begin{pmatrix} 0 & 1 \\ 0 & 0 \end{pmatrix}$$

$$D^{\frac{1}{2}}(J_-) = \begin{pmatrix} \langle +|J_-|+\rangle & \langle +|J_-|-\rangle \\ \langle -|J_-|+\rangle & \langle -|J_-|-\rangle \end{pmatrix} = \begin{pmatrix} 0 & 0 \\ 1\langle -|-\rangle & 0 \end{pmatrix} = \begin{pmatrix} 0 & 0 \\ 1 & 0 \end{pmatrix}$$

Now, it is easy to calculate the next matrices:

$$D^{\frac{1}{2}}(J_x) = \frac{1}{2}\left(D^{\frac{1}{2}}(J_+) + D^{\frac{1}{2}}(J_-)\right) = \frac{1}{2}\begin{pmatrix} 0 & 1 \\ 1 & 0 \end{pmatrix} = \frac{1}{2}\sigma_x$$

$$D^{\frac{1}{2}}(J_y) = \frac{1}{2i}\left(D^{\frac{1}{2}}(J_+) - D^{\frac{1}{2}}(J_-)\right) = \frac{1}{2}\begin{pmatrix} 0 & -i \\ i & 0 \end{pmatrix} = \frac{1}{2}\sigma_y$$

The matrix of the spinor transformation will be given by

$$\hat{U}_2(\alpha,\beta,\gamma) = \begin{pmatrix} e^{-i\frac{\alpha}{2}} & 0 \\ 0 & e^{i\frac{\alpha}{2}} \end{pmatrix}\begin{pmatrix} \cos\frac{\beta}{2} & -\sin\frac{\beta}{2} \\ \sin\frac{\beta}{2} & \cos\frac{\beta}{2} \end{pmatrix}\begin{pmatrix} e^{-i\frac{\gamma}{2}} & 0 \\ 0 & e^{i\frac{\gamma}{2}} \end{pmatrix}$$

$$= \begin{pmatrix} \cos\frac{\beta}{2}e^{-i\frac{\alpha}{2}}e^{-i\frac{\gamma}{2}} & -\sin\frac{\beta}{2}e^{-i\frac{\alpha}{2}}e^{i\frac{\gamma}{2}} \\ \sin\frac{\beta}{2}e^{i\frac{\alpha}{2}}e^{-i\frac{\gamma}{2}} & \cos\frac{\beta}{2}e^{i\frac{\alpha}{2}}e^{i\frac{\gamma}{2}} \end{pmatrix} \tag{1.111}$$

In the above equations of the $SU(2)$ group, one can replace the symbol J by S from spin.

1.9.3 WIGNER–ECKART THEOREM

In quantum mechanics applications, we often encounter the problem of two subsystems that have both common and distinct operators. An example is a two-particle system, where the kinetic energy operator is distinct for each particle but the potential energy operator actions simultaneously on both particles. Another example is the addition of angular momenta (spin-orbital or spin–spin).

Let us consider a system with two parts in a weak interaction. If we ignore the interaction, then the conservation of angular momentum is true for each of them. The total angular momentum L of the system is the sum of angular momenta L_1 and L_2. In terms of operators, there are J_1 and J_2. They operate separately onto each subsystem. The law of addition of angular momenta is obvious.

We consider a set of four operators $\hat{J}_1^2, \hat{J}_2^2, \hat{J}_{1z}$, and $\hat{J}_{2z}$ to completely describe the state of the system. We suppose to know the eigenfunctions of both the square angular momentum and the projection on the z axis (they are common). We can then write the equations of eigenvalues:

$$\hat{J}_{1z}\left|j_1m_1\right\rangle = m_1\left|j_1m_1\right\rangle$$

$$\hat{J}_{1+}\left|j_1m_1\right\rangle = \sqrt{(j_1 - m_1)(j_1 + m_1 + 1)}\left|j_1m_1 + 1\right\rangle$$

$$\hat{J}_{1-}\left|j_1m_1\right\rangle = \sqrt{(j_1 + m_1)(j_1 - m_1 + 1)}\left|j_1m_1 - 1\right\rangle$$

$$\hat{J}_{2z}\left|j_2m_2\right\rangle = m_2\left|j_2m_2\right\rangle$$

$$\hat{J}_{2+}\left|j_2m_2\right\rangle = \sqrt{(j_2 - m_2)(j_2 + m_2 + 1)}\left|j_2m_2 + 1\right\rangle$$

$$\hat{J}_{2-}\left|j_2m_2\right\rangle = \sqrt{(j_2 + m_2)(j_2 - m_2 + 1)}\left|j_2m_2 - 1\right\rangle$$

Thus, the wavefunction of the total system can be written as a product of the subsystem functions:

$$\left|j_1m_1; j_2m_2\right\rangle = \left|j_1m_1\right\rangle\left|j_2m_2\right\rangle$$

and they are orthonormated too:

$$\langle j_1'm_1'; j_2'm_2' | j_1''m_1''; j_2''m_2''\rangle = \langle j_1'm_1' | j_2'm_2'\rangle\langle j_1''m''_1 | j_2''m_2''\rangle = \delta_{j_1'j_2'}\delta_{m_1'm_2'}\delta_{j_1''j_2''}\delta_{m_1''m_2''}$$

There are $(2j_1 + 1)(2j_2 + 1)$ such functions of the entire system.

Let us now apply the addition law of the angular momentum: $\hat{J} = \hat{J}_1 + \hat{J}_2$. It is also valuable for the components of the angular momentum. The problem is to find a set of linear combinations of wavefunctions that are simultaneously eigenfunctions of both z projection and square angular momentum of the entire system.

We apply the z projection operator to the total function:

$$\begin{aligned}\hat{J}_z\left|j_1m_1; j_2m_2\right\rangle &= (\hat{J}_{1z} + \hat{J}_{2z})\left|j_1m_1\right\rangle\left|j_2m_2\right\rangle = \left(\hat{J}_{1z}\left|j_1m_1\right\rangle\right)\left|j_2m_2\right\rangle + \left(\hat{J}_{2z}\left|j_2m_2\right\rangle\right)\left|j_1m_1\right\rangle \\ &= m_1\left|j_1m_1\right\rangle\left|j_2m_2\right\rangle + m_2\left|j_1m_1\right\rangle\left|j_2m_2\right\rangle = (m_1 + m_2)\left|j_1m_1\right\rangle\left|j_2m_2\right\rangle\end{aligned}$$

Thus, $|j_1m_1\rangle|j_2m_2\rangle$ is the eigenfunction of the z projection operator and corresponds to the eigenvalue $m = m_1 + m_2$. We must find the linear combination of functions that is simultaneously the eigenfunction of both the

z projection operator and the square angular momentum, and this can be written as follows:

(1.112)

$$|jm\rangle = \sum_{m_1}\sum_{m_2}\langle j_1m_1;j_2m_2|j_1j_2;jm\rangle|j_1m_1\rangle|j_2m_2\rangle, \text{ where } m = m_1 + m_2$$

The coefficients of this linear combination $\langle j_1m_1;\ j_2m_2|j_1j_2;\ jm\rangle$ are named *Clebsch–Gordan coefficients.* You can find different notations for Clebsch–Gordan coefficients (Landau and Lifshitz 1991).

We note here two main properties (see proofs and details in Sakurai and Napolitano 2011):

1. The Clebsch–Gordan coefficients *vanish* unless $m = m_1 + m_2$.
2. The Clebsch–Gordan coefficients *vanish* unless $|j_1 - j_2| \leq j \leq j_1 + j_2$.

The main problem just begins: how do we calculate the Clebsch–Gordan coefficients? It is not the purpose of this book to address this. We mention that there are a lot of tables, programs, and reduction and summation formulae to work with. We note here that if you know one coefficient, you can obtain others by using the recursion relations derived by putting the J_+ or J_- operators between states.

In spectroscopy, the main subject is the interaction between electromagnetic field (represented by the measured radiation) and atoms or molecules. During emission or absorption processes, something is changed in the atom and this is not only the energy but also the angular momentum. This implies to evaluate the matrix elements of a tensor operator using the eigenfunctions of angular momentum. A classic example is a spherical harmonic $Y_l^m(\theta,\varphi)$, which often appears in integrals as describing a transition. We can define a vector $\vec{v}$ to give the space orientation instead of (θ, φ) angles. The result is a spherical tensor:

$$Y_l^m(v) \equiv T_m^{(l)}$$

The spherical tensor is also irreducible as spherical harmonic is.

Thus, during a radiative transition, there is an irreducible tensor that is changing between two stationary states characterized by angular quantum numbers j and m and other quantum numbers. Therefore, we must calculate the matrix elements:

$$\langle \alpha jm|T_{m2}^{(j2)}|\beta j_1m_1\rangle$$

where α and β represent other quantum numbers that are not important now.

The matrix elements of irreducible tensor with respect to angular momentum eigenstates are given by the relationship as follows:

$$\left\langle \alpha jm \middle| T_{m2}^{(j2)} \middle| \beta j_1 m_1 \right\rangle = \left\langle j_1 m_1 ; j_2 m_2 \middle| j_1 j_2 ; jm \right\rangle \frac{\left\langle \alpha j \middle\| T^{(j2)} \middle\| \beta j_1 \right\rangle}{\sqrt{2j_1 + 1}} \tag{1.113}$$

This is the *Wigner–Eckart theorem*. The first term on the right side is represented by the Clebsch–Gordan coefficients.

To be more explicit, we take as an example to calculate an integral as follows:

$$\int Y_{l1}^{m1} Y_{l2}^{m2} Y_{l3}^{m3} d\tau = \left\langle l_1 m_1 \middle| Y_{l2}^{m2} \middle| l_3 m_3 \right\rangle = C_{m1m2m3}^{l1l2l3} \frac{\left\langle l_3 \middle\| Y_{l2} \middle\| l_1 \right\rangle}{\sqrt{2l_3 + 1}}$$

where C_{m1m2m3}^{l1l2l3} is a symbol of the Clebsch–Gordan coefficients. Because the reduced matrix element $\left\langle l_3 \middle\| Y_{l2} \middle\| l_1 \right\rangle$ does not depend on m values, we can reduce the time to calculate it by replacing $m_1 = m_2 = m_3 = 0$ in the above equation. After the calculation of the reduced matrix element, we come back to the original task taking the coefficients from tables.

1.10 EXAMPLES OF SYMMETRY

1.10.1 CARBON TETRACHLORIDE

Carbon tetrachloride molecule—CCl_4. It is also known as tetrachloromethane. The last name suggests its structure is similar to methane (NIST 2013a,b). Figure 1.76 illustrates the spatial configuration of the molecule.

The molecule of carbon tetrachloride has a high symmetry as a result of similarity to the methane molecule. Therefore, it belongs to the $\boldsymbol{T}_d$ point group and has five classes of symmetry operations E, $8C_3$, $3C_2$, $6\,\Sigma_d$,

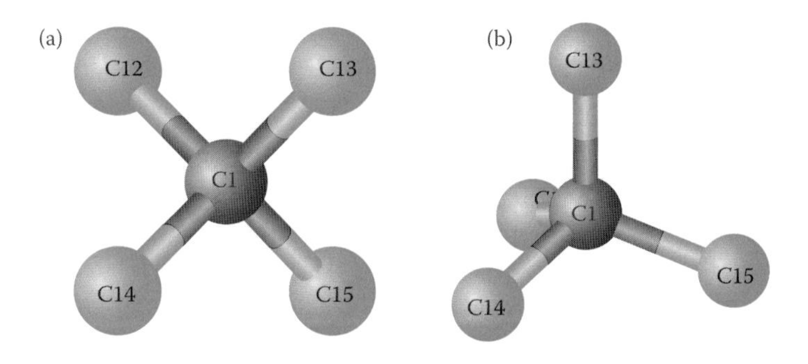

FIGURE 1.76 (a) Spatial configuration of the carbon tetrachloride molecule. (b) The molecule of carbon tetrachloride belongs to the $\boldsymbol{T}_d$ point group.

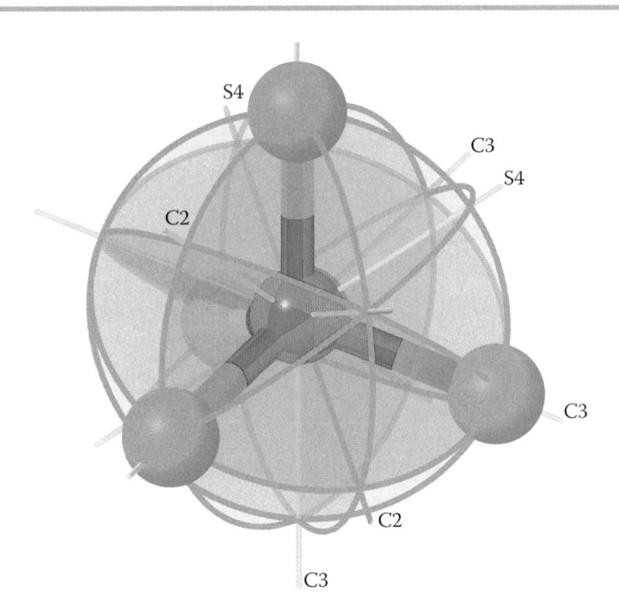

FIGURE 1.77 Symmetry elements of the carbon tetrachloride molecule. The principal rotation axis is one of the four equivalent threefold rotation axes passing through each chlorine atom and the carbon central atom.

and $6S_4$. For a better view of the molecule symmetry, we prefer to draw the molecule in the right position of the tetrahedron (Figure 1.76b) and then we illustrate the symmetry elements in Figure 1.77.

1.10.2 ORGANOMETALLIC MOLECULES

Let us now take a more complex example of an organometallic molecule. The most important journal of the field *Organometallics* defines organometallic as a compound "in which there is a bonding interaction (ionic or covalent, localized or delocalized) between one or more carbon atoms of an organic group or molecule and a main group, transition, lanthanide, or actinide metal atom (or atoms)" (Organometallics 2013). These compounds are strongly investigated because of their spectacular properties with many important applications "in particular in optoelectronics, conductivity and superconductivity, charge transfer and magnetism, nanoporous materials and biomimetic materials" (Braga et al. 2001). We have already presented the ferrocene molecule, one of the most well-known organometallic molecule, and its symmetry.

Nickel tetracarbonyl is a compound known for a long time because it is present as an intermediate product in the purification of nickel metal. Two molecular configurations have been proposed for the $Ni(CO)_4$ molecule: one planar and the other tetrahedral. The planar model is not consistent with the point symmetry because it would show fourfold symmetry that is not observed. Its $\boldsymbol{T_d}$ symmetry (Ladell et al. 1952) is similar to that of methane (Figure 1.78).

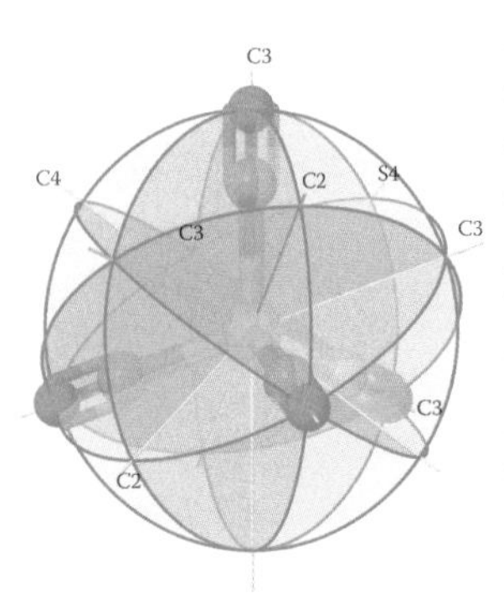

FIGURE 1.78 Symmetry elements of nickel tetracarbonyl molecule, which belongs to the $\boldsymbol{T_d}$ symmetry group.

1.10.3 HYDROXYAPATITE

Hydroxyapatite is an inorganic calcium phosphate-based mineral that compose bones and teeth. The hydroxyapatite crystals account for the rigidity of bones and teeth. The bones and teeth have an organic matrix filled with small crystals of hydroxyapatite. The formula of the molecule is $[Ca_5(PO_4)_3(OH)]$. Looking for crystalized form, it is better to take into consideration the stoichiometric formula $[Ca_{10}^{2+}(PO_4)_6^{2-}(OH)_2^{-}]$ of the unit cell. The phosphate ion $(PO_4)^{2-}$ has a tetrahedral symmetry similar to carbon tetrachloride. To simplify the spatial representation of the hydroxyapatite crystal, we will take only the phosphor ion as representative for its ionic complex. The hydroxyl ion $(OH)^-$ will also be reduced to oxygen representation. The arrangement of ions in the crystal of hydroxyapatite viewed from the top is shown in Figure 1.79a (Shi 2006). The top view (or plane projection) is useful to see the hexagonal symmetry of the unit cell. The positions of calcium ions can be classified into two types: Ca1 and Ca2. There are six positions of Ca2 located at the corners of the hexagon around the OH^- ion. However, the six Ca2 ions are not located in the same plane as you can see in this projection but they are arranged in two triangular planes (Figure 1.79b). The other four positions Ca1 (marked by black color) named columnar calcium are located outside of the phosphate ions.

Both Ca2 and Ca1 ions form two coaxial hexagonal channels, specific to the hydroxyapatite crystal (Figure 1.80). They assure the rigidity of the crystal together with phosphate ions. The crystallization system is hexagonal and corresponds to the space group $P6_3/m$. The principal symmetry element is the screw axis 6_3, which is imposed by the six Ca2 ions arranged in two stacked opposite triangles. It means the 60° rotation is followed by a translation from 0.25 to 0.75 from the *c* axis of the unit cell. This axis is perpendicular to the horizontal mirror plane.

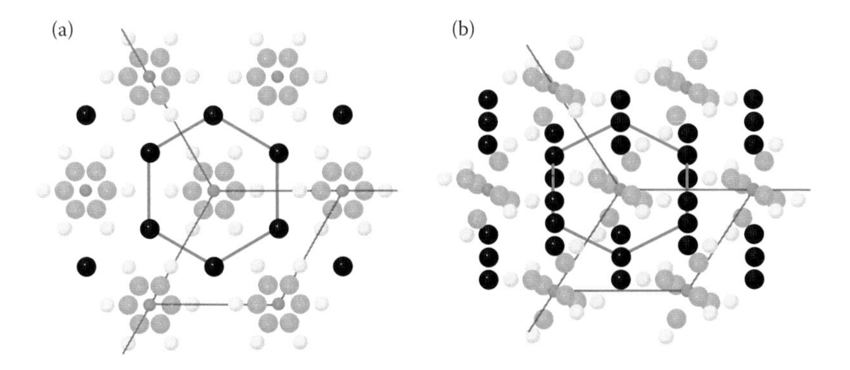

FIGURE 1.79 (a) Planar arrangement of ions in a hydroxyapatite crystal shows the hexagonal symmetry. *c*-axis is perpendicular to the plane determined by the three *a* axes illustrated. (b) The spatial configuration (rotated) shows that the Ca1 ions forms vertical columns and the Ca2 ions are arranged in two triangular planes.

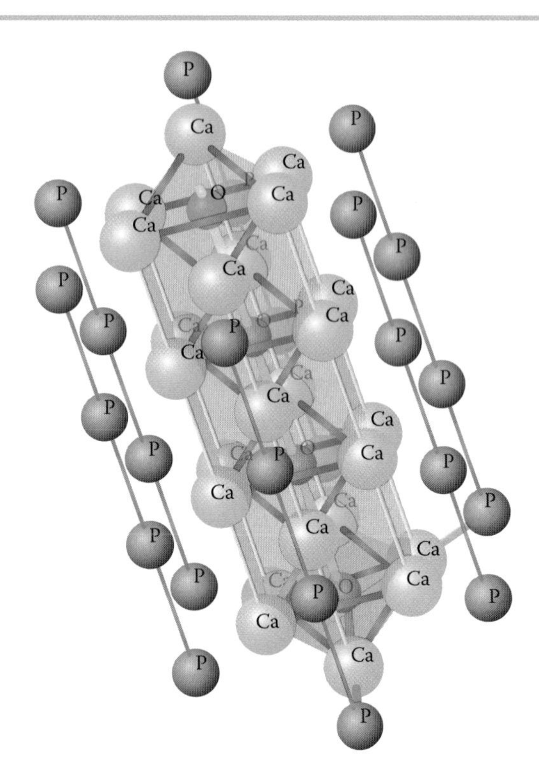

FIGURE 1.80 Both Ca1 and Ca2 ions form two coaxial hexagonal channels.

1.10.4 Ti^{3+}:SAPPHIRE

Ti^{3+}:sapphire is the best active material for the new class of ultrashort pulsed lasers. The sapphire (host crystal) is a crystal made from aluminum oxide (Al_2O_3). The pure crystal is colorless and can have different colors depending on the transition metal ions present as impurities. Sapphire has a hexagonal unit cell, which means the basic unit of the cell is a hexagonal prism (Figure 1.81). The voids represented by small black balls are Al^{3+} ion vacancies with respect to the electrical neutrality of the crystal. We note here that the presence of voids is a source of stress for electrical field in the crystal and the geometry of the crystal structure (angles and atomic distances not equal) is not perfect, as we will present it.

The symmetry elements of the unit cell are:

- A threefold rotoinversion axis (Figure 1.82a)
- Three rotation axes of the second order perpendicular to main axis (Figure 1.82b)
- Three mirror planes perpendicular to the axis of the second order and intersecting the sixfold axis
- A center of inversion

The symmetry of the unit cell belongs to the $\bar{3}2/m$ group in Hermann–Maugain notation of the point group. The corresponding space group is $R\bar{3}c$. It means the shape is rhombohedral (trigonal). The crystal main axis is a threefold rotoinversion and it also has a glide parallel to the *c* axis.

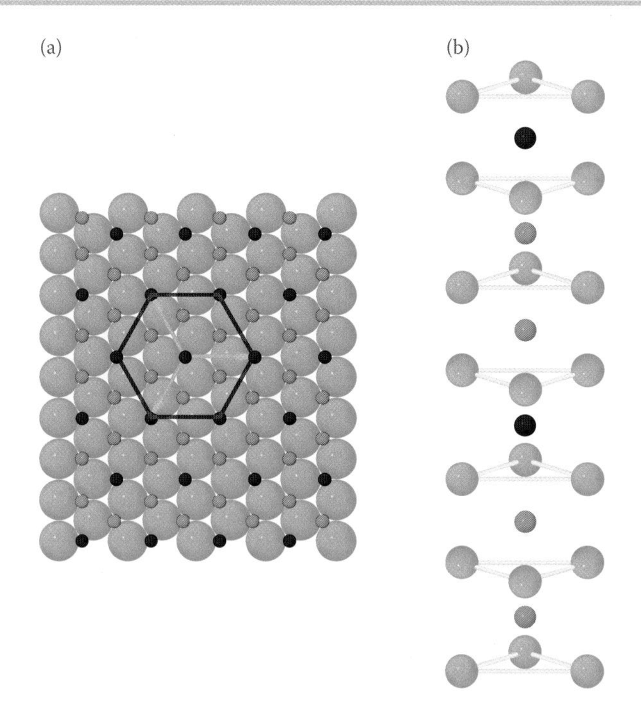

FIGURE 1.81 (a) Two-dimensional illustration of Al^{3+} ions arrangement (small gray spheres) between two layers of O^{2-} (big spheres) in sapphire crystal. The upper oxygen layer is not represented. The small black spheres are voids of metal ions. The arrows represent the three crystallographic axes related to hexagonal cell. (b) Vertical arrangement of the ions.

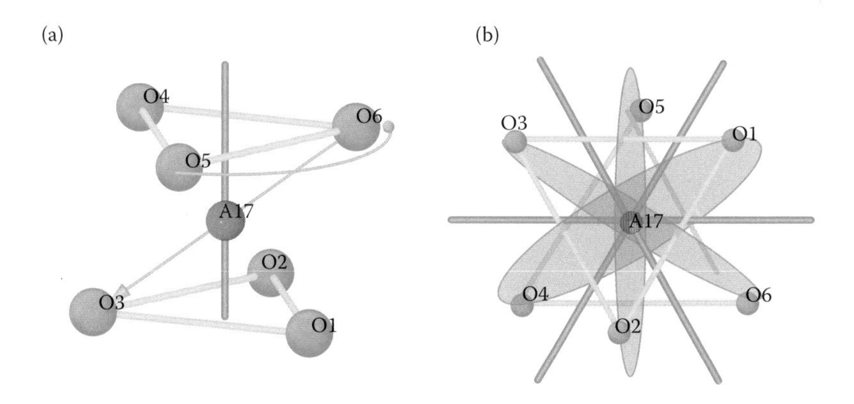

FIGURE 1.82 Illustration of symmetry elements: (a) rotoinversion axis; (b) three twofold axes and three mirror planes.

1.10.5 SILICON

The chemical element silicon is second in abundance in the Earth's crust (Encyclopedia Britannica 2013). The most common natural form is SiO_2 present in soil, atmosphere, and even tissues of animals. The silicon crystal for the electronic industry is produced starting from sands rich in

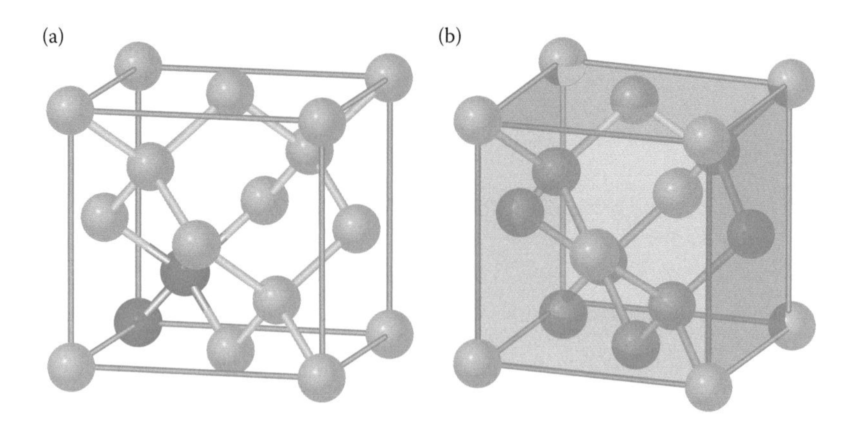

FIGURE 1.83 Unit cell of a silicon crystal.

SiO_2 at the end of a long industrial processing. The computer used to write this book contains a very small quantity of "smart" silicon processed crystal.

The silicon crystal has a diamond-like structure and belongs to the space group $Fd\bar{3}m$. It means the crystal cell symmetry can be described by a cube with a lattice point at each of its corners (Figure 1.83a). The symmetry is called face-centered cubic because there is a lattice point on each of the six faces (Figure 1.83b). d means diamond glide reflection and translation $(a + b + c)/4$. There are also four threefold rotoinversion axes on the direction of each principal diagonal (Figure 1.84a) and six mirror planes (Figure 1.84b). The unit cell illustrated here shows well the symmetry of the crystal but it is not the smallest cell of the crystal (primitive cell) that can reproduce the entire crystal by simple repetition. The primitive cell is composed of only two atoms darker marked in Figure 1.83a.

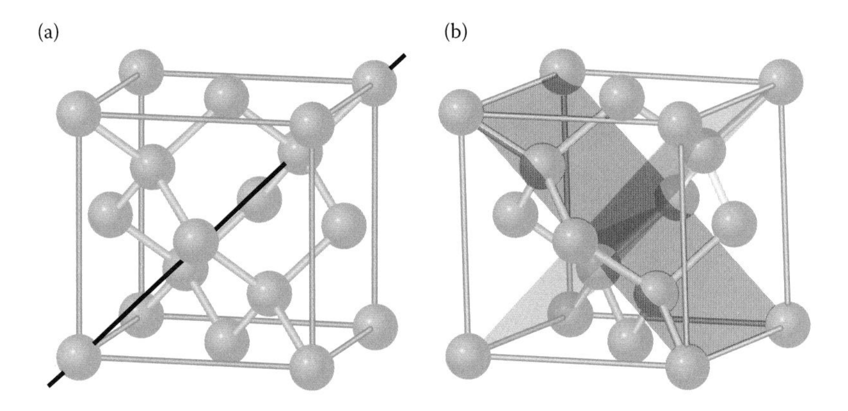

FIGURE 1.84 Elements of symmetry of a silicon crystal: (a) four threefold rotoinversion axes on the direction of each principal diagonal (only one represented here); (b) six mirror planes (only two represented here).

STUDY QUESTIONS

1.1 Complete the multiplication table for the point group $\boldsymbol{S}_4$ having the set of elements: $\boldsymbol{S}_4 = \{E, S_4^1, S_4^2 = C_2, S_4^3\}$:

Right \ Left	E	S_4^1	C_2	S_4^3
E	E	S_4^1	C_2	S_4^3
S_4^1	S_4^1			
C_2	C_2			
S_4^3	S_4^3			

1.2 Find the symmetry elements of the square planar complex $[AuCl_4]^-$.

1.3 For the above molecular complex show that C_4 and σ_v do not commute, but C_4 and σ_h commute.

1.4 Construct a multiplication table for the symmetry operations of water molecule.

1.5 Figure out all axes of rotation that can be found for benzene.

1.6 Identify all symmetry elements in PCl_5 molecule, which is used as a regular chlorinating agent. The shape of phosphorus pentachloride is trigonal bipyramidal.

1.7 Identify all symmetry operations of PCl_5 molecule. What point group does it belong to?

1.8 Identify all symmetry operations of difluoroethylene molecule ($C_2H_2F_2$). The molecule is planar. Both hydrogen atoms bond symmetrically to C1 atom at an angle of 118°. Both fluorine atoms bond symmetrically to C2 atom at an angle of 109°. What point group does it belong to?

1.9 Identify all symmetry operations of ethane molecule (C_2H_6). What point group does it belong to?

1.10 Identify all symmetry operations of HCl molecule. What point group does it belong to?

1.11 Fill in the multiplication table of $\boldsymbol{D}_{3h}$ group shown in Table 1.9 of the book.

1.12 The point group $\boldsymbol{C}_{3v}$ has the following set of symmetry operations $\{E, 2C_3, 3\Sigma_v\}$. Find its irreducible representations and characters using the rules derived from the great theorem of orthogonality.

1.13 How many ions are there in a single unit cell of CsCl?

REFERENCES

Braga, D., L. Maini, M. Polito, L. Scaccianoce, G. Cojazzi, and F. Grepioni. 2001. Design of organometallic molecular and ionic materials. *Coordination Chemistry Reviews* 216–217: 225–248.

Chaplin, M. 2013. Water Structure and Science. http://www.lsbu.ac.uk/water/escs.html and http://www.lsbu.ac.uk/water/clusters.html#java.

Chuvilin, A., U. Kaiser, E. Bichoutskaia, N. A. Besley, and A. N. Khlobystov. 2010. Direct transformation of graphene to fullerene. *Nature Chemistry* 2: 450–453.

Cotton, A. F. 1990. *Chemical Applications of Group Theory*, 3rd edition. New York: John Wiley & Sons. Inc.

Darvas, G. 2012. What Is Symmetry? http://symmetry.hu/definition.html.

Encyclopedia Britannica. 2013. Silicon. http://www.britannica.com/EBchecked/topic/544301/silicon-Si.

Evans, R. C. 1964. *An Introduction to Crystal Chemistry*, 2nd edition. Cambridge: Cambridge University Press.

Fowler, P. W., D. E. Manolopoulos, D. B. Redmond, and R. P. Ryan. 1993. Possible symmetries of fullerene structures. *Chemical Physics Letters* 202: 371–378.

IUPAC. 1971. Commission on Nomenclature of Inorganic Chemistry, ed. R. M. Adams, *Nomenclature of Inorganic Boron Compounds*, London: Butterworths, p. 694.

Jensen, W. B. 2007. The origin of the s, p, d, f orbital labels. *Journal of Chemical Education* 84: 757–758.

Manthey, D. 2013. Orbital Viewer. www.orbitals.com/orb.

Ladell, J., B. Post, and I. Fankuchen. 1952. The crystal structure of nickel carbonyl, $Ni(CO)_4$. *Acta Crystallographica* 5: 795–800.

Landau, L. D. and E. M. Lifshitz, reprinted 1991. *Quantum Mechanics*, 3rd edition. Oxford: Pergamon Press.

NIST. 2013a. Carbon Tetrachloride. http://webbook.nist.gov/cgi/cbook.cgi?ID=C56235 &Mask=4.

NIST. 2013b. Nickel Tetracarbonyl. http://webbook.nist.gov/cgi/cbook.cgi?Name=nickel+tetracarbonyl&Units=SI.

Nobel Prize in Chemistry. 1996. *Nobelprize.org*. Nobel Media AB 2013. Web. 12 July 2013. http://www.nobelprize.org/nobel_prizes/chemistry/laureates/1996/.

Nobel Prize in Physics. 1915. *Nobelprize.org*. Nobel Media AB 2013. Web. 29 July 2013 http://www.nobelprize.org/nobel_prizes/physics/laureates/1915/.

Nordman, C. E. and D. L. Schmitkons. 1965. Phase transition and crystal structures of adamantane. *Acta Crystallographica* 18: 764–767.

O'Keeffe, M. and B. G. Hyde. 1996. *Crystal Structures I: Patterns and Symmetry*. Washington D.C.: Mineralogical Society of America (available at http://www.public.asu.edu/~rosebudx/okeeffe.htm).

Organometallics. 2013. Author Guidelines. http://pubs.acs.org/paragonplus/submission/orgnd7/orgnd7_authguide.pdf.

Rupp, B. 2013. Space Group Decoder. http://www.ruppweb.org/xray/comp/space_instr.htm.

Sakurai, J. J. and J. Napolitano. 2011. *Modern Quantum Mechanics*, 2nd edition. San Francisco: Addison-Wesley.

Shi, D. (ed.). 2006. *Introduction to Biomaterials*. Singapore: World Scientific Publication, pp. 13–28.

Sivaev, I. B., V. I. Bregadze, and N. T. Kuznetsov. 2002. Derivatives of the *closo*-dodecaborate anion and their application in medicine. *Russian Chemical Bulletin* 51: 1362–1374.

Thaler, J. 2001. Classical and Quantum Mechanics on SU(2). http://v1.jthaler.net/physics/notes/matrix/SU(2).pdf.

Crystal Field Theory

2

2.1 STATES AND ENERGIES OF FREE ATOMS AND IONS

2.1.1 CRYSTAL FIELD THEORY VERSUS LIGAND FIELD THEORY

Group theory helps us to understand optical spectra of crystals doped with metal ions. The description of optical spectra of metal ions introduced into a lattice is more qualitative but is quick to determine and good enough. The practical idea is that metal ion energy levels are influenced by the surrounding host ions. The central metal cation and the nearest surrounding anions form a *molecular complex*. There are more ways to treat this molecular complex theoretically:

1. *Crystal field theory.* This theory was proposed by the physicist Hans Bethe in 1929 (Bethe 1929). The molecular complex is like an "ionic" molecule (Ballhausen 1962). Ions of the molecular complex are seen as point charges and the interaction between the central ion and surrounding ions is a pure electrostatic interaction between point charges. Thus, the valence orbitals of metal ions are influenced and just deformed in the direction of neighboring ions. The electrons of metal ions *are not allowed to mix* with electrons of the surrounding ions. As a consequence, their orbitals will have a lower symmetry inside the complex than the spherical symmetry as free ions.
2. *Ligand field theory* is a modified version of the crystal field theory proposed by J. H. Van Vleck in 1935 to allow for some covalency in the interactions (Van Vleck and Sherman 1935). It assumes that the interaction is not purely electrostatic between two point charges. Thus, the orbitals of the metal and neighboring ions *overlap* partly. It supposes that the interaction is electrostatic as well as covalent. If the covalence is small, we obtain the same results as by using the crystal theory.
3. *Molecular orbital theory.* When the covalent interaction is *strong,* molecular orbital theory is preferred but the symmetry considerations remain the same as in crystal field theory.

Before we start to study the influence of neighboring ions, we must understand a little about the free ion state. For this purpose, we need to review atomic physics.

2.1.2 VECTOR MODEL OF ATOM

We are interested in how to find the energy of n electrons moving in a central field. The treatment was well developed in the classic book by Condon and Shortley, *Theory of Atomic Spectra* (Condon and Shortley 1935).

We start filling with electrons the basic orbitals of hydrogen-like orbitals *s*, *p*, *d*, and *f*. In filling orbitals with electrons, we must take into consideration the Pauli exclusion principle that states that no two electrons in an atom may be at the same time in the same state. The functions describing orbitals *s*, *p*, *d*, and *f* are solutions of the Schrödinger equation $\hat{H}\Psi = E\Psi$, where the Hamiltonian contains only two terms, the kinetic and potential energy of the orbital motion of the electron in the central field of the nucleus:

$$\hat{H}_0 = -\frac{\hbar^2}{2m}\nabla^2 - \frac{Ze^2}{r} \tag{2.1}$$

The solutions of the Schrödinger equation depend on three quantum numbers: principal, orbital, and magnetic (n, ℓ, and m_ℓ). The eigenfunction Ψ_{nlm} is the product between the radial function and the angular function (Equation 1.92). The radial function determines the eigenvalues of energy that depend only on the principal quantum number $E_n \propto -1/n^2$. The angular function is a measure of the probability of finding the electron and determines the shape of orbitals that we discussed in Section 1.7.4. A value of energy corresponds to one or more eigenfunctions and we say the energy is degenerated $(2\ell + 1)$ times.

In the case of multielectron atoms (j electrons), the Hamiltonian operator changes from that given by Equation 2.1 because of two new phenomena: electrostatic repulsion between electrons and the screening effect of nuclear charge produced by the inner shells:

$$\hat{H} = -\frac{\hbar^2}{2m}\sum_j \nabla_j^2 - \sum_j \frac{Z_j e^2}{r_j} + \sum_{j,k} \frac{e^2}{r_{jk}} = \hat{H}_0 + \hat{H}_1 \tag{2.2}$$

where ∇_j^2 is the Laplacian operator of the j electron, Z_j is the effective nuclear charge "seen" by the j electron, and r_{jk} is the distance between the j and k electrons. $\hat{H}_0$ is the zero-order Hamiltonian operator due to the central field and determines the gross structure of energy levels that now depends on n and ℓ quantum numbers. $\hat{H}_1$ is the Hamiltonian of electrostatic correction (first perturbation) and will split the energy levels into different *terms*. The energy does not depend on m_ℓ and every value is $(2\ell + 1)$ times degenerated; therefore, there are $(2\ell + 1)$ states for every energy value. (Attention: Always do the difference between energy levels and states!)

The outer shell electrons of the metal ion belong to either d orbitals (transition metals) or f orbitals (lanthanides and actinides). According to the vector model, the orbital quantum number associated with these orbitals is $\ell = 2$ and 3, respectively. In this model, we reduce the electron to the discrete charged particle moving around the nucleus and this orbital motion is characterized by the angular momentum. The angular momentum vector is perpendicular to the orbital plane. In quantum mechanics, the angular momentum has only a few discrete values reported to a reference direction (which can be the direction of a magnetic field). In the case of *d* electrons, there are only five values of projection of the angular momentum on the reference direction shown in Figure 2.1. These values are described by the magnetic quantum number m_ℓ that can take values: 2, 1, 0, −1, and −2. In terms of wavefunctions, it means that there are five orbitals of *d* type described by five different wavefunctions (see Figure 1.60). This state is denoted by the spectral symbol ^2D where the superscript is the spin multiplicity $(2 \cdot s + 1 = 2 \cdot 1/2 + 1 = 2)$. (In a particular case of the single-electron atom, the symbol is ^2d.) In this state, the electron can be found in any *d* orbital, but it has only one value of energy. The energy level is fivefold degenerate.

In terms of group theory, the free atom has the highest symmetry, the *spherical symmetry*. In this point group, there are only rotations and one inversion as the symmetry operations. Rotations leave unchanged the angular momentum in the isotropic space (no privileged direction). The inversion also leaves unchanged the wavefunction describing *d* orbitals because we previously established that all orbitals "*d*" are symmetric to the inversion. All filled shells are closed and they have spherical symmetry. So, the symmetry of the atom is finally given only by the partially filled shell.

For *multi-electron atom,* we must apply the Russell–Saunders coupling scheme, also called the *L–S* coupling. If every electron has an orbital angular momentum and a spin angular momentum, then the total orbital angular momentum $\vec{L}$ of the atom is the sum of all orbital momenta $\vec{L} = \sum \vec{l}_i$ and the total spin angular momentum $\vec{S}$ of the atom is the sum of all spin momenta $\vec{S} = \sum \vec{s}_i$. In terms of quantum numbers, the relations are similar without the sign of the vector. Finally, the total angular momentum is the sum of both the orbital and spin momentum $\vec{J} = \vec{L} + \vec{S}$. Here, we note the use of capital letters as symbols for quantum numbers of the entire electronic configuration. In the case of multielectron atoms such as transition metals, the second-order Hamiltonian $\hat{H}_2$ is due to the spin–orbit interaction. It will further split the terms leading to the fine structure of energy levels.

We now take as an example an *atom with two d electrons.* In this case, the electrons can be in the same shell (equivalent electrons) or in different shells (nonequivalent electrons). If the electrons are nonequivalent (excited atom), each of them has the orbital momentum $\ell_1 = 2$ and $\ell_2 = 2$, respectively. The total orbital momentum L may have values from $\ell_1 + \ell_2$

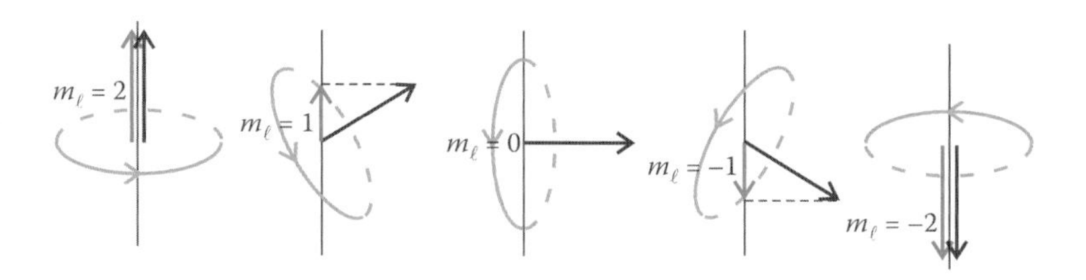

FIGURE 2.1 Illustration of the five value projections of angular momentum on a reference direction for one *d* electron.

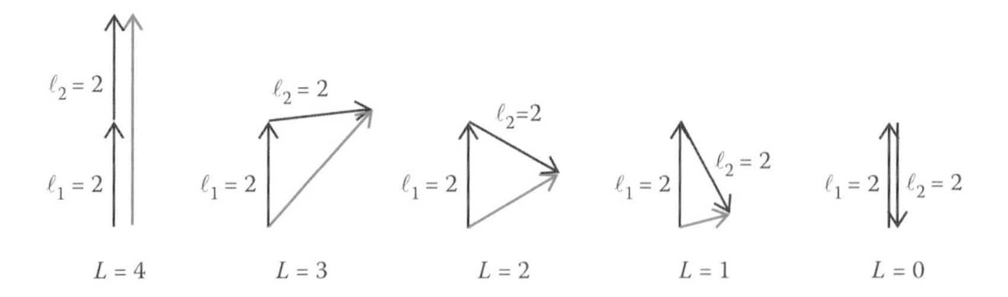

FIGURE 2.2 The addition of orbital momenta of two *d* nonequivalent electrons ($nd^1 - n'd^1$) may yield five different results.

to $\ell_1-\ell_2$; therefore, for two *d* non-equivalent electrons *L* may have values 4, 3, 2, 1, and 0 as shown in the Figure 2.2. Each value of the total orbital momentum gives a new state labeled by *G*, *F*, *D*, *P*, and *S* as with the *f*, *d*, *p*, and *s* for a single electron (Table 2.1). The values of *L*, *S*, and *J* specify *term* symbol according to the form

$$^{2S+1}L_J \qquad (2.3)$$

The superscript symbol called total spin multiplicity (given by $2S + 1$) is calculated from the number of unpaired electrons and the correlation between the value of *S* and the name of spin multiplicity is also presented in Table 2.1. The subscript is the value of the total angular momentum (*J*).

Each electron has a spin momentum that equals 1/2. Thus, the total spin momentum *S* may have two values: 1 or 0. Therefore, the total angular momentum of the atom may be any combination of the total orbital momentum and total spin momentum. It means *J* may take values from $L + S$ to $L - S$. All new states obtained as a result of all combinations are presented in Table 2.2.

If both electrons are in the same shell (*equivalent electrons*), we must respect the Pauli exclusion principle that states that no two electrons in an atom may be at the same time in the same state. In other words, no two electrons can have the same set of quantum numbers. Therefore, when $L = 4$, 2, and 0, the magnetic quantum number is the same as shown in

TABLE 2.1 Relation between *S* and *L* Values and Name of Spin Multiplicity and *L* Value and Term Symbol, Respectively

S Value	$2S + 1$	Name of Spin Multiplicity	*L* Value	Term Symbol
0	1	Singlet	0	*S*
1/2	2	Doublet	1	*P*
1	3	Triplet	2	*D*
3/2	4	Quartet	3	*F*
2	5	Quintet	4	*G*
5/2	6	Sextet	5	*H*

TABLE 2.2 All Possible States of the Atom with Two Nonequivalent *d* Electrons ($nd^1 - n'd^1$)

$L = 0$	$L = 1$	$L = 2$	$L = 3$	$L = 4$	
$^1S_0$	$^1P_1$	$^1D_2$	$^1F_3$	$^1G_4$	$S = 0$
$^3S_1$	$^3P_2$	$^3D_3$	$^3F_4$	$^3G_5$	
	$^3P_1$	$^3D_2$	$^3F_3$	$^3G_4$	$S = 1$
	$^3P_0$	$^3D_1$	$^3F_2$	$^3G_3$	

Figure 2.3 and, as a consequence of the Pauli principle, the two electrons must have only opposite spins (+1/2 and −1/2) and the total spin momentum may be only 0.

The first moment is to find all possible distinct microstates of the nd^2 electronic configuration, taking into consideration the Pauli principle, as shown in Table 2.3. We arrange electrons in each cell, which corresponds to all permitted values of the magnetic quantic number m_ℓ. Arrows indicate electrons in different spin states. The total number of possible combinations given by the degeneracy formula is 45.

Then we may summarize the results of Table 2.3 by writing all possible terms that we have obtained for the nd^2 configuration as shown in Table 2.4.

Finally, we can write the available states of the free atom with d^2 configuration as shown in Table 2.5.

As a conclusion, the Russell–Saunders coupling applied to all d^n configurations of the free atom gives the terms as presented in Table 2.6.

Notes on Table 2.6:

1. The number of energy levels equals the number of terms.
2. The number of microstates (all possible combinations of M_L and M_S numbers) depends on the number of electrons and the maximum number of positions for electrons. For example, the d^2 configuration has $(10!)/[2! \cdot (10 - 2)!] = (9 \cdot 10)/2 = 45$ quantum microstates. The d^3 configuration has $(10!)/[3! \cdot (10 - 3)!] = (8 \cdot 9 \cdot 10)/6 = 120$ quantum microstates.
3. The ground term is the most stable (has the lowest energy) and is given by Hund's rules:
   a. Maximum S value.

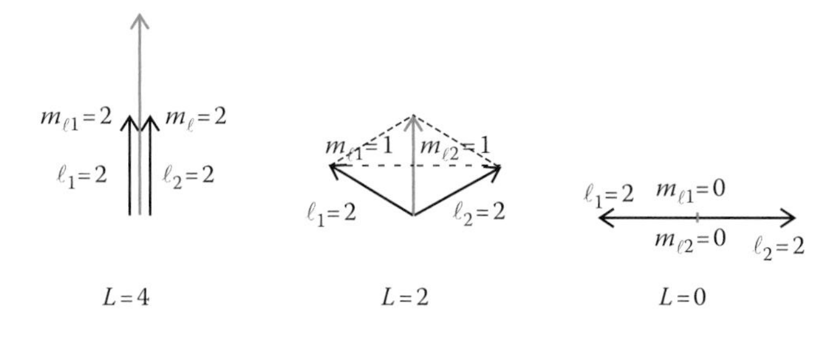

FIGURE 2.3 Illustration of cases in which the quantum numbers n, ℓ, and m_ℓ are the same for the nd^2 configuration.

TABLE 2.3 All Possible Distinct Microstates of the nd^2 Electronic Configuration

	$m_l = 2$	1	0	−1	−2	M_L	M_S	Term
1	↑↓					4	0	^{1}G
2	↑	↓				3	0	^{1}G
3	↑	↑				3	1	^{3}F
4	↑		↓			2	0	^{1}G
5	↑		↑			2	1	^{3}F
6	↑			↓		1	0	^{1}G
7	↑			↑		1	1	^{3}F
8	↑				↓	0	0	^{1}G
9	↑				↑	0	1	^{3}F
10		↑↓				2	0	^{3}F
11		↑	↓			1	0	^{3}F
12		↑	↑			1	1	^{3}P
13		↑		↓		0	0	^{3}F
14		↑		↑		0	1	^{3}P
15		↑			↓	−1	0	^{1}G
16		↑			↑	−1	1	^{3}F
17			↑↓			0	0	^{3}P
18			↑	↓		−1	0	^{3}F
19			↑	↑		−1	1	^{3}P
20			↑		↓	−2	0	^{1}G
21			↑		↑	−2	1	^{3}F
22				↑↓		−2	0	^{3}F
23				↑	↓	−3	0	^{1}G
24				↑	↑	−3	1	^{3}F
25					↑↓	−4	0	^{1}G
	↓↑							
26	↓	↑				3	0	^{3}F
27	↓	↓				3	−1	^{3}F
28	↓		↑			2	0	^{1}D
29	↓		↓			2	−1	^{3}F
30	↓			↑		1	0	^{1}D
31	↓			↓		1	−1	^{3}F
32	↓				↑	0	0	^{1}D
33	↓				↓	0	−1	^{3}F
		↓↑						
34		↓	↑			1	0	^{3}P
35		↓	↓			1	−1	^{3}P
36		↓		↑		0	0	^{1}S
37		↓		↓		0	−1	^{3}P
38		↓			↑	−1	0	^{1}D
39		↓			↓	−1	−1	^{3}F
			↓↑					
40			↓	↑		−1	0	^{3}P
41			↓	↓		−1	−1	^{3}P
42			↓		↑	−2	0	^{1}D
43			↓		↓	−2	−1	^{3}F
				↓↑				
44				↓	↑	−3	0	^{3}F
45				↓	↓	−3	−1	^{3}F
					↓↑			

TABLE 2.4 All Possible Terms of nd^2 Configuration Summarized

$L = 4$	$M_L = 4$	3	2	1	0	−1	−2	−3	−4	^1G
$S = 0$	$M_S = 0$									
$L = 3$	$M_L = 3$	2	1	0	−1	−2	−3			^3F
$S = 1$	$M_S = 1$	0	−1							
$L = 2$	$M_L = 2$	1	0	−1	−2					^1D
$S = 0$	$M_S = 0$									
$L = 1$	$M_L = 1$	0	−1							^3P
$S = 1$	$M_S = 1$	0	−1							
$L = 0$	$M_L = 0$									^1S
$S = 0$	$M_S = 0$									

TABLE 2.5 Possible States of the Free Atom with nd^2 Configuration

$L = 0$	$L = 1$	$L = 2$	$L = 3$	$L = 4$	
$^1S_0$		$^1D_2$		$^1G_4$	$S = 0$
	$^3P_2$		$^3F_4$		$S = 1$
	$^3P_1$		$^3F_3$		
	$^3P_0$		$^3F_2$		

TABLE 2.6 Term Symbols of All $3d^n$ Configurations of Free Atom (Ion) States

Configuration	Number of Micro States	Number of Energy Levels	Ground Term	Excited Terms
d^1, d^9	10	1	^2D	—
d^2, d^8	45	5	^3F	$^3P, ^1G, ^1D, ^1S$
d^3, d^7	120	8	^4F	$^4P, ^2H, ^2G, ^2F, ^2D, ^2D, ^2P$
d^4, d^6	210	16	^5D	$^3H, ^3G, ^3F, ^3F, ^3D, ^3P, ^3P, ^1I, ^1G, ^1G, ^1F, ^1D, ^1D, ^1S, ^1S$
d^5	252	16	^6S	$^4G, ^4F, ^4D, ^4P, ^2I, ^2H, ^2G, ^2G, ^2F, ^2F, ^2D, ^2D, ^2D, ^2P, ^2S$

b. Maximum L value.
c. Smallest J value if the subshell is less than half filled or the largest J value if the subshell is more than half filled.

For example, for the d^3 configuration, there are three electrons and five orbitals. The electrons may be arranged in spins to have the maximum total spin number. Thus, the maximum value of S equals 3/2 and the spin multiplicity equals $2 \cdot 3/2 + 1 = 4$. As a consequence, the three electrons may be placed in three different orbitals to respect the Pauli principle. So, the maximum L value will be obtained for the arrangement as shown in Figure 2.4 (Bransden and Joachain 1983).

Orbital	$Y_{22} = d_{x^2-y^2}$	$Y_{21} = d_{xz}$	$Y_{20} = d_{z^2}$	$Y_{2-1} = d_{yz}$	$Y_{2-2} = d_{xy}$
	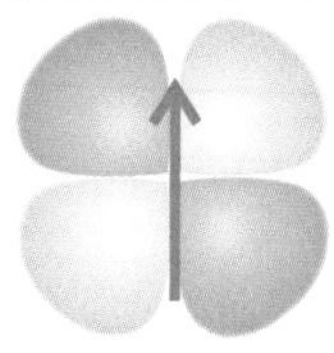	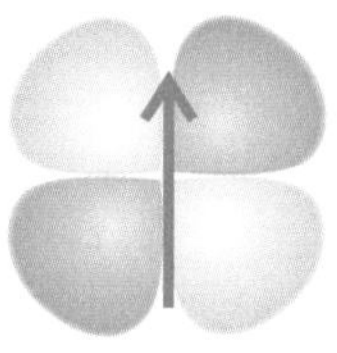		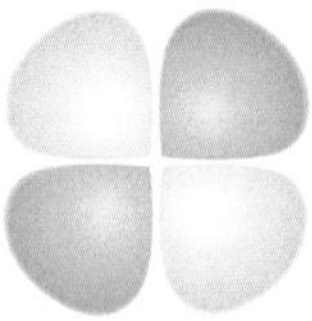	
$L = 3$	2 +	1 +	0		
$J =$	$3 + 3/2 = 9/2; 7/2; 5/2;$ **$3/2$**; (shell is less than half filled)				
Term			${}^4F_{\frac{3}{2}}$		

FIGURE 2.4 The ground term arrangement of three equivalent electrons (d^3 configuration). The electrons are placed in three different orbitals to maximize the orbital number. The arrow indicates the electron spin.

2.1.3 ENERGIES OF THE *LS* TERMS

We will concentrate on evaluating the positions on the energy scale of the terms previously determined for the d^n configuration. The Hamiltonian of the system is

$$\hat{H} = \hat{H}_0 + \hat{H}_1 + \hat{H}_2 + \hat{H}_3 \tag{2.4}$$

where $\hat{H}_0$ is the zero-order Hamiltonian operator due to the central field, $\hat{H}_1$ is the Hamiltonian of electrostatic repulsion, $\hat{H}_2$ is the Hamiltonian operator due to the spin–orbit momentum coupling, and $\hat{H}_3$ is the Hamiltonian due to the spin–spin momentum coupling. The order of terms is determined by their decreasing contribution to the main term and is valuable for most of the transition metals.

We assume that the electrostatic repulsion is stronger than the spin–orbit interaction; thus, the Hamiltonian has only two terms. Using the hydrogen-like wave functions, it is possible to evaluate the matrix elements of Coulomb repulsion. This was performed by Slater as integrals of the radial part of wave functions for two electrons:

$$F^k(n_il_i, n_jl_j) = e^2 \int_0^\infty \int_0^\infty \frac{r_<^k}{r_>^{k+1}} R^*_{n_il_i}(r_1) R^*_{n_jl_j}(r_2) R_{n_il_i}(r_1) R_{n_jl_j}(r_2) r_1^2 dr_1 r_2^2 dr_2 \tag{2.5}$$

where k equals 2 or 4 (Slater 1960). (The reader can find details of the Slater parameters evaluation and other results presented here in the excellent reference book of Ballhausen *Introduction to Ligand Field Theory*.) It was shown later that it is more convenient to express the energies of free ions in terms of three parameters A, B, and C called Racah's parameters, which are linear combinations of the Slater parameters, namely $B = F_2 - 5F_4$ and $C = 35F_4$. In Table 2.7, we present the expressions of spectroscopic term

TABLE 2.7 Energies of d^2 and d^3 Configurations Expressed in Terms of Racah's Parameters

	${}^3F = A - 8B$
	${}^3P = A + 7B$
$n = 2$	${}^1G = A + 4B + 2C$
	${}^1D = A - 3B + 2C$
	${}^1S = A + 14B + 7C$
$n = 3$	${}^4F = 3A - 15B$
	${}^4P = 3A$
	${}^2H = {}^2P = 3A - 6B + 3C$
	${}^2G = 3A - 11B + 3C$
	${}^2F = 3A + 9B + 3C$
	${}^2D = 3A + 5B + 5C \pm \sqrt{193B^2 + 8BC + 4C^2}$

energies of free ions with electronic configurations d^2 and d^3, respectively. Any textbook of atomic physics may contain the complete table.

This table contains all that you need to go further; therefore, it contains all states of free transition metal ions calculated with respect to the electrostatic repulsion as the first correction. We are not interested in their absolute value, but only in the relative distance between them, which will be reflected in the energy of transition we can measure. The first term of each expression is the same for a given type of ion. Thus, it will vanish when we will make the difference. Only the parameters B and C remain important for state energy from the point of view of spectra and they were determined by an experiment, by fitting the theoretical dependence on the B/C ratio to the measured spectrum as shown in Figure 2.5.

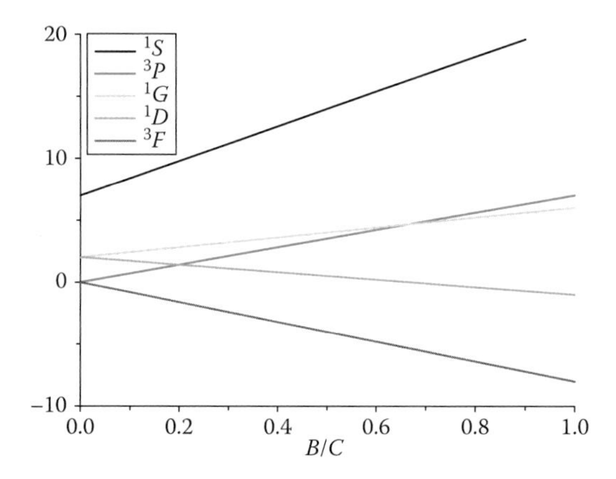

FIGURE 2.5 Theoretical dependence of free ion states with d^2 configuration on Racah's parameters ratio.

2.2 OPTICAL SPECTRA OF IONIC CRYSTALS

Before inserting the metal ion in the crystal lattice, we should discuss about the optical properties of pure crystals to make a distinction between the crystal contribution and that of the ion on the measured spectrum.

We can easily measure the absorption spectrum of a crystal at room temperature or lower temperature. The method of measurement is presented in Chapter 5 of this book. Most of the ionic crystals are transparent in the visible (Vis) range if they are pure and perfect. They absorb light in the ultraviolet (UV) range and in the infrared (IR) range too. The mechanism of light absorption in IR is different from the mechanism of absorption in the visible and UV ranges. We are now interested only in the light absorption at higher energy (shorter wavelengths). This feature is also present in the construction of instruments dedicated to measuring the absorption spectrum. They are divided into IR instruments and UV–Vis instruments. Several UV–Vis instruments have extended their work range into the near-infrared region (NIR).

In the absorption spectrum of a pure crystal, we can see three classes of absorption bands:

1. Fundamental absorption band
2. Exciton bands
3. Color center bands

2.2.1 FUNDAMENTAL ABSORPTION

The ionic crystals are a category of very-well-studied crystals. The best examples of ionic crystals are the alkali halides: LiF, NaCl, KCl, KBr, and CsI. We can say they are “pure” ionic crystals. Other examples of ionic crystals are MgF_2, CdF_2, ZnF_2, and so on. What is characteristic for ionic crystals? They are formed by *ions with filled electronic configurations* (noble gas configuration). Let us take NaCl as an example: Na^+ has the complete configuration of Ne ($1s^22s^22p^6$) and Cl^- has the complete configuration of Ar ($1s^22s^22p^63s^23p^6$). All these crystals are transparent in the visible range, but the measured spectrum shows a very large spectral range free of absorption from IR to vacuum UV. You can imagine that all spectroscopic instruments for the IR range were built many decades ago based on light dispersion by prisms made of alkali halides. The most popular prisms were from NaCl or KCl. The CsI crystal was very good and was the only material available for far IR. However, LiF was a very good material for prisms used in vacuum UV, where the common quartz absorbs. What is the origin of this large transparency range? The ionic bond (nearly pure electrostatic) between positive ions and the surrounding negative ions is very strong. In solid-state physics, this property means that the valence band is completely filled with electrons and the distance between the valence band and the conduction band (called “energy gap”) is very large. Therefore, the energy gap is very large (from 6.4 eV for CsI to 12.5 eV for LiF) and electrons cannot pass into the conduction band without external high-energy input. The valence band originates in the filled np valence orbitals of the halide ion (e.g., $3p^6$ of Cl^-)

while the conduction band originates in the first empty *ns* orbital of the alkali ion (in our example, that $3s^0$ of Na^+). The very large energy gap results in other properties of ionic crystals, namely that they are very good insulators. If we convert the energy gap *Eg* (eV) of the crystal into wave number $\tilde{\nu}$ (cm^{-1}) or wavelength λ(nm) of the light, we may obtain the position in the spectrum where the light absorption of the ionic crystal begins (Figure 2.6). This value is called the *threshold wavelength*. Therefore, the radiation is absorbed if the wavelength is shorter than the threshold wavelength, and is not absorbed (transmitted) if the wavelength is longer than the threshold wavelength. We used the energy conversion formula 1 eV = 8065.54 cm^{-1} and the relationship $\lambda(\text{nm}) = 10^7/\tilde{\nu}(\text{cm}^{-1})$.

Energy of light refers to the energy of a photon and not to the intensity of the light beam. A high-intensity light beam with a wavelength longer than the threshold value does not produce absorption. (In a special condition of an ultra-high-intensity-focused laser beam, the problem is completely different!) The term "threshold wavelength" is used due to its similar meaning to the threshold wavelength for the photoelectric effect in metals. It is the internal photoelectric effect in our case, which means that absorption of a photon results in the excitation of an electron from the valence band to the conduction band. Therefore, the electron leaves a halide ion in ionic crystals. This phenomenon was intensively studied in semiconductors and has many practical applications such as photovoltaic cells. There are as yet no practical applications of this effect in dielectrics. The other term used in spectroscopy is (optical) *fundamental absorption edge*, but it does not specify the physical quantity referred to. Is it possible to determine the optical band gap from transmittance or absorbance measurement? In other words, what does an absorption spectrum look like in this range? Looking at the theoretical model presented in Figure 2.6, the absorption spectrum will show a very sharp edge

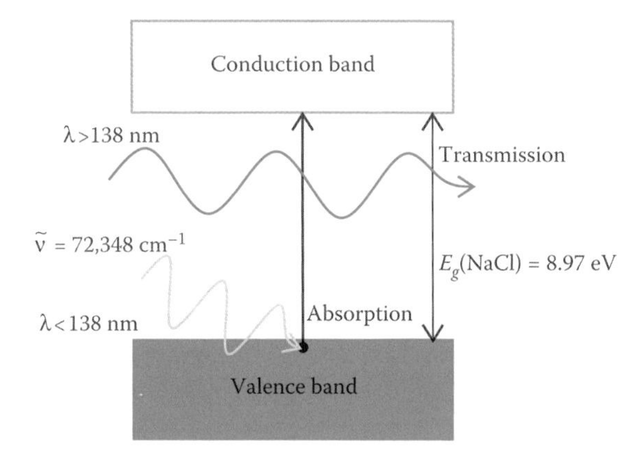

FIGURE 2.6 The absorption of light begins when the energy of light is larger than the gap energy of the ionic crystal (wavelength is shorter than the threshold wavelength). The value of the energy gap at room temperature was taken from the literature. (From Strehlow W. H., and E. L. Cook. 1973. Compilation of energy band gaps in elemental and binary compound semiconductors and insulators. *Journal of Physical and Chemical Reference Data* 2. http://www.nist.gov/data/PDFfiles/jpcrd22.pdf.)

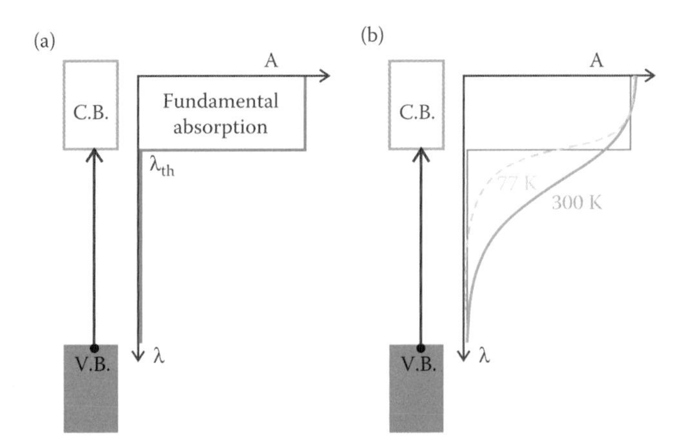

FIGURE 2.7 The (simulated) absorption spectrum of the NaCl crystal near the fundamental absorption edge: (a) theoretical shape (the fundamental absorption edge is very sharp) and (b) real shape of the spectrum at room temperature (continuous line) and low temperature (dotted line).

near the threshold wavelength as shown in Figure 2.7a. The real shape of the absorption spectrum as shown in Figure 2.7b does not have a sharp edge even at a lower temperature. The dependence on the temperature of the spectrum shape suggests that the effect is due to the vibrations of the crystal lattice, whose amplitude depends on temperature and never stop. Thus, it is not possible to determine the exact value of the threshold wavelength even at liquid helium temperature, but the error can be much reduced at this temperature. We note that this is the case of ultra-high purity (no chemical impurities) and a perfect (no structural defects) crystal or, in other words, an ideal crystal. The precise measurement of the fundamental absorption edge of high-energy gap crystals is a difficult experimental task. It is a challenge to have a very-good-quality spectrometer (very low level of stray light), very-low-temperature equipment (4 K), and a crystal of very good quality.

2.2.2 EXCITON SPECTRA

During measurement of the fundamental absorption of a pure and perfect crystal at low temperature, two or three sharp bands could be seen near the fundamental absorption edge of the longer wavelengths (Figure 2.8). These are called *exciton bands*. There is a special requirement to see exciton bands: to have a very thin crystal layer to reduce the fundamental absorption as much as possible.

What are excitons? We begin to underline that they are not impurities or structural defects. They are electronic excitations in insulating solids. Remember the electronic excitation in a free atom. The excited electron could receive enough energy to become free (ionization process). If the absorbed energy is a little bit lower than the ionization energy, then the excited electron remains weakly bonded to its nucleus. The process is similar inside the crystal lattice. The difference is that the crystal site left by the

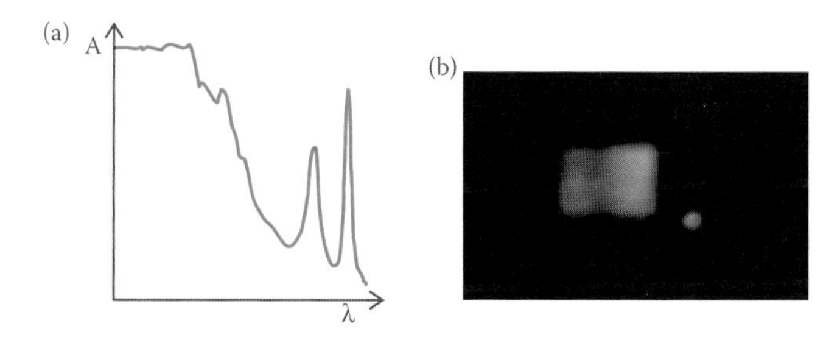

FIGURE 2.8 The exciton bands are located near the fundamental absorption onto the longer wavelengths. The first two peaks are visible only at liquid nitrogen temperature. (a) The absorption spectrum and (b) the image that can be seen at the exit of the spectrograph.

electron is equivalent to a positive charge, called a *hole* (similar to semiconductors). Between the electron and the hole, there is a weak electrostatic interaction in such a way that the coupled electron–hole pair has an energy-level structure similar to that of the hydrogen atom. The electron moves around the hole (Figure 2.9). The interaction radius of the exciton is determined by the atomic and electronic structures of the lattice. The radius can be bigger than the ion–ion distance and the attractive electrostatic force between the electron and the hole will be weakened by the dielectric constant of the crystal. Thus, inside the crystal, where the local electron and the hole interact with several surrounding atoms, the excitons are collective electronic excitations of the lattice. Excitons can also migrate through the crystal, but they do not transport effective electric charge. So, they do not have a contribution to the photoconduction currents in the crystal. The measurement of the exciton optical absorption requires low temperatures (77 K—liquid nitrogen recommended) and a high-resolution spectrometer.

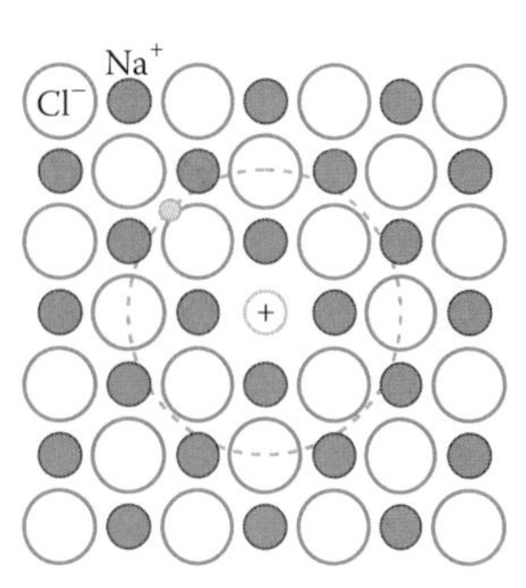

FIGURE 2.9 The model of an exciton in ionic crystals.

2.2.3 COLOR CENTERS

An exciting property of ionic crystals is that they become colored under X-ray irradiation. The NaCl crystal appears yellow and the KBr crystal appears blue after irradiation. It means that visible light is absorbed in some spectral range by centers that appear in crystals under irradiation (we suppose pure crystals). They are called *color centers.* We present below, the color centers in an alkali halide, but many of their properties could also be transferred to other types of dielectric crystals.

The measured absorption spectrum of the NaCl crystal can look like that presented in Figure 2.10a. A strong absorption band appears in the blue region of the visible range. The crystal of KCl shows the maximum absorption of color centers in the green region (Figure 2.10b). The weaker the electrostatic interaction, the lower is the energy of color center absorption. It is evidence of the structural origin of color centers.

It is obvious that the absorption is due to some electronic transitions inside the energy gap. If the crystal is pure, it means that the structure of the crystal was affected by X-ray irradiation where photons have high energy. This phenomenon can also be produced by γ-rays irradiation whose

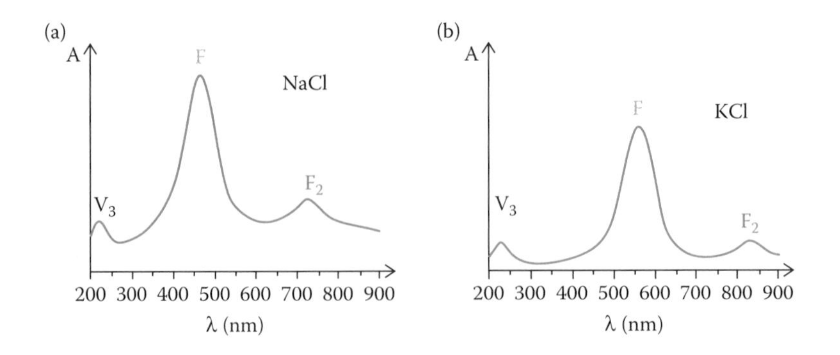

FIGURE 2.10 The absorpton spectrum in the UV–Vis range of an irradiated crystal of NaCl (a) and KCl (b). The spectra are similar, but the F and F_2 bands are shifted to the longer wavelength in the KCl crystal.

photons have much more energy than the X-ray photons. When the crystal is highly colored, the absorption is mainly due to the so called *F centers*. These defects have been mostly investigated by various spectroscopic techniques.

The main idea is that no crystal is perfect. During the crystal growth from melting when the temperature is high (lower or higher than 1000°C for alkali halides), some ions fail to take the right site. Empty sites appear and they are called *vacancies*. An anionic vacancy (e.g., a missing Cl^- ion) is equivalent to a positive local charge. A cationic vacancy (e.g., a missing Na^+ ion) is equivalent to a negative local charge (Figure 2.11a). We underline that a *vacancy is different from a hole*. A vacancy is a missing ion while a hole is a halide (e.g., Cl^-) ion present but without one electron. Both defects are equivalent to a positive charge. There are many such defects in the crystal as grown but they do not interact with light because they do not have electrons; so, they do not manifest in the absorption spectrum. Under certain conditions, "free" electrons can be created in the crystal. Therefore, an electron from the valence band can receive enough energy (more than the energy gap) to pass in the conduction band. It may be trapped by an anionic vacancy and forms an F-type center (Figure 2.11b). A hole can be

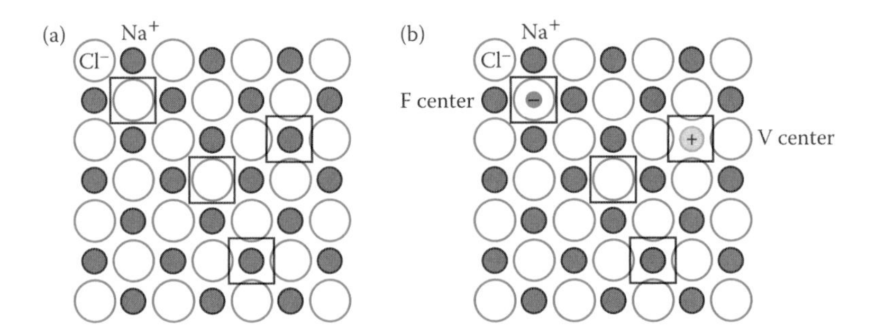

FIGURE 2.11 (a) A missing ion is a vacancy equivalent to an electric charge of the opposite sign. (b) A trapped electron on an anionic vacancy forms an F-type color center and a hole trapped on a cationic vacancy forms a V-type color center.

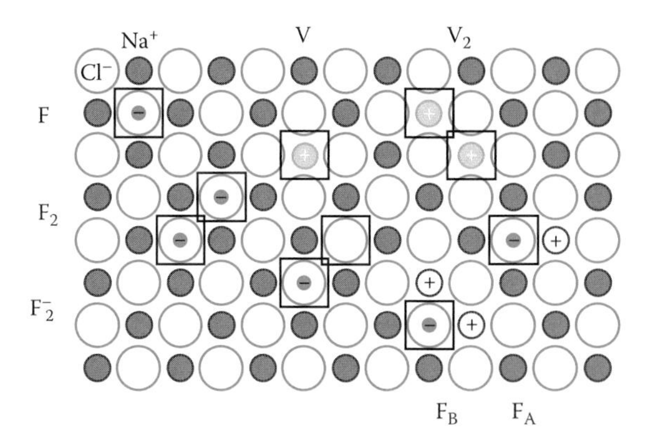

FIGURE 2.12 Different types of color centers present in ionic crystals.

trapped by a cationic vacancy and forms a V-type center. An F center is like an atomic entity with energy levels inside the energy gap of the crystal. It can absorb light and the F center band appears in the visible range (Figure 2.10). The F band amplitude decreases in time after a long irradiation with light of F band wavelength and the crystal loses its color and will become transparent again. It means the color centers are not stable. The electron can receive enough energy to pass again in the conduction band and to be stronger trapped on the other side. It is the case of color centers created by X-ray irradiation. The γ-ray irradiation produces stable color centers.

There are several types of color centers because many types of defects can be produced during crystal growth (Figure 2.12):

- F centers appear when an electron is trapped by an anionic vacancy.
- F_2 and F_3 centers appear when two or three F centers agglomerate.
- F^+ is an F center that loses one electron and F_2^- is made of two anionic vacancies with three trapped electrons.
- V centers appear when a hole is trapped by a cationic vacancy (metal missing).
- V_2 and V_3 appear due to the agglomeration process of cationic vacancies.
- F_A and F_B centers appear when the F center has one or two neighbor cations different from host cations. These types of F centers are interesting for lasers based on color centers.

Stable F centers can also be formed by heating an alkaline halide crystal in an atmosphere of specific metal vapor. The other method to produce stable color centers is to apply a high voltage between the opposite faces of the crystal heated at a high temperature. The injected electrons will be trapped on vacancies. The color depends on the size of the defect hole, which in turn depends on which halide was removed. The electrolytic method may be relatively easily used in a student laboratory. All you need are (Figure 2.13):

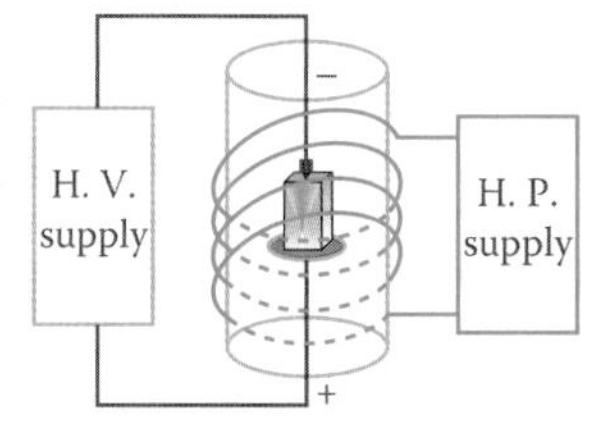

FIGURE 2.13 The device for electrolytic coloration of alkali halide crystals.

1. A furnace made from a quartz tube and a resistive coil of Kanthal flat wire. The temperature depends on the crystal type but varies between 400°C and 500°C, below the melting temperature of the crystal.

2. Two metal electrodes, a plate of nickel (+) and a tip of tungsten (–).
3. A variable alternate current (ac) power supply for the furnace (low voltage–high current) and a high-voltage direct current (dc) power supply for electrodes (400–1000 V, low current). A smaller crystal means a lower–high voltage.

Using this method, it is easy to obtain crystals with a high concentration of F centers but with nonuniform distribution. The uniformity can be improved using many electrode tips. The crystals are stable and present a strong and large absorption band in the visible range. This method has the advantage of real-time observation of the changing color of crystals, and it is easy to be controlled.

We have already mentioned that F centers and other types are optically active in the visble spectral range or UV range and it is easy to measure their contribution to the crystal absorption because of the large energy gap. A theoretical model of F centers is based on the idea that there is an electrostatic interaction between the cationic vacancy and the trapped electron. The presence of a vacancy is equivalent to a square well of potential where the electron can move and may have one energy level in classical mechanics and several discrete energy levels in quantum mechanics. The quantum model is correct because the interatomic distances in the crystal are of the order of an angstrom. It is similar to the hydrogen atom. So, there are electronic levels of energy. This pseudoatomic system is not isolated like the hydrogen atom but is connected to the crystal lattice by strong interactions. Also, the crystal lattice vibrates with frequency characteristics to the crystal periodicity and therefore to the crystal type. The lattice vibrations modify the dimension of the quantum well where the trapped electron is moving. This interaction results in a representation of energy levels similar to the diatomic molecule in the harmonic oscillator approximation where both ground (G) and excited (E) energy level of each electron is split into several equidistant vibrational energy levels (Figure 2.14). In the ground state, the electron is in the lowest possible vibrational energy level (1). Then the electron absorbs light and passes into an excited state corresponding to the absorbed photon energy. The corresponding vibrational level (2) of the excited state is always different from the lowest energy. This is a consequence of the Franck–Condon principle, which states that the electronic transition is faster than nuclear motions because of the large mass difference. The second moment is the nonradiative relaxation ($2 \rightarrow 3$) from the vibrational excited level 2 to the lowest vibrational level into the excited electronic state. The energy difference is taken by the crystal lattice through the collective vibrations. The third moment is the radiative transition by light emission from the excited state into the ground state ($3 \rightarrow 4$). The final moment is again the nonradiative transition from the excited vibrational level (4) into the lowest vibrational level (1) where the crystal lattice takes the rest of the absorbed photon energy. It is time for the following question: Why does the nonradiative transition ($2 \rightarrow 3$) occur before radiative transition ($3 \rightarrow 4$)? It is because the relaxation time of the lattice (phonon lifetime) is orders of magnitude shorter than the radiative lifetime (10^{-12} s compared to 10^{-9} s).

What can we see? The crystal with color centers can absorb light with specific energy (blue absorption band on the left side of Figure 2.14) and

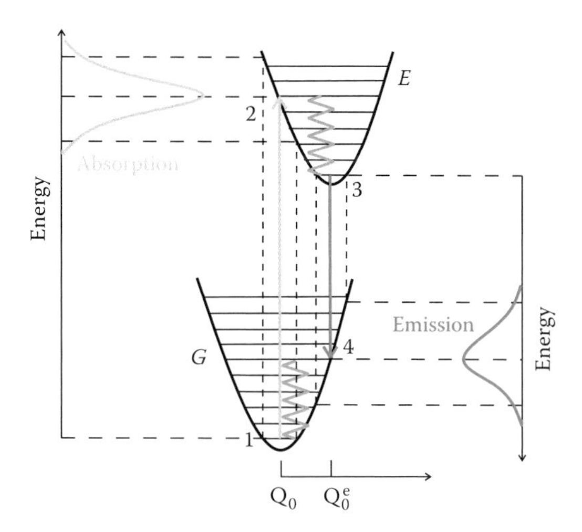

FIGURE 2.14 Simplified energy diagram of optical transitions between two vibrational levels of F centers explains their photoluminescence. Q_0 is the coordinate of the ground state at equilibrium and Q_0^e is the coordinate of the excited state at equilibrium, too. (From Di Bartolo, B. 1978. *Luminescence of Inorganic Solids*. New York: Plenum Press.)

can also emit light with lower energy (red emission band on the right side of Figure 2.14). In terms of wavelength, it means that the crystal emits light with a wavelength longer than that of absorbed light (*Stokes shift*). This phenomenon is called *photoluminescence* in solid-state physics and is a nonresonant phenomenon. It is similar to the *fluorescence* phenomenon produced by molecules and is largely used in biomedical research. Therefore, spectrofluorimetry, the method of materials characterization by measurements of photoluminescence or fluorescence, is a branch of optical spectroscopy. In this field, the absorbed light is usually called excitation light. Here, we note that the light emission continues for some time even after the excitation light is turned off. The after-glow time (or the fluorescence decay time) is not usually observable by the naked eye because it is in the range of nanoseconds or microseconds. Both the position of the emission band and decay time of photoluminescence are sensitive to the surroundings of the emitting system and therefore to the local symmetry of the crystal.

The photoluminescence of color centers has some important characteristics:

- The Stokes shift (the spectral distance between the excitation and emission bands) is very large, by hundreds of nanometers compared to tens of nanometers in molecules.
- The emission band is broad because many vibrational levels are involved (Figure 2.14).

We observe that the energy diagram presented in Figure 2.14 is similar to the four-level laser scheme (the most popular scheme) and the broad band emission also suggests the possibility of building a *tunable solid-state laser* similar to the well-known dye lasers based on organic molecules in solutions. Such a laser is a very useful tool in (molecular) spectroscopy where you need a controlled high intensity to a specific wavelength that you can change. Another advantage is given by the first property (a large or very large Stokes shift), which means that the laser can emit in NIR, the spectral range where we have a few spectral sources. With an appropriate choice of crystal, it is possible to obtain emission at any wavelength in the NIR range (0.8–3 μm). This was an interesting research topic for decades, but the properties of color centers are still interesting in the new field of nanoparticles.

We mention here an interesting and useful phenomenon due to the F centers called *thermoluminescence* (McKeever 1985). The process consists of light emission by the crystal during sample heating. The mechanism involves the supply of thermal energy to the crystal, which allows the trapped electrons on *F* centers to become free. It means that electrons pass into the conduction band. Then the electrons can recombine with other electron-deficient defects and the recombination energy is released as visible light. The emitted light intensity is proportional to the freed electrons number. It is a good method for the dating of archeological materials and objects (Aitken 1985).

The thermoluminescence dating technique has been important in determining timelines in archeology and geology in the last 500,000 years for various materials, such as volcanic material, meteoritic craters, faults caused by earthquakes, and anthropogenic materials. Since the 1970s, the thermoluminescence technique began to be used for dating sediments, sand dunes, and loess that have been exposed to the sun.

The thermoluminescence technique is based on the accumulation over time of irradiation defects in the crystal lattice of the mineral grains due to natural radiation. Quartz and feldspar are the most common minerals used by thermoluminescence dating because they are easily found, are easily separated from the samples, and show intense luminescence signals.

It supposes a zero moment when the crystalline material (e.g., feldspar) has no empty defects (electron free). It could be the moment after ceramic burning in a furnace to over 400°C. In the dark (sample material covered by the other material), the high-energy radiation (electromagnetic particles) that exists on the Earth penetrates the covering material and produces new stable defects. These defects are kept for a long time (thousands of years). In the laboratory, the finely controlled temperature of the sample is increased, and free electrons are created and recombine with empty defects. The emitted light intensity is proportional to the number of stable defects that is proportional to the dose of high-energy radiation absorbed by the crystal. We must compare the thermoluminescence unknown signal to the signal after a controlled high-energy irradiation (Figure 2.15). If we suppose that the annual natural radiation dose is constant and we can determine the natural dose per year, then we may calculate the age of the object. Other information may be given by the temperature at which the light is emitted. A lower temperature indicates the presence of trapping centers with energy near the conduction band. A higher temperature means that the trapping centers are deeper in the energy gap of the material. The absence of low-depth centers is a sign indicating that the object was reheated (reburned) during its history.

The thermoluminescence method has many interesting applications: geological dating using the zircon mineral (Secu et al. 2007), dosimetry on tooth enamel (Secu et al. 2011), detection of irradiated food (Cutrubinis

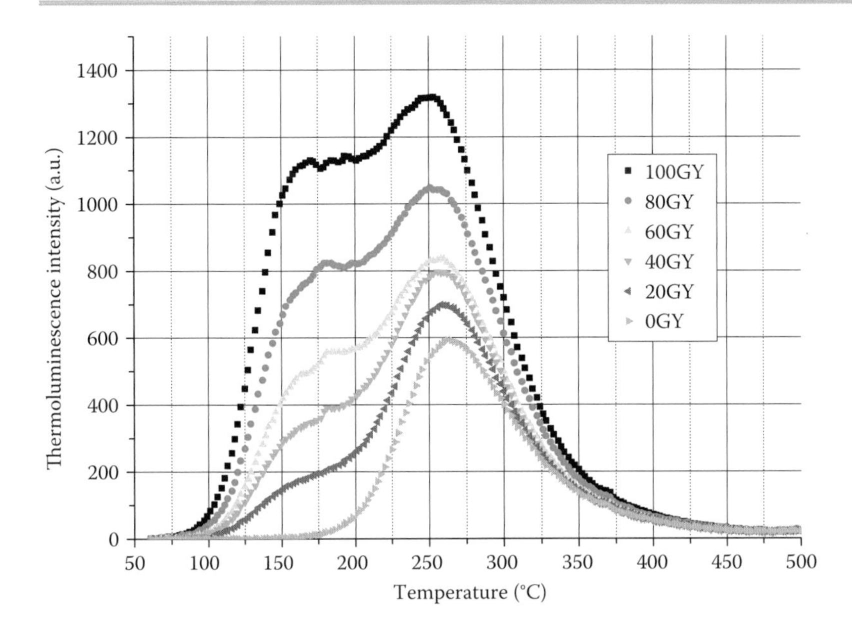

FIGURE 2.15 The compared thermoluminescence signals of quartz extracted from a sedimentary rock sample at different γ-irradiation doses. (Acknowledgments to Dr. Mihail Secu.)

et al. 2007), and detection of irradiation defects in materials (Secu et al. 2008).

2.3 IMPURITIES IN CRYSTAL LATTICE: SPLITTING OF LEVELS AND TERMS IN LATTICE SYMMETRY

2.3.1 IMPURITY TYPES

We just finished presenting the principal characteristics of the absorption spectrum of alkali halide crystals. The three main causes that produce measurable absorption bands in the absorption spectrum (fundamental absorption, excitons, and color centers) are also common to other classes of materials such as metal oxides or complex oxides, vanadates, tungstates, borates, and so on. Diamond is a covalent crystal with similar properties. All of them are insulators and have a large energy gap (5–15 eV). Therefore, from the point of view of optics, the pure insulator materials are transparent, if we can control the quality of the crystal lattice well. Transparency does not necessarily mean being clear materials in the visible range. UV and NIR are also spectral ranges of great interest. However, we like and use color materials for different purposes such as jewelry, wine bottles, and even for windshields instead of the transparent material. The difference between a ruby and a sapphire is given only by the low-concentration impurity that is optically active (chromium and iron, respectively). The crystalline matrix is the same (Al_2O_3). There are a lot of practical applications of doped crystals, but much of the research in the field in the last 50 years was due to the discovery and development of lasers.

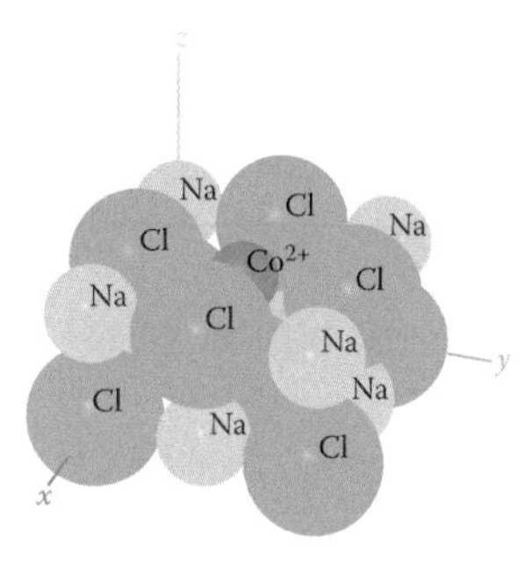

FIGURE 2.16 The Co^{2+} ion takes the place of the Na^+ ion during the crystal growth process.

Thus, our problem is to understand what happens when known impurities are introduced in a controlled manner in a crystal lattice. Therefore, we are finally interested in how impurities generate absorption bands and also manifest emission in the transparency range of the host matrix. It is not the purpose of this book to describe methods of crystal doping. However, we mention here for the bulk crystals that the doping is performed at high temperature, usually when the crystal is pulled out from the molten state or during the crystallization process. Let us imagine that a quantity of ultrapure powder of sodium chloride with added traces (<1%) of cobalt chloride is melted in a glass or quartz ampoule and then cooled very slowly. When the temperature decreases under the melting temperature, the crystallization process begins. The host ions (Na^+ and Cl^-) take their places one by one, forming the crystal lattice already shown in Figure 1.61. The guest ions of Co^{2+} that are 100 times less frequent than Na^+ ions are also pushed from the melt to the solid state. It is natural for them to accommodate in sites of sodium ions that are being surrounded by negative ions of chlorine (Figure 2.16). When the crystal is cooled, it remains captured in the cationic sites. The growing process in proper conditions can assure a relatively uniform distribution of such ions in the crystal host matrix. When the temperature is lower than the melting temperature but high enough to have a strong thermal motion, the impurity ions can migrate through the crystal and agglomerate to form aggregates. These are usually seen as defects. The guest ion by itself is a defect (a factor of stress for the crystal lattice) if its electrical charge is not equal to the electrical charge of the host cation replaced. The crystal lattice must accommodate this supplementary positive electrical charge, creating a structural defect equivalent to a negative charge.

What can the guest ion "see" in the crystal host? The guest ion replaces a host ion, for example, sodium. We have already shown in Figure 1.62 that six chlorine ions are placed in octahedron vertices around the Na^+ ion. If the Co^{2+} ion takes its place, it will have the same neighbors, even if the excess of positive charge should stress the local lattice. In this situation, the Co^{2+} ion is called *substitutional impurity*, that is, a host ion was substituted by an impurity ion. The six chlorine ions in our example are called *ligands*. Using the new terms, we can say that the impurity ion is placed in the center of an octahedron formed by ligands (Figure 2.17). The new symmetry is octahedral, more precisely $\boldsymbol{O}_h$.

Most of the optically active impurities are ions of transition metals. Examples of the most common ions are presented in Table 2.8.

We note that Cu and Ag, which belong to the transition metals category, are two important exceptions to the general rule of filling the shells and have the electronic configuration [Ar] $3d^{10}\,4s^1$ and [Kr] $4d^{10}\,5s^1$, respectively. They can easily lose an electron to become Cu^+ and Ag^+, but the new ions have a filled last shell and are not optically active. However, the Cu^{2+} ion still exists and belongs to the d^9 configuration.

The fluorescence of transition metal complexes has a new application: live cell imaging with luminescent metal complexes (Baggaley et al. 2012). Using such complexes is attractive for the reason of a longer lifetime of complexes (microseconds) compared to organic fluorophores currently used in biology (nanoseconds). This property can increase the contrast in fluorescence lifetime imaging microscopy to eliminate the autofluorescence of the

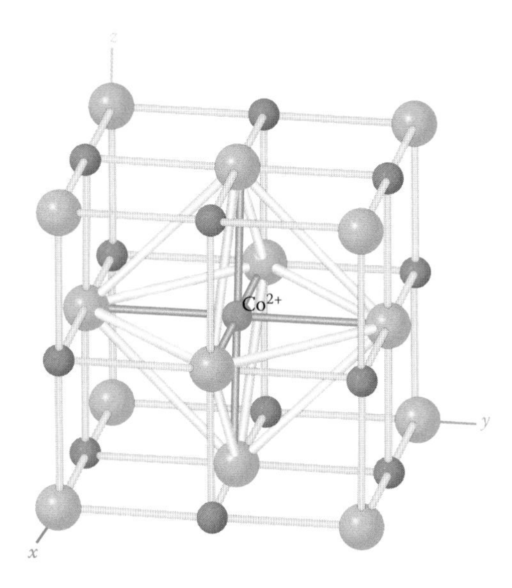

FIGURE 2.17 A substitutional impurity in the alkali halide crystal has $\boldsymbol{O}_h$ symmetry of the ligand field.

TABLE 2.8 Transition Metal Ions Commonly Used as Impurities (Dopants) in Ion Crystals Originate from the First Row of the Periodic Table ($3d^n$ Configuration)

Electronic Configuration	Ion*
d^1	**Ti^{3+}**, Vn^{4+}
d^2	**V^{3+}**, Cr^{4+}, and Mn^{5+}
d^3	V^{2+}, **Cr^{3+}**, and Mn^{4+}
d^4	Cr^{2+}, **Mn^{3+}**
d^5	**Mn^{2+}**, **Fe^{3+}**
d^6	**Fe^{2+}**, Co^{3+}
d^7	Fe^{+}, **Co^{2+}**, and Ni^{3+}
d^8	**Ni^{2+}**
d^9	**Cu^{2+}**

* Ions in bold are the most common states.

cell (Koo et al. 2009). Another opportunity offered by transition metal complexes is the high sensitivity of the emission lifetime to the microenvironment and therefore to the various ions present in the live cell.

Other ions intensively studied are rare earths, otherwise referred to as the 15 lanthanides. They are dopants in many types of matrices for different applications as shown in Table 2.9. Technologically, the most important are neodymium- and erbium-doped gain media for lasers.

2.3.2 COUPLING SCHEMES

Both TM and RE ions introduced in the crystal lattice change the absorption spectrum of the crystal host. Both classes of ions change their

TABLE 2.9 Examples of Rare Earth Ions Commonly Used as Impurities (Dopants) in Ion Crystals in Order of Their Importance as Laser-Active Media

Electronic Configuration	Ion	Common Host Media
[Xe] $4f^3$	Neodymium Nd^{3+}	YAG, YVO_4, YLF, and silica
[Xe] $4f^{13}$	Ytterbium (Yb^{3+})	YAG, tungstates, and silica
[Xe] $4f^{11}$	Erbium (Er^{3+})	YAG, silica
[Xe] $4f^{12}$	Thulium (Tm^{3+})	YAG, silica, and fluoride glasses
[Xe] $4f^2$	Praseodymium (Pr^{3+})	Silica, fluoride glasses
[Xe] $4f^{10}$	Holmium (Ho^{3+})	YAG, YLF, and silica

absorption and emission spectra with respect to the crystal host. However, there is an important difference between them: TM ions spectra strongly depend on the crystal host symmetry while RE ions spectra are influenced only slowly. Looking at the electronic configuration, we can see that the optically active electrons of TM ions are placed in $3d$ orbitals that are in an outer shell while those electrons of RE ions are placed in $4f$ orbitals that are in an inner shell. This difference in spatial position becomes important when the ion is introduced in an external (electrostatic) field as in a crystalline matrix.

Now, we return to the Schrödinger equation of the ion introduced in an external field to determine the possible energies (eigenvalues). The Hamiltonian of the free ion given by Equation 2.4 will be completed by a new term as follows:

$$\hat{H} = \hat{H}_0 + \hat{H}_{ee} + \hat{H}_{LS} + \hat{H}_{SS} + \hat{H}_{CF} \tag{2.6}$$

We just changed the subscripts so that the significance of each term can be more evident:

$$\hat{H}_0$$

is the Hamiltonian of the central field,

$$\hat{H}_{ee} = \frac{1}{2}\sum_{i>j=1}\frac{e^2}{r_{ij}}$$

is the Hamiltonian of electrostatic repulsion between electrons,

$$\hat{H}_{LS} = \sum_{i=1}\xi_i \vec{l}_i \cdot \vec{s}_i$$

is the Hamiltonian of the spin–orbit interaction,

$$\hat{H}_{SS} = \sum_{i>j=1}\zeta \vec{s}_i \cdot \vec{s}_j$$

is the Hamiltonian of the spin–spin interaction, and

$$\hat{H}_{CF} = \sum_{i=1} eV_i$$

is the Hamiltonian of the interaction between each electron and the crystal field.

Before to apply the theory of perturbations, it is important to establish the order of perturbations and we have three schemes in terms of the position of the crystal field reported to electrostatic repulsion and spin–orbit coupling:

1. *Weak crystalline field scheme.* It means that the Hamiltonian of the interaction with the crystal field is weaker than the electrostatic repulsion but stronger than the spin–orbit coupling:

$$\hat{H}_{ee} > \hat{H}_{CF} > \hat{H}_{LS}$$

   It is the case of most complexes of transition metal ions.
2. *Strong crystalline field scheme.* It means that the Hamiltonian of the interaction with the crystal field is stronger than the electrostatic repulsion and the spin–orbit coupling:

$$\hat{H}_{CF} > \hat{H}_{ee} > \hat{H}_{LS}$$

   It is the case of a few complexes of transition metal ions (Fe and Co).
3. *Rare-earths scheme.* It means that the Hamiltonian of the interaction with the crystal field is weaker than the spin–orbit coupling:

$$\hat{H}_{ee} > \hat{H}_{LS} > \hat{H}_{CF}$$

As the name says, it is the case of all rare-earth ions. The valence orbitals are somewhat screened toward the neighbor ions. Thus, they keep the energetic levels of the free state.

We underline that there is clear distinction between the three schemes. Each scheme just suggests the way to treat a specific problem.

2.4 WEAK CRYSTALLINE FIELD OF OCTAHEDRAL SYMMETRY

2.4.1 PREDICTION OF 3*d* ORBITALS SPLITTING IN $\boldsymbol{O}_h$ SYMMETRY BASED ON GROUP THEORY

We begin to study the optical properties of crystals doped with transition metal ions, that is, to apply weak crystalline field scheme without actually solving the Schrödinger equation. We need to know:

- The strength of the supplementary electrostatic field created by ligands (crystal field) in the site occupied by the doping ion

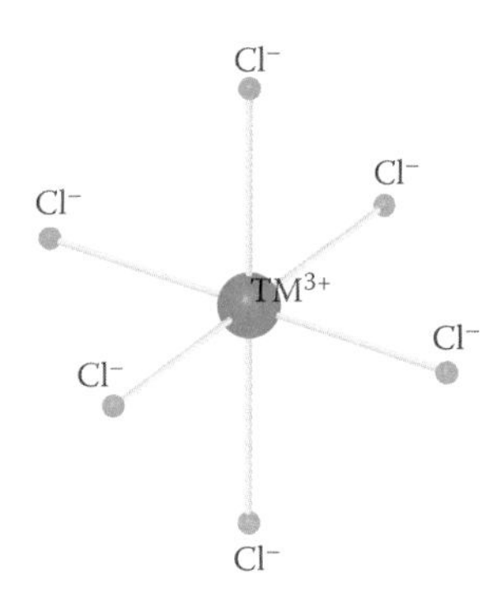

FIGURE 2.18 The transition metal ion is surrounded by six negative point charges in $\boldsymbol{O}_h$ symmetry.

- How the orbitals of the central ion are affected by the anisotropy of the crystal field in terms of symmetry
- To evaluate the energetic changes produced by the crystal field to orbitals

The scheme will be applied to octahedral symmetry that is typical of *d*-metal complexes where every ligand ion is approximated to a point charge (Figure 2.18).

Let us see how useful the group theory is before performing an extensive calculation. We must just look at the Appendix to this book containing the character table of the $\boldsymbol{O}_h$ group. The *d* orbitals ($\ell = 2$) are described by five spherical harmonics Y_{22}, Y_{21}, Y_{20}, Y_{2-1}, and Y_{2-2}. In Cartesian coordinates, they correspond to $x^2 - y^2$, xz, $2z^2 - x^2 - y^2$, yz, and xy, respectively (Figure 1.60). In the character table of the $\boldsymbol{O}_h$ group, the pair of functions $(2z^2 - x^2 - y^2, x^2 - y^2)$ appears to be the second-order basis of the two-dimensional representation E_g. The triplet of functions (xz, yz, and xy) also appears to be the second-order basis of the three-dimensional representation T_{2g}. Using the single-electron notation, we can denote both representations as lowercase letters e_g and t_{2g}. We draw a quick conclusion that the five orbitals of *d*-type will have two different values of energy in the octahedral symmetry of the crystal instead of one value of energy in the spherical symmetry of the free ion. We observe that they are either grouped as orbitals with lobes directed along the axes or as orbitals with lobes directed between the axes (Figure 2.19). It means that if an electron belongs to the orbitals d_{z^2} or $d_{x^2-y^2}$, the repulsion electron–anion

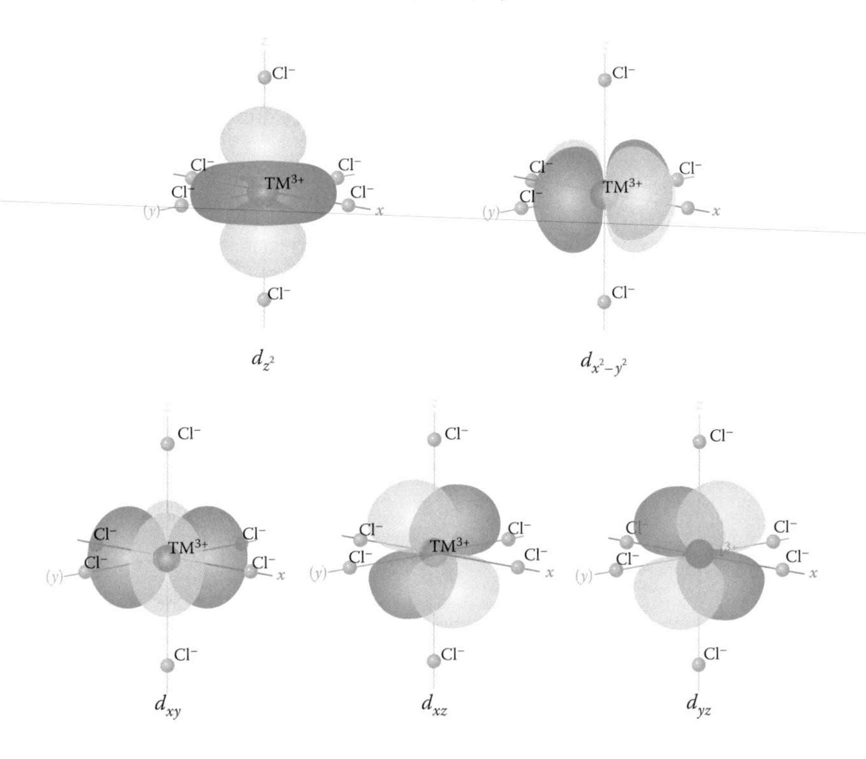

FIGURE 2.19 Spatial orientation of *d* orbitals with respect to ligand anion positions in $\boldsymbol{O}_h$ symmetry.

(chloride) will be stronger than if the electron belongs to one of the orbitals d_{xz}, d_{yz}, and d_{xy}. Therefore, in terms of energy, the level e_g will be upper level and the level t_{2g} will be lower level (Figure 2.20). In other words, we can say that the state of the free ion will split into two states in the octahedral field. The levels represented by several distinct lines in Figure 2.20 should be seen as represented by only one line because they have the same energy, therefore: the five lines of 3*d* as one line, the two lines of e_g leaves one, and the three lines of t_{2g} leaves one, too. If the symmetry changes, then the procedure is similar. This is an example of how powerful group theory can be. However, group theory cannot provide any quantitative information about how large the split is.

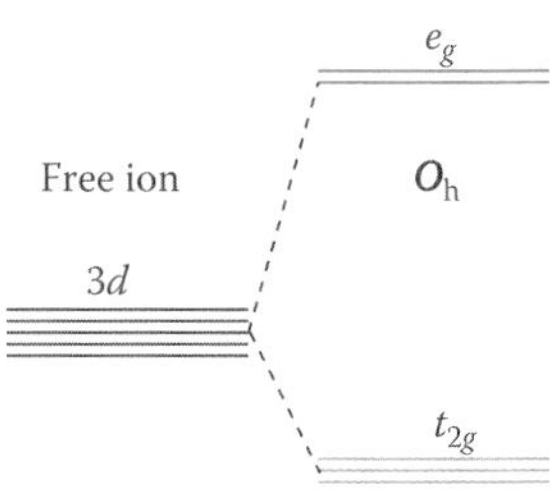

FIGURE 2.20 Splitting of 3*d* level into e_g and t_{2g} in octahedral symmetry.

2.4.2 CRYSTAL FIELD STRENGTH OF OCTAHEDRAL SYMMETRY

The central issue of theory is to find an expression for the electrostatic field produced by several points charges that possess a given symmetry. We assume that the point charges are located on coordinate axes at equal distances from the central ion, thus giving an octahedral environment.

The electric potential created by the *i*-th point charge with coordinates (x_0, y_0, z_0) in a point of coordinates (x, y, z) as shown in Figure 2.21 is given by

$$V_i = \frac{e}{|\vec{r} - \vec{r}_0|} \tag{2.7}$$

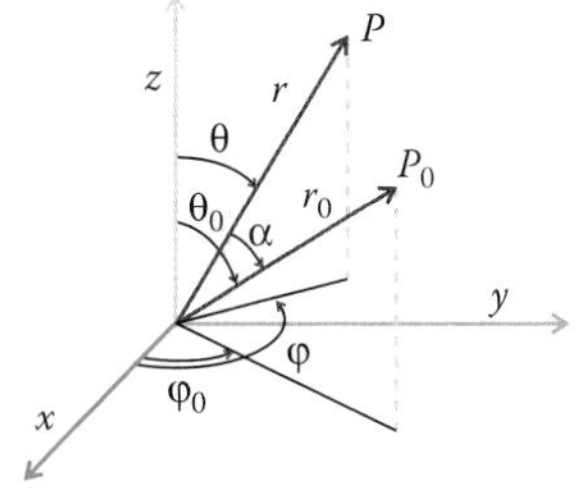

FIGURE 2.21 Illustration of a general case where a point charge located in P_0 generates a potential in *P*.

This potential will be applied to the orbitals of the central ion. We suppose that the crystal field is weaker than both the central field of the nucleus and electrostatic repulsion between electrons. Thus, the crystal field effect becomes a perturbation of the free ion state. The orbitals of the free ion are described by known eigenfunctions that are products between the radial wavefunction and the angular wavefunction. We suppose that the radial function is not affected by the crystal field. We will focus on the spatial orientation of orbitals in a crystal field. The orbitals are described by spherical harmonics that can be separated into two functions on each angle θ and φ, respectively:

$$Y_{lm}(\theta,\varphi) = \frac{1}{\sqrt{2\pi}}\Theta_{lm}(\theta)e^{im\varphi} \tag{2.8}$$

The equation of angular momentum

$$\left[-\frac{1}{\sin\theta}\frac{\partial}{\partial\theta}\left(\sin\theta\frac{\partial}{\partial\theta}\right) + \frac{m}{\sin^2\theta}\right]\Theta_{lm}(\theta) = l(l+1)\Theta_{lm}(\theta) \tag{2.9}$$

suggests that the solutions of Θ_{lm} could be expressed in terms of Legendre polynomials of l order $P_l(\cos\theta)$. The Legendre polynomials have the important property of orthogonality that we ask for in spherical harmonics. Then we change the Legendre polynomials into the Legendre-associated functions by a general relationship:

$$P_l^m(x) = (1-x^2)^{\frac{m}{2}}\frac{d^m}{dx^m}P_l(x) \quad \text{where } m = 0, 1, 2, 3, \ldots, l \tag{2.10}$$

The Legendre-associated functions also obey the orthogonality condition. Thus, the function on θ angle can be written in terms of Legendre functions as follows:

$$\Theta_{lm}(\theta) = (-1)^m \left[\frac{(2l+1)(l-m)!}{2(l+m)!}\right]^{1/2} P_l^m(\cos\theta) \quad \text{if } m \geq 0$$

$$\Theta_{lm}(\theta) = (-1)^{|m|}\,\Theta_{l|m|}(\theta) \quad \text{if } m < 0$$

Finally, we can write the expression of spherical harmonics as

$$Y_{lm}(\theta,\varphi) = (-1)^m \left[\frac{(2l+1)(l-m)!}{2\pi(l+m)!}\right]^{1/2} P_l^m(\cos\theta)e^{im\varphi} \quad \text{if } m < 0 \tag{2.11}$$

$$Y_{l,-m}(\theta,\varphi) = (-1)^m Y^*_{lm}(\theta,\varphi) \quad \text{if } m < 0$$

The spherical harmonics also obey the orthonormality condition:

$$\iint Y^*_{l'm'}(\theta,\varphi)Y_{lm}(\theta,\varphi)\sin\theta d\theta d\varphi = \delta_{ll'}\delta_{mm'} \tag{2.12}$$

In conclusion, the angular part of eigenfunctions of the free atom is based on the Legendre polynomials. This part gives the shape of the corresponding orbital and we are interested in how the orbital shape influences the position (direction) of orbitals in the external field of ligands. The rearrangement of orbitals in this anisotropic field can change the corresponding energy of orbitals. However, we know that the electrostatic repulsion between electrons in free atoms $1/r_{ij}$ is also expanded on the Legendre polynomials. The idea is to expand the potential of the ligand in terms of the Legendre polynomial P_k where α is the angle between $\vec{r}$ and $\vec{r}_0$ (Figure 2.21) as follows:

$$V_i = \frac{e}{|\vec{r}-\vec{r}_0|} = \frac{e}{r_0}\sum_{k=0}^{\infty}\left(\frac{r}{r_0}\right)^k P_k(\cos\alpha) \tag{2.13}$$

Each Legendre polynomial can be expanded into spherical harmonics separately for the ligand and electrons of the central ion:

$$\begin{aligned} P_k(\cos\alpha) &= \frac{4\pi}{2k+1}\sum_{m=-k}^{k} Y_{km}(\theta,\varphi)Y^*_{km}(\theta_0,\varphi_0) \\ &= \frac{4\pi}{2k+1}\sum_{m=-k}^{k}\frac{1}{\sqrt{2\pi}}\Theta_{km}(\theta)\Phi_m(\varphi)\frac{1}{\sqrt{2\pi}}\Theta^*_{km}(\theta_0)\Phi^*_m(\varphi_0) \\ &= P_k(\cos\theta_0)P_k(\cos\theta) + 2\sum_{m=-k}^{k}\frac{(k-m)!}{(k+m)!}P_k^m(\cos\theta_0)P_k^m(\cos\theta) \\ &\quad \times \cos[m(\varphi-\varphi_0)] \end{aligned} \tag{2.14}$$

Thus, the final expression of the potential will be

$$V_i = \frac{e}{r_0}\sum_{k=0}^{\infty}\left(\frac{r}{r_0}\right)^k \left\{P_k(\cos\theta_0)P_k(\cos\theta) + 2\sum_{m=-k}^{k}\frac{(k-m)!}{(k+m)!}P_k^m(\cos\theta_0)P_k^m(\cos\theta)\cos[m(\varphi-\varphi_0)]\right\} \tag{2.15}$$

Using the expressions of first Legendre polynomials and functions given in Table 2.10, we can calculate the contribution of each ligand to the total potential. Let us take the ligand located on the positive branch of the x axis. Its position is described by the spherical coordinates: r_0, $\varphi_0 = 0$, and $\theta_0 = 90°$. Thus, its contribution to the potential is given by

$$V_i(x) = \frac{e}{r_0}\sum_{k=0}^{\infty}\left(\frac{r}{r_0}\right)^k \left\{P_k(\cos\theta)P_k(0) + 2\sum_{m=-k}^{k}\frac{(k-m)!}{(k+m)!}P_k^m(\cos\theta)P_k^m(0)\cos(m\varphi)\right\} \tag{2.16}$$

Looking into Table 2.10, we see that the terms $P_1(0), P_3(0), P_5(0),\ldots,$ $P_2^1(0), P_3^2(0),\ldots$ vanish. We denote

$$\frac{(k-m)!}{(k+m)!} = \beta_{km}^2 \tag{2.17}$$

Thus,

$$\begin{aligned} V(x) = \frac{e}{r_0}\Bigg\{&1 + \frac{r}{r_0}2\beta_{11}^2P_1^1(\cos\theta)\cos\varphi \\ &+ \left(\frac{r}{r_0}\right)^2\left[-\frac{1}{2}P_2(\cos\theta) + 6\beta_{22}^2P_2^2(\cos\theta)\cos 2\varphi\right] \\ &+ \left(\frac{r}{r_0}\right)^3 2\left[-\frac{3}{2}\beta_{31}^2P_3^1(\cos\theta)\cos\varphi + 15\beta_{33}^2 2P_3^3(\cos\theta)\cos 3\varphi\right] \\ &+ \left(\frac{r}{r_0}\right)^4 2\left[\begin{matrix}\frac{3}{8}P_4(\cos\theta) - 15\beta_{42}^2P_4^2(\cos\theta)\cos 2\varphi \\ + 210\beta_{44}^2P_4^4(\cos\theta)\cos 4\varphi\end{matrix}\right] + \cdots\Bigg\} \end{aligned} \tag{2.18}$$

All four ligands placed in (x, y) plane give similar expressions of the potential. There is only one difference concerning the φ_0 angle. The ligand located on the negative branch of the x axis has $\varphi_0 = 180°$ and it changes trigonometric functions as follows: $\cos(\varphi - 180°) = -\cos\varphi$,

TABLE 2.10 Expressions of First Legendre Polynomials and Associated Legendre Functions

k	$P_k(x)$	$P_k^m(x)$
0	1	
1	x	$P_1^1(x) = (1 - x^2)^{1/2}$
2	$\frac{1}{2}(3x^2 - 1)$	$P_2^0(x) = \frac{1}{2}(3x^2 - 1)$
		$P_2^1(x) = 3(1 - x^2)^{1/2}x$
		$P_2^2(x) = 3(1 - x^2)$
		$P_2^{-1}(x) = -\frac{1}{6}P_2^1(x)$
		$P_2^{-2}(x) = \frac{1}{24}P_2^2(x)$
3	$\frac{1}{2}(5x^3 - 3x)$	$P_3^1(x) = \frac{3}{2}(1 - x^2)^{1/2}(5x^2 - 1)$
		$P_3^2(x) = 15x(1 - x^2)$
		$P_3^2(x) = 15(1 - x^2)^{3/2}$
4	$\frac{1}{8}(35x^4 - 30x^2 + 3)$	$P_4^0(x) = \frac{1}{8}(35x^4 - 30x^2 + 3)$
		$P_4^1(x) = -\frac{5}{2}(7x^3 - 3x)(1 - x^2)^{1/2}$
		$P_4^2(x) = \frac{15}{2}(7x^2 - 1)(1 - x^2)$
		$P_4^3(x) = -105x(1 - x^2)^{3/2}$
		$P_4^4(x) = 105(1 - x^2)^2$
		$P_4^{-1}(x) = -\frac{1}{20}P_4^1(x)$
		$P_4^{-2}(x) = \frac{1}{360}P_4^2(x)$
		$P_4^{-3}(x) = -\frac{1}{5040}P_4^3(x)$
		$P_4^{-4}(x) = \frac{1}{40320}P_4^4(x)$
5	$\frac{1}{8}(63x^5 - 70x^3 + 15x)$	…
6	$\frac{1}{16}(231x^6 - 315x^4 + 105x^2 - 5)$	…

$\cos[2(\varphi - 180°)] = \cos(2\varphi)$, $\cos[3(\varphi - 180°)] = -\cos(3\varphi)$, and $\cos[4(\varphi - 180°)] = \cos(4\varphi)$. The potential becomes

$$V(-x) = \frac{e}{r_0}\left\{1 - \frac{r}{r_0}2\beta_{11}^2 P_1^1(\cos\theta)\cos\varphi + \left(\frac{r}{r_0}\right)^2\left[-\frac{1}{2}P_2(\cos\theta) + 6\beta_{22}^2 P_2^2(\cos\theta)\cos 2\varphi\right] + \left(\frac{r}{r_0}\right)^3 2\left[\frac{3}{2}\beta_{31}^2 P_3^1(\cos\theta)\cos\varphi - 15\beta_{33}^2 2P_3^3(\cos\theta)\cos 3\varphi\right] + \left(\frac{r}{r_0}\right)^4 2\left[\frac{3}{8}P_4(\cos\theta) - 15\beta_{42}^2 P_4^2(\cos\theta)\cos 2\varphi + 210\beta_{44}^2 P_4^4(\cos\theta)\cos 4\varphi\right] + \cdots\right\} \quad (2.19)$$

Thus, the potential produced by two ligands on the x axis, obtained by the addition of the previous expressions (Equations 2.18 and 2.19), is more simple:

$$V(\pm x) = \frac{e}{r_0}\left\{2 + 2\left(\frac{r}{r_0}\right)^2\left[-\frac{1}{2}P_2(\cos\theta) + 6\beta_{22}^2 P_2^2(\cos\theta)\cos 2\varphi\right] + 2\left(\frac{r}{r_0}\right)^4 2\left[\frac{3}{8}P_4(\cos\theta) - 15\beta_{42}^2 P_4^2(\cos\theta)\cos 2\varphi + 210\beta_{44}^2 P_4^4(\cos\theta)\cos 4\varphi\right] + \cdots\right\}$$

The ligands located on the y axis have $\varphi_0 = 90°$ and $270°$, respectively. So, we finally obtain the contribution of the four ligands located in the (x,y) plane as follows:

$$V(\pm x, \pm y) = \frac{e}{r_0}\left\{4 - 2\left(\frac{r}{r_0}\right)^2 P_2(\cos\theta) + \frac{3}{2}\left(\frac{r}{r_0}\right)^4 P_4(\cos\theta) + \frac{1}{48}\left(\frac{r}{r_0}\right)^4 P_4^4(\cos\theta)\cos 4\varphi + \cdots\right\} \quad (2.20)$$

The other two ligands located on the z axis ($\theta_0 = 0$ and $180°$, respectively; $\varphi_0 = 0$) will give the potential:

$$V(\pm z) = \frac{e}{r_0}\left\{2 + 2\left(\frac{r}{r_0}\right)^2 P_2(\cos\theta) + 2\left(\frac{r}{r_0}\right)^4 P_4(\cos\theta) + \cdots\right\} \quad (2.21)$$

Now, we are able to write the expression of the *potential created in a point of spherical coordinates* r, θ, and φ *by six point charges located in octahedral positions:*

$$V(\pm x, \pm y, \pm z) = \frac{e}{r_0}\left\{6 + \left(\frac{r}{r_0}\right)^4\left[\frac{7}{2}P_4(\cos\theta) + \frac{1}{48}P_4^4(\cos\theta)\cos 4\varphi + \ldots\right]\right\} \Rightarrow$$

(2.22)

$$V(\pm x,\pm y,\pm z) = \frac{6e}{r_0} + \left(\frac{35e}{4r_0^5}\right)\left[\frac{2}{5}r^4 P_4(\cos\theta) + \frac{1}{420}r^4 P_4^4(\cos\theta)\cos 4\varphi\right] + \cdots$$

Discussions

- The first term of Equation 2.22 is a constant $6e/r_0$ and represents the ligand field in the center of the octahedron ($r = 0$).
- We denote the *potential constant* by the symbol D:

(2.23)
$$D = \left(\frac{35e}{4r_0^5}\right)$$

- The square brackets of Equation 2.22 can be rewritten in terms of Cartesian coordinates (x, y, z) of the d electron as

$$\left[x^4 + y^4 + z^4 - \frac{3}{5}r^4\right]$$

- In the perturbation method, when we apply the field potential to calculate the matrix elements $\langle \Psi_i | V | \Psi_j \rangle$, we will find that all vanish except for $k \leq 4$ (from the orthonormality property of spherical harmonics). This condition is valuable for $3d$-electrons where $l_1 = l_2 = 2$ that restricts the value of k to 0–4. For $4s$ electrons, $l_1 = l_2 = 0$ leaving only $k = 0$.

As a conclusion the short expression of the *electrostatic potential in octahedral symmetry that acts on the 3d-electron* of the central TM ion can be written as

(2.24)
$$V = \frac{6e}{r_0} + D\left[x^4 + y^4 + z^4 - \frac{3}{5}r^4\right]$$

The first term is the potential in the center of the octahedron (origin of coordinates). It is totally symmetric and it elevates all the energy levels of the atom by the same amount. The second term depends on the geometrical arrangement of the point charges and the geometry of orbitals and will split the energy levels as will be shown later. It is this term that we will take into account.

How strong is this crystal field? Let us write the expression of the potential at a point located at half the distance between one vertex and the center of the octahedron:

(2.25)
$$V(r_0/2) = \left(\frac{35e}{4r_0^5}\right)\left[\left(\frac{r_0}{2}\right)^4 - \frac{3}{5}\left(\frac{r_0}{2}\right)^4\right] = \frac{7e}{32r_0}$$

What is the electrostatic field intensity? We can obtain the expression of the field intensity from the potential expression:

(2.26)
$$\frac{\partial V}{\partial x} = \frac{\partial}{\partial x} D\left[x^4 + y^4 + z^4 - \frac{3}{5}r^4\right] = D\left(4x^3 - \frac{3}{5}4x^3\right) = D\frac{8}{5}x^3$$

Its value at a point located at half the distance between one vertex and the center of the octahedron will be

(2.27)
$$\left.\frac{\partial V}{\partial x}\right|_{r_0/2} = D\frac{8}{5}\left(\frac{r_0}{2}\right)^3 = \frac{7e}{4r_0^2}$$

For $e = 1.6 \cdot 10^{-19}$ C and $r_0 = 10^{-10}$ m, we use a correction factor of $9 \cdot 10^9$, in order to simplify the formula of the potential in Gaussian system of units (Gaussian units are still used due to the simple expressions of the main electromagnetic formulas), and by doing so, we will obtain the values:

(2.28)
$$V(r_0/2) = 9 \cdot 10^9 \cdot \frac{7 \cdot 1.6 \cdot 10^{-19}}{32 \cdot 10^{-10}} = 3.15 \text{ V}$$

(2.29)
$$\left.\frac{\partial V}{\partial x}\right|_{r0/2} = 9 \cdot 10^9 \cdot \frac{7 \cdot 1.6 \cdot 10^{-19}}{4 \cdot \left(10^{-10}\right)^2} = 252 \cdot 10^9 \frac{\text{V}}{\text{m}} = 2.52 \cdot 10^9 \frac{\text{V}}{\text{cm}}$$

The value of the potential is relevant enough. If we express the energy of the electron in this potential, we can find 3.15 eV, which is much less than the energy gap in ionic crystals. It is approximately the energy of a violet photon.

Comments

Ballhausen presented in his wonderful book another way to express the potential created by the octahedral crystalline field. His method is mainly based on group theory and we will present it here in short. He starts from the idea that the potential has two components: the first one is isotropic with spherical symmetry and the second one is weaker and anisotropic with octahedral symmetry. Only the anisotropic component called V_O may partially lift the degeneracy of d orbitals. The total potential produced by six charges located in the vertices of the octahedron must have the same symmetry as produced by ligands. It means the potential function is invariant to any symmetry operation of the $\boldsymbol{O}_h$ group. Since the first term is totally symmetric, results in V_O are also invariant to all symmetry operations; therefore, it must transform as a_{1g}.

The expansion of the anisotropic term of the potential can be done using the Legendre polynomials or even spherical harmonics, because they are related. The spherical harmonics are part of the wavefunctions, and thus they may be more useful. For reasons we already discussed, the maximum order of harmonics used in the expansion is 4. The Y_{4m} harmonics can be used as bases to generate a ninefold-reducible representation of the octahedral group. Then this representation can be decomposed into irreducible representations of the octahedral group. One of them must be the totally symmetric representation a_{1g}. Thus, there is a linear combination of spherical harmonics that will transform as a_{1g} and we have to find it.

We apply rotations C_4, C_2, and C_3 to the set of harmonics. Any α angle rotation around the z axis will transform only the $e^{im\varphi}$ term into $e^{im(\varphi+\alpha)} = e^{im\alpha}e^{im\varphi}$. For the particular rotations, we have the multiplication coefficients equal to $e^{im\pi/2}$ and $e^{im\pi}$, respectively. The C_3 axis is not collinear to the fourfold axis (z axis) and we cannot apply this simple result to it. It will be treated separately. You may also consider the C_3 axis as the main axis and finally must obtain the same result. Therefore, the effect of the C_4 rotation around the z axis may be written as follows:

$$C_4 \begin{bmatrix} Y_{44} \\ Y_{43} \\ Y_{42} \\ Y_{41} \\ Y_{40} \\ Y_{4-1} \\ Y_{4-2} \\ Y_{4-3} \\ Y_{4-4} \end{bmatrix} = \begin{bmatrix} Y_{44} \\ -iY_{43} \\ -Y_{42} \\ iY_{41} \\ Y_{40} \\ -iY_{4-1} \\ -Y_{4-2} \\ iY_{4-3} \\ Y_{4-4} \end{bmatrix}$$

because $e^{im\alpha} = \cos\alpha + i\sin\alpha$; so, the effect of rotations C_4 and C_2 will be given in Table 2.11.

Since only the functions Y_{44}, Y_{40}, and Y_{4-4} rest unchanged, results in their linear combination have the symmetry a_{1g}. Thus, the expansion of the anisotropic term may be written as

$$V_O = Y_{40} + aY_{44} + bY_{4-4}$$

TABLE 2.11 Effect of C_4 and C_2 Rotations on Spherical Harmonics

	Y_{44}	Y_{43}	Y_{42}	Y_{41}	Y_{40}	Y_{4-1}	Y_{4-2}	Y_{4-3}	Y_{4-4}
$e^{im\pi/2}$	1	$-i$	-1	i	1	$-i$	-1	i	1
$e^{im\pi}$	1				1				1

We apply the C_2 rotation to the set of these three harmonics:

$$C_2\begin{bmatrix} Y_{44} \\ Y_{40} \\ Y_{4-4} \end{bmatrix} = \begin{bmatrix} Y_{44} \\ Y_{40} \\ Y_{4-4} \end{bmatrix}$$

It results in $a = b$ and $V_O = Y_{40} + a(Y_{44} + Y_{4-4})$.
The term V_O also remains unchanged to the action of the C_3 rotation:

$$C_3V_O = V_O \quad \Rightarrow \quad C_3[Y_{40} + a(Y_{44} + Y_{4-4})] = Y_{40} + a(Y_{44} + Y_{4-4})$$

We express spherical harmonics in Cartesian coordinates starting from Equation 2.11:

$$\begin{aligned} Y_{44}(\theta,\varphi) &= (-1)^4\left[\frac{(2\cdot 4+1)(4-4)!}{4\pi(4+4)!}\right]^{1/2} P_4^4(\cos\theta)e^{i4\varphi} \\ &= \sqrt{\frac{9}{4\pi\cdot 8!}}105(1-\cos^2\theta)^2e^{4i\varphi} = \sqrt{\frac{9}{4\pi}}\sqrt{\frac{35}{128}}(\sin^2\theta)^2e^{4i\varphi} \\ &= \sqrt{\frac{9}{4\pi}}\sqrt{\frac{35}{128}}\frac{(r\sin\theta e^{i\varphi})^4}{r^4} = \sqrt{\frac{9}{4\pi}}\sqrt{\frac{35}{128}}\frac{(x+iy)^4}{r^4} \end{aligned}$$

$$Y_{4-4}(\theta,\varphi) = (-1)^4Y_{44}^*(\theta,\varphi) = \sqrt{\frac{9}{4\pi}}\sqrt{\frac{35}{128}}\frac{(x-iy)^4}{r^4}$$

and

$$\begin{aligned} Y_{40}(\theta,\varphi) &= (-1)^0\left[\frac{(2\cdot 4+1)(4-0)!}{4\pi(4+0)!}\right]^{1/2} P_4^0(\cos\theta)e^{i0\varphi} \\ &= \sqrt{\frac{9}{4\pi}}\frac{1}{8}(35\cos^4\theta - 30\cos^2\theta + 3) \\ &= \sqrt{\frac{9}{4\pi}}\frac{1}{8}\frac{(35r^4\cos^4\theta - 30r^2\cos^2\theta r^2 + 3r^4)}{r^4} \\ &= \sqrt{\frac{9}{4\pi}}\frac{1}{8}\frac{(35z^4 - 30z^2r^2 + 3r^4)}{r^4} \end{aligned}$$

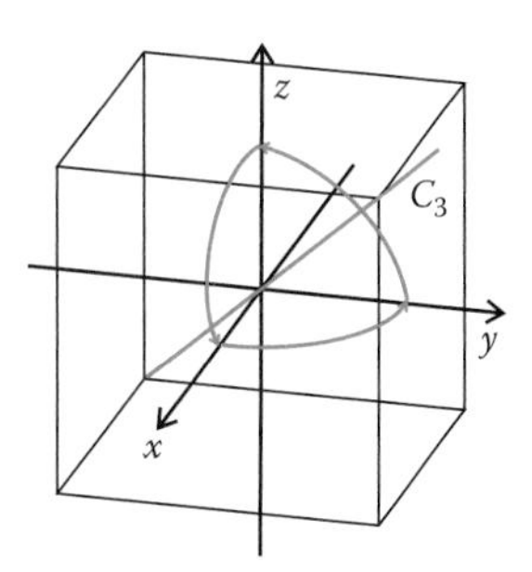

FIGURE 2.22 C_3 rotation around the principal diagonal of the cube (1, 1, 1) transforms coordinates as follows: $x \rightarrow y$, $y \rightarrow z$, and $z \rightarrow x$, respectively.

It is now easy to write the effect of the C_3 rotation around the principal diagonal of the cube on Cartesian coordinates only by simple inspection of Figure 2.22:

$$C_3\begin{bmatrix} x \\ y \\ z \end{bmatrix} = \begin{bmatrix} y \\ z \\ x \end{bmatrix}$$

We apply the effect of rotation on Cartesian coordinates into the octahedral potential term as follows:

$$C_3[Y_{40} + a(Y_{44} + Y_{4-4})] = C_3\left\{\sqrt{\frac{9}{4\pi}}\frac{1}{8}\frac{(35z^4 - 30z^2r^2 + 3r^4)}{r^4}\right.$$

$$\left. + a\left[\sqrt{\frac{9}{4\pi}}\sqrt{\frac{35}{128}}\frac{(x+iy)^4}{r^4} + \sqrt{\frac{9}{4\pi}}\sqrt{\frac{35}{128}}\frac{(x-iy)^4}{r^4}\right]\right\}$$

$$= \sqrt{\frac{9}{4\pi}}\frac{1}{8}\frac{(35x^4 - 30x^2r^2 + 3r^4)}{r^4}$$

$$+ a\sqrt{\frac{9}{4\pi}}\sqrt{\frac{35}{128}}\left[\frac{(y+iz)^4}{r^4} + \frac{(y-iz)^4}{r^4}\right]$$

$$\equiv \sqrt{\frac{9}{4\pi}}\frac{1}{8}\frac{(35z^4 - 30z^2r^2 + 3r^4)}{r^4}$$

$$+ a\sqrt{\frac{9}{4\pi}}\sqrt{\frac{35}{128}}\left[\frac{(x+iy)^4}{r^4} + \frac{(x-iy)^4}{r^4}\right]$$

Therefore,

$$\sqrt{\frac{9}{4\pi}}\frac{1}{8}3z^4 + a\sqrt{\frac{9}{4\pi}}\sqrt{\frac{35}{128}}\left[2z^4\right] \equiv \sqrt{\frac{9}{4\pi}}\frac{1}{8}(35z^4 - 30z^4 + 3z^4)$$

$$\Rightarrow \frac{1}{8}3 + a\sqrt{\frac{35}{128}}2 \equiv \frac{1}{8}8 \Rightarrow a = \sqrt{\frac{5}{14}}$$

Thus,

$$V_O = Y_{40} + \sqrt{\frac{5}{14}}(Y_{44} + Y_{4-4})$$

The last method appears to be simpler, but it does not provide information about the coefficient of the term, even if this cannot be determined and remains as a parameter to be determined experimentally also for the first method. The expansion of the potential in terms of spherical harmonics is useful during the next part where one calculates the contribution of the potential to the energy of each ionic state that is described by spherical harmonics. Their orthogonality property helps us to simplify the calculus. Using the expression of spherical harmonics, the potential can also be converted into Cartesian coordinates.

2.4.3 MATRIX ELEMENTS FOR $3d^1$ CONFIGURATION IN $\boldsymbol{O}_h$ SYMMETRY

The ground state of the free ion with $3d^1$ configuration was found to be ^2d in a single-electron notation or ^2D in a multielectron notation (Table

2.6). We know the potential expression of the crystal field. We also know that the crystal field contributes to electron energy lower than other terms of the free ion; so, the Hamiltonian of the crystal field (H_{CF}) can be considered as a small perturbation to the free ion Hamiltonian. We can calculate the contribution of H_{CF} to electron energy by means of the perturbation theory. The ligand field formalism follows a first-order perturbation treatment. Employing crystal field theory requires calculations of matrix elements between wavefunctions:

$$H_{ij} = \langle \Psi_i | H_{CF} | \Psi_j \rangle = \langle \Psi_i | eV | \Psi_j \rangle \tag{2.30}$$

We take the free ion wavefunctions of $3d^1$ configuration (nonperturbed wavefunctions) as the starting wavefunctions:

$$\Psi i = R_{3d}(r_i) Y_{2m}(\theta_i, \varphi_i) \tag{2.31}$$

where m_i takes values 2, 1, 0, –1, and –2. Thus, we may now proceed to find the Hamiltonian matrix elements for the interaction of the ion with the surrounding ligands. The matrix elements will be calculated as

$$\begin{aligned} H_{ij} &= \langle \Psi_i | V | \Psi_j \rangle = \iiint \Psi_i^* eV \Psi_j \, d\tau \\ &= \iiint R_{3d}(r) Y_{2mi}^*(\theta,\varphi) eV R_{3d}(r) Y_{2mj}(\theta,\varphi) r^2 dr \sin\theta \, d\theta d\varphi \end{aligned} \tag{2.32}$$

We leave out the first term of V_{CF} which gives an equal contribution to all levels.

$$\begin{aligned} H_{ij} &= \langle \Psi_i | V | \Psi_j \rangle = \iiint \Psi_i^* eV \Psi_j d\tau \\ &= \iiint R_{3d}^2(r) Y_{2mi}^*(\theta,\varphi) eD \left[\frac{2}{5} r^4 P_4(\cos\theta) + \frac{1}{420} r^4 P_4^4(\cos\theta)\cos 4\varphi \right] \\ &\quad \times Y_{2mj}(\theta,\varphi) r^2 dr \sin\theta \, d\theta d\varphi \end{aligned}$$

Then we treat the radial part separately from the angular part:

$$\begin{aligned} H_{ij} &= eD \int R_{3d}^2(r) r^4 r^2 dr \iint \left[\frac{2}{5} P_4(\cos\theta) + \frac{1}{420} P_4^4(\cos\theta)\cos 4\varphi \right] \\ &\quad \times Y_{2mi}^* Y_{2mj} \sin\theta \, d\theta d\varphi \end{aligned} \tag{2.33}$$

We do not know the exact expression of radial functions; so, we denote by symbol q the entire radial integral:

$$q = \frac{2e}{105} \int R_{3d}^2(r) r^4 r^2 \, dr = \frac{2e\overline{r^4}}{105} \tag{2.34}$$

and then incorporate Equation 2.34 into Equation 2.33:

(2.35)

$$H_{ij} = Dq\iint \frac{105}{2}\left[\frac{2}{5}P_4(\cos\theta) + \frac{1}{420}P_4^4(\cos\theta)\cos 4\varphi\right]Y_{2m_i}^{*}Y_{2m_j}\sin\theta\, d\theta d\varphi$$

We remember the orthonormality condition of spherical harmonics given by Equation 2.12. The conclusion of the orthonormality condition is that the matrix elements vanish, except those where $m_i = m_j$. Thus, we can apply the one-electron spherical harmonics given by Equation 2.11 only 5 times: $m_i = m_j = 2, 1, 0, -1, -2$.

We start to evaluate the matrix element for $m_i = m_j = 1$. The spherical harmonics will be

(2.36)

$$Y_{21}(\theta,\varphi) = (-1)^1\left[\frac{(2\cdot 2+1)(2-1)!}{4\pi(2+1)!}\right]^{1/2} P_2^1(\cos\theta)e^{i\varphi} = -\left[\frac{5}{24\pi}\right]^{1/2} P_2^1(\cos\theta)e^{i\varphi}$$

Thus, the matrix element will be written as

(2.37)

$$H_{11} = Dq\iint \frac{105}{2}\left[\frac{2}{5}P_4(\cos\theta) + \frac{1}{420}P_4^4(\cos\theta)\cos 4\varphi\right]\left[\frac{5}{24\pi}\right]\left[P_2^1(\cos\theta)\right]^2$$
$$\times \sin\theta\, d\theta d\varphi$$

Then the quantity will be written as the sum of two integrals:

(2.38)

$$H_{11} = Dq\iint \frac{105}{2}\left[\frac{2}{5}P_4(\cos\theta)\right]\left[\frac{5}{24\pi}\right]\left[P_2^1(\cos\theta)\right]^2\sin\theta\, d\theta d\varphi$$
$$+ Dq\iint \frac{105}{2}\left[\frac{1}{420}P_4^4(\cos\theta)\cos 4\varphi\right]\left[\frac{5}{24\pi}\right]\left[P_2^1(\cos\theta)\right]^2\sin\theta\, d\theta d\varphi$$
$$= Dq\frac{105}{24\pi}\int_0^{\pi} P_4(\cos\theta)\left[P_2^1(\cos\theta)\right]^2\sin\theta\, d\theta\int_0^{2\pi} d\varphi$$
$$+ Dq\frac{5}{12\pi}\int_0^{\pi} P_4^4(\cos\theta)\left[P_2^1(\cos\theta)\right]^2\sin\theta\, d\theta\int_0^{2\pi}\cos 4\varphi\, d\phi$$

The second term equals zero because $\int_0^{2\pi}\cos 4\varphi\, d\varphi = 0$. To evaluate the first term, we must look for expressions of the Legendre polynomials and functions in Table 2.10 and we can write

(2.39)

$$H_{11} = Dq\frac{105}{24\pi}2\pi\int_0^{\pi}\frac{1}{8}(35\cos^4\theta - 30\cos^2\theta + 3)\left[3(1-\cos^2\theta)^{1/2}\cos\theta\right]^2\sin\theta\,d\theta$$

$$= Dq\frac{105}{96}\cdot 9\int_0^{\pi}(35\cos^4\theta - 30\cos^2\theta + 3)(1-\cos^2\theta)\cos^2\theta\sin\theta\,d\theta$$

$$= -Dq\frac{105}{96}\cdot 9\int_1^{-1}(-35x^8 + 65x^6 - 33x^4 + 3x^2)dx$$

We used a suitable change of variable $\cos\theta = x$, and thus we can find the matrix element:

(2.40)
$$H_{11} = -Dq\frac{105}{96}\cdot 9\cdot\frac{128}{9\cdot 35} = -4Dq$$

This is an important result in what follows.

We now evaluate the matrix element for $m_i = m_j = 0$:

(2.41)

$$H_{00} = Dq\iint\frac{105}{2}\left[\frac{2}{5}P_4(\cos\theta) + \frac{1}{420}P_4^4(\cos\theta)\cos 4\varphi\right]Y_{20}^*Y_{20}\sin\theta\,d\theta d\varphi$$

where

(2.42)

$$Y_{20}(\theta,\varphi) = (-1)^0\left[\frac{(2\cdot 2+1)(2-0)!}{4\pi(2+0)!}\right]^{\frac{1}{2}}P_2^0(\cos\theta)e^{i0\varphi} = \left[\frac{5}{4\pi}\right]^{\frac{1}{2}}P_2(\cos\theta)$$

Thus,

(2.43)

$$H_{00} = Dq\iint\frac{105}{2}\left[\frac{2}{5}P_4(\cos\theta) + \frac{1}{420}P_4^4(\cos\theta)\cos 4\varphi\right]\frac{5}{4\pi}\frac{1}{4}\left[3\cos^2\theta - 1\right]^2$$
$$\times\sin\theta\,d\theta d\varphi$$

$$= Dq\frac{105\cdot 2\cdot 5}{2\cdot 5\cdot 16\pi}\int_0^{\pi}P_4(\cos\theta)\left[3\cos^2\theta - 1\right]^2\sin\theta\,d\theta\int_0^{2\pi}d\varphi$$

$$+ Dq\frac{105\cdot 5}{2\cdot 420\cdot 16\pi}\int_0^{\pi}P_4^4(\cos\theta)\left[3\cos^2\theta - 1\right]^2\sin\theta\,d\theta\int_0^{2\pi}\cos 4\varphi\,d\varphi$$

The second term is zero as shown previously. Then

(2.44)

$$\begin{aligned} H_{00} &= Dq\frac{105}{8}\int_0^{\pi}\frac{1}{8}(35\cos^4\theta - 30\cos^2\theta + 3)\left[3\cos^2\theta - 1\right]^2 \sin\theta\, d\theta \\ &= Dq\frac{105}{64}\int_0^{\pi}(315\cos^8\theta - 480\cos^6\theta + 242\cos^4\theta - 48\cos^2\theta + 3)\sin\theta\, d\theta \\ &= -Dq\frac{105}{64}\int_1^{-1}(315x^8 - 480x^6 + 242x^4 - 48x^2 + 3)dx \end{aligned}$$

Finally, the matrix element will be given by the expression:

(2.45) $$\mathrm{H}_{00} = 6Dq$$

We suggest to the reader to find all matrix elements to be nonzero by repeating the previous treatment and we present here just the results:

(2.46)
$$\begin{cases} H_{22} = H_{-2-2} = Dq \\ H_{2-2} = H_{-22} = 5Dq \\ H_{11} = H_{-1-1} = -4Dq \\ H_{00} = 6Dq \end{cases}$$

The product Dq is a parameter of the crystal field, called by Van Vleck the *field strength parameter*, which will always be semiempirically determined and not calculated, because we do not know exactly the radial function R_{3d}.

Using the results obtained earlier (Equation 2.46), we can already write the Hamiltonian for one 3*d*-electron moving in the potential produced by six surrounding point charges located in octahedral positions. The *perturbation matrix of the weak crystalline field of* $\mathbf{O}_h$ *symmetry* can be written as follows:

(2.47)

m	2	1	0	−1	−2
2	Dq	0	0	0	$5Dq$
1	0	$-4Dq$	0	0	0
0	0	0	$6Dq$	0	0
−1	0	0	0	$-4Dq$	0
−2	$5Dq$	0	0	0	Dq

2.4.4 ENERGY SPLITTING IN O_h SYMMETRY OF 3*d* ORBITALS OF ONE-ELECTRON CONFIGURATION

The next moment is to solve the *secular equation* or the characteristic equation that is $\det(\hat{H} - \varepsilon\hat{E}) = 0$, where $\hat{E}$ is the unit matrix with elements δ_{ij}, to find the eigenvalues of the Hamiltonian:

$$\begin{vmatrix} Dq-\varepsilon & 0 & 0 & 0 & 5Dq \\ 0 & -4Dq-\varepsilon & 0 & 0 & 0 \\ 0 & 0 & 6Dq-\varepsilon & 0 & 0 \\ 0 & 0 & 0 & -4Dq-\varepsilon & 0 \\ 5Dq & 0 & 0 & 0 & Dq-\varepsilon \end{vmatrix} = 0 \tag{2.48}$$

Since the matrix has nondiagonal terms connecting the states with $m_i = +2$, $m_j = -2$, we have to apply the diagonalization procedure to rearrange the determinant in the smaller blocks on the diagonal:

$$\begin{vmatrix} Dq-\varepsilon & 5Dq & 0 & 0 & 0 \\ 5Dq & Dq-\varepsilon & 0 & 0 & 0 \\ 0 & 0 & 6Dq-\varepsilon & 0 & 0 \\ 0 & 0 & 0 & -4Dq-\varepsilon & 0 \\ 0 & 0 & 0 & 0 & -4Dq-\varepsilon \end{vmatrix} = 0 \tag{2.49}$$

Thus, the energy values will be given by:

$$\begin{aligned} (Dq-\varepsilon)^2 - (5Dq)^2 = 0 &\Rightarrow \varepsilon_1 = 6Dq \text{ and } \varepsilon_2 = -4Dq \\ 6Dq - \varepsilon = 0 &\Rightarrow \varepsilon_3 = 6Dq \\ (-4Dq-\varepsilon)^2 = 0 &\Rightarrow \varepsilon_{4,5} = -4Dq \end{aligned} \tag{2.50}$$

So, the solutions are $6Dq$ (2 times) and $-4Dq$ (3 times).

It is time to remember that the expression we have deduced for the potential created by six point charges, which are located in octahedral positions, has two terms (Equations 2.22 and 2.24). The first term is a constant $6e/r_0$ and represents the ligand field in the center of the octahedron. It is totally symmetric and *elevates all the energy levels of the atom by the same amount*. Thus, the matrix elements we have calculated do not contain the contribution of the first term.

Finally, the perturbation energy will have two values that may be written as

$$\begin{aligned} \varepsilon_1 &= \varepsilon_0 + 6Dq \quad \text{(twofold degenerate)} \\ \varepsilon_2 &= \varepsilon_0 - 4Dq \quad \text{(threefold degenerate)} \end{aligned} \tag{2.51}$$

where ε_0 is the part of energy origins in the first term that is common.

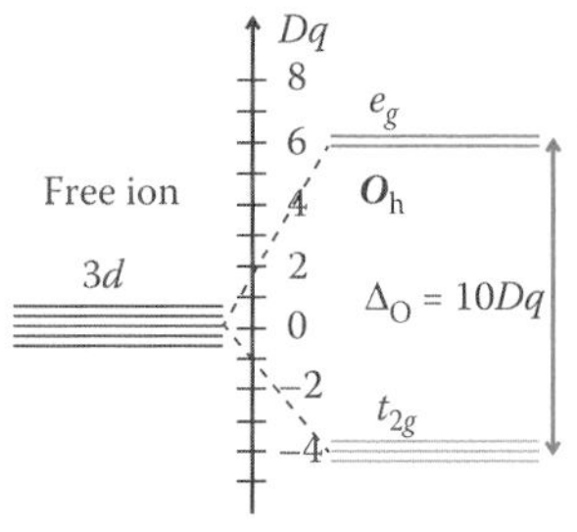

FIGURE 2.23 The energy level of one *d* electron in the free ion will be split into two energy levels by the crystalline field with $\boldsymbol{O}_h$ symmetry.

Which is the main conclusion? The energy level of one d electron in the free ion, which is fivefold degenerate, will be split into two energy levels; the lower level is threefold degenerate and the higher level is twofold degenerate, when the ion is introduced into an electric field with octahedral symmetry (Figure 2.23). The illustration from Figure 2.23 is similar to that from Figure 2.20, obtained only by group theory considerations. Quantum mechanics offers supplementary information on the relative position of two levels (10*Dq*) but not on an exact value of it! The extent of splitting in energies is represented by the *splitting energy* Δ_O (or parameter of the crystalline field). The subscript is employed to indicate the geometry of the surrounding point charges, for example, "O" for octahedral.

What orbitals correlate to these energy values? The group theory uses character tables to show that the orbitals d_{xz}, d_{yz}, and d_{xy} form a basis of t_{2g} representation, while orbitals $d_{x^2-y^2}$ and $d_{2z^2-x^2-y^2}$ form a basis of e_g representation. There are also valuable considerations related to electrostatic repulsion with anionic neighbors discussed in Section 2.4.1. The crystalline field generated by six point charges located in the octahedron vertices *partly removes the fivefold degeneracy* of the free ion state: the lower state has a threefold degeneracy and the higher state has a twofold degeneracy.

The first term of the potential expression (Equation 2.24), which has a spherical symmetry, elevates all levels with the same quantity (ε_0). Its contribution is larger than the contribution of the second term coming from the octahedral disposition of six negative charges. Thus, the total effect of the crystalline field is illustrated in Figure 2.24. There are three orbitals with the same energy ε_2. They can be filled with six electrons and the total energy will be

$$6(\varepsilon_0 - 4Dq) = 6\varepsilon_0 - 24Dq \quad (2.52a)$$

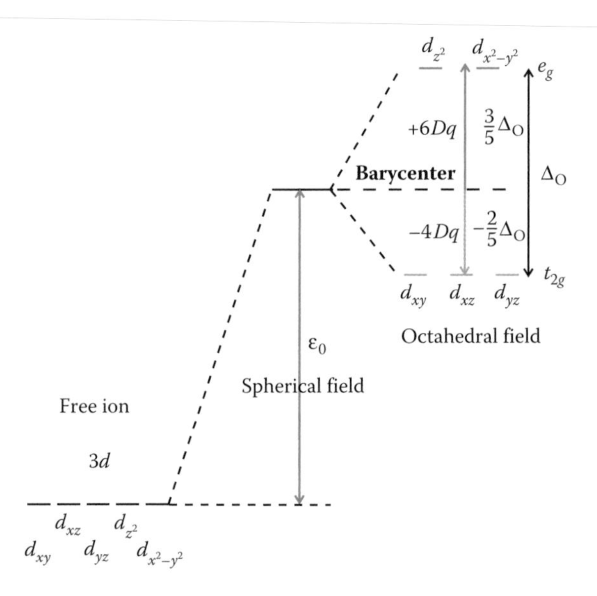

FIGURE 2.24 The total effect of the $\boldsymbol{O}_h$ crystalline field on *d* orbitals.

The two orbitals with higher energy ε_1 can be filled with four electrons and their total energy will be

(2.52b) $$4(\varepsilon_0 + 6Dq) = 4\varepsilon_0 + 24Dq$$

The total energy of the system will be obtained by the addition of the two values (Equations 2.52a and 2.52b):

(2.52c) $$6(\varepsilon_0 - 4Dq) + 4(\varepsilon_0 + 6Dq) = 10\varepsilon_0$$

Thus, the splitting will be done in such a manner that the position of the barycenter (the average energy of orbitals in the complex) is not changed during splitting (the *rule of barycenter conservation*). This rule is valuable for all symmetries of crystalline fields.

2.4.5 SPLITTING OF ATOMIC ORBITALS IN OCTAHEDRAL SYMMETRY

We have just solved the problem of splitting in the energy level of the *d* orbitals in octahedral symmetry. What happens with other types of orbitals? We discuss again the case of $\boldsymbol{O}_h$ symmetry because it is the highest symmetry we can find in ionic complexes.

The general method is to find the characters of representation generated by each type of orbital. Then this representation will be reduced to irreducible representations of the group and that is finished. The irreducible representations give the number and names of the new levels obtained by splitting of the original level in the crystalline field.

To find characters of the matrices is to apply each symmetry operation to each wavefunction describing the orbital type of work. This way is very long and difficult, but fortunately, group theory offers shortcuts.

First of all, we must look at the character table of the $\boldsymbol{O}_h$ group given by Table 2.12. There are 10 rows: the first five rows correspond to even representations (gerade) and the last five rows correspond to the same

TABLE 2.12 Character Table for the O_h Point Group

O_h	E	$8C_3$	$6C_2$	$6C_4$	$3C_2 = C_4^2$	I	$8S_6$	$6\Sigma_d$	$6S_4$	$3\Sigma_h$	Linear, Rotations	Quadratic
A_{1g}	1	1	1	1	1	1	1	1	1	1		$x^2 + y^2 + z^2$
A_{2g}	1	1	−1	−1	1	1	1	−1	−1	1		
E_g	2	−1	0	0	2	2	−1	0	0	2		$(2z^2 - x^2 - y^2, x^2 - y^2)$
T_{1g}	3	0	−1	1	−1	3	0	−1	1	−1	(R_x, R_y, R_z)	
T_{2g}	3	0	1	−1	−1	3	0	1	−1	−1		(xz, yz, xy)
A_{1u}	1	1	1	1	1	−1	−1	−1	−1	−1		
A_{2u}	1	1	−1	−1	1	−1	−1	1	1	−1		
E_u	2	−1	0	0	2	−2	1	0	0	−2		
T_{1u}	3	0	−1	1	−1	−3	0	1	−1	1	(x, y, z)	
T_{2u}	3	0	1	−1	−1	−3	0	−1	1	1		

but odd representations (ungerade). So, we will concentrate on the gerade representation. The characters of the first five columns are equal to the characters of the last five columns. It means that we can confine our work to the area marked in gray. It is the character table of the ***O*** group. It contains only rotations and we know how to find the matrix of a rotation.

Second, the wavefunction describing any orbital can be written as the product of three functions that depends on only one coordinate: $R(r)$, $\Theta(\theta)$, and $\Phi(\varphi)$. The radial function is invariant to any symmetry operation. If we choose a proper orientation of the coordinate system in such a way that the rotation will also be done around the z axis, the function $\Theta(\theta)$ will be invariant. Thus, we reduce working with the function $\Phi(\varphi) = e^{im\varphi}$.

In conclusion, we will apply each rotation of the ***O*** group to the function Φ to find the characters of the generated representation.

Let us start with the *orbital s*. The shape of s orbitals is spherical and any symmetry operation leaves it unchanged. Thus, all matrices of the generated representation will be equal to the unit matrix and all characters equal 1. Moreover, the s orbital wavefunction is even (gerade). Therefore, the wavefunction of the s orbital is a basis of the totally symmetric representation called a_{1g} in terms of the single-electron notation. The conclusion is that the *orbital s* is not split in the octahedral field. We may extrapolate that the orbital s is not split in any symmetry field, because the representation a_{1g} (also called a_1 or a) is present in all symmetry groups. The character table of the point group that you can find in any book always contains this information in the last column: the function $x^2 + y^2 + z^2$ (or shorter $x^2 + y^2$), which is the sphere equation in Cartesian coordinates, forms a basis of the totally symmetric irreducible representation.

We may continue with the *orbital p*. Just looking over the character table, we know that the set of all p orbitals (p_x, p_y, p_z) forms a basis for the t_{1u} irreducible representation (written again in a single-electron notation). The p orbitals are odd (ungerade) and will be found in the ungerade part of the table. It is better to apply the method of character calculation as an exercise. The orbital number of the p electron is $l = 1$. Then the magnetic number will have three values: 1, 0, and −1. The functions corresponding to all three orbitals form a one-column matrix. The rotation of the angle α around the z axis will transform the one-column matrix of functions corresponding to all three orbitals as follows:

$$\text{Initial state}\begin{bmatrix} e^{i\varphi} \\ e^{0} \\ e^{-i\varphi} \end{bmatrix} \xrightarrow{\alpha \text{ angle rotation around } z \text{ axis}} \begin{bmatrix} e^{i(\varphi+\alpha)} \\ e^{0} \\ e^{-i(\varphi+\alpha)} \end{bmatrix}\text{final state}$$

What is the matrix of this transformation?

$$\begin{bmatrix} a_{11} & a_{12} & a_{13} \\ a_{21} & a_{22} & a_{23} \\ a_{31} & a_{32} & a_{33} \end{bmatrix}\begin{bmatrix} e^{i\varphi} \\ e^{0} \\ e^{-i\varphi} \end{bmatrix} = \begin{bmatrix} e^{i(\varphi+\alpha)} \\ e^{0} \\ e^{-i(\varphi+\alpha)} \end{bmatrix}$$

$$\begin{bmatrix} a_{11}e^{i\varphi} + a_{12} + a_{13}e^{-i\varphi} \\ a_{21}e^{i\varphi} + a_{22} + a_{23}e^{-i\varphi} \\ a_{31}e^{i\varphi} + a_{32} + a_{33}e^{-i\varphi} \end{bmatrix} = \begin{bmatrix} e^{i(\varphi+\alpha)} \\ e^{0} \\ e^{-i(\varphi+\alpha)} \end{bmatrix}$$

$$\begin{aligned} a_{11}e^{i\varphi} + a_{12} + a_{13}e^{-i\varphi} &= e^{i(\varphi+\alpha)} \\ a_{21}e^{i\varphi} + a_{22} + a_{23}e^{-i\varphi} &= 1 \\ a_{31}e^{i\varphi} + a_{32} + a_{33}e^{-i\varphi} &= e^{-i(\varphi+\alpha)} \end{aligned}$$

(2.53)
$$R(\alpha) = \begin{bmatrix} e^{i\alpha} & 0 & 0 \\ 0 & e^{i0} & 0 \\ 0 & 0 & e^{-i\alpha} \end{bmatrix}$$

By simple extrapolation, the matrix corresponding to the orbital quantum number l will have the $(2l + 1) \times (2l + 1)$ dimension as follows:

(2.54)
$$R(\alpha) = \begin{bmatrix} e^{il\alpha} & 0 & \ldots & 0 & 0 \\ 0 & e^{i(l-1)\alpha} & \ldots & 0 & 0 \\ \ldots & \ldots\ldots & \ldots & \ldots & \ldots \\ 0 & 0 & \ldots & e^{-i(l-1)\alpha} & 0 \\ 0 & 0 & \ldots & 0 & e^{-il\alpha} \end{bmatrix}$$

Then the character of the rotation matrix will be given by the sum of the geometric progression:

(2.55)
$$\chi(\alpha) = \frac{\sin\left[(2l+1)\dfrac{\alpha}{2}\right]}{\sin\dfrac{\alpha}{2}}, \quad \text{where } \alpha \neq 0$$

The character of the matrix corresponding to identity operation ($\alpha = 0$) equals $(2l + 1)$. Relationship (2.55) will be applied for the particular cases of ***O*** group rotations and the results are shown in Table 2.13.

$$\chi_p(\alpha) = \frac{\sin\dfrac{3\alpha}{2}}{\sin\dfrac{\alpha}{2}} \Rightarrow \chi_p(C_3) = \frac{\sin\dfrac{3\cdot 120}{2}}{\sin\dfrac{120}{2}} = 0;$$

$$\chi_p(C_2) = \frac{\sin\dfrac{3\cdot 180}{2}}{\sin\dfrac{180}{2}} = -1; \quad \chi_p(C_4) = \frac{\sin\dfrac{3\cdot 90}{2}}{\sin\dfrac{90}{2}} = 1$$

TABLE 2.13 Characters of Rotation Matrices of the *O* Group

Orbital Type	*l*	$\chi(E)$	$\chi(C_3)$	$\chi(C_2)$	$\chi(C_4)$	$\chi(C_2)$
s	**0**	1	1	1	1	1
p	**1**	3	0	–1	1	–1
d	**2**	5	–1	1	–1	1
f	**3**	7	1	–1	–1	–1
g	**4**	9	0	1	1	1
h	**5**	11	–1	–1	1	–1
i	**6**	13	1	1	–1	1

$$\chi_d(\alpha) = \frac{\sin\frac{5\alpha}{2}}{\sin\frac{\alpha}{2}} \Rightarrow \chi_d(C_3) = \frac{\sin\frac{5\cdot 120}{2}}{\sin\frac{120}{2}} = -1;$$

$$\chi_d(C_2) = \frac{\sin\frac{5\cdot 180}{2}}{\sin\frac{180}{2}} = 1;\quad \chi_d(C_4) = \frac{\sin\frac{5\cdot 90}{2}}{\sin\frac{90}{2}} = -1$$

$$\chi_f(\alpha) = \frac{\sin\frac{7\alpha}{2}}{\sin\frac{\alpha}{2}} \Rightarrow \chi_f(C_3) = \frac{\sin\frac{7\cdot 120}{2}}{\sin\frac{120}{2}} = 1;$$

$$\chi_f(C_2) = \frac{\sin\frac{7\cdot 180}{2}}{\sin\frac{180}{2}} = -1;\quad \chi_f(C_4) = \frac{\sin\frac{7\cdot 90}{2}}{\sin\frac{90}{2}} = -1$$

$$\chi_g(\alpha) = \frac{\sin\frac{9\alpha}{2}}{\sin\frac{\alpha}{2}} \Rightarrow \chi_g(C_3) = \frac{\sin\frac{9\cdot 120}{2}}{\sin\frac{120}{2}} = 0;$$

$$\chi_g(C_2) = \frac{\sin\frac{9\cdot 180}{2}}{\sin\frac{180}{2}} = 1;\quad \chi_g(C_4) = \frac{\sin\frac{9\cdot 90}{2}}{\sin\frac{90}{2}} = 1$$

$$\chi_h(\alpha) = \frac{\sin\frac{11\alpha}{2}}{\sin\frac{\alpha}{2}} \Rightarrow \chi_h(C_3) = \frac{\sin\frac{11\cdot 120}{2}}{\sin\frac{120}{2}} = -1;$$

$$\chi_h(C_2) = \frac{\sin\frac{11\cdot 180}{2}}{\sin\frac{180}{2}} = -1;\quad \chi_h(C_4) = \frac{\sin\frac{11\cdot 90}{2}}{\sin\frac{90}{2}} = 1$$

$$\chi_i(\alpha) = \frac{\sin\frac{13\alpha}{2}}{\sin\frac{\alpha}{2}} \Rightarrow \chi_i(C_3) = \frac{\sin\frac{13 \cdot 120}{2}}{\sin\frac{120}{2}} = 1;$$

$$\chi_i(C_2) = \frac{\sin\frac{13 \cdot 180}{2}}{\sin\frac{180}{2}} = 1; \quad \chi_i(C_4) = \frac{\sin\frac{13 \cdot 90}{2}}{\sin\frac{90}{2}} = -1$$

Table 2.12 shows that the representations of the $\boldsymbol{O}_\text{h}$ group have dimensions equal to 1, 2, and 3. Thus, any generated representation with dimension >3 is reducible to the irreducible representations of the group. The decomposition will show the splitting of the corresponding level in the crystalline field of octahedral symmetry. Therefore, we must now apply Equation 1.77 to find the number of irreducible representations Γ_i that appears in a reducible representation.

Let us check the method of decomposition for the *d* orbitals, where we already know the results. The order of the $\boldsymbol{O}$ group is $h = 24$, given by the number of symmetry operations. The number of times each irreducible representation is included in the reducible representation generated by the effect of rotations on *d* orbitals will be

$$a_1 = \frac{1}{24}[1 \cdot (1)(5) + 8 \cdot (1)(-1) + 6 \cdot (1)(1) + 6 \cdot (1)(-1) + 3 \cdot (1)(1)] = 0$$

$$a_2 = \frac{1}{24}[1 \cdot (1)(5) + 8 \cdot (1)(-1) + 6 \cdot (-1)(1) + 6 \cdot (-1)(-1) + 3 \cdot (1)(1)] = 0$$

$$e = \frac{1}{24}[1 \cdot (2)(5) + 8 \cdot (-1)(-1) + 6 \cdot (0)(1) + 6 \cdot (0)(-1) + 3 \cdot (2)(1)] = 1$$

$$t_1 = \frac{1}{24}[1 \cdot (3)(5) + 8 \cdot (0)(-1) + 6 \cdot (-1)(1) + 6 \cdot (1)(-1) + 3 \cdot (-1)(1)] = 0$$

$$t_2 = \frac{1}{24}[1 \cdot (3)(5) + 8 \cdot (0)(-1) + 6 \cdot (1)(1) + 6 \cdot (-1)(-1) + 3 \cdot (-1)(1)] = 1$$

Therefore, taking into account the symmetry to inversion of the *d* orbitals (gerade), the representation of *d* can be decomposed into irreducible representations as follows:

$$\gamma_d = e_g + t_{2g}$$

We will repeat the procedure for *f orbitals*:

$$a_1 = \frac{1}{24}[1 \cdot (1)(7) + 8 \cdot (1)(1) + 6 \cdot (1)(-1) + 6 \cdot (1)(-1) + 3 \cdot (1)(-1)] = 0$$

TABLE 2.14 Splitting of Orbitals in the O_h Symmetry of a Weak Field

Orbital Type	l	Orbital Degeneracy	Irreducible Representations
s	0	1	a_{1g}
p	1	3	t_{1u}
d	2	5	$e_g + t_{2g}$
f	3	7	$a_{2u} + t_{1u} + t_{2u}$
g	4	9	$a_{1g} + e_g + t_{1g} + t_{2g}$
h	5	11	$e_u + 2t_{1u} + t_{2u}$
i	6	13	$a_{1g} + a_{2g} + e_g + t_{1g} + 2t_{2g}$

$$a_2 = \frac{1}{24}[1 \cdot (1)(7) + 8 \cdot (1)(1) + 6 \cdot (-1)(-1) + 6 \cdot (-1)(-1) + 3 \cdot (1)(-1)] = 1$$

$$e = \frac{1}{24}[1 \cdot (2)(7) + 8 \cdot (-1)(1) + 6 \cdot (0)(-1) + 6 \cdot (0)(-1) + 3 \cdot (2)(-1)] = 0$$

$$t_1 = \frac{1}{24}[1 \cdot (3)(7) + 8 \cdot (0)(1) + 6 \cdot (-1)(-1) + 6 \cdot (1)(-1) + 3 \cdot (-1)(-1)] = 1$$

$$t_2 = \frac{1}{24}[1 \cdot (3)(7) + 8 \cdot (0)(1) + 6 \cdot (1)(-1) + 6 \cdot (-1)(-1) + 3 \cdot (-1)(-1)] = 1$$

Therefore, taking into account the f orbitals that are antisymmetric with respect to inversion (ungerade), the representation generated by rotation of the f orbitals can be decomposed into odd irreducible representations as follows:

$$\gamma_f = a_{2u} + t_{1u} + t_{2u}$$

As an exercise, the reader can obtain the decomposition for g, h, and i orbitals. Even though we do not use the orbitals of g, h, and i type in practice, the decomposition results will be used later for the multielectron states of G, H, and I type. The splitting of each type of orbital in the octahedral field is shown in Table 2.14.

It should be noted that the sum of dimensions of all irreducible representations is equal to the total orbital degeneracy. This is a way to control whether the decomposition is complete.

2.5 EFFECT OF A WEAK CRYSTALLINE FIELD OF LOWER SYMMETRIES

2.5.1 EFFECT OF LOWER SYMMETRY ON IRREDUCIBLE REPRESENTATIONS

The octahedron is the most common shape of a molecular complex. The highest symmetry is that of the $\boldsymbol{O}_h$ group ($m3m$) with order 48 (the total number of symmetry operations). The next symmetry elements belong to an octahedron:

- Three c_4 axes that pass through opposite vertices of the octahedron (Figure 2.25a). They generate six C_4 rotations and $3C_2(= C_4^2)$ rotations.
- Four c_3 axes that pass through opposite faces of the octahedron (Figure 2.25b). They generate eight C_3 rotations.
- Six c_2 axes that pass through the midpoint of opposite edges of the octahedron (Figure 2.25b). They generate six C_2 rotations.
- Inversion center i.
- Three s_4 axes coincide with the c_4 axes (Figure 2.25a).
- Four s_6 axes coincide with the c_3 axes (Figure 2.25b).
- Three σ_h planes. Each plane contains four vertices of the octahedron and is perpendicular to a c_4 axis (Figure 2.25c).
- Six planes called σ_d, each of them containing a c_3 axis (Figure 2.25d).

All symmetry operations of the $\boldsymbol{O}_h$ group are shown in Table 2.12. We can lower the symmetry of the complex by distortion of the local geometry and the $\boldsymbol{O}_h$ group will be reduced to one of its subgroups. The order of every subgroup must be the divisor of 48:24, 16, 12, 8, 6, 4, 3, 2, and 1. This will be the control key for the next task. The subgroups of $\boldsymbol{O}_h$ are generated by removing certain symmetry elements.

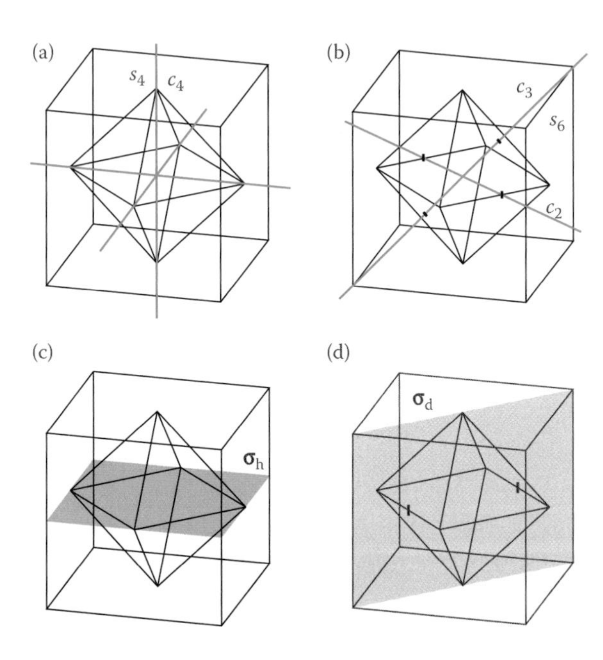

FIGURE 2.25 The main symmetry elements of the $\boldsymbol{O}_h$ group. Inversion center is the point where all rotation axes and planes intersect. Some of the rotation axes (c_3 and c_2 type) and planes (σ_h and σ_d) were not represented to simplify the illustration.

1. **O** subgroup. We have already used it. It is obtained from O_h by removing the inversion center and mirror planes. Thus, the O group keeps all rotations of the O_h group and its order equals 24 (Table 2.13).
2. T_d ($\bar{4}3m$) is the second symmetry important in molecular complexes. There are only four anion neighbors placed in opposite vertices of the cube. The CH_4 molecule is the most frequently used example. This group was analyzed in Chapter 1 (Section 1.3.4). The tetrahedron lost all c_4 axes, the inversion center, the improper rotation axis s_6, and all planes of σ_h. See illustrations of Figure 2.26 and compare them to those of Figure 2.25a, b, and d. The T_d order equals 24.
3. T_h and T are uncommon groups. T_h is derived from T_d by removing the s_4 and σ_d elements but has the inversion center, improper s_6, and σ_h. T contains all rotations, that is, the c_3 and c_2 elements of T_d. The cubic symmetry ends here.
4. The S_6 and S_4 subgroups are related to certain organometallic molecules. The groups contain improper rotations.
5. D_{3d} ($\bar{3}m$) and D_{2d}. The objects of antiprismatic shape belong to the D_{3d} group and they keep one c_3 axis as the main axis, inversion center, three c_2 axes perpendicular to the main axis, and three dihedral planes. The main axis is also an improper axis of s_6.
6. D_{4h} ($4/m\ mm$) and D_{2h}. The D_{4h} group will be discussed in detail later. You may imagine the octahedron distorted in the z direction. Therefore, the octahedron is compressed or stretched in the z direction. The symmetry becomes lower as you can see in Figure 2.27. We must compare Figure 2.26 with Figure 2.27. It completely loses the c_3 rotation axes and s_6 improper rotation axes. It keeps only one c_4 axis and the corresponding c_2 axis in the z direction. The other two c_4 axes change into two c_2' axes. Two (= C_4^2) axes also change into c_2' axes. The six c_2 axes change into two c_2'' axes. The three horizontal planes σ_h (in fact, they are vertical planes but each plane is perpendicular to one main axis) change as follows: one remains as σ_h (it is perpendicular to the main rotation axis) and two become vertical planes σ_v (they contain the main axis).
7. D_4 (422), D_3, and D_2. The dihedral group D_4 is a group that has a fourfold axis of rotation ($2C_4$ and C_2 operations) and two sets

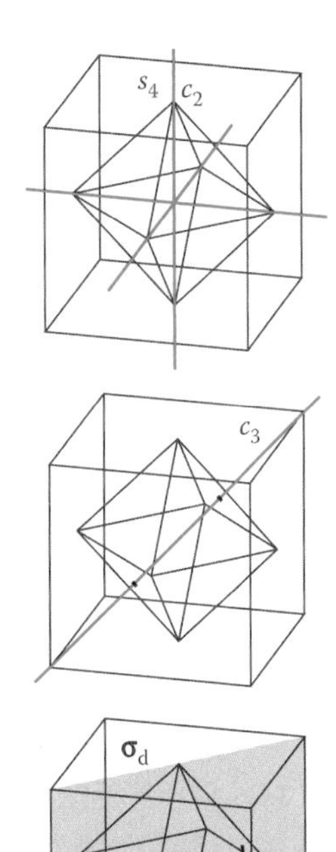

FIGURE 2.26 Illustrations of symmetry elements of the tetrahedron group T_d that are common to the octahedron group O_h.

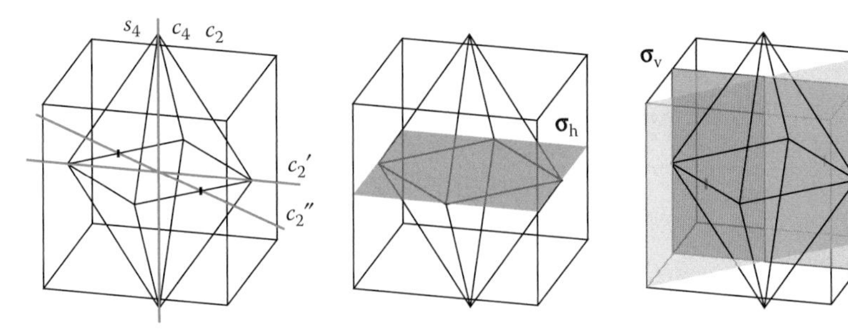

FIGURE 2.27 The symmetry elements of the D_{4h} group (octahedron stretched in the z axis direction).

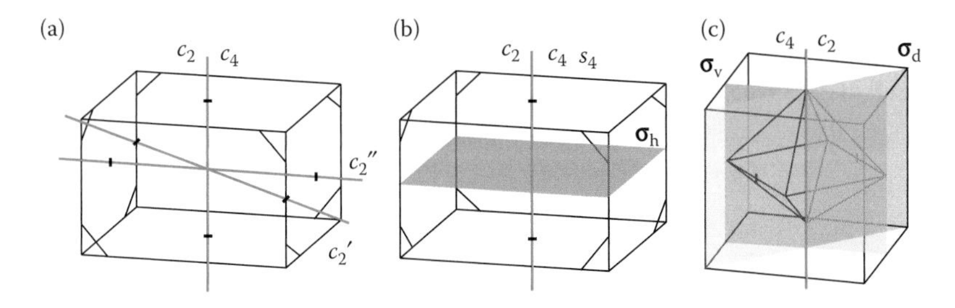

FIGURE 2.28 Illustration of symmetry elements of the dihedral subgroup $\boldsymbol{D}_4$ (a), $\boldsymbol{C}_{4h}$ (b), and $\boldsymbol{C}_{4v}$ (c).

of twofold axes perpendicular to it (Figure 2.28a). The dihedral group $\boldsymbol{D}_4$ is the rotation group of a regular prism.

8. $\boldsymbol{C}_{4h}$ (4/*m*), $\boldsymbol{C}_{3h}$, and $\boldsymbol{C}_{2h}$. The main axis of $\boldsymbol{C}_{4h}$ is a fourfold rotation, twofold rotation, and fourfold improper rotation axis at the same time (Figure 2.28b). It also has a horizontal plane. This group is called the prismatic group.
9. $\boldsymbol{C}_{4v}$ (4*mm*), $\boldsymbol{C}_{3v,}$ and $\boldsymbol{C}_{2v}$. If in the molecular complex of octahedron symmetry $\boldsymbol{O}_h$, an anion is substituted by another type (e.g., the chlorine by hydroxyl), the symmetry of the new object will be drastically lowered to $\boldsymbol{C}_{4v}$ (Figure 2.28c). The new group loses most of the symmetry elements c_3, c_2, horizontal planes, and improper rotation axes. It keeps only the main vertical axis, that is, both the c_4 and c_2 rotation axis and the vertical and dihedral planes.
10. $\boldsymbol{C}_4$ (4), $\boldsymbol{C}_3$, and $\boldsymbol{C}_2$. The uniaxial group $\boldsymbol{C}_4$ contains only one rotation c_4 axis that generates both C_4 and C_2 rotations.
11. $\boldsymbol{C}_s$. The group $\boldsymbol{C}_s$ has one symmetry element that is a vertical plane.
12. $\boldsymbol{S}_2 = \boldsymbol{C}_i$ is reduced to the inversion center.
13. $\boldsymbol{E}$. Each group has the subgroup containing only the identity operation.

    We have now as the starting data:

    a. The correspondence between the symmetry operations of each subgroup and the main group (Table 2.15)
    b. The character tables of all groups (Appendix)
    c. The theorem of decomposition of any representation into irreducible representations

So, we can begin to lower the symmetry from the $\boldsymbol{O}_h$ group to groups we often meet: $\boldsymbol{T}_d$, $\boldsymbol{D}_{3d}$, and $\boldsymbol{D}_{4h}$.

2.5.1.1 CORRESPONDENCE $\boldsymbol{O}_h \rightarrow \boldsymbol{T}_d$

We have to compare the character tables of $\boldsymbol{T}_d$ with the corresponding part of $\boldsymbol{O}_h$ (marked areas in Table 2.16). The differences between the two tables are that the subscripts *g* and *u* disappear in the $\boldsymbol{T}_d$ group as a consequence of the lack of inversion operation. They also have the same dimension of irreducible representations. Therefore, the lower symmetry $\boldsymbol{T}_d$ does not lead to further splitting. We observe that the gray cells have the same characters in both tables. We conclude that when the symmetry is lower, the

TABLE 2.15 Symmetry Operations of Point Subgroups of the O_h Group

Point Group	Symmetry Operations																Order
O_h	E		$8C_3$		$6C_2$	$6C_4$	$3C_2(C_4^2)$	I		$8S_6$			$6S_4$		$6\Sigma_d$	$3\Sigma_h$	48
O	E		$8C_3$		$6C_2$	$6C_4$	$3C_2(C_4^2)$										24
T_d	E		$8C_3$				$3C_2$						$6S_4$		$6\Sigma_d$		24
T_h	E	$4C_3$		$4C_3^2$	$3C_2$			I	$4S_6$		$4S_6^5$					$3\Sigma_h$	24
T	E	$4C_3$		$4C_3^2$	$3C_2$												12
S_6	E	C_3		C_3^2				I	S_6		S_6^5						6
S_4	E				C_2							S_4		S_4^3			4
D_{3d}	E		$2C_3$		$3C_2$			I		$2S_6$					$3\Sigma_d$		12
D_{2d}	E				C_2	$2C_2'$							$2S_4$		$2\Sigma_d$		8
D_{4h}	E		$2C_2'$		$2C_2''$	$2C_4$	C_2	I		$2\Sigma_v$			$2S_4$		$2\Sigma_d$	Σ_h	16
D_{2h}	E		C_2'		C_2''		C_2'''	I		Σ'					Σ''	Σ'''	8
C_{4h}	E					$2C_4$	C_2	I					$2S_4$			Σ_h	8
C_{4v}	E					$2C_4$	C_2			$2\Sigma_v$					$2\Sigma_d$		8
D_4	E		$2C_2'$		$2C_2''$	$2C_4$	C_2										8
C_4	E				C_4^3	C_4	C_2										4
C_i	E							I									2
C_s	E															Σ_h	2
C_1	E																1

TABLE 2.16 Character Tables to Be Compared When Lowering the Symmetry from O_h to T_d

O_h	E	$8C_3$	$6C_2$	$6C_4$	$3C_2 = C_4^2$	I	$8S_6$	$6S_4$	$6\Sigma_d$	$3\Sigma_h$
A_{1g}	1	1	1	1	1	1	1	1	1	1
A_{2g}	1	1	−1	−1	1	1	1	−1	−1	1
E_g	2	−1	0	0	2	2	−1	0	0	2
T_{1g}	3	0	−1	1	−1	3	0	1	−1	−1
T_{2g}	3	0	1	−1	−1	3	0	−1	1	−1
A_{1u}	1	1	1	1	1	−1	−1	−1	−1	−1
A_{2u}	1	1	−1	−1	1	−1	−1	1	1	−1
E_u	2	−1	0	0	2	−2	1	0	0	−2
T_{1u}	3	0	−1	1	−1	−3	0	−1	1	1
T_{2u}	3	0	1	−1	−1	−3	0	1	−1	1

T_d	E	$8C_3$	$3C_2$	$6S_4$	$6\Sigma_d$
A_1	1	1	1	1	1
A_2	1	1	1	−1	−1
E	2	−1	2	0	0
T_1	3	0	−1	1	−1
T_2	3	0	−1	−1	1

representations will change as follows: $a_{1g} \rightarrow a_1$, $a_{2g} \rightarrow a_2$, $e_g \rightarrow e$, $t_{1g} \rightarrow t_1$, and $t_{2g} \rightarrow t_2$. We used capital letters for representations in character tables like those in any textbook and small letters in our comments because we are still discussing the single-electron occupancy of orbitals.

The next question is what happens to odd (ungerade) representations? Their characters are not equal to $\boldsymbol{T}_\mathrm{d}$ group. We have to reduce the ungerade representations of the $\boldsymbol{O}_\mathrm{h}$ group to the irreducible representations of the $\boldsymbol{T}_\mathrm{d}$ group. The number of times each irreducible representation of $\boldsymbol{T}_\mathrm{d}$ is included in the "reducible" a_{1u} representation of $\boldsymbol{O}_\mathrm{h}$ will be given by

$$a_1 = \frac{1}{24}[1 \cdot (1)(1) + 8 \cdot (1)(1) + 3 \cdot (1)(1) + 6 \cdot (1)(-1) + 6 \cdot (1)(-1)] = 0$$

$$a_2 = \frac{1}{24}[1 \cdot (1)(1) + 8 \cdot (1)(1) + 3 \cdot (1)(1) + 6 \cdot (-1)(-1) + 6 \cdot (-1)(-1)] = 1$$

We do not need to continue because the a_{1u} representation is one-dimensional. Thus, a_{1u} will be "reduced" to a_2.

We continue with a_{2u}.

$$a_1 = \frac{1}{24}[1 \cdot (1)(1) + 8 \cdot (1)(1) + 3 \cdot (1)(1) + 6 \cdot (1)(1) + 6 \cdot (1)(1)] = 1$$

We stop here to search the decomposition of a_{2u}. Thus, $a_{2u} \rightarrow a_1$.

Decomposition of e_u. We observe that the characters of e_u are equal to the corresponding characters of e. Thus, $e_u \rightarrow e$.

Decomposition of t_{1u}:

$$a_1 = \frac{1}{24}[1 \cdot (1)(3) + 8 \cdot (1)(0) + 3 \cdot (1)(-1) + 6 \cdot (1)(-1) + 6 \cdot (1)(1)] = 0$$

$$a_2 = \frac{1}{24}[1 \cdot (1)(3) + 8 \cdot (1)(0) + 3 \cdot (1)(-1) + 6 \cdot (-1)(-1) + 6 \cdot (-1)(1)] = 0$$

$$e = \frac{1}{24}[1 \cdot (2)(3) + 8 \cdot (-1)(0) + 3 \cdot (2)(-1) + 6 \cdot (0)(-1) + 6 \cdot (0)(1)] = 0$$

$$t_1 = \frac{1}{24}[1 \cdot (3)(3) + 8 \cdot (0)(0) + 3 \cdot (-1)(-1) + 6 \cdot (1)(-1) + 6 \cdot (-1)(1)] = 0$$

$$t_2 = \frac{1}{24}[1 \cdot (3)(3) + 8 \cdot (0)(0) + 3 \cdot (-1)(-1) + 6 \cdot (-1)(-1) + 6 \cdot (1)(1)] = 1$$

Decomposition of t_{2u}:

$$a_1 = \frac{1}{24}[1 \cdot (1)(3) + 8 \cdot (1)(0) + 3 \cdot (1)(-1) + 6 \cdot (1)(1) + 6 \cdot (1)(-1)] = 0$$

TABLE 2.17 Decomposition of Representations of the O_h Group When Symmetry Decreases in the T_d Group

O_h		T_d
a_{1g}	→	a_1
a_{2g}	→	a_2
e_g	→	e
t_{1g}	→	t_1
t_{2g}	→	t_2
a_{1u}	→	a_2
a_{2u}	→	a_1
e_u	→	e
t_{1u}	→	t_2
t_{2u}	→	t_1

$$a_2 = \frac{1}{24}[1 \cdot (1)(3) + 8 \cdot (1)(0) + 3 \cdot (1)(-1) + 6 \cdot (-1)(1) + 6 \cdot (-1)(-1)] = 0$$

$$e = \frac{1}{24}[1 \cdot (2)(3) + 8 \cdot (-1)(0) + 3 \cdot (2)(-1) + 6 \cdot (0)(1) + 6 \cdot (0)(-1)] = 0$$

$$t_1 = \frac{1}{24}[1 \cdot (3)(3) + 8 \cdot (0)(0) + 3 \cdot (-1)(-1) + 6 \cdot (1)(1) + 6 \cdot (-1)(-1)] = 1$$

$$t_2 = \frac{1}{24}[1 \cdot (3)(3) + 8 \cdot (0)(0) + 3 \cdot (-1)(-1) + 6 \cdot (-1)(1) + 6 \cdot (1)(-1)] = 0$$

The conclusion of how representations are reduced when one lowers the symmetry from $\boldsymbol{O}_h$ to $\boldsymbol{T}_d$ is summarized in Table 2.17.

2.5.1.2 CORRESPONDENCE $\boldsymbol{O}_h \rightarrow \boldsymbol{D}_{3d}$

We have to follow the procedure described in the previous example. We start to compare the character tables of the $\boldsymbol{D}_{3d}$ group with the corresponding part of $\boldsymbol{O}_h$ (marked areas in Table 2.18). The cells with the gray background have the same characters in both tables. Thus, the symmetry representations will change as follows: $a_{1g} \rightarrow a_{1g}$, $a_{2g} \rightarrow a_{2g}$, $e_g \rightarrow e_g$, $a_{1u} \rightarrow a_{1u}$, $a_{2u} \rightarrow a_{2u}$, and $e_u \rightarrow e_u$.

The threefold representations of the $\boldsymbol{O}_h$ group are not equivalent in the $\boldsymbol{D}_{3d}$ group and must be decomposed by the known procedure.

Decomposition of t_{1g}:

$$a_{1g} = \frac{1}{12}[1 \cdot (1)(3) + 2 \cdot (1)(0) + 3 \cdot (1)(-1) + 1 \cdot (1)(3) + 2 \cdot (1)(0) + 3 \cdot (1)(-1)] = 0$$

TABLE 2.18 Character Tables to Be Compared When Symmetry Lowers from O_h to D_{3d}

O_h	E	$8C_3$	$6C_2$	$6C_4$	$3C_2(C_4^2)$	I	$8S_6$	$6S_4$	$6\Sigma_d$	$3\Sigma_h$
A_{1g}	1	1	1	1	1	1	1	1	1	1
A_{2g}	1	1	−1	−1	1	1	1	−1	−1	1
E_g	2	−1	0	0	2	2	−1	0	0	2
T_{1g}	3	0	−1	1	−1	3	0	1	−1	−1
T_{2g}	3	0	1	−1	−1	3	0	−1	1	−1
A_{1u}	1	1	1	1	1	−1	−1	−1	−1	−1
A_{2u}	1	1	−1	−1	1	−1	−1	1	1	−1
E_u	2	−1	0	0	2	−2	1	0	0	−2
T_{1u}	3	0	−1	1	−1	−3	0	−1	1	1
T_{2u}	3	0	1	−1	−1	−3	0	1	−1	1

D_{3d}	E	$2C_3$	$3C_2$	I	$2S_6$	$3\Sigma_d$
A_{1g}	1	1	1	1	1	1
A_{2g}	1	1	−1	1	1	−1
E_g	2	−1	0	2	−1	0
A_{1u}	1	1	1	−1	−1	−1
A_{2u}	1	1	−1	−1	−1	1
E_u	2	−1	0	−2	1	0

$$a_{2g} = \frac{1}{12}[1\cdot(1)(3) + 2\cdot(1)(0) + 3\cdot(-1)(-1) + 1\cdot(1)(3) + 2\cdot(1)(0) + 3\cdot(-1)(-1)] = 1$$

$$e_g = \frac{1}{12}[1\cdot(2)(3) + 2\cdot(-1)(0) + 3\cdot(0)(-1) + 1\cdot(2)(3) + 2\cdot(-1)(0) + 3\cdot(0)(-1)] = 1$$

Here, we stop searching the decomposition of t_{1g}. Thus, $t_{1g} \rightarrow a_{2g} + e_g$. Decomposition of t_{2g}:

$$a_{1g} = \frac{1}{12}[1\cdot(1)(3) + 2\cdot(1)(0) + 3\cdot(1)(1) + 1\cdot(1)(3) + 2\cdot(1)(0) + 3\cdot(1)(1)] = 1$$

$$a_{2g} = \frac{1}{12}[1\cdot(1)(3) + 2\cdot(1)(0) + 3\cdot(-1)(1) + 1\cdot(1)(3) + 2\cdot(1)(0) + 3\cdot(-1)(1)] = 0$$

$$e_g = \frac{1}{12}[1\cdot(2)(3) + 2\cdot(-1)(0) + 3\cdot(0)(1) + 1\cdot(2)(3) + 2\cdot(-1)(0) + 3\cdot(0)(1)] = 1$$

Here, we stop searching the decomposition of t_{2g}. Thus, $t_{2g} \rightarrow a_{1g} + e_g$.

In a similar way, we can find the decomposition of the other two representations to be: $t_{1u} \rightarrow a_{2u} + e_u$ and $t_{2u} \rightarrow a_{1u} + e_u$.

2.5.1.3 CORRESPONDENCE $\boldsymbol{O}_h \rightarrow \boldsymbol{D}_{4h}$

Lowering the symmetry from $\boldsymbol{O}_h$ to $\boldsymbol{D}_{4h}$ is frequent in molecules and crystals. You simply imagine the distortion of the molecular complex in one coordinate direction (e.g., z axis). We start to compare the character tables of the $\boldsymbol{D}_{4h}$ subgroup with the character table of $\boldsymbol{O}_h$. It is crucial to establish the exact correspondence between symmetry operations as shown in Table 2.19, because the characters of corresponding representations in $\boldsymbol{D}_{4h}$ are given by the operations before lowering the symmetry. The corresponding new representations $\Gamma(E_g)$, $\Gamma(T_{1g})$, and $\Gamma(T_{2g})$ are not part of the irreducible representations of the $\boldsymbol{D}_{4h}$ group.

We now have almost all the information to start comparing the character tables of both groups (Table 2.20). The simple inspection of tables shows that the maximum dimension of irreducible representations of the $\boldsymbol{D}_{4h}$ group is 2. Thus, lower symmetry means supplementary splitting. The cells with a gray background have the same characters in both tables. Another direct observation is that the symmetry representations will change as follows: $a_{1g} \rightarrow a_{1g}$, $a_{2g} \rightarrow b_{1g}$, $a_{1u} \rightarrow a_{1u}$, and $a_{2u} \rightarrow b_{1u}$.

The two- and threefold representations of the $\boldsymbol{O}_h$ group are not equivalent among irreducible representations of the $\boldsymbol{D}_{4h}$ group, and the new corresponding representations $\Gamma(E_g)$, $\Gamma(T_{1g})$, and $\Gamma(T_{2g})$ must be decomposed by the known procedure.

Decomposition of $\gamma(e_g)$:

$$\begin{aligned} a_{1g} &= \frac{1}{16}[1 \cdot (1)(2) + 2 \cdot (1)(2) + 2 \cdot (1)(0) + 2 \cdot (1)(0) + 1 \cdot (1)(2) \\ &\quad + 1 \cdot (1)(2) + 2 \cdot (1)(2) + 2 \cdot (1)(0) + 2 \cdot (1)(0) + 1 \cdot (1)(2)] = 1 \end{aligned}$$

TABLE 2.19 Correspondence between Symmetry Operations of the O_h Group and Symmetry Operations of the D_{4h} Subgroup

O_h	E	$8C_3$	$6C_2$	$6C_4$	$3C_2(=C_4^2)$	I	$8S_6$	$6S_4$	$6\Sigma_d$	$3\Sigma_h$
E_g	2	−1	0	0	2	2	−1	0	0	2
T_{1g}	3	0	−1	1	−1	3	0	1	−1	−1
T_{2g}	3	0	1	−1	−1	3	0	−1	1	−1
D_{4h}	E	$2C_2'$	$2C_2''$	$2C_4$	$C_2(=C_4^2)$	I	$2\Sigma_v$	$2S_4$	$2\Sigma_d$	Σ_h
$\Gamma(E_g)$	2	2	0	0	2	2	2	0	0	2
$\Gamma(T_{1g})$	3	−1	−1	1	−1	3	−1	1	−1	−1
$\Gamma(T_{2g})$	3	−1	1	−1	−1	3	−1	−1	1	−1

TABLE 2.20 Character Tables to Be Compared When Symmetry Lowers from O_h to D_{4h}

O_h	E	$8C_3$	$6C_2$	$6C_4$	$3C_2 = C_4^2$	I	$8S_6$	$6S_4$	$6\Sigma_d$	$3\Sigma_h$
A_{1g}	1	1	1	1	1	1	1	1	1	1
A_{2g}	1	1	−1	−1	1	1	1	−1	−1	1
E_g	2	−1	0	0	2	2	−1	0	0	2
T_{1g}	3	0	−1	1	−1	3	0	1	−1	−1
T_{2g}	3	0	1	−1	−1	3	0	−1	1	−1
A_{1u}	1	1	1	1	1	−1	−1	−1	−1	−1
A_{2u}	1	1	−1	−1	1	−1	−1	1	1	−1
E_u	2	−1	0	0	2	−2	1	0	0	−2
T_{1u}	3	0	−1	1	−1	−3	0	−1	1	1
T_{2u}	3	0	1	−1	−1	−3	0	1	−1	1
D_{4h}	E	$2C_2'$	$2C_2''$	$2C_4$	$C_2(C_4^2)$	I	$2\Sigma_v$	$2S_4$	$2\Sigma_d$	Σ_h
A_{1g}	1	1	1	1	1	1	1	1	1	1
A_{2g}	1	−1	−1	1	1	1	−1	1	−1	1
B_{1g}	1	1	−1	−1	1	1	1	−1	−1	1
B_{2g}	1	−1	1	−1	1	1	−1	−1	1	1
E_g	2	0	0	0	−2	2	0	0	0	−2
A_{1u}	1	1	1	1	1	−1	−1	−1	−1	−1
A_{2u}	1	−1	−1	1	1	−1	1	−1	1	−1
B_{1u}	1	1	−1	−1	1	−1	−1	1	1	−1
B_{2u}	1	−1	1	−1	1	−1	1	1	−1	−1
E_u	2	0	0	0	−2	−2	0	0	0	2
$\Gamma(E_g)$	2	2	0	0	2	2	2	0	0	2
$\Gamma(T_{1g})$	3	−1	−1	1	−1	3	−1	1	−1	−1
$\Gamma(T_{2g})$	3	−1	1	−1	−1	3	−1	−1	1	−1

$$
\begin{aligned}
b_{1g} &= \frac{1}{16}[1\cdot(1)(2) + 2\cdot(1)(2) + 2\cdot(-1)(0) + 2\cdot(-1)(0) + 1\cdot(1)(2) \\
&\quad + 1\cdot(1)(2) + 2\cdot(1)(2) + 2\cdot(-1)(0) + 2\cdot(-1)(0) + 1\cdot(1)(2)] = 1
\end{aligned}
$$

Here, we stop searching the decomposition of $\gamma(e_g)$. Thus, $e_g \to a_{1g} + b_{1g}$.
Decomposition of $\gamma(t_{1g})$:

$$
\begin{aligned}
a_{2g} &= \frac{1}{16}[1\cdot(1)(3) + 2\cdot(-1)(-1) + 2\cdot(-1)(-1) + 2\cdot(1)(1) + 1\cdot(1)(-1) \\
&\quad + 1\cdot(1)(3) + 2\cdot(-1)(-1) + 2\cdot(1)(1) + 2\cdot(-1)(-1) + 1\cdot(1)(-1)] = 1
\end{aligned}
$$

$$
\begin{aligned}
e_g &= \frac{1}{16}[1\cdot(2)(3) + 2\cdot(0)(-1) + 2\cdot(0)(-1) + 2\cdot(0)(1) + +1\cdot(-2)(-1) \\
&\quad + 1\cdot(2)(3) + 2\cdot(0)(-1) + 2\cdot(0)(1) + 2\cdot(0)(0) + 1\cdot(-2)(-1)] = 1
\end{aligned}
$$

TABLE 2.21 Correlation Table of the O_h Irreducible Representations with the Irreducible Representations of Some of Its Subgroups

O_h	T_d	D_{4h}	D_{3d}	D_{2d}	C_{4v}	D_4	C_{3v}	C_{2h}
a_{1g}	a_1	a_{1g}	a_{1g}	a_1	a_1	a_1	a_1	a_g
a_{2g}	a_2	b_{1g}	a_{2g}	b_1	b_1	b_1	a_2	b_g
e_g	e	$a_{1g}+b_{1g}$	e_g	a_1+b_1	a_1+b_1	a_1+b_1	e	a_g+b_g
t_{1g}	t_1	$a_{2g}+e_g$	$a_{2g}+e_g$	a_2+e	a_2+e	a_2+e	a_2+e	a_g+2b_g
t_{2g}	t_2	$b_{2g}+e_g$	$a_{1g}+e_g$	b_2+e	b_2+e	b_2+e	a_1+e	$2a_g+b_g$
a_{1u}	a_2	a_{1u}	a_{1u}	b_1	a_2	a_1	a_2	a_u
a_{2u}	a_1	b_{1u}	a_{2u}	a_1	b_2	b_1	a_1	b_u
e_u	e	$a_{1u}+b_{1u}$	e_u	a_1+b_1	a_2+b_2	a_1+b_1	e	a_u+b_u
t_{1u}	t_2	$a_{2u}+e_u$	$a_{2u}+e_u$	b_2+e	a_1+e	a_2+e	a_1+e	a_u+2b_u
t_{2u}	t_1	$b_{2u}+e_u$	$a_{1u}+e_u$	a_2+e	b_1+e	b_2+e	a_2+e	$2a_u+b_u$

Note: Find a complete collection of correlation tables in Table X-14 from Wilson et al. 1955.

Here, we stop searching the decomposition of $\gamma(t_{1g})$. Thus, $t_{1g} \rightarrow a_{2g} + e_g$.
Decomposition of $\gamma(t_{2g})$:

$$b_{2g} = \frac{1}{16}[1 \cdot (1)(3) + 2 \cdot (-1)(-1) + 2 \cdot (1)(1) + 2 \cdot (-1)(-1) + 1 \cdot (1)(-1) + 1 \cdot (1)(3) + 2 \cdot (-1)(-1) + 2 \cdot (-1)(-1) + 2 \cdot (1)(1) + 1 \cdot (1)(-1)] = 1$$

$$e_g = \frac{1}{16}[1 \cdot (2)(3) + 2 \cdot (0)(-1) + 2 \cdot (0)(1) + 2 \cdot (0)(-1) + 1 \cdot (-2)(-1) + 1 \cdot (2)(3) + 2 \cdot (0)(-1) + 2 \cdot (0)(-1) + 2 \cdot (0)(1) + 1 \cdot (-2)(-1)] = 1$$

Here, we stop searching the decomposition of $\gamma(t_{2g})$. Thus, $t_{2g} \rightarrow b_{2g} + e_g$.

In a similar way, we can find the decomposition of the other two representations to be: $t_{1u} \rightarrow a_{2u} + e_u$ and $t_{2u} \rightarrow a_{1u} + e_u$.

The results obtained here and much more are presented in Table 2.21, which may be used in your future work.

2.5.2 SPLITTING OF ATOMIC ORBITALS IN A LOWER SYMMETRY FIELD

What is good in Table 2.21 about correlations between irreducible representations of the $\boldsymbol{O}_h$ group and those of its subgroups? We have already expressed the importance of determining the manner in which the atomic orbitals will be split by the field of a particular environment. We calculated the splitting of orbitals in the $\boldsymbol{O}_h$ symmetry field and the results are summarized in Table 2.14. We can continue the previous work just to find the splitting of orbitals in a particular lower symmetry. There are two ways:

1. To repeat all the procedures described in Section 2.4.6
2. To use the results of Table 2.21

Let us take as an example the point group $\boldsymbol{D}_{4h}$.

1. First way: To determine the representation for which each set of wavefunctions of the free ion forms a basis, and then to reduce these representations into irreducible representations of the group. For the $\boldsymbol{D}_{4h}$ group, we must first calculate the elements of the matrices that express the effect on the wavefunctions of each class of the symmetry operations in the group. We know to calculate the characters of representations generated as a result of rotation action on different orbitals by using Equation 2.55. They will be applied for the particular cases of the $\boldsymbol{D}_4$ subgroup (marked area in Table 2.22) composed of only rotations and the results are shown in Table 2.23.

$$\chi_p(\alpha) = \frac{\sin\frac{3\alpha}{2}}{\sin\frac{\alpha}{2}} \Rightarrow \chi_p(C_4) = \frac{\sin\frac{3\cdot 90}{2}}{\sin\frac{90}{2}} = 1;$$

$$\chi_p(C_2) = \frac{\sin\frac{3\cdot 180}{2}}{\sin\frac{180}{2}} = -1$$

$$\chi_d(\alpha) = \frac{\sin\frac{5\alpha}{2}}{\sin\frac{\alpha}{2}} \Rightarrow \chi_d(C_4) = \frac{\sin\frac{5\cdot 90}{2}}{\sin\frac{90}{2}} = -1;$$

$$\chi_d(C_2) = \frac{\sin\frac{5\cdot 180}{2}}{\sin\frac{180}{2}} = 1$$

TABLE 2.22 Character Table of the D_4 Subgroup (Pure Rotations) Is Restricted to the Marked Area of the D_{4h} Group

D_{4h}	E	$2C_2'$	$2C_2''$	$2C_4$	$C_2(C_4^2)$	I	$2\Sigma_v$	$2S_4$	$2\Sigma_d$	Σ_h
A_{1g}	1	1	1	1	1	1	1	1	1	1
A_{2g}	1	−1	−1	1	1	1	−1	1	−1	1
B_{1g}	1	1	−1	−1	1	1	1	−1	−1	1
B_{2g}	1	−1	1	−1	1	1	−1	−1	1	1
E_g	2	0	0	0	−2	2	0	0	0	−2
A_{1u}	1	1	1	1	1	−1	−1	−1	−1	−1
A_{2u}	1	−1	−1	1	1	−1	1	−1	1	−1
B_{1u}	1	1	−1	−1	1	−1	−1	1	1	−1
B_{2u}	1	−1	1	−1	1	−1	1	1	−1	−1
E_u	2	0	0	0	−2	−2	0	0	0	2

TABLE 2.23 Characters of Rotation Matrices of the D_4 Group for Atomic Orbitals

Orbital Type	l	$\chi(E)$	$\chi(C_2')$	$\chi(C_2'')$	$\chi(C_4)$	$\chi(C_2)$
s	0	1	1	1	1	1
p	1	3	−1	−1	1	−1
d	2	5	1	1	−1	1
f	3	7	−1	−1	−1	−1
g	4	9	1	1	1	1
h	5	11	−1	−1	1	−1
i	6	13	1	1	−1	1

$$\chi_f(\alpha) = \frac{\sin\frac{7\alpha}{2}}{\sin\frac{\alpha}{2}} \Rightarrow \chi_f(C_4) = \frac{\sin\frac{7\cdot 90}{2}}{\sin\frac{90}{2}} = -1;$$

$$\chi_f(C_2) = \frac{\sin\frac{7\cdot 180}{2}}{\sin\frac{180}{2}} = -1$$

$$\chi_h(\alpha) = \frac{\sin\frac{11\alpha}{2}}{\sin\frac{\alpha}{2}} \Rightarrow \chi_h(C_4) = \frac{\sin\frac{11\cdot 90}{2}}{\sin\frac{90}{2}} = 1;$$

$$\chi_h(C_2) = \frac{\sin\frac{11\cdot 180}{2}}{\sin\frac{180}{2}} = -1$$

The maximum dimension of irreducible representations of the $\boldsymbol{D}_4$ subgroup equals 2. Thus, from Table 2.23, except *s* orbitals, all others will be split in the $\boldsymbol{D}_4$ symmetry field. The decomposition will show the splitting of the corresponding level in the crystalline field of this symmetry.

The order of the $\boldsymbol{D}_4$ group is $h = 8$. The number of times each irreducible representation is included in the reducible representation generated by the effect of rotations on *p* orbitals will be

$$a_1 = \frac{1}{8}[1\cdot(1)(3) + 2\cdot(1)(-1) + 2\cdot(1)(-1) + 2\cdot(1)(1) + 1\cdot(1)(-1)] = 0$$

$$a_2 = \frac{1}{8}[1\cdot(1)(3) + 2\cdot(-1)(-1) + 2\cdot(-1)(-1) + 2\cdot(1)(1) + 1\cdot(1)(-1)] = 1$$

$$b_1 = \frac{1}{8}[1\cdot(1)(3) + 2\cdot(1)(-1) + 2\cdot(-1)(-1) + 2\cdot(-1)(1) + 1\cdot(1)(-1)] = 0$$

$$b_2 = \frac{1}{8}[1 \cdot (1)(3) + 2 \cdot (1)(-1) + 2 \cdot (-1)(-1) + 2 \cdot (-1)(1) + 1 \cdot (1)(-1)] = 0$$

$$e = \frac{1}{8}[1 \cdot (2)(3) + 2 \cdot (0)(-1) + 2 \cdot (0)(-1) + 2 \cdot (0)(1) + 1 \cdot (-2)(-1)] = 1$$

Thus, the representation of p orbitals will be decomposed into irreducible representations as follows:

$$\gamma_p = a_{2u} + e_u$$

Concerning d orbitals, the decomposition will give

$$a_1 = \frac{1}{8}[1 \cdot (1)(5) + 2 \cdot (1)(1) + 2 \cdot (1)(1) + 2 \cdot (1)(-1) + 1 \cdot (1)(1)] = 1$$

$$a_2 = \frac{1}{8}[1 \cdot (1)(5) + 2 \cdot (1)(1) + 2 \cdot (-1)(1) + 2 \cdot (-1)(-1) + 1 \cdot (1)(1)] = 0$$

$$b_1 = \frac{1}{8}[1 \cdot (1)(5) + 2 \cdot (1)(1) + 2 \cdot (-1)(1) + 2 \cdot (-1)(-1) + 1 \cdot (1)(1)] = 1$$

$$b_2 = \frac{1}{8}[1 \cdot (1)(5) + 2 \cdot (-1)(1) + 2 \cdot (1)(1) + 2 \cdot (-1)(-1) + 1 \cdot (1)(1)] = 1$$

$$e = \frac{1}{8}[1 \cdot (2)(5) + 2 \cdot (0)(1) + 2 \cdot (0)(1) + 2 \cdot (0)(-1) + 1 \cdot (-2)(1)] = 1$$

Therefore, the representation of d can be decomposed into irreducible representations as follows:

$$\gamma_d = a_{1g} + b_{1g} + b_{2g} + e_g$$

So, the d orbitals of fivefold degeneracy in the free ion split into three levels of onefold degeneracy and one level of twofold degeneracy in the $\boldsymbol{D}_{4h}$ symmetry crystalline field. We recommend you to exercise for other orbitals.

Finally, we can complete Table 2.24 on how the single-electron orbitals will split when the ion will be introduced in the $\boldsymbol{D}_{4h}$ symmetry field.

Again, we note that the results obtained for single-electron orbitals are also valuable for multielectron orbitals.

2. The second method is to use Table 2.21 of the correlation between irreducible representations when the symmetry decreases. We have to use the orbitals splitting in $\boldsymbol{O}_h$ symmetry presented in Table 2.14 and apply the effect of lower symmetry to representations. The final results are presented in Table 2.25. We may observe by simple inspection that the results are the same as in Table 2.24.

TABLE 2.24 Splitting of Atomic Orbitals in the D_{4h} Symmetry of a Weak Field

Orbital Type	l	Orbital Degeneracy	Irreducible Representations	Total Dimension
s	0	1	a_{1g}	1
p	1	3	$a_{2u} + e_u$	$1 + 2 = 3$
d	2	5	$a_{1g} + b_{1g} + b_{2g} + e_g$	$1 + 1 + 1 + 2 = 5$
f	3	7	$a_{2u} + b_{1u} + b_{2u} + 2e_u$	$1 + 1 + 1 + 2 \cdot 2 = 7$
g	4	9	$2a_{1g} + a_{2g} + b_{1g} + b_{2g} + 2e_g$	$2 \cdot 1 + 1 + 1 + 1 + 2 \cdot 2 = 9$
h	5	11	$a_{1u} + 2a_{2u} + b_{1u} + b_{2u} + 3e_u$	$1 + 2 \cdot 1 + 1 + 1 + 3 \cdot 2 = 11$
i	6	13	$2a_{1g} + a_{2g} + 2b_{1g} + 2b_{2g} + 3e_g$	$2 \cdot 1 + 1 + 2 \cdot 1 + 2 \cdot 1 + 3 \cdot 2 = 13$

TABLE 2.25 Splitting of Atomic Orbitals of a Free Ion as an Effect of Lowering the Symmetry from O_h to D_{4h}

Orbital Type	l	Splitting in O_h Symmetry	Splitting in D_{4h} Symmetry
s	0	a_{1g}	a_{1g}
p	1	t_{1u}	$a_{2u} + e_{2u}$
d	2	$e_g + t_{2g}$	$a_{1g} + b_{1g} + b_{2g} + e_g$
f	3	$a_{2u} + t_{1u} + t_{2u}$	$b_{1u} + a_{2u} + e_u + b_{2u} + e_u$
g	4	$a_{1g} + e_g + t_{1g} + t_{2g}$	$a_{1g} + a_{1g} + b_{1g} + a_{2g} + e_g + b_{2g} + e_g$
h	5	$e_u + 2t_{1u} + t_{2u}$	$a_{1u} + b_{1u} + 2a_{2u} + 2e_u + b_{2u} + e_u$
i	6	$a_{1g} + a_{2g} + e_g + t_{1g} + 2t_{2g}$	$a_{1g} + b_{1g} + a_{1g} + b_{1g} + a_{2g} + e_g + 2b_{2g} + 2e_g$

TABLE 2.26 Splitting of Atomic Orbitals of a Free Ion in Several Particular Symmetries

Orbital Type	O_h	T_d	D_{3d}	D_{4h}	C_{3v}
s	a_{1g}	a_1	a_{1g}	a_{1g}	a_1
p	t_{1u}	t_2	$a_{2u} + e_u$	$a_{2u} + e_u$	$a_1 + e$
d	$e_g + t_{2g}$	$e + t_2$	$a_{1g} + 2e_g$	$a_{1g} + b_{1g} + b_{2g} + e_g$	$a_1 + 2e$
f	$a_{2u} + t_{1u} + t_{2u}$	$a_2 + t_1 + t_2$	$a_{1u} + 2a_{2u} + 2e_u$	$a_{2u} + b_{1u} + b_{2u} + 2e_u$	$a_1 + 2a_2 + 2e$
g	$a_{1g} + e_g + t_{1g} + t_{2g}$	$a_1 + e + t_1 + t_2$	$2a_{1g} + a_{2g} + 3e_g$	$2a_{1g} + a_{2g} + b_{1g} + b_{2g} + 2e_g$	$2a_1 + a_2 + 3e$
h	$e_u + 2t_{1u} + t_{2u}$	$e + 2t_1 + t_2$	$a_{1u} + 2a_{2u} + 4e_u$	$a_{1u} + 2a_{2u} + b_{1u} + b_{2u} + 3e_u$	$a_1 + 2a_2 + 4e$
i	$a_{1g} + a_{2g} + e_g + t_{1g} + 2t_{2g}$	$a_1 + a_2 + e + t_1 + 2t_2$	$3a_{1g} + 2a_{2g} + 4e_g$	$2a_{1g} + a_{2g} + 2b_{1g} + 2b_{2g} + 3e_g$	$3a_1 + 2a_2 + 4e$

As a conclusion, we may build Table 2.26 about the splitting of different sets of orbitals with single-electron occupancy in some particular symmetries, which we may encounter more frequently.

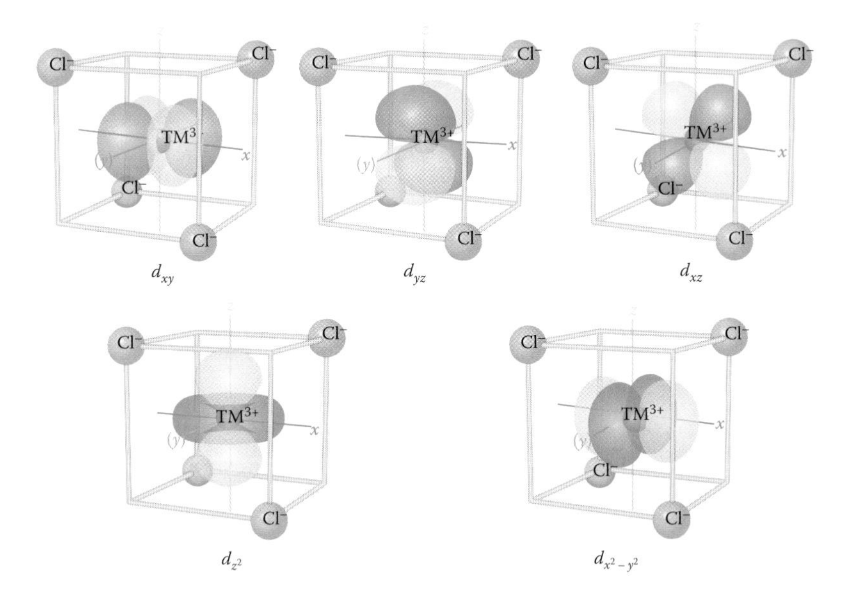

FIGURE 2.29 Spatial orientation of *d* orbitals with respect to ligand anion positions in $\boldsymbol{T}_\mathrm{d}$ symmetry.

2.5.3 ENERGY SPLITTING OF 3*d* ORBITALS IN $\boldsymbol{T}_\mathrm{d}$ SYMMETRY

The order of splitting in $\boldsymbol{T}_\mathrm{d}$ symmetry may be estimated by simple inspection of the *d* orbitals symmetry (Figure 2.29). They may be grouped either as orbitals with lobes between axes or as orbitals with lobes directed along the axes. The ligand ions are arranged in opposite vertices of the cube and not on the axes as in the case of an octahedron. Ligands do not directly point to any of the *d* orbitals but are closer to the d_{xy}, d_{yz}, and d_{xz} orbitals. Thus, if an electron belongs to the orbitals d_{xy}, d_{yz}, and d_{xz}, the repulsion electron–anion (chloride) will be stronger than if the electron belongs to the orbitals d_{z^2} or $d_{x^2-y^2}$. The lobes of the last two orbitals are oriented so that they are equally placed between chloride ligands and are thus well balanced.

However, the electrostatic potential produced by the negative point charges positioned at the four opposite vertices of a cube will be lower than the potential produced by six negative charges positioned in the vertices of the octahedron inscribed in the cube (Figure 2.30).

Therefore, the state of the free ion will split into two states in the tetrahedral field in the same way as in the octahedral field but in the reverse order (Figure 2.31). The level *e* will be lower placed and the level t_2 will be higher placed in terms of energy. Note that the subscript *g* is missing because tetrahedral geometry has no center of symmetry. The splitting energy in tetrahedral symmetry is lower than the splitting energy in octahedral symmetry $\Delta_\mathrm{T} < \Delta_\mathrm{O}$ because there are four ligands instead of six and they are located farther than in octahedron symmetry (Figure 2.30).

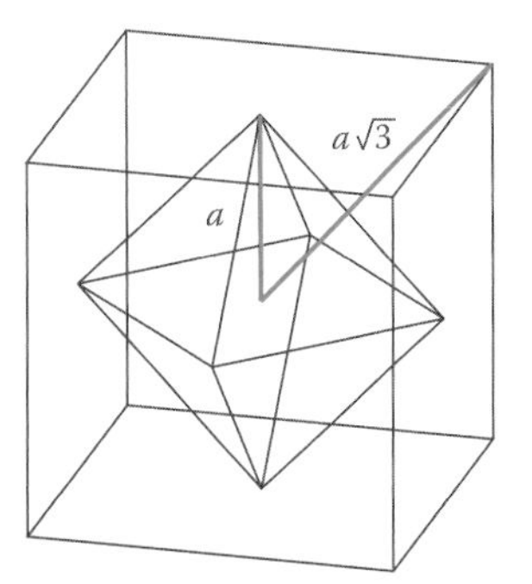

FIGURE 2.30 The distance from the central ion to ligands is greater from the tetrahedron than from the octahedron.

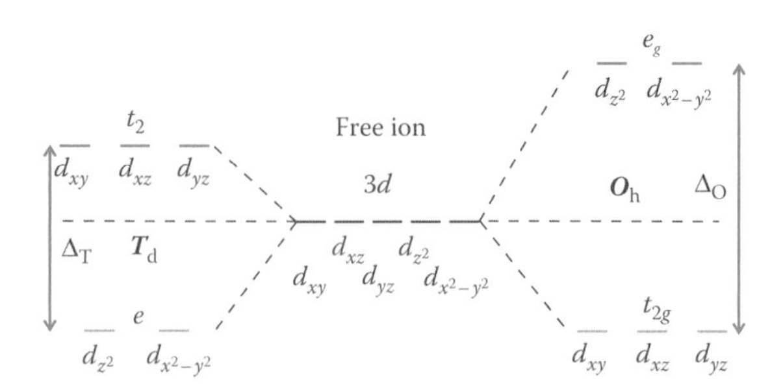

FIGURE 2.31 The energy splitting in tetrahedral symmetry is lower and reversed compared to that in octahedral symmetry.

Proceeding in a similar way to that described in Section 2.4.3, the potential created by four point charges located in the vertices of the tetrahedron can be expanded as follows:

$$V_T = \frac{4e}{r_0} + D\left[x^4 + y^4 + z^4 - \frac{3}{5}r^4\right] + \frac{20e}{\sqrt{3}r_0^4}xyz \tag{2.56}$$

where the potential constant given by

$$D = \left(\frac{35e}{9r_0^5}\right) \tag{2.57}$$

is different from that of the octahedral potential.

The first term in the potential expansion has a spherical symmetry and elevates all levels with the same quantity. Its contribution is much larger than the contribution of the second term coming from the tetrahedral disposition of four negative charges. The third term is odd and does not contribute to matrix elements because all wavefunctions of d orbitals are even. The splitting of energy will be produced only by the second term of Equation 2.56.

As we have already stated based on simple inspection, the tetrahedral field is weaker than the octahedral field, but now we can exactly calculate the ratio:

$$\frac{D_T}{D_O} = -\frac{4}{9} \tag{2.58}$$

How can we check the correctness of this result? The most common example is the experimental measurement of splitting energy for the $[Ti(H_2O)_6]^{3+}$ complex. The Ti^{3+} ion is a typical example of the d^1 electronic configuration and the octahedral complex with six equidistant water molecules is not difficult to be manipulated in the laboratory. (We recommend the use of 0.1 M aqueous solution.) The water molecules are not ionized but have a small electric dipole, and hence the ligand field is weak. The color of the solution appears to be purple, which means that the complex absorbs greenish-yellow light. Indeed, the measured absorption spectrum shows a very large absorption band

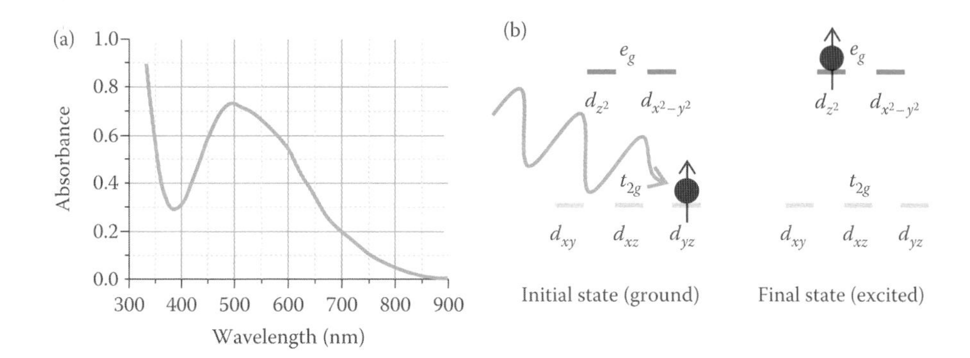

FIGURE 2.32 (a) The absorption spectra of the Ti^{3+} ion in octahedral symmetry $[Ti(H_2O)_6]^{3+}$. (b) The green absorption band is due to electronic *d–d* transition from ground state $t_{2g}^1e_g^0$ to excited state $t_{2g}^0e_g^1$.

with a maximum around 495 nm (green), which means in energy units 20,200 cm^{-1} (Figure 2.32).

From the theory presented here, we expect a single weak band corresponding to the electronic transition $t_{2g} \rightarrow e_g$ to appear. It is called *d–d* transition and the observed color is a measure of the energy splitting produced by the presence of ligands around the central ion. The transition metal ions are not colored as free ions (gaseous state) and only splitting of *d* orbitals will give color. The band position on the energy scale equals $10Dq$, and thus the corresponding $Dq = 2020$ cm^{-1}. The shoulder of the absorption band to longer wavelength (lower energy) is due to the Jahn–Teller effect (for details, see Section 2.7). The octahedron is compressed in the *z* direction and all orbitals with lobes in the *z* direction will have higher energy compared to that without *z*, that is, d_{xz} and d_{yz} compared to d_{xy} and d_{z^2} compared to $d_{x^2-y^2}$. Thus, both t_{2g} and e_g will split. The strong absorption that starts in UV is due to the charge transfer from the central ion to the ligand. The same spectrum can be obtained by measuring the absorption of the Ti:sapphire crystal. It is the best-known *active medium* for lasers used to generate *ultrashort pulses*. The crystal appears to be pink colored because the concentration of titanium ions is much lower than in solution (0.1–0.01%). In Ti:Al_2O_3, the Ti^{3+} ions substitute for Al^{3+} ions during the crystal growth from melting. Thus, the 3*d*-electron interacts with electrostatic charges of six surrounding oxygen ions (the other difference vs. solution). The crystal field splits the energy level of the free ion into lower T_{2g} and upper E_g. The large absorption band of Ti:sapphire is centered at 490 nm as given by two crystal-producer companies (FocTek 2013; KryLight 2013; Kusuma et al. 2010), which makes it suitable for a variety of laser pump sources (argon ion lasers, frequency-doubled Nd:YAG lasers, frequency-doubled diode lasers, or flash lamps). So, the splitting results in light absorption with the maximum located around 20,400 cm^{-1}. The small difference between splitting energy for both cases comes from the difference between the nature of ligands (H_2O molecules vs. O ions). The emission wavelength of the laser depends on the active (lasing) ion as well as the crystal host into which the ion is incorporated as an impurity. For the transition metal ions with an incomplete 3*d* shell (Cr^{3+}, Ti^{3+}, Co^{2+}, Ni^{2+}, and V^{2+}), the lasing transition is between the ion energy levels as split by the crystal field.

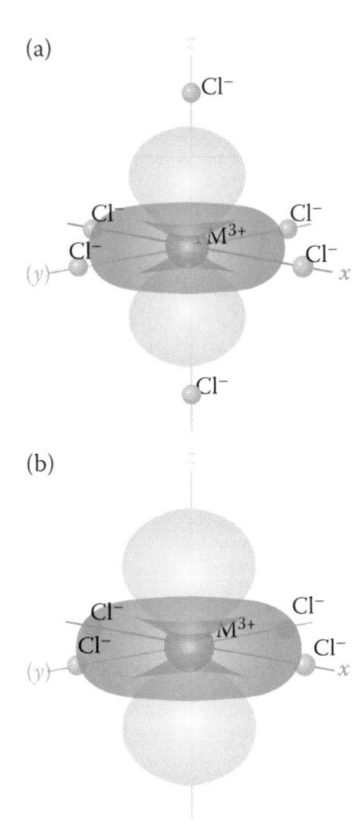

FIGURE 2.33 Removing z ligands results in the energy of d_{z^2} orbital falls the most.

Concerning tetrahedral symmetry, we are tempted to look for the $TiCl_4$ compound that is tetrahedral indeed, but the titanium ion is here tetravalent (d^2 electronic configuration). The titanium3+ ion (d^1 configuration) and chlorine ions form only $TiCl_3$ compound. As an example of the d^1 configuration splitting in tetrahedral symmetry, we may take vanadium tetrachloride VCl_4, which that is an inorganic bright-red liquid. The V^{4+} ion is isoelectronic with the Ti^{3+} ion.

2.5.4 *d* ORBITALS SPLITTING IN TETRAGONAL SYMMETRY

A square planar complex has a tetragonal symmetry. It can be imagined when two ligands on the z axis of the octahedron are removed. The four negative ligands in the xy plane keep the square shape but they move a little bit closer to the central cation. Thus, the symmetry of the complex belongs to the $\boldsymbol{D}_{4h}$ group.

What is the effect on orbitals? It is simple to imagine seeing the illustration of Figure 2.33. All orbitals oriented on the z axis are the most affected. The electrostatic repulsion between electrons concentrated in z lobes and negative ligands on the z axis disappears and results in lower energy. The energy of the d_{z^2} orbital will decrease the most because its lobe in the z direction is 2 times bigger than any other orbital lobe.

While in octahedral symmetry, the lobes of both d_{z^2} and $d_{x^2-y^2}$ orbitals point exactly toward the point charges of ligands and this similarity results in both orbitals having the same energy; in tetragonal symmetry, they have different orientations related to the four ligands. The lobes of the $d_{x^2-y^2}$ orbital interact more strongly with ligands that are closer (Figure 2.34) and its energy will be higher. Consequently, the two orbitals d_{z^2} and $d_{x^2-y^2}$ will have different positions on the energy scale. It means that the degeneracy of the e_g state is totally removed.

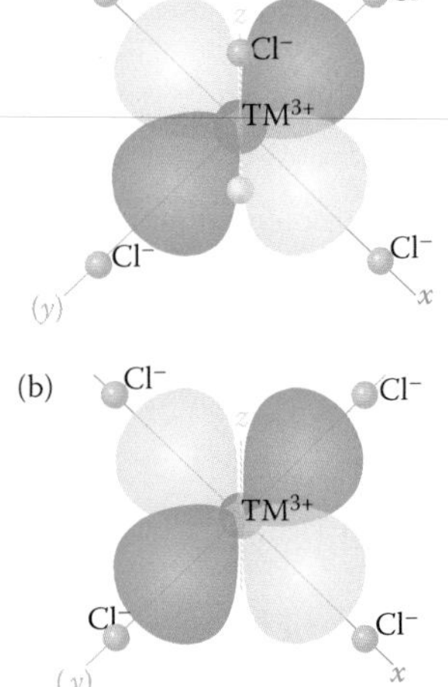

FIGURE 2.34 The $d_{x^2-y^2}$ orbital interacts stronger with xy ligands in tetragonal symmetry compared to octahedral symmetry and its energy will increase.

Regarding the other three orbitals d_{xy}, d_{xz}, and d_{yz}, their behavior follows the rules above: while the orbitals with a z component (d_{xz} and d_{yz}) will decrease in energy, the orbitals with lobes in the xy plane (d_{xy}) will increase in energy. The diagram of splitting in the $\boldsymbol{D}_{4h}$ symmetry shown in Figure 2.35 is more complex than that in octahedral and tetrahedral symmetry, but the degeneracy is not yet completely removed.

The qualitative survey of splitting of d orbitals when the symmetry lowers from $\boldsymbol{O}_h$ to $\boldsymbol{D}_{4h}$ can be immediately seen in Table 2.21:

$$\begin{cases} e_g \rightarrow a_{1g} + b_{1g} \\ t_{2g} \rightarrow b_{2g} + e_g \end{cases}$$

Therefore, the two levels of $\boldsymbol{O}_h$ symmetry will be split into four levels in $\boldsymbol{D}_{4h}$ symmetry and one of them is twofold degenerated. This table gives no information about the relative position of the new levels. However, for spectroscopy, it is important to know that the single-band spectrum obtained in $\boldsymbol{O}_h$ symmetry will change to a multiband spectrum in $\boldsymbol{D}_{4h}$ symmetry. It is not mandatory for all possible transitions to occur but more than one band spectrum suggests the presence of lower symmetry.

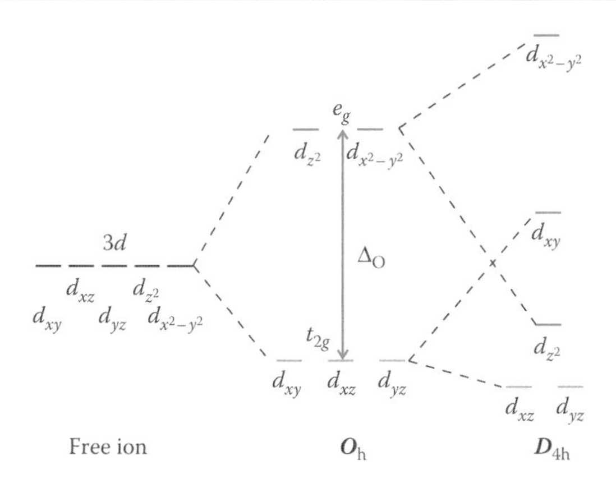

FIGURE 2.35 The splitting of *d* orbitals is more complex in $\boldsymbol{D}_{4h}$ symmetry than in $\boldsymbol{O}_h$ symmetry and the fivefold degeneracy of the free ion state is almost removed by the lower symmetry.

2.5.5 FACTORS GOVERNING THE SPLITTING ENERGY (SPECTROCHEMICAL SERIES)

Why is the color of $[TiCl_6]^{3-}$ orange and $[Ti(urea)_6]^{3+}$ blue, whereas $[Ti(H_2O)_6]^{3+}$ is purple? The difference between the three complexes is the ligand type. The complex $[Fe(H_2O)_6]^{2+}$ has a different color from the complex $[Fe(H_2O)_6]^{3+}$. The color is directly related to the value of splitting energy.

Values of Δ_O are typically in the range 7000–30,000 cm^{-1} for the weak field. The splitting energy, that is, Δ_O in octahedral symmetry, depends on the following factors (order of presentation is not necessarily related to their importance):

1. *The charge of the metal ion (oxidation state).* The higher charge of the central ion means a stronger interaction with ligands. Therefore, the ligands are more strongly pulled toward the metal ion. Thus, *splitting energy increases with increasing ionic charge* of the central metal ion. An example is Ru in the same octahedral environment in water: $\Delta_O = 19{,}800$ cm^{-1} for $[Ru(H_2O)_6]^{2+}$, $\Delta_O = 28{,}600$ cm^{-1} for $[Ru(H_2O)_6]^{3+}$.
2. *Nature of the metal ion.* Therefore, the splitting energy depends on whether the metal is a 3*d*, 4*d*, or 5*d* element: *Δ increases* for analogous complexes within a given group in the order 3*d*< 4*d*< 5*d*. As a consequence, the transition metal ions with electronic configuration 4*d* and 5*d* will be mostly in the low-spin state than 3*d* ions.
3. *Number and geometry of ligands.* The splitting energy Δ_T for tetrahedral complexes is roughly 50% of that for octahedral complexes: An example is the same ion of V^{4+} in a different environment: $\Delta_O = 15{,}400$ cm^{-1} for $[VCl_6]^{2-}$; $\Delta_T = 7900$ cm^{-1} for VCl_4.
4. *Nature of the ligands.* The transition metal complexes exhibit different colors that are determined by the value of Δ_O. For the same central ion and different types of ligands in the same

TABLE 2.27 Spectrochemical Series of Several Ligands

$I^- < Br^- < S^{2-} < Cl^- < F^- < OH^- < O^{2-} < H_2O < NH_3 < NO_2^- < CN^- < CO < NO$	
← Weaker field ligands	Stronger field ligands →
← Longer wavelength absorption	Shorter wavelength absorption →

symmetry, one may observe that the maximum of absorption bands corresponding to the *d–d* transition slightly changes its position according to the nature of ligands. By changing the ligand from iodide to fluoride, the maximum moves to higher energy as it is normal. Using the position of the absorption band, the halogen ions were arranged in a series called the spectrochemical series as follows: $I^- < Br^- < Cl^- < F^-$. A smaller ligand in size means increased ability to split orbitals. Ligands are ranked according to the extent to which they split the *d* electrons. Strong field ligands cause the *d* orbitals to be split by a large amount. Weak field ligands, on the other hand, cause the *d* orbitals to be split only by a small amount. By studying a large number of complexes with different types of ligands and not only ions (they can also be dipolar molecules), the *spectrochemical series* was extended as shown in Table 2.27.

A ligand such as CO is called a strong field ligand, whereas a ligand such as Cl^- is considered a weak field ligand. The presence of a negative charge on the ligand is not necessarily a strong field. The neutral (but polar) water molecule H_2O is stronger than OH^-. This and others are problems in explaining the trends in the spectrochemical series with crystal field theory. On the basis of crystal field theory, all anionic ligands should come later in the series (at stronger field strength), but most of them exhibit weaker field strengths. CO is neutral but gives the strongest field. All observations mentioned have led to the development of the ligand field theory from crystal field theory that supposes only pure electrostatic interaction between point charges. Ligand field theory allows covalent interactions between the central metal ion and ligands.

2.6 SPLITTING OF MULTIELECTRON d^n CONFIGURATIONS IN THE CRYSTALLINE FIELD

2.6.1 SPLITTING OF MULTIELECTRON TERMS

Most of the transition metals or lanthanide ions have more than one electron in the valence orbital. What happens to more than one electron in *d* orbitals? What happens to the states of the free ion in the crystal field? *The results obtained for single-electron occupancy of various types of orbitals also apply to the terms arising from multielectron occupancy.*

For multielectron configuration, the electron–electron interactions must be taken into account but *no symmetry operation changes the spin.*

As a free ion, the single-electron state called by the symbol ^2d (d from $l = 2$ and the superscript 2 is the spin multiplicity $2 \cdot 1/2 + 1$) is described as a set of five wavefunctions corresponding to the five values of the magnetic quantic number m. The state is fivefold degenerated. The multielectron ion has an analogous behavior. The Φ factor of the wavefunction also depends on the azimuthal angle as the function $e^{-iM\varphi}$, where M is the total magnetic number. The M number also takes five values 2, 1, 0, –1, and –2 for the D term. (We reviewed the theory of the multielectron ion in Section 2.1.2.) Thus, the symmetry operations will have the same effect on both the D state arising from any group of electrons and the d state of a single electron. Therefore, the fivefold-degenerated D term will split in the crystalline field exactly like the single-electron d orbital.

Remember: The small letter denotes the single-electron terms (e.g., ^2d) and the capital letter denotes multielectron terms (^1D).

The same relation exists between s orbitals and S states, p orbitals and P states, and so on. So, the results given in Table 2.28 for the splitting of different one-electron orbitals also apply to the splitting of analogous Russell–Saunders terms.

Let us take as an example the ion with *two equivalent d electrons* (the *d^2 electron configuration*). The degeneracy of d orbitals is partially lifted by the electrostatic repulsion between electrons. The possible terms of the free ion are ^1S, ^3P, ^1D, ^3F, and ^1G. They may cause a supplementary split when the ion is inserted into a crystalline field with particular symmetry as presented in Table 2.28. We underline that the *ligand field has no effect on the electron spin*. Thus, the spin multiplicity of the free ion term remains unchanged.

Important note:

Knowing the symmetry of the ligand field

- Enables us to *predict the degeneracies* of the derived states
- *Gives no information about their relative energies* (energy order of the new states).

TABLE 2.28 Splitting of Terms of the nd^2 Configuration in Several Symmetries

Term of Free Ion	Terms Splitting in the Crystal Field		
	O_h	T_d	D_{4h}
^1S	$^1A_{1g}$	$^1A_1$	$^1A_{1g}$
^1G	$^1A_{1g} + {}^1E_g + {}^1T_{1g} + {}^1T_{2g}$	$^1A_1 + {}^1E + {}^1T_1 + {}^1T_2$	$2{}^1A_{1g} + {}^1A_{2g} + {}^1B_{1g} + {}^1B_{2g} + 2{}^1E_g$
^3P	$^3T_{1g}$	$^3T_1$	$^3A_{2g} + {}^3E_g$
^1D	$^1E_g + {}^1T_{2g}$	$^1E + {}^1T_2$	$^1A_{1g} + {}^1B_{1g} + {}^1B_{2g} + {}^1E_g$
^3F	$^3A_{2g} + {}^3T_{1g} + {}^3T_{2g}$	$^3A_2 + {}^3T_1 + {}^3T_2$	$^3A_{2g} + {}^3B_{1g} + {}^3B_{2g} + 2{}^3E_g$

In Section 2.4.5, we have seen that the ^2d term of the free ion is split into e_g and t_{2g} states in $\boldsymbol{O}_\mathrm{h}$ symmetry. We have also seen that t_{2g} is lower than e_g, and we have calculated the separation energy between them. From Table 2.28, we may see that the splitting is the same $(e + t_2)$ in $\boldsymbol{T}_\mathrm{d}$ symmetry, but we have no information about the order of the new states and the value of the splitting energy.

2.6.2 ELECTRON FILLING OF e_g AND t_{2g} ORBITALS

We know the fivefold degeneracy term of the free ion split into two terms in octahedral symmetry as shown in Figure 2.36. Let us consider that the ion may have d^1, d^2, d^3, … , d^{10} electronic configurations. How will the electrons be arranged?

Remember: Electrons fill orbitals starting with the lowest-state level, according to Hund's rule and Pauli's exclusion principle:

1. *Hund's rule*: The most stable arrangement of electrons (for a ground-state electron configuration) has the maximum number of unpaired electrons, all with the same spin.
2. *Pauli's exclusion principle*: No two electrons in the same atom can be in the same quantum state. If quantum numbers n, l, and m of two electrons are the same, their spins must be antiparallel.

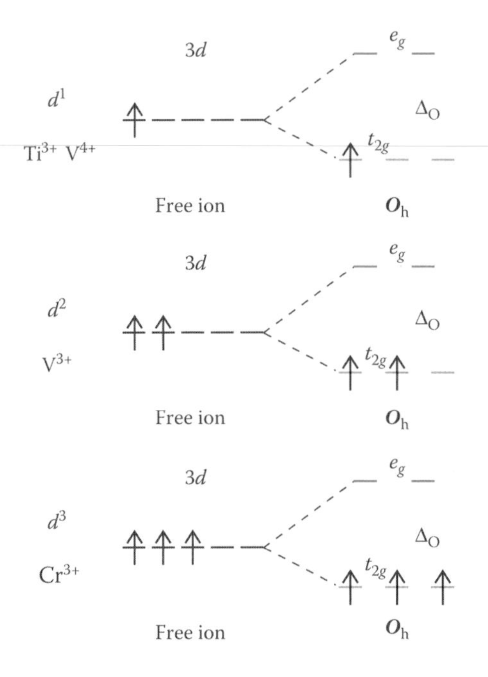

FIGURE 2.36 For d^1–d^3 configurations, the electrons will not pair and occupy the t_{2g} set, one electron in each orbital.

We start to fill the threefold-degenerated t_{2g} orbitals by adding electrons with spins parallel as long as Pauli's principle allows. It means that for d^1–d^3 systems, the electrons will not pair and occupy the t_{2g} set, one electron on each orbital.

For the d^4 configuration, there are two possibilities when the set of t_{2g} orbitals is half-filled:

1. Either put an additional electron in the t_{2g} set and it must pair to occupy one of the orbitals of the degenerate set (Figure 2.37a). The total spin will be 1 and the case is called the *low-spin case.*
2. Alternatively, put the additional electron in the e_g set but the electrons do not pair (Figure 2.37b). The total spin will be 2 and the case is called the *high-spin case.*

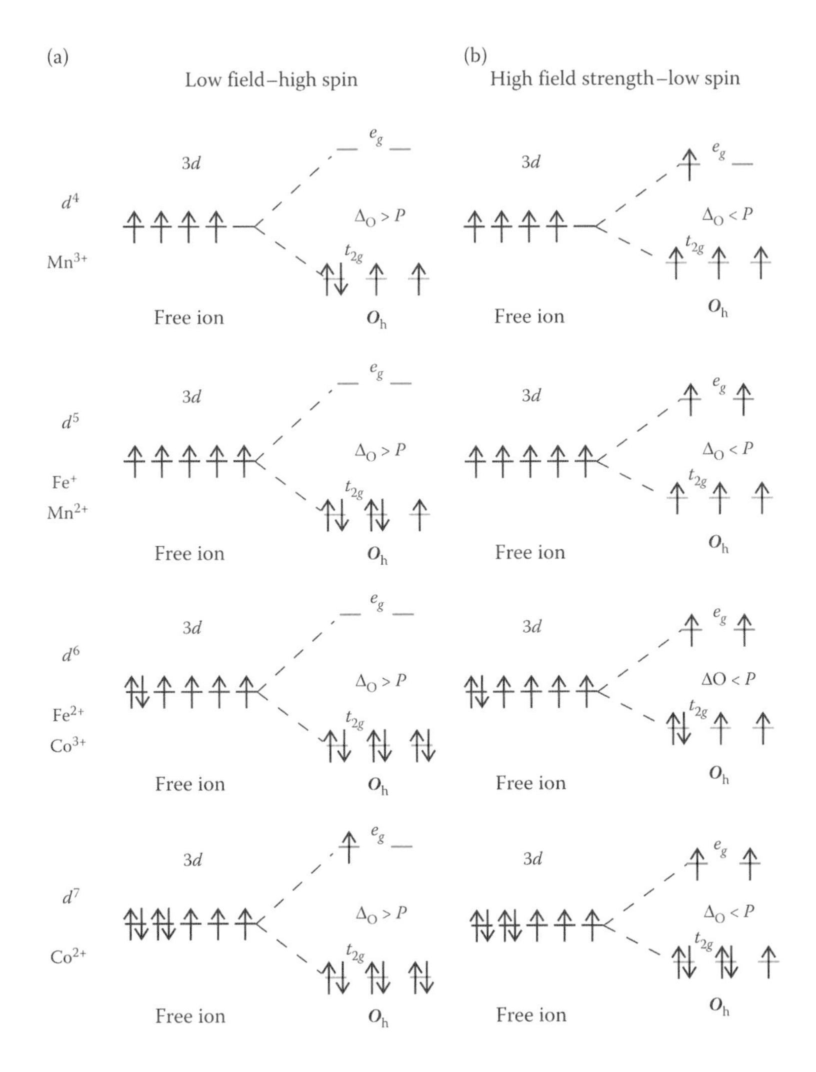

FIGURE 2.37 The occupation of 3*d* orbitals in the free ion, low field, and high field for d^4–d^7 complexes.

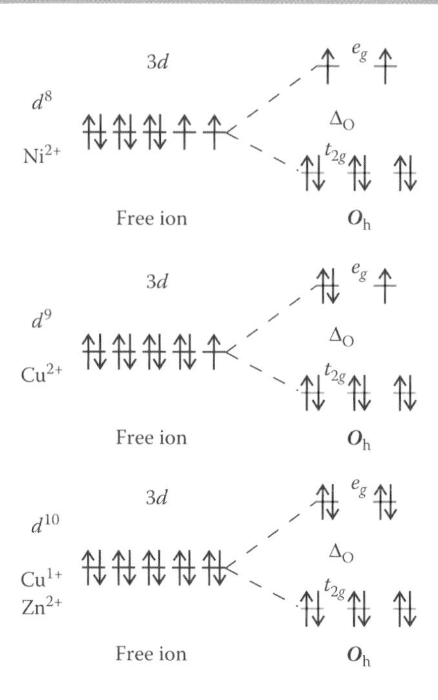

FIGURE 2.38 The occupation of 3*d* orbitals in the free ion and octahedral field for d^8–d^{10} complexes.

What is the criterion to produce one case or the other? There are two parameters to consider:

- The *pairing energy P*
- The *splitting energy* Δ

If the pairing energy P is greater than the splitting energy Δ_O, putting the electron in the higher-level e_g results in a *lower energy state.* For configurations d^4–d^7, there are two possible filling schemes depending on the values of P and Δ_O (Figure 2.37). For a given ion, the pairing energy P is constant but the splitting energy Δ depends on the environment.

There is only one ground state for configurations d^8–d^{10} (Figure 2.38). Owing to the full t_{2g} orbitals, the added electron has the only option of going into the e_g orbital.

2.6.3 SPECTRAL TERMS IN OCTAHEDRAL SYMMETRY (NAMING OF ELECTRONIC STATES)

In Section 2.1.2, we have seen that two electrons can be arranged in *d* orbitals in many ways, resulting in many possible states of the free ion. Then the electronic transitions that give absorption and emission light spectra will imply these states. The states are called starting from the old spectral observation (*s*, *p*, *d*, and *f*), the total spin, and the total orbital momentum. In the crystal field of octahedral symmetry, the *d* orbitals always split into t_{2g} and e_g energy levels. The arrangement of many

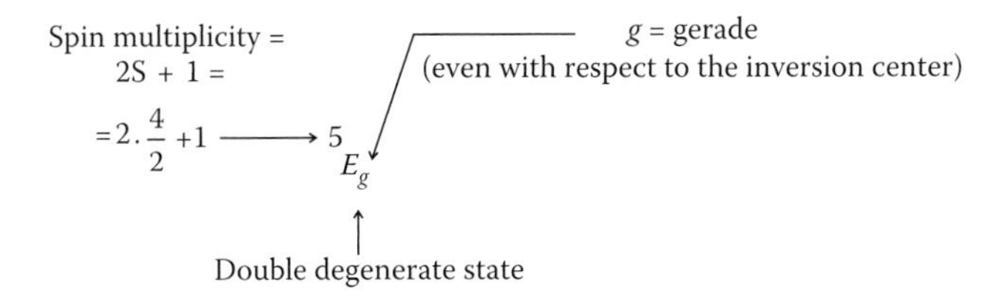

FIGURE 2.39 The state name contains information about the symmetry representation, the existence of the symmetry center, and the total spin. An example is for high-spin $t_{2g}^3e_g^1$ configuration (parallel spins).

electrons on such energy levels will result in many possible multielectron states, so called to contain the symmetry information and the total spin, as in the following example shown in Figure 2.39.

2.6.4 CRYSTAL FIELD STABILIZATION ENERGY

In Figure 2.24, we have seen that the energy of d orbitals of a metal ion becomes higher by placing the ion in the crystal field generated by a set of ligands. This is due to electrostatic repulsion between electrons of the metal ion and the negative charge of the ligands. In the first approximation, the crystal field mainly has a spherical symmetry and the energy becomes much higher (20–40 eV). In the second moment, the specific position (symmetry) of the ligands splits the d orbitals, some of them with lower energy with respect to the barycenter. As a result of the arrangement of electrons in the lower level (e.g., t_{2g} in $\boldsymbol{O}_\text{h}$ symmetry), the metal ion becomes more stable relative to the energy in a spherical field. Therefore, each electron placed on a lower-level t_{2g} will have a contribution of $-4Dq$ ($-0.4\ \Delta_\text{O}$) to the stabilization of the molecular complex. Conversely, each electron placed on a higher-level e_g will reduce by $+6Dq$ ($+0.6\ \Delta_\text{O}$) the energy of stabilization of the molecular complex. Thus, one can define the *crystal field stabilization energy (CFSE)* as the total energy difference relative to the barycenter:

$$\text{CFSE} = \left(-0.4 \cdot n_{t2g} + 0.6 \cdot n_{eg}\right)\Delta_O \tag{2.59}$$

Starting with the d^4 configuration, we must also take into account the *pairing energy* when two electrons occupy the same orbital. The pairing energy will reduce the value of CFSE; therefore,

$$\text{CFSE} = \left(-0.4 \cdot n_{t2g} + 0.6 \cdot n_{eg}\right)\Delta_O + n \cdot P \tag{2.60}$$

where P is the mean pairing energy and n is the number of electron pairs created. Sometimes, CFSE is called the ligand field stabilization energy.

Similarly, for a tetrahedral complex, CFSE will be given by

$$\text{CFSE} = \left(+0.4 \cdot n_{t2g} - 0.6 \cdot n_{eg}\right)\Delta_T + n \cdot P \tag{2.61}$$

In Table 2.29, the calculated values of CFSE are shown for octahedral and tetrahedral symmetry. We note that the *pairing energy is constant* for a given metal ion. It does not depend on the ligand nature, but it depends on the oxidation state of the metal ion. The marked cells show that the d^3 and d^8 configurations benefit the most from the ligand field stabilization energy in octahedral symmetry. However, in tetrahedral symmetry, the d^2 and d^7 configurations benefit the most from the ligand field stabilization energy. When you compare CFSE in octahedral symmetry versus tetrahedral symmetry, keep in mind that the tetrahedral parameter Δ_T is much smaller than the octahedral parameter Δ_O. Thus, the CFSE of the d^2 configuration is greater as an absolute value in octahedral symmetry than in tetrahedral symmetry, because $0.8\Delta_O > 1.2\Delta_T$ ($\approx 1.2 \cdot 4/9 \cdot \Delta_O$). So, the octahedral complex is more stable than the tetrahedral complex. When you compare the values on the column of CFSE in octahedral symmetry, you may see that it increases in absolute value; therefore, the complex becomes more stable, from d^1 to d^3 configuration. However, also remember that the value of Δ_O depends on the nature and the valence state of the metal ion. So, a d^3 complex is not necessarily more stable than a d^2 complex, even for the same element (e.g., V^{2+} vs. V^{3+}). The pairing energy appears to be a destabilization factor because it reduces the total stabilization energy with its positive contribution. Therefore, the electron–electron pairing disfavors the directional arrangement of ligands. The splitting energy in tetrahedral symmetry is relatively small and results in the tetrahedral complexes with low spin not allowed.

TABLE 2.29 Crystal Field Stabilization Energies in Low Field (High Spin) of O_h and T_d Symmetries

Electron Configuration of Free Ion	Octahedral Configuration	CFSE	Tetrahedral Configuration	CFSE
d^0	$t_{2g}^0e_g^0$	$0\ \Delta_O$	$e^0t_2^0$	$0\ \Delta_T$
d^1	$t_{2g}^1e_g^0$	$-0.4\ \Delta_O$	$e^1t_2^0$	$-0.6\ \Delta_T$
d^2	$t_{2g}^2e_g^0$	$-0.8\ \Delta_O$	$e^2t_2^0$	$-1.2\ \Delta_T$
d^3	$t_{2g}^3e_g^0$	$-1.2\ \Delta_O$	$e^2t_2^1$	$-0.8\ \Delta_T$
d^4	$t_{2g}^3e_g^1$	$-0.6\ \Delta_O$	$e^2t_2^2$	$-0.4\ \Delta_T$
d^5	$t_{2g}^3e_g^2$	0	$e^2t_2^3$	$0\ \Delta_T$
d^6	$t_{2g}^4e_g^2$	$-0.4\ \Delta_O + 1P$	$e^3t_2^3$	$-0.6\ \Delta_T + 1P$
d^7	$t_{2g}^5e_g^2$	$-0.8\ \Delta_O + 2P$	$e^4t_2^3$	$-1.2\ \Delta_T + 2P$
d^8	$t_{2g}^6e_g^2$	$-1.2\ \Delta_O + 3P$	$e^4t_2^4$	$-0.8\ \Delta_T + 3P$
d^9	$t_{2g}^6e_g^3$	$-0.6\ \Delta_O + 4P$	$e^4t_2^5$	$-0.4\ \Delta_T + 4P$
d^{10}	$t_{2g}^6e_g^4$	$0\ \Delta_O + 5P$	$e^4t_2^6$	$0\ \Delta_T + 5P$

TABLE 2.30 Comparative Crystal Field Stabilization Energies in High and Low Fields of O_h Symmetries

Electron Configuration of Free Ion	High Field versus Low Field	CFSE	Example of Molecular Complexes	Spin State
d^4	$t_{2g}^4e_g^0$	$-1.6\,\Delta_O + 1P$	$[Mn(CN)_6]^{3-}$	Low spin
	$t_{2g}^3e_g^1$	$-0.6\,\Delta_O$	$[Mn(H_2O)_6]^{3+}$	High spin
d^5	$t_{2g}^5e_g^0$	$-2.0\,\Delta_O + 2P$	$[Fe(CN)_6]^{3-}$	Low spin
	$t_{2g}^3e_g^2$	0	$[Fe(H_2O)_6]^{3+}$	High spin
d^6	$t_{2g}^6e_g^0$	$-2.4\,\Delta_O + 3P$	$[Fe(CN)_6]^{4-}$	Low spin
	$t_{2g}^4e_g^2$	$-0.4\,\Delta_O + 1P$	$[Fe(H_2O)_6]^{2+}$	High spin
d^7	$t_{2g}^6e_g^1$	$-1.8\,\Delta_O + 3P$	$[CoF_6]^{3-}$	Low spin
	$t_{2g}^5e_g^2$	$-0.8\,\Delta_O + 2P$	$[Co(H_2O)_6]^{2+}$	High spin

Table 2.30 shows a comparison between the CFSE of the same metal ion both in weak and strong fields. As long as the pairing energy has a small value compared to splitting energy, the low-spin complexes tend to be more stable than the high-spin complexes. If P is comparable to Δ, the CFSE favors the formation of high-spin complexes.

The above formula of CFSE (Equations 2.60 and 2.61) reflects the limitation of crystal field theory. The presence of pairing energy is evidence to support covalence bonding within transition metal complexes.

2.7 JAHN–TELLER EFFECT

The Jahn–Teller effect may occur when *all six ligands* of the molecular complex *are the same* and the complex is distorted in one direction (two opposite ligands are closer or farther away from the central ion). The theorem states:

The nonlinear symmetric molecular complexes which are in a degenerate electronic state tend to lower their symmetry by distortion and thereby lower their energy and lift the degeneracy.

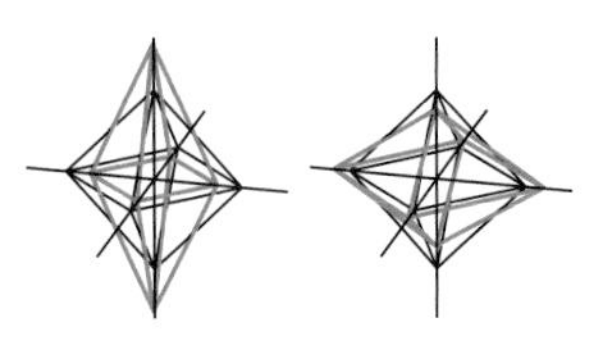

FIGURE 2.40 An elongated or compressed octahedron changes the symmetry to tetragonal.

As an example, the elongated or compressed octahedron illustrated in Figure 2.40 changes the symmetry to tetragonal. Hence, we may treat the Jahn–Teller effect as a particular case of Section 2.5.4. For an octahedral complex, $d_{x^2-y^2}$ and d_{z^2} orbitals have the same energy. Imagine that one electron occupies the d_{z^2} orbital. Thus, the greater electrostatic repulsion between the negative charge of the electron concentrated in the d_{z^2} orbital and two ligands on the z axis will push the ligands up and down (Figure 2.41a). If two ligands increase their distance along the z axis, the energy of the d_{z^2} orbital decreases due to the electrostatic repulsion reduces. We suppose that the elongation on the z axis is counterbalanced by the compression on the x and y axes to keep the total energy constant. The compression in the xy plane results in stronger repulsion and the energy of the $d_{x^2-y^2}$ orbital increases (Figure 2.41b). The orbitals containing the z

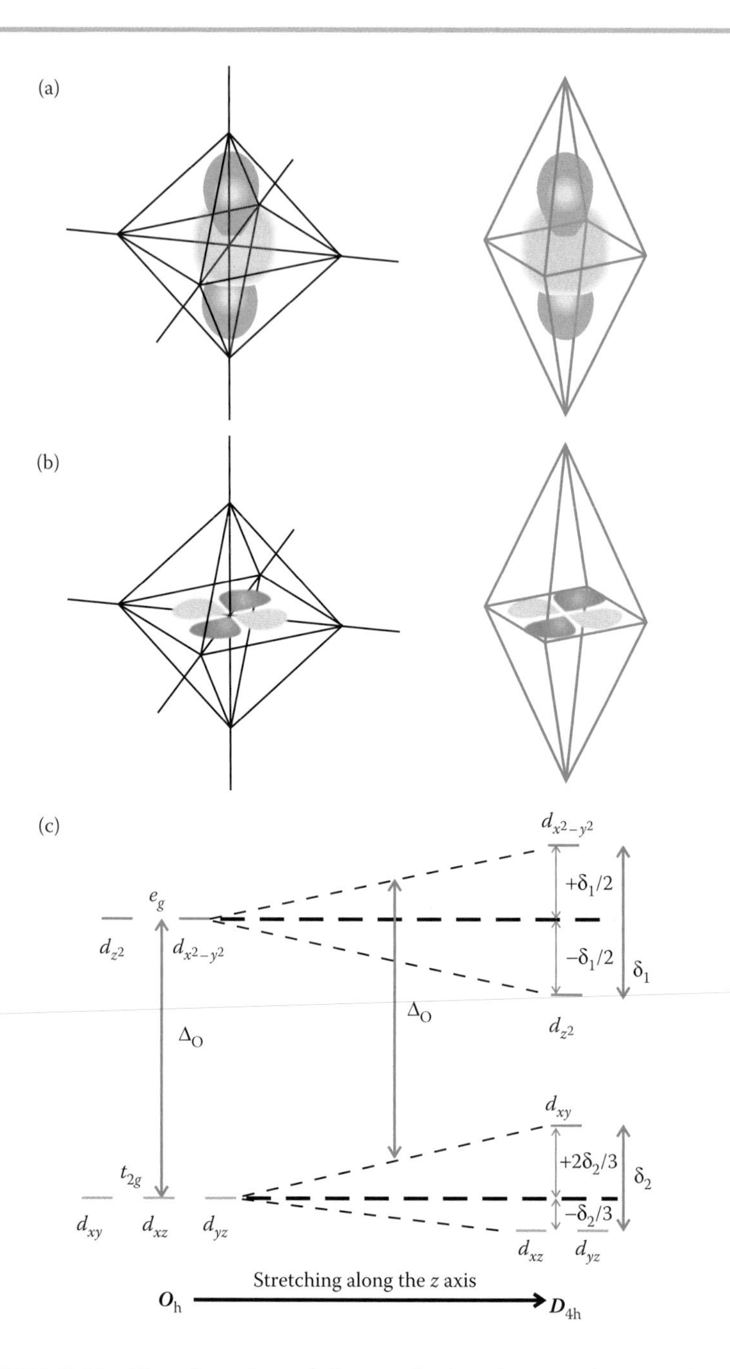

FIGURE 2.41 The distortion of the octahedron by stretching in the z direction results in weaker repulsion on the d_{z^2} orbital and stronger repulsion on the $d_{x^2-y^2}$ orbital.

coordinates d_{xz} and d_{yz} will be affected similar to d_{z^2}; therefore, their energy will be lower. The energy of the orbital located in the xy plane d_{xy} will be higher. The further splitting of the t_{2g} and e_g levels must respect the rule of barycenter conservation (Figure 2.41c). The upper e_g level splits equally into two nondegenerated levels called a_{1g} and b_{1g}. The lower t_{2g} level splits into one double-degenerated new e_g level and the other nondegenerated b_{2g} level. b_{2g} is located 2 times higher with respect to the barycenter than e_g has decreased. The new states were denoted using the correlation table (Table 2.21) because the new symmetry of the complex is $\boldsymbol{D}_{4h}$. The splitting energy δ_1 is larger than δ_2 because both the d_{z^2} and $d_{x^2-y^2}$ orbitals have lobes directed toward the ligands. Both orbitals located in the xy plane are equally affected by compressed ligands, so they have the same energy slope (increase with the same amount). This may be written as

$$\frac{\delta_1}{2} = 2\frac{\delta_2}{3} \Rightarrow \delta_1 = \frac{4}{3}\delta_2$$

The value of splitting energies of both t_{2g} and e_g levels depends on the stretching value in the z direction. Higher elongation means higher splitting energy.

Conversely, the occupation of the $d_{x^2-y^2}$ orbital will induce elongation on the xy plane and compression on the z axis; thus, the splitting of both the t_{2g} and e_g levels will be reversed.

The Jahn–Teller distortion is mostly observed in octahedral environments. The distortion is not possible in all nondegenerate configurations: d^3, d^8, d^{10}, d^5 high spin, and d^6 low spin (Figure 2.42).

Theoretically, all other degenerate electronic configurations in octahedral symmetry may distort, but the Jahn–Teller distortion is usually observed in

- d^4 high spin ($t_{2g}^3e_g^1$) configuration, for example, Cr^{2+}, Mn^{3+}
- d^7 low spin ($t_{2g}^6e_g^1$) configuration, for example, Co^{2+}, Ni^{3+}
- d^9 ($t_{2g}^6e_g^3$) configuration, for example, Cu^{2+}

This happens because of e_g orbitals (i.e., d_{z^2} and $d_{x^2-y^2}$) that are directed toward the ligands.

Figure 2.32 illustrated the absorption spectra of the Ti^{3+} ion in octahedral symmetry $[Ti(H_2O)_6]^{3+}$, where the green absorption band (maximum located at around 500 nm) was supposed to be due to the electronic d–d transition from the ground state $t_{2g}^1e_g^0$ to the excited state $t_{2g}^0e_g^1$. The absorption band is highly asymmetric toward longer wavelength and shows a distinct shoulder. It means that another weaker absorption band appears that is not well separated from the first absorption band at room temperature. It is because this time the Jahn–Teller distortion will remove

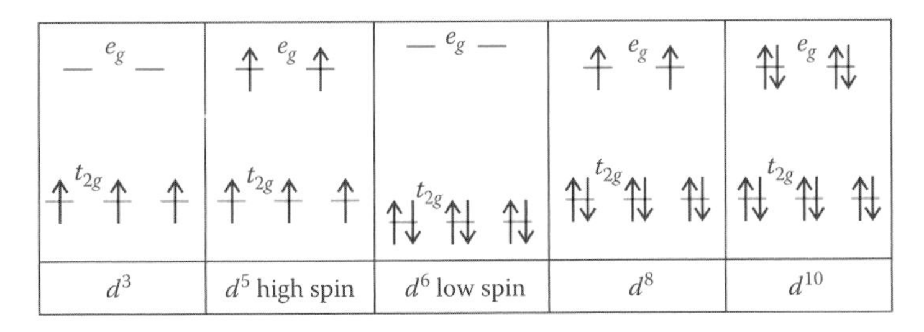

FIGURE 2.42 The nondegenerate configurations do not exhibit the Jahn–Teller distortion.

the degeneracy of the *excited state* and not of the ground state. The excited state will be split into two excited states, one for each orbital d_{z^2} and $d_{x^2-y^2}$, respectively. Thus, two transitions are possible that lead to the splitting of the green absorption band. The phenomenon is also present in Ti:Al_2O_3 and is very useful because it allows us to use different optical pumping systems in Ti:Sapphire lasers (a 514 nm argon ion laser or a 532 nm neodymium frequency-doubled laser).

2.8 CONSTRUCTION OF ENERGY-LEVEL DIAGRAMS

We have seen in Table 2.28 the splitting of L–S terms of d^2 configuration in the octahedral weak field. Group theory shows an easy way to know what the splitting does for each term of the free ion, but it does not directly show:

- The relative energy of the new terms and
- How the energy of the terms depends on the Δ_O value

In other words, we know that the ground level ^3F splits into $^3A_{2g}$, $^3T_{1g}$, and $^3T_{2g}$, but we do not know their order on their energy scale for different values of crystal field strength.

We try to describe two different procedures to obtain the correct order without solving the secular equations. Thus, we can obtain only qualitative information useful to recognize the absorption bands of an ion complex.

2.8.1 METHOD OF DESCENDING SYMMETRY

Here, we will describe how to find the ion states in a strong crystal field.

The method of descending symmetry was elaborated by Bethe. It is rigorous but too long. The main advantage is that the method can be applied to all systems.

Let us take as an example d^2 configuration. We have previously seen that the d orbitals of the free ion split into t_{2g} and e_g in octahedral symmetry. Filling with two electrons t_{2g} and e_g orbitals, we may obtain the following configurations: $t_{2g}^2e_g^0$, $t_{2g}^1e_g^1$, and $t_{2g}^0e_g^2$. It is simpler to start by showing how to use the procedure for the last configuration.

1. What are the states of $t_{2g}^0e_g^2$ configuration in $\boldsymbol{O}_{\mathrm{h}}$ symmetry? The steps of the procedure are shown in Table 2.31.
   The multiplicity of new states will be found by lowering the symmetry.
   Remember that the total degeneracy of each term is the product of its spin degeneracy and its orbital degeneracy $(2S + 1) \cdot (2L + 1)$. As a consequence, the degeneracy of the e_g term equals $(2 \cdot 1/2 + 1)\ (2) = 4$.
   What are the states of $t_{2g}^0e_g^2$ configuration by lowering the symmetry to $\boldsymbol{D}_{4\mathrm{h}}$?

TABLE 2.31 Procedure to Find the States of $t_{2g}^0e_g^2$ Configuration in O_h Symmetry

1. Finding the reducible representation generated by e_g^2

e_g	2	−1	0	0	2	2	−1	0	0	2
$e_g^2 = e_g \times e_g$	4	1	0	0	4	4	1	0	0	4

2. Reducing the corresponding representation to irreducible representations (multielectron notation):

A_{1g}
$$\frac{1}{48}[1\cdot(1)(4) + 8\cdot(1)(1) + 6\cdot(1)(0) + 6\cdot(1)(0) + 3\cdot(1)(4)$$
$$+ 1\cdot(1)(4) + 8\cdot(1)(1) + 6\cdot(1)(0) + 6\cdot(1)(0) + 3\cdot(1)(4)] = 1$$

A_{2g}
$$\frac{1}{48}[1\cdot(1)(4) + 8\cdot(1)(1) + 6\cdot(-1)(0) + 6\cdot(-1)(0) + 3\cdot(1)(4)$$
$$+ 1\cdot(1)(4) + 8\cdot(1)(1) + 6\cdot(-1)(0) + 6\cdot(-1)(0) + 3\cdot(1)(4)] = 1$$

E_g
$$\frac{1}{48}[1\cdot(1)(4) + 8\cdot(1)(1) + 6\cdot(1)(0) + 6\cdot(1)(0) + 3\cdot(1)(4)$$
$$+ 1\cdot(1)(4) + 8\cdot(1)(1) + 6\cdot(1)(0) + 6\cdot(1)(0) + 3\cdot(1)(4)] = 1$$

3. The decomposition of e_g^2 results in irreducible representations of the $\boldsymbol{O}_h$ group

$e_g^2 = e_g \times e_g = \quad A_{1g} + A_{2g} + E_g$

The e_g orbitals split into a_{1g} and b_{1g} when the symmetry decreases to tetrahedral $\boldsymbol{D}_{4h}$ (Figure 2.43). From the character table of $\boldsymbol{D}_{4h}$, it results in that d_{z^2} has a_{1g} symmetry and $d_{x^2-y^2}$ has b_{1g} symmetry. We can put electrons on the new levels as shown in Figure 2.43. Table 2.32 shows the new configurations, the direct product, and the spin multiplicities of the new states.

The symmetry operations do not apply to the electron spin; Pauli's exclusion principle must be applied and the new spin multiplicity results from Figure 2.43.

Now, by simple inspection of the correlation Table 2.21, we may establish the correspondence between the states of d^2 configuration from $\boldsymbol{O}_h$ to $\boldsymbol{D}_{4h}$ as follows in Table 2.33.

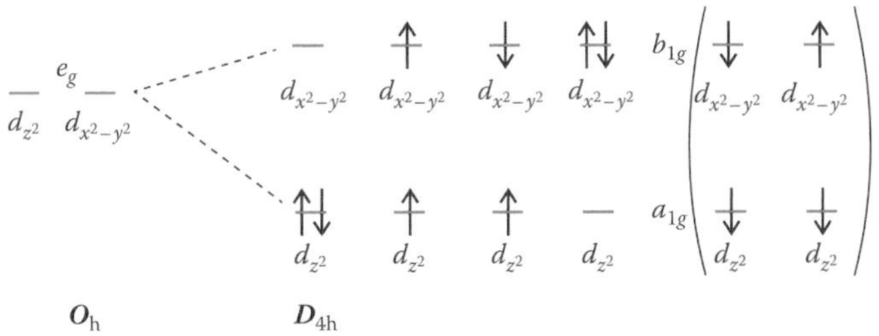

FIGURE 2.43 There are six possibilities to put two electrons on the levels a_{1g} and b_{1g} when the symmetry decreases from octahedral to tetrahedral $\boldsymbol{D}_{4h}$, but only four are distinct.

TABLE 2.32 States of $t_{2g}^0 e_g^2$ Configuration in Octahedral Symmetry Will Generate Four Different States by Decreasing the Symmetry to Tetrahedral

Configuration	Direct Product	Spin Multiplicity
$a_{1g}^2 b_{1g}^0$	A_{1g}	${}^1A_{1g}$
$a_{1g}^1 b_{1g}^1$	B_{1g}	${}^3B_{1g}$ ${}^1B_{1g}$
$a_{1g}^0 b_{1g}^2$	A_{1g}	${}^1A_{1g}$

Taking into consideration that spin multiplicity is not changed by lowering of symmetry, we are able to write the multiplicity of states in $\boldsymbol{O}_h$ symmetry as shown above.

2. We continue to find the states of $t_{2g}^2 e_g^0$ *configuration* by the same procedure as shown in Table 2.34. The representation generated by this configuration has a ninefold dimension and must be reduced to irreducible representations.

The result means that this configuration has two states that are threefold degenerate, one state that is twofold degenerate, and one state that is nondegenerate.

Looking at correlation Table 2.21, lowering the symmetry from $\boldsymbol{O}_h$ to $\boldsymbol{D}_{4h}$ can be seen, but the degeneracy is still present with the twofold-degenerate e_g states. $\boldsymbol{C}_{2h}$ is the lower symmetry of the complex where all irreducible representations are nondegenerate. The $\boldsymbol{C}_{2v}$ group is also useful for this procedure.

The states of $t_{2g}^2 e_g^0$ configuration in octahedral symmetry will generate nine different states by decreasing the symmetry to tetrahedral as shown in Figure 2.44 and has spin multiplicity as shown in Table 2.35.

Now, by simple inspection of the correlation Table 2.21, we may establish the correspondence between the states of d^2 configuration from $\boldsymbol{O}_h$ to $\boldsymbol{C}_{2h}$ as follows. We also observe that there is only one triplet ${}^3A_g$ and this can correspond to T_{1g}. Then the spin multiplicity is obviously assigned as in Table 2.36.

TABLE 2.33 Setting of Spin Multiplicity by Using the Correspondence between States of d^2 Configuration in O_h to D_{4h}

O_h	D_{4h}	Spin correspondence $D_{4h} \rightarrow O_h$	
A_{1g}	A_{1g}	${}^1A_{1g}$ →	${}^1A_{1g}$
A_{2g}	B_{1g}	${}^3B_{1g}$ →	${}^3A_{2g}$
E_g	$A_{1g} + B_{1g}$	${}^1A_{1g} + {}^1B_{1g}$ →	${}^1E_g$

TABLE 2.34 Procedure to Find the States of $t_{2g}^2 e_g^0$ Configuration in O_h Symmetry

1. Finding the reducible representation generated by t_{2g}^2

t_{2g}	3	0	1	−1	−1	3	0	1	−1	−1
$t_{2g}^2 = t_{2g} \times t_{2g}$	9	0	1	1	1	9	0	1	1	1

2. Reducing the corresponding representation to irreducible representations (multielectron notation):

A_{1g} $$\frac{1}{48}[1\cdot(1)(9) + 8\cdot(1)(0) + 6\cdot(1)(1) + 6\cdot(1)(1) + 3\cdot(1)(1) + 1\cdot(1)(9) + 8\cdot(1)(0) + 6\cdot(1)(1) + 6\cdot(1)(1) + 3\cdot(1)(1)] = 1$$

A_{2g} $$\frac{1}{48}[1\cdot(1)(9) + 8\cdot(1)(0) + 6\cdot(-1)(1) + 6\cdot(-1)(1) + 3\cdot(1)(1) + 1\cdot(1)(9) + 8\cdot(1)(0) + 6\cdot(-1)(1) + 6\cdot(-1)(1) + 3\cdot(1)(1)] = 0$$

E_g $$\frac{1}{48}[1\cdot(2)(9) + 8\cdot(-1)(0) + 6\cdot(0)(1) + 6\cdot(0)(1) + 3\cdot(2)(1) + 1\cdot(2)(9) + 8\cdot(-1)(0) + 6\cdot(0)(1) + 6\cdot(0)(1) + 3\cdot(2)(1)] = 1$$

T_{1g} $$\frac{1}{48}[1\cdot(3)(9) + 8\cdot(0)(0) + 6\cdot(-1)(1) + 6\cdot(1)(1) + 3\cdot(-1)(1) + 1\cdot(3)(9) + 8\cdot(0)(0) + 6\cdot(-1)(1) + 6\cdot(1)(1) + 3\cdot(-1)(1)] = 1$$

T_{2g} $$\frac{1}{48}[1\cdot(3)(9) + 8\cdot(0)(0) + 6\cdot(1)(1) + 6\cdot(-1)(1) + 3\cdot(-1)(1) + 1\cdot(3)(9) + 8\cdot(0)(0) + 6\cdot(1)(1) + 6\cdot(-1)(1) + 3\cdot(-1)(1)] = 1$$

3. The decomposition of t_{2g}^2 results in irreducible representations of the $\boldsymbol{O}_h$ group

$t_{2g}^2 = t_{2g} \times t_{2g}$ $A_{1g} + E_g + T_{1g} + T_{2g}$

3. Concerning the states of the $t_{2g}^1 e_g^1$ configuration, we know from the beginning that each state can be a triplet and a singlet because the two electrons will never be on the same level and Pauli's exclusion principle does not apply. The reader may take this as an exercise to apply all of the above procedures.

The energy diagram will be built after we present the second method.

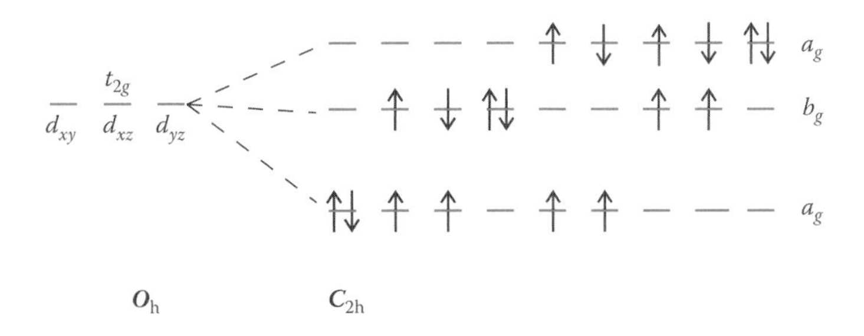

FIGURE 2.44 Filling with electrons of levels obtained by supplementary splitting in the lower symmetry $\boldsymbol{C}_{2h}$. (The energy position of new levels is arbitrarily chosen.)

TABLE 2.35 Spin Multiplicity of Each Microstate of the $t_{2g}^2e_g^0$ Configuration

Configuration	Direct Product	Spin Multiplicity
$a_g^2b_g^0a_g^0$	A_g	^1A_g
$a_g^1b_g^1$	B_g	^3B_g ^1B_g
$a_g^0b_g^2$	A_g	^1A_g
$a_g^1a_g^1$	A_g	^3A_g ^1A_g
$b_g^1a_g^1$	B_g	^3B_g ^1B_g
$a_g^0b_g^0a_g^2$	A_g	^1A_g

TABLE 2.36 Setting of Spin Multiplicity of $t_{2g}^2e_g^0$ Configuration States by Reducing the Symmetry from O_h to C_{2h}

O_h	C_{2h}	Spin Correspondence $C_{2h} \to O_h$
A_g	A_g	$^1A_{1g} \longrightarrow {}^1A_{1g}$
E_g	A_g+B_g	$^1A_{1g} + {}^1B_{1g} \longrightarrow {}^1E_g$
T_{1g}	$A_g+B_g+B_g$	$^3A_g + {}^3B_g + {}^3B_g \longrightarrow {}^3T_{1g}$
T_{2g}	$A_g+B_g+B_g$	$^1A_g + {}^1A_g + {}^1B_g \longrightarrow {}^1T_{2g}$

2.8.2 METHOD OF CORRELATION OF THE WEAK FIELD–STRONG FIELD

The other way to build the energy diagram is to construct the correlation diagram between the weak field and the strong field. The main idea is that in the weak field, we know the exact splitting and in the strong field, we can know the order on the energy scale. Thus, the correlation between two extreme situations could help us to obtain more information about ion terms in the crystal field without performing difficult calculus.

We may suppose:

1. The interactions between electrons of different orbitals are negligible in the strong field.
2. Each state on the weak field must be retrieved on the strong field (*biunivocal correspondence rule*).
3. The correlation lines of states with the same symmetry and spin do not cross (*noncrossing rule*).

So, for the d^2 *configuration*, we know the free ion terms (as a result of the spin–orbit coupling) and their relative position on the energy scale (Table 2.7 and Figure 2.5) and they are shown on the left side of Figure 2.45. When the ion is placed into a molecular complex, the ligand field will partly remove the degeneracy of free ion terms by additionally splitting them into 11 terms (Table 2.26), conserving the spin multiplicity as shown in Table 2.37.

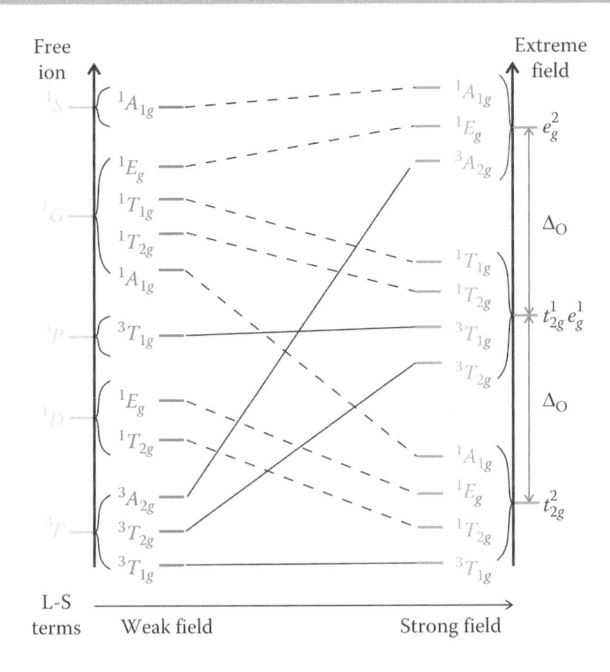

FIGURE 2.45 The correlation diagram between the weak field and strong field states of the d^2 electronic configuration in $\boldsymbol{O}_h$ symmetry. The solid line was used to connect weak field–strong field states involved in transitions that give stronger absorption (see later selection rules).

When the crystal field of octahedral symmetry is extremely strong, the splitting of d orbitals is so high that there are only three configurations possible in increasing order of their energy: $t_{2g}^2e_g^0$, $t_{2g}^1e_g^1$, and $t_{2g}^0e_g^2$. They are represented on the right side of Figure 2.45. Let us imagine reducing the interaction between the ion and ligands. The two d electrons of the central ion begin to interact with each other and Pauli's principle must be applied. Each configuration will generate a set of new states. The procedure to find the new states was described previously: we decompose into irreducible representations specific to the symmetry group the representation

TABLE 2.37 Weak Field of O_h Symmetry Partly Removes the Degeneracy of Free Ion Terms of d^2 Configuration by Supplementary Splitting (5 Terms to 11 Terms)

Term of Free Ion	Splitting in O_h	11 States
^1S	$^1A_{1g}$	$2 \times A_{1g}$
^1G	$^1A_{1g} + {}^1E_g + {}^1T_{1g} + {}^1T_{2g}$	$1 \times A_{2g}$
^3P	$^3T_{1g}$	$2 \times E_g$
^1D	$^1E_g + {}^1T_{2g}$	$3 \times T_{1g}$
^3F	$^3A_{2g} + {}^3T_{1g} + {}^3T_{2g}$	$3 \times T_{2g}$

TABLE 2.38 Strong Field of O_h Symmetry Also Removes the Degeneracy of Free Ion Terms of d^2 Configuration by Splitting 5 Terms into 11 Terms

Splitting in Strong Field of O_h Symmetry	Extreme Field Configurations	11 States
$A_{1g} + A_{2g} + E_g$	$t_{2g}^0e_g^2$	$2\times A_{1g}$, $1\times A_{2g}$, $2\times E_g$,
$^1T_{1g} + {}^1T_{2g}$ and $^3T_{1g} + {}^3T_{2g}$	$t_{2g}^1e_g^1$	$3\times T_{1g}$, $3\times T_{2g}$
$A_{1g} + E_g + T_{1g} + T_{2g}$	$t_{2g}^2e_g^0$	

corresponding to each configuration, that is, $t_{2g} \times t_{2g}$, $t_{2g} \times e_g$, and $e_g \times e_g$. The total number of new states equals 11, as shown in Table 2.38.

The correlation diagram between the weak and strong fields helps us to order the new terms based on the noncrossing rule.

It is easier to start with unsplit terms: $^1S \rightarrow {}^1A_{1g}$. On the right side, there are two terms $^1A_{1g}$, the one from the higher configuration $t_{2g}^0e_g^2$ and the other from the lower configuration $t_{2g}^2e_g^0$. It is obvious from the noncrossing rule that the correspondence will be to the first term. Thus, the term $^1A_{1g}$ from ^1G will correspond to $^1A_{1g}$ from the $t_{2g}^2e_g^0$ configuration. The higher term $^3T_{1g}$ from ^3P corresponds to $^3T_{1g}$ from the higher $t_{2g}^1e_g^1$ and the lower $^3T_{1g}$ from ^3F corresponds to $^3T_{1g}$ from the lower term $t_{2g}^2e_g^0$. The higher ^1E_g from ^1G corresponds to ^1E_g from the higher $t_{2g}^0e_g^2$ and the lower ^1E_g from ^1D corresponds to ^1E_g from the lower $t_{2g}^2e_g^0$, and so on. The connecting lines can only be drawn as shown in Figure 2.45.

2.8.3 CORRELATION DIAGRAM AND HOLE FORMALISM

We should look again at Table 2.6 that presents all terms of transition metal free ions for all possible configurations. We may see that: The *terms for complementary configurations* d^n *and* d^{10-n} *are the same*. That is, for example, the filling of five *d* orbitals with *eight electrons* results in six paired electrons and *two electrons* that are free to participate in transitions. We may also look to the d^8 electron configuration like to a d^2 positive charges configuration (holes), as in semiconductor physics. This way of thinking, called the *hole formalism* (Cotton 1990), can be used to find terms in the strong field configuration. It means that the *behavior of configuration* d^{10-n} will be the same as that for d^n configuration, but the energies of interaction will have opposite signs, because the electric charge of *d* orbitals of the central ion is formally *positive* for holes.

The simplest example is to compare the d^1 to d^9 configuration. While the d^1 configuration splits into T_{2g} and E_g terms, the d^9 configuration splits into E_g and T_{2g} terms (Figure 2.46). (We used the same multielectron notation for both configurations to avoid confusion.) The absorption of light is due to the electron transition from the ground state $^2T_{2g}$ to the excited state ^2E_g for d^1 configuration of the $[Ti(H_2O)_6]^{3+}$ complex. For the d^9 configuration, the electron transition from t_{2g} to e_g orbital is equivalent to the hole transition from e_g to t_{2g} orbital; therefore, the light absorption is due

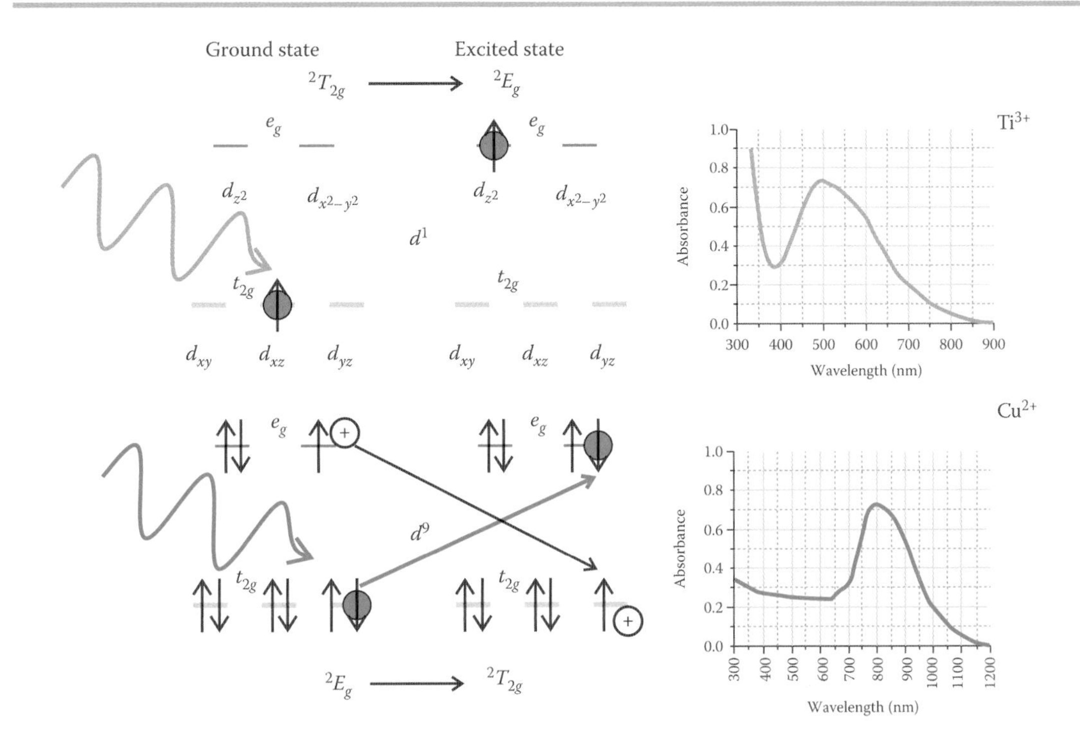

FIGURE 2.46 The light absorption for the d^1 configuration represented by the Ti^{3+} ion is due to electron transition from the ground state $^2T_{2g}$ to the excited state ^2E_g. For the d^9 configuration represented by the Cu^{2+} ion, the absorption is formally explained by the hole transition from ^2E_g to excited state $^2T_{2g}$.

to the transition from the ground state ^2E_g to the excited state $^2T_{2g}$ for the d^9 configuration of the $[Cu(H_2O)_6]^{2+}$ complex. The difference between the peak positions of the light absorption, 500 nm for Ti^{3+} versus 800 nm for Cu^{2+}, is due to the difference between atomic species of the central ion.

Here, we note a new evidence of the Jahn–Teller distortion effect shown in the absorption spectrum of $[Cu(H_2O)_6]^{2+}$ (Figure 2.46). The presence of the shoulder in the single band to longer wavelength is due to the splitting of the ^2E_g ground state into two distinct states with two transition energies to the $^2T_{2g}$ excited state (slightly split too).

The consequence of hole formalism is, for example, for the d^8 configuration, that the extreme field configurations are in the order of increasing energy: $t_{2g}^{6e}e_g^{2e} = e_g^{2p}t_{2g}^{0p}$, $t_{2g}^{5e}e_g^{3e} = e_g^{1p}t_{2g}^{1p}$, and $t_{2g}^{4e}e_g^{4e} = e_g^{0p}t_{2g}^{2p}$ (superscript *e* means electrons and *p* means a positive charge). The correlation diagram between the weak and strong field states of the d^8 electronic configuration is illustrated in Figure 2.47.

2.8.4 EXTENSION OF THE HOLE FORMALISM RELATIONSHIP

Can we more simplify our work based on group theory results?

We have already seen that the tetrahedral environment splits *d* orbitals into two states such as the octahedral environment but is reversed: e_g lower and t_{2g} higher. This is exactly to the behavior of complementary configuration (of hole formalism) in the octahedral environment just

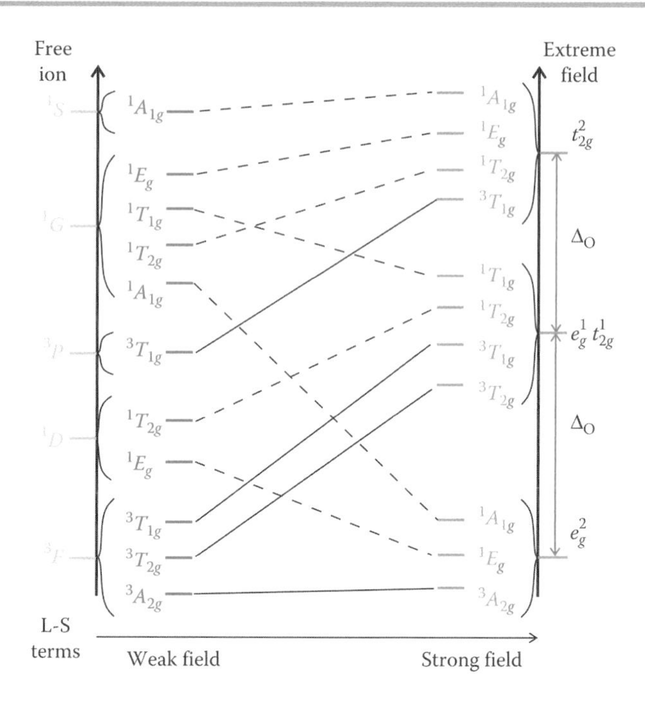

FIGURE 2.47 The correlation diagram between weak field and strong field states of the d^8 electronic configuration in $\boldsymbol{O}_h$ symmetry.

TABLE 2.39 Equivalence of Correlation Diagrams for the O_h and T_d Groups

Octahedron		Tetrahedron	
$1/d^9$	d^1	d^9	$1/d^1$
$1/d^8$	d^2	d^8	$1/d^2$
$1/d^7$	d^3	d^7	$1/d^3$
$1/d^6$	d^4	d^6	$1/d^4$
$1/d^5$	d^5	d^5	$1/d^5$

presented above. Therefore, we must know the correlation diagram only for d^1, d^2, d^3, d^4, and d^5 in the octahedron and we have both the diagram of complementary configuration in the octahedron and the diagram of the same configuration in the tetrahedron (Table 2.39).

2.9 TANABE–SUGANO DIAGRAMS

2.9.1 SELECTION RULES AND POLARIZATIONS

Most transition metal compounds are highly colored. The color of transition metals arises from a split in the energies of the d orbitals caused by the ligand field. The color is due to electronic transitions. There are two important types of transitions for TM ions in a ligand field (crystalline or molecular complex):

- *Charge transfer transitions*. They are often called ligand-to-metal charge transitions (LMCT) and consist of electron jumps from the ligand orbital to the metal orbital. LMCT supposes this happens due to the interaction between ligand molecular orbitals and atomic orbitals of the metal ion. So, they are not objects of the crystal field theory, if you remember the start suppositions of this theory. Anyway, the charge transitions take place inside the molecular complex. LMCTs are not restricted by any rule. They can happen if the light energy (meaning the photon energy and not the beam intensity) is great enough to cover the energy difference between the initial state and the final state. As a consequence, LMCT produces extremely intensive and large absorption bands located in the UV range (≈200 nm) or vacuum UV (<180 nm). Such transition often saturates the measurement capacity of the spectrophotometer. Metal-to-ligand charge transitions (MLCTs) are also possible. In the molecular complexes, the charge transitions can produce from the complex to the solvent. In the case of TM ions-doped crystals, the charge transfer implies that the metal ions of the host crystal (e.g., K in the KCl crystal) and electron transition produces from the complex (e.g., $(CoCl_4)^{2-}$) to the metal ion (K^+) (Trutia 2005). To measure the intensity of the charge transition band, the crystal sample must be very thin to reduce the band intensity under the maximum value of the spectrophotometer range. In some cases, they can also be obscured by the fundamental absorption of the crystal.
- *d–d transitions*. They mean the electron jumps from one *d* orbital to another *d* orbital of the metal ion. This type of transition happens because the crystal field splits the *d* orbitals creating new levels of energy possible for *d* electrons. Any transition between two new terms in the crystalline field is a *d–d* transition and it implies only the central ion.

The intensity of *d–d* transitions can vary greatly from one band to another. They are governed by selection rules.

2.9.1.1 LAPORTE'S RULE

The intensity of the measured band depends on many factors, one of which is the *probability of transition*. In quantum mechanics, the probability of transition is given by

$$\int_{-\infty}^{+\infty} \Psi^*_f \hat{\mu} \Psi_i \, d\tau$$

where Ψ_i is the wave function of the initial state, Ψ_f is the wave function of the final state, and $\hat{\mu}$ is the transition moment operator. If this quantity equals zero, the transition is forbidden. Thus, we do not need to calculate this integral if we have a way to predict when it vanishes, and group theory is a strong tool based on symmetry considerations. The *integral over the whole space* (i.e., over a symmetric range) of an *odd function equals zero.* In the first approximation, we may suppose that the transition is given by the electric dipole that has a formula $e \cdot \vec{r}$; therefore, it depends on coordinates that are *antisymmetric relative to the inversion center.* The wavefunctions of the initial and final states in the case of the transition metal ions are linear combinations of *d* orbital wavefunctions. We know that *all d wavefunctions are symmetric relative to the inversion center,* and thus the wavefunctions of the initial and final states are symmetric too. The product of two symmetric functions and one antisymmetric function is an antisymmetric (*odd*) function (+ × − × + = −). As a consequence, the integral vanishes and the transition is forbidden in a *centrosymmetric environment*! This happens for antisymmetric wavefunctions too (− × − × − = −).

Laporte's rule states: in the centrosymmetric complex, the electric dipole transitions are forbidden (p–p, d–d, and f–f transitions).

The rule applies to free atoms and ions (spherical symmetry has an inversion center) in the octahedral environment.

However, the TM compounds are colored. Why? That means some mechanisms can relax the rule. The most important mechanism is the *vibronic coupling.* At room temperature, the complex vibrations can distort the octahedral symmetry so as to remove the center of symmetry. Thus, the electronic transitions are weakly allowed and the spectrum will present absorption bands that are "vibronically" allowed. They are several orders of magnitude weaker than Laporte allowed bands. The vibrations are also responsible for the large width of absorption bands. When temperature of a crystal sample decreases the width of absorption band decreases and other neighbor small absorption bands could become visible in a spectrum.

Later, we will discuss in detail the vibronic coupling and its consequences on spectral behavior.

We note that the same ion presents stronger absorption bands in a tetrahedral environment (no inversion center) than in an octahedral environment. This is a method to investigate the local symmetry of a sample.

2.9.1.2 SPIN RULE

The *spin selection rule* states:

> *Transitions between states of different multiplicities are forbidden.*
> (or Transitions between states of the same multiplicities are allowed.)

This rule is also relaxed by molecular vibrations and spin–orbit coupling. The relaxation of the spin rule can produce spin-forbidden bands with the intensity of one order of magnitude lower than that of spin allowed bands.

The extinction coefficient ε (will be discussed later) measured in $l \cdot mol^{-1} \cdot cm^{-1}$ can give an indication of the type of electronic transition (Adamson 1969) as shown in Table 2.40.

TABLE 2.40 Extinction Coefficient Can Give an Indication of Transition Type

Transition Type	ε ($l \cdot mol^{-1} \cdot cm^{-1}$)	Example
Spin forbidden	≈1	d^5 configuration
d–d Laporte forbidden	10–100	O_h complexes
d–d noncentrosymmetric	≈500	T_d complexes
Charge transfer	10^3–10^5	

2.9.2 RELATION BETWEEN DIAGRAM AND OPTICAL SPECTRA

Until now, we have learned what happens when the metal ion is placed inside a symmetric structure (crystal or molecule), how to find the new states, and spectral terms. The anisotropic environment partly removes the degeneracy of initial states of the free ion. By using the correlation diagrams, we are able to order the new terms on the energy scale, but group theory does not give any information about the relative position of new terms. Therefore, it does not provide the value of light energy absorbed during the electron transition from the ground state to the excited state. Such information helps us to recognize the nature of the absorbent specimen. This is the real task of spectroscopy: from the measurement of the spectrum, we can have information about the matter that emits or absorbs light.

The exact quantitative calculus of ion energy levels in the crystal field is difficult to be performed. The imprecision comes from the quantum mechanics method applied to the free ion. Nowadays, the higher precision results need a long time for parallel computation on an expensive computer cluster.

Table 2.7 of this chapter presents energies of a few d^n configurations expressed in terms of Racah's parameters. The different free ion terms for a given electron configuration have different energies due to the *variations in electron–electron repulsion*. The term "energy" can be expressed by using a number of electrostatic parameters *A*, *B*, and *C*. They are unsolved integrals related to the strength of electron–electron repulsion. The *A* Racah's parameter does not play any role in spectroscopy because only the energy difference between levels is important in electron transition and *A* is roughly the same for any metal center.

For example, the energies of the d^2 configuration free ion terms are given in Table 2.41.

Thus, the transition from the ground state ^3F to the excited state ^3P needs the energy $(A + 7B) - (A - 8B) = 15B$. The *B* and *C* Racah's parameters are determined by an experiment, by fitting the theoretical dependence on the B/C ratio to the measured spectrum of light absorption. We underline that the *B* and *C* parameters depend on the nature (and size) of the ion, but most ions have the ratio C/B around 4.

In the crystalline environment, including a molecular complex, a new parameter appears that is splitting energy Δ. It is the product between the potential constant *D* and *q*, the symbol of the entire radial integral. We do not know the exact expression of radial functions, so Δ remains a parameter to be experimentally determined too. Thus, the energy of each

TABLE 2.41 Energies of the Free Ion Terms of d^2 Configuration Expressed in Racah's Parameters

$E(^1S) = A + 14B + 7C$
$E(^1G) = A + 4B + 2C$
$E(^1D) = A - 3B + 2C$
$E(^3P) = A + 7B$
$E(^3F) = A - 8B$

spectral term in the crystal field depends on the B, C, and Δ parameters. The sets of secular equations of each configuration in the octahedral environment were solved in terms of B, C, and Δ for each representation and the obtained expressions for energy of all configurations are very well summarized in Mc Clure's book (Mc Clure 1959). For example, the terms of d^2 configuration have the following expressions:

$$E(^3T_{1g}) = 7.5B - 3Dq - 0.5\sqrt{225B^2 + 100(Dq)^2 + 180DqB}$$

$$E(^3T_{2g}) = -8Dq + 9B + 2C - 12B^2/10Dq$$

$$E(^1E_g) = -8Dq + 9B + 2C - 6B^2/10Dq$$

$$E(^1A_{1g}) = -8Dq + 18B + 5C - 108B^2/10Dq \tag{2.62}$$

$$E(^3T_{2g}) = +2Dq$$

$$E(^3T_{1g}) = 7.5B - 3Dq + 0.5\sqrt{225B^2 + 100(Dq)^2 + 180DqB}$$

$$E(^3A_{2g}) = +12Dq$$

The most important Racah's parameter is B. It is a measure of electrostatic repulsion between electrons of the ion. Its value is obtained from spectroscopic measurements of ions in the gas phase. The value of the B parameter decreases in the crystal field due to the *ligand presence that reduces the electron–electron repulsion* (called the *nephelauxetic effect*). The dependence of the relative position of the new spectral terms on splitting energy Δ scaled by Racah's parameter B for all electronic configurations in an octahedral symmetry was represented thus long ago and is called the *Tanabe–Sugano diagrams (TS)* (Tanabe and Sugano 1954). The difference between the TS diagrams and the weak field–strong field correlation diagrams is that the first diagram illustrates the relative energies of states in a quantitative manner. They suppose that the ratio B/C equals 4, and the ground state is always on the x axis (it is a reference).

The TS diagrams are important for practical activity because they can be used to determine both B and Δ of a given ion in a specific symmetry environment. These diagrams also obey the rule of noncrossing: the electronic states with the same symmetry cannot cross and they always mix. It means that if two states of the *same symmetry* are likely to cross by increasing the field parameter, they will mix to avoid crossing (Figure 2.48). Thus, the lines become curved.

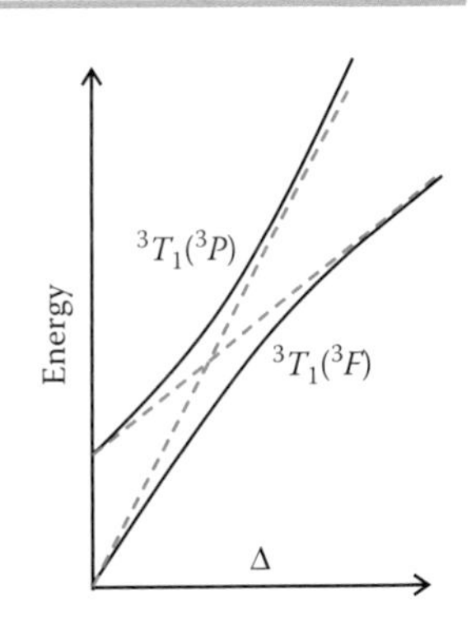

FIGURE 2.48 Illustration of noncrossing rule: two states of the same symmetry will mix to avoid crossing (particular case of d^8 configuration).

2.9.2.1 TS DIAGRAM OF THE d^2 CONFIGURATION IN $\boldsymbol{O}_h$ SYMMETRY

An example of a TS diagram is illustrated in Figure 2.49 for the simplest configuration d^2 in $\boldsymbol{O}_h$ symmetry. On the left side are all terms of the free ion. The ground state is ^3F. The terms ^3F, ^1D, and ^1G will split in the crystalline field according to Table 2.28. From Equation 2.62, we know that

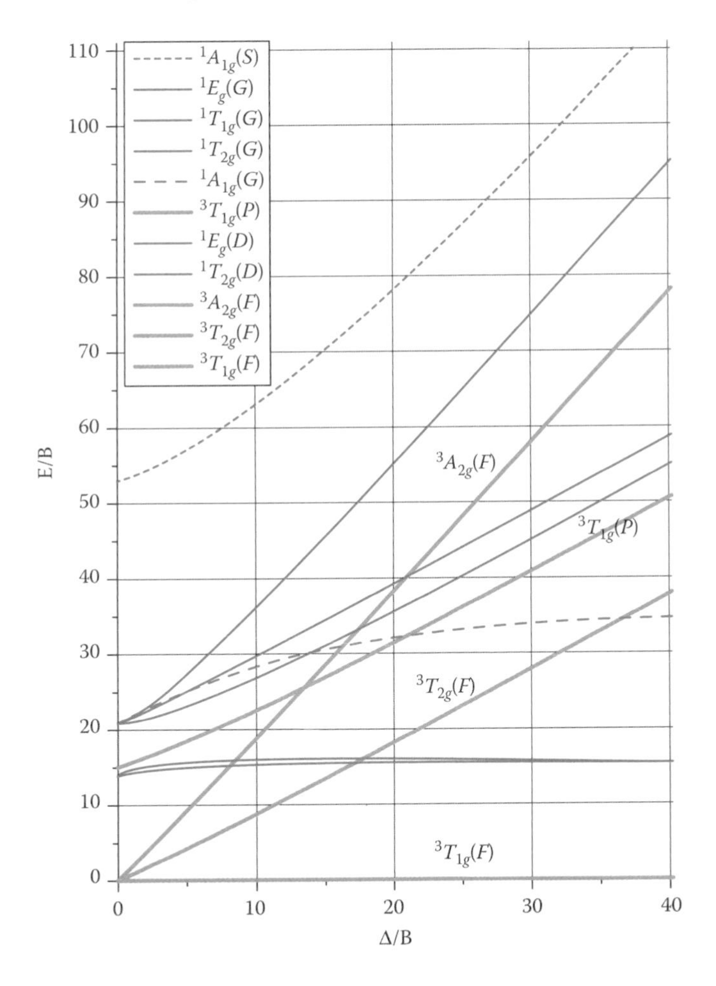

FIGURE 2.49 The TS diagram of d^2 configuration in $\boldsymbol{O}_h$ symmetry ($C/B = 4.42$). The spin-allowed transitions can take place between triplet states that are highlighted by thick lines. (The representation was possible by courtesy of Professor Robert Lancashire http://wwwchem.uwimona.edu.jm/courses/Tanabe-Sugano/TSspread.html)

$^3T_{1g}$ has the lowest energy and it will be represented by a horizontal line as a reference in energy. Each level will be represented by the difference between its energy and the ground term energy that is in fact the energy of transition between the two terms. We take into consideration only terms with triplet multiplicity, that is, $^3T_{1g}$, $^3T_{2g}$, $^3A_{2g}$, and $^3T_{1g}$ (originally from P).

Now, we divide each expression by B, we denote $10Dq = \Delta_O$, and we assume $C/B = 4.42$; so, we may write:

$$E(^3T_{1g})/B = 0$$

$$\frac{E(^3T_{2g})}{B} = 0.5\left(\frac{\Delta}{B} - 15 + \sqrt{225 + \left(\frac{\Delta}{B}\right)^2 + 18\frac{\Delta}{B}}\right)$$

$$\frac{E(^3A_{2g})}{B} = 0.5\left(3\frac{\Delta}{B} - 15 + \sqrt{225 + \left(\frac{\Delta}{B}\right)^2 + 18\frac{\Delta}{B}}\right)$$

$$\frac{E(^3T_{1g})}{B} = \sqrt{225 + \left(\frac{\Delta}{B}\right)^2 + 18\frac{\Delta}{B}}$$

Finally, we are able to make a representation of the energy versus (Δ/B) parameter.

How do we make use of the TS diagrams in our laboratory work? We describe below a *suggestion of the experimental procedure* (most of this procedure can be completely performed by a computer with proper software. The steps described here could be used to design the algorithm before writing the software):

1. Carefully prepare the sample (liquid or solid). It is easier to prepare liquid samples and control the concentration. For didactical reasons, it is better to use pure reagents to be sure that you have only one type of ion, for example, V^{3+}, in one type of symmetry, for example, six H_2O molecules. For the V^{3+} ion in $\boldsymbol{O}_h$ symmetry (a medium strong field), we expect to obtain three transitions from the ground state as represented in Figure 2.50: $^3T_{2g}(F) \leftarrow {}^3T_{1g}(F)$, $^3T_{1g}(P) \leftarrow {}^3T_{1g}(F)$, and $^3A_{2g}(F) \leftarrow {}^3T_{1g}(F)$. They are represented separately for didactical reasons to precisely see the initial and final states. They normally have the same position on the abscissa because the crystal field does not change during electron transition, which is faster than any possible small rearrangement of ions. We underline that this stage is very important in getting accurate results and depends on your skills.
2. Record the UV/Vis spectrum of the sample. The spectrum of a solution containing $[V(H_2O)_6]^{3+}$ is illustrated in Figure 2.51. The instrument was a standard spectrophotometer with a spectral range of 200–800 nm. For most of the molecular complexes,

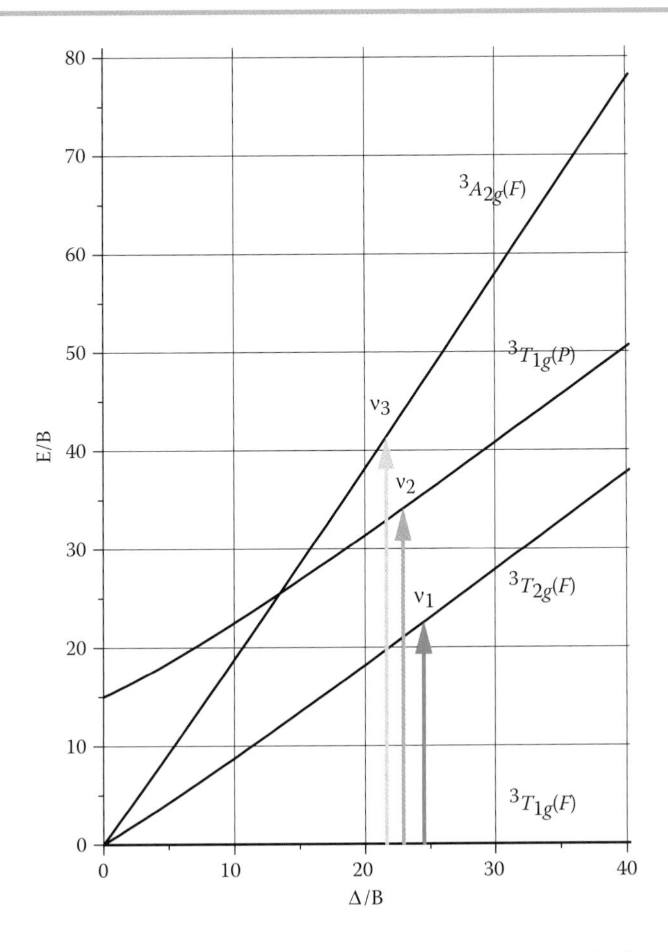

FIGURE 2.50 Three transitions are spin allowed in any ion with d^2 configuration in a given $\boldsymbol{O}_h$ symmetry environment.

you do not need to use more expensive instruments with an extended spectral range either to a shorter wavelength (180 nm) or to a longer wavelength (2500 nm). A few $\boldsymbol{O}_h$ complexes present the first absorption band in NIR. At a shorter wavelength (higher energies), it is usual for an extremely strong absorption to appear due to the charge transfer from the ligand to the central ion. Two bands are observed at 554 and 382 nm. The first two bands in the V^{3+} spectrum probably involve transitions to the $^3T_{2g}$ and $^3T_{1g}$ terms. No absorption for the $^3A_{2g}(F) \leftarrow {}^3T_{1g}(F)$ transition is observed. This highest energy absorption band is obscured by the "tail" of charge transfer absorption.

3. Convert the peak position from wavelength (nm) to wave number (cm^{-1}), which is an energy-proportional quantity. The wave numbers are $\tilde{\nu}_1 = 18{,}050\ \text{cm}^{-1}$ and $\tilde{\nu}_2 = 26{,}178\ \text{cm}^{-1}$ for the vanadium example.
4. Calculate the experimental ratio of their transition energies: $\tilde{\nu}_2/\tilde{\nu}_1 = 1.45$.

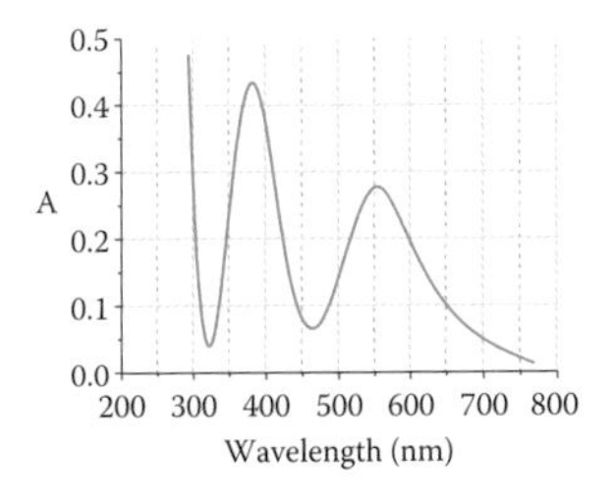

FIGURE 2.51 The absorption spectrum of the $[V(H_2O)_6]^{3+}$ molecular complex in the spectral range 200–800 nm.

TABLE 2.42 Determination of the Average Value of Racah's *B* Parameter from Absorption Frequencies

$^3T_{2g}(F)$	$\tilde{\nu}_1/B = 29.7$	$B = 606\ \text{cm}^{-1}$	$\bar{B} = 609\ \text{cm}^{-1}$
$^3T_{1g}(P)$	$\tilde{\nu}_2/B = 42.7$	$B = 613\ \text{cm}^{-1}$	

5. Locate on the TS diagram where the ratio matches that of the experimental value. When $\Delta/B = 20$, the diagram gives the ratio $\tilde{\nu}_2/\tilde{\nu}_1 = 1.79$. When $\Delta/B = 30$, the diagram gives the ratio $\tilde{\nu}_2/\tilde{\nu}_1 = 1.47$. When $\Delta/B = 40$, the diagram gives the ratio $\tilde{\nu}_2/\tilde{\nu}_1 = 1.33$. Assuming a linear dependence of ratio on Δ/B by interpolation, we obtain $\Delta/B = 31.4$ for a 1.45 ratio.
6. Find the values of $\tilde{\nu}_1/B$ and $\tilde{\nu}_2/B$ from the intersection of diagram lines for $^3T_{2g}(F)$ and $^3T_{1g}(P)$ with a vertical line at $\Delta/B = 31.4$.
7. Using the experimental values of $\tilde{\nu}_1$ and $\tilde{\nu}_2$, calculate an average value of Racah *B*. The value obtained for the Racah *B* parameter is smaller than that of the free ion where $B = 860\ \text{cm}^{-1}$. This behavior was expected because the presence of ligands reduces the repulsion of electron–electron in the metal ion (nephelauxetic effect). All values presented here are summarized in Table 2.42.
8. Calculate Δ_O based on the value of the Racah *B* parameter. If $\Delta_O/B = 31.4$ and $B = 609\ \text{cm}^{-1}$, then the splitting energy $\Delta_O = 19138\ \text{cm}^{-1}$.

The values of Δ/B and *B* can be used to calculate the energy $\tilde{\nu}_3$ of the third transition (unobserved) from the TS diagram, in our example 36,720 cm^{-1} (272 nm).

We note here two important sources of error:

- The absorption bands of transition metal ions are generally very large and it is possible to make a small error in reading their positions, which this will change the experimental ratio $\tilde{\nu}_2/\tilde{\nu}_1$.
- Reading on the TS diagram, the intersection point between the *y* axis and a line with a big slope (e.g., $^3A_{2g}$) could also induce a supplementary error.

2.9.2.2 TS DIAGRAM OF THE d^3 CONFIGURATION IN $\boldsymbol{O}_h$ SYMMETRY

The TS diagram for ions with d^3 configuration (e.g., Cr^{3+}) is obtained and used in a similar way to that with d^2 configuration. Table 2.3 shows that the ground-state term of the free ion is ^4F and, as a consequence, our interest will be focused on the higher ^4P term, where transition is spin allowed. In the $\boldsymbol{O}_h$ crystalline field, they present the same splitting as the terms of d^2 configuration but in reverse order: $^4F = {}^4A_{2g} + {}^4T_{2g} + {}^4T_{1g}$ and $^4P = {}^4T_{1g}$. We expect three transitions too: $^4T_{2g}(F) \leftarrow {}^4A_{2g}(F)$, $^4T_{1g}(F) \leftarrow {}^4A_{2g}(F)$, and $^4T_{1g}(P) \leftarrow {}^4A_{2g}(F)$. The above-described procedure can be applied without any problems. The third transition (highest energy) will give a hidden absorption band too.

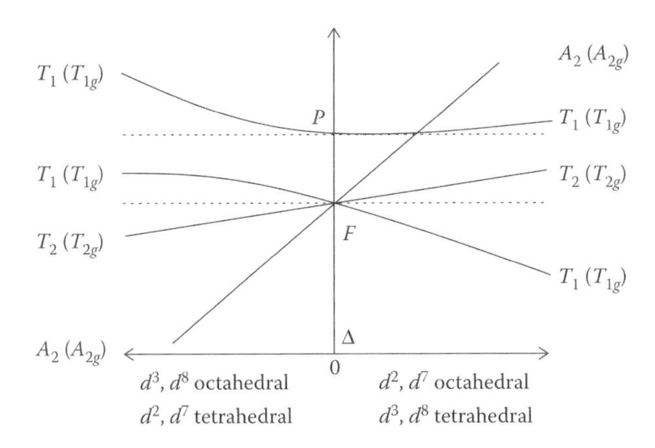

FIGURE 2.52 Orgel's diagram for d^2, d^3, d^7 (high spin), and d^8 configurations in octahedral and tetrahedral symmetries.

2.9.2.3 TS DIAGRAMS OF d^7 AND d^8 CONFIGURATIONS IN O_h SYMMETRY

The TS diagrams for ions with d^7 configuration (e.g., Co^{2+}) and d^8 configuration (e.g., Ni^{2+}) are obtained and used in a way similar to that with d^3 configuration and d^8 configuration, respectively, but with a reverse splitting order due to the hole formalism. The prediction of three spin-allowed transitions is still correct and they may be seen in the absorption spectra (in the case of Ni and Co) because the splitting energy has a much lower value (e.g., 8600 cm^{-1} of Ni^{2+} vs. 18,000 cm^{-1} of V^{3+}). We note that this is the reason why the prediction is correct for the d^7 (*high-spin*) configuration! From an inspection of the TS diagram of d^7 results a dramatically changing plot for the Δ/B ratio of larger than 22. The ground state changes from $^4T_1$ to ^2E and all terms are plotted relative to the energy of the ground state. For a large ligand field strength, a high-spin-to-low-spin transition occurs.

The simplified Orgel diagram presented in Figure 2.52 illustrates the conclusion of the above discussion. It is useful to easily predict the number of spin-allowed absorption bands expected and their symmetry-state designations. We note that the energy values cannot be obtained from them.

2.9.2.4 TS DIAGRAMS OF d^4 AND d^6 CONFIGURATIONS IN O_h SYMMETRY

The metal ions with d^4 and d^6 configurations have a ^5D ground term. All other terms have a spin multiplicity lower than 5. The ground term of such an ion placed in an octahedral field splits into $^5T_{2g}$ and ^5E_g. In the tetrahedral field, the order will be reversed. Thus, these ions present only one spin-allowed transition ($^5T_{2g} \leftarrow {}^5E_g$) that can generate one absorption band that is easy to be seen (Figure 2.53). Both configurations may be high-spin and low-spin cases for the same ion (e.g., Mn^{3+} and Fe^{2+}) depending on the strength of the crystal field. There is a limit between

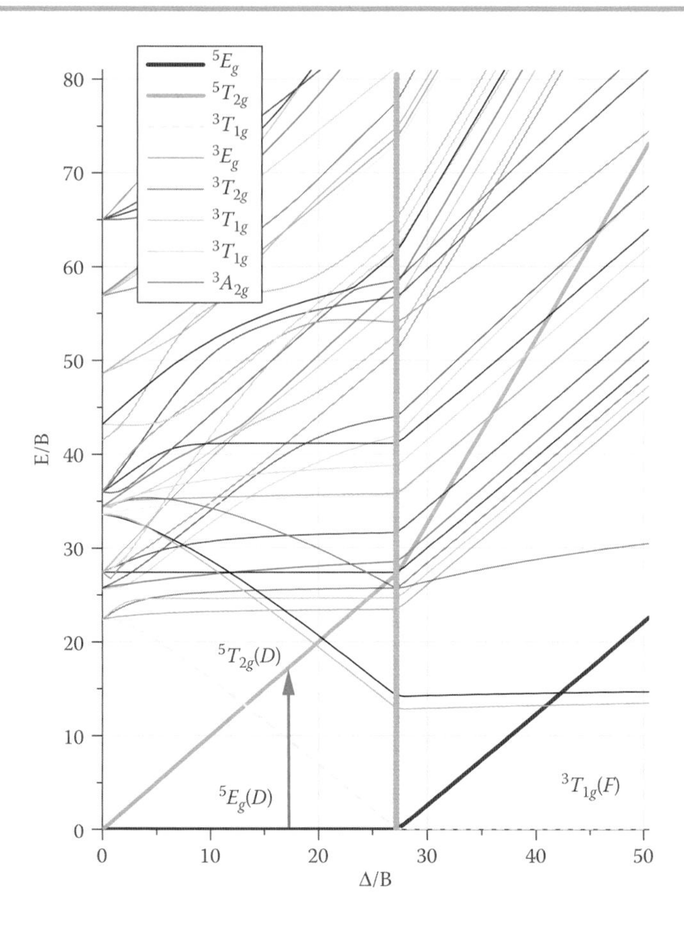

FIGURE 2.53 TS diagram of d^4 configuration in $\mathbf{O}_h$ symmetry ($C/B = 4.6$). There is only one spin-allowed transition in high-spin state (left side). (The representation was made by using data available by courtesy of Professor Robert Lancashire.)

high spin and low spin in TS diagrams of both configurations where each diagram drastically changes because the ground term changes and its spin multiplicity changes too. On the right side of Figure 2.53, the triplet term $^3T_{1g}$ becomes the ground term and the spin rule will allow many transitions instead of one transition in a high-spin case. Most of the molecular complexes are in a high-spin case. In the tetrahedral field that is weaker than the octahedral field, this is the normal case.

We also mention that the d^1 and d^9 configurations have a ^2D term that splits into $^2T_{2g}$ and ^2E_g and the absorption spectrum will have only one band too.

The simplified description of the spin-allowed transition is illustrated by the Orgel diagram in Figure 2.54.

The d^5 configuration is a special case due to the nondegenerate term ^6S that does not split in the crystal field and no transition, is spin allowed. There are several spin-forbidden transitions, but they generate very weak absorption bands (1–2 orders of magnitude). The usual case of such an ion (e.g., Mn^{2+}) is high spin.

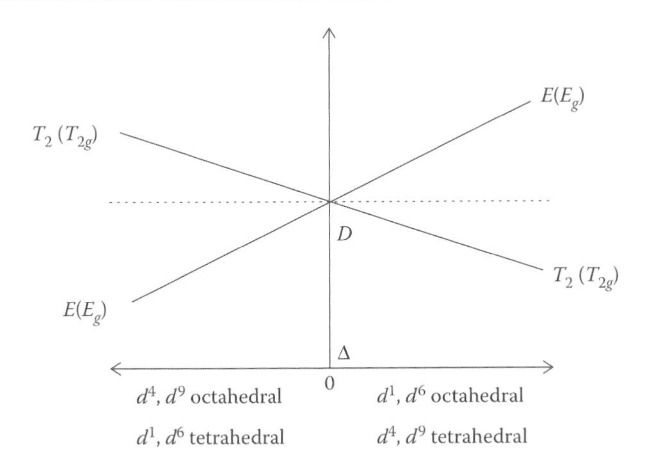

FIGURE 2.54 Orgel's simplified diagram for d^1, d^4 (high spin), d^6 (high spin), and d^9 configurations in octahedral and tetrahedral symmetries.

2.9.3 SPLITTING OF F TERMS IN OCTAHEDRAL SYMMETRY

We must consider one f electron in the octahedral environment of six ligands (Ballhausen 1972). The wavefunctions describing f orbitals

$$4f_{y^3}, 4f_{x^3}, 4f_{z^3}, 4f_{x(z^2-y^2)}, 4f_{y(z^2-x^2)}, 4f_{z(x^2-y^2)}, 4f_{xyz}$$

form a basis for reducible representation (see Table 2.13)

Orbital Type	*l*	χ(*E*)	χ(C₃)	χ(C₂)	χ(C₄)	χ(C₂)
f	3	7	1	−1	−1	−1

that can be reduced to irreducible representations: $\gamma_f = a_{2u} + t_{1u} + t_{2u}$.

The multielectron terms ^3F (from d^2 and d^8 configurations) or ^4F (from d^3 and d^7 configurations) formed due to the many-electron filling of d orbitals split similar to f orbitals in the $\boldsymbol{O}_h$ field in $A_{2g} + T_{1g} + T_{2g}$. We underline that the F terms of transition metal ions originate from d orbitals and their symmetry is *gerade*, arising from d orbitals symmetry.

What is the right order of new spectral terms and what is the energy splitting?

To determine the energies of new terms A_2, T_1, and T_2, we will use the perturbation theory where the wavefunctions of the ^3F term will be used as nonperturbed functions to calculate the matrix elements of the secular equation.

The total spin multiplicity rests unchanged; so, we are interested only in all combinations of two electrons whose spins are parallel as shown in Table 2.43.

As a result, there are states described by pure wavefunctions from ^3F and states described by a linear combination of wavefunctions from ^3F to ^3P as follows in Table 2.44 where we noted the two-electron function

TABLE 2.43 Combinations of Two Electrons Whose Spins Are Parallel

m_l						
2	**1**	**0**	**−1**	**−2**	M_L	**Term**
↑	↑				3	^3F
↑		↑			2	^3F
↑			↑		1	^3F
↑				↑	0	^3F
	↑	↑			1	^3P
	↑		↑		0	^3P
	↑			↑	−1	^3P
		↑	↑		−1	^3F
		↑		↑	−2	^3F
			↑	↑	−3	^3F

by $\left\langle 2^+,1^+ \right\rangle = (1/\sqrt{2})\,[u_1(2^+)\cdot u_2(1^+) - u_1(1^+)\cdot u_2(2^+)]$. The function u is a single-electron orbital function. It contains the radial part and spherical harmonics too.

Using these functions, we must calculate all matrix elements (49) from H_{33} to H_{-3-3}, where

$$H_{ij} = \left\langle \Psi_i \middle| eV \middle| \Psi_j \right\rangle = \iiint \Psi^*_{MLMS} eV \Psi_{ML'MS'}\, d\tau_1 d\tau_2$$

TABLE 2.44 Wavefunctions of States that Arise from ^3F Expressed as Linear Combinations of Wavefunctions from ^3F and ^3P

Ψ_{31}	$\left\langle 2^+,1^+ \right\rangle$
Ψ_{21}	$\left\langle 2^+,0^+ \right\rangle$
Ψ_{11}	$\sqrt{\frac{3}{5}}\left\langle 2^+,-1^+ \right\rangle + \sqrt{\frac{2}{5}}\left\langle 1^+,0^+ \right\rangle$
Ψ_{01}	$\sqrt{\frac{1}{5}}\left\langle 2^+,-2^+ \right\rangle + \sqrt{\frac{4}{5}}\left\langle 1^+,-1^+ \right\rangle$
Ψ_{-11}	$\sqrt{\frac{3}{5}}\left\langle 1^+,-2^+ \right\rangle + \sqrt{\frac{2}{5}}\left\langle 0^+,-1^+ \right\rangle$
Ψ_{-21}	$\left\langle 0^+,-2^+ \right\rangle$
Ψ_{-31}	$\left\langle -1^+,-2^+ \right\rangle$

Here, the potential is given by $V = V_1 + V_2$ and the matrix element calculation can be reduced to single-electron integrals:

$$\iiint \Psi^*_{MLMS} e(V_1 + V_2)\Psi_{MLMS} d\tau_1\, d\tau_2 = \iiint u_1^* eV_1 u_1\, d\tau_1 + \iiint u_2^* eV_2 u_2\, d\tau_2$$

and

$$\iiint \Psi^*_{MLMS} e(V_1 + V_2)\Psi_{ML'MS'}\, d\tau_1 d\tau_2 = (-1)^\tau \iiint u_\lambda^* eV_1 u_{\lambda'}\, d\tau_1$$

when wavefunctions differ by only one single-electron function. τ is the parity of permutation between both functions.

The matrix element vanishes if the wavefunctions Ψ_{MLMS} and $\Psi_{ML'MS'}$ differ by more than one single-electron function.

In the following equations, we will use the results previously obtained for matrix elements of a single-electron function (Equations 2.32 through 2.46) summarized below.

$$\begin{cases} H_{22} = H_{-2-2} = Dq \\ H_{2-2} = H_{-22} = 5Dq \\ H_{11} = H_{-1-1} = -4Dq \\ H_{00} = 6Dq \end{cases}$$

$$H_{33} = \iiint \Psi^*_{31} e(V_1 + V_2)\Psi_{31}\, d\tau_1 d\tau_2 = \iiint \langle 2^+, 1^+ \rangle e(V_1 + V_2) \langle 2^+, 1^+ \rangle\, d\tau_1 d\tau_2$$
$$= \iiint 2^* eV2\, d\tau_1 + \iiint 1^* eV1\, d\tau_2 = Dq + (-4Dq) = -3Dq$$

$$H_{32} = \iiint \Psi^*_{31} e(V_1 + V_2)\Psi_{21}\, d\tau_1 d\tau_2$$
$$= \iiint \langle 2^+, 1^+ \rangle e(V_1 + V_2) \langle 2^+, 0^+ \rangle\, d\tau_1 d\tau_2 = 0$$

$$H_{31} = \iiint \langle 2^+, 1^+ \rangle e(V_1 + V_2) \left[\sqrt{\frac{3}{5}} \langle 2^+, -1^+ \rangle + \sqrt{\frac{2}{5}} \langle 1^+, 0^+ \rangle \right] d\tau_1 d\tau_2 = 0$$

$$H_{30} = \iiint \langle 2^+, 1^+ \rangle e(V_1 + V_2) \left[\sqrt{\frac{1}{5}} \langle 2^+, -2^+ \rangle + \sqrt{\frac{4}{5}} \langle 1^+, -1^+ \rangle \right] d\tau_1 d\tau_2 = 0$$

$$H_{3-1} = \iiint \langle 2^+, 1^+ \rangle e(V_1 + V_2) \left[\sqrt{\frac{3}{5}} \langle 1^+, -2^+ \rangle + \sqrt{\frac{2}{5}} \langle 0^+, -1^+ \rangle \right] d\tau_1 d\tau_2$$
$$= -\sqrt{15} Dq$$

$$H_{3-2} = \iiint \langle 2^+,1^+ \rangle e(V_1 + V_2) \langle 0^+,-2^+ \rangle d\tau_1 d\tau_2 = 0$$

$$H_{3-3} = \iiint \langle 2^+,1^+ \rangle e(V_1 + V_2) \langle -1^+,-2^+ \rangle d\tau_1 d\tau_2 = 0$$

$$H_{23} = \iiint \langle 2^+,0^+ \rangle e(V_1 + V_2) \langle 2^+,1^+ \rangle d\tau_1 d\tau_2 = (-1)^{\tau} \iiint 0^{*} eV1\, d\tau_1 = 0$$

$$H_{22} = \iiint \langle 2^+,0^+ \rangle e(V_1 + V_2) \langle 2^+,0^+ \rangle d\tau_1 d\tau_2 = 7Dq$$

$$H_{21} = \iiint \langle 2^+,0^+ \rangle e(V_1 + V_2) \left[\sqrt{\frac{3}{5}} \langle 2^+,-1^+ \rangle + \sqrt{\frac{2}{5}} \langle 1^+,0^+ \rangle \right] d\tau_1 d\tau_2 = 0$$

$$H_{20} = \iiint \langle 2^+,0^+ \rangle e(V_1 + V_2) \left[\sqrt{\frac{1}{5}} \langle 2^+,-2^+ \rangle + \sqrt{\frac{4}{5}} \langle 1^+,-1^+ \rangle \right] d\tau_1 d\tau_2 = 0$$

$$H_{2-1} = \iiint \langle 2^+,0^+ \rangle e(V_1 + V_2) \left[\sqrt{\frac{3}{5}} \langle 1^+,-2^+ \rangle + \sqrt{\frac{2}{5}} \langle 0^+,-1^+ \rangle \right] d\tau_1 d\tau_2 = 0$$

$$H_{2-2} = \iiint \langle 2^+,0^+ \rangle e(V_1 + V_2) \langle 0^+,-2^+ \rangle d\tau_1 d\tau_2 = -5Dq$$

$$H_{2-3} = \iiint \langle 2^+,0^+ \rangle e(V_1 + V_2) \langle -1^+,-2^+ \rangle d\tau_1 d\tau_2 = 0$$

$$H_{13} = \iiint \left[\sqrt{\frac{3}{5}} \langle 2^+,-1^+ \rangle + \sqrt{\frac{2}{5}} \langle 1^+,0^+ \rangle \right] e(V_1 + V_2) \langle 2^+,1^+ \rangle d\tau_1 d\tau_2 = 0$$

$$H_{12} = \iiint \left[\sqrt{\frac{3}{5}} \langle 2^+,-1^+ \rangle + \sqrt{\frac{2}{5}} \langle 1^+,0^+ \rangle \right] e(V_1 + V_2) \langle 2^+,0^+ \rangle d\tau_1 d\tau_2 = 0$$

$$H_{11} = \iiint \left[\sqrt{\frac{3}{5}} \langle 2^+,-1^+ \rangle + \sqrt{\frac{2}{5}} \langle 1^+,0^+ \rangle \right] e(V_1 + V_2) \left[\sqrt{\frac{3}{5}} \langle 2^+,-1^+ \rangle \right]_2$$
$$+ \sqrt{\frac{2}{5}} \langle 1^+,0^+ \rangle d\tau_1 d\tau = -Dq$$

$$H_{10} = \iiint \left[\sqrt{\frac{3}{5}} \langle 2^+,-1^+ \rangle + \sqrt{\frac{2}{5}} \langle 1^+,0^+ \rangle \right] e(V_1 + V_2)$$
$$\times \left[\sqrt{\frac{1}{5}} \langle 2^+,-2^+ \rangle + \sqrt{\frac{4}{5}} \langle 1^+,-1^+ \rangle \right] d\tau_1 d\tau_2 = 0$$

$$H_{1-1} = \iiint \left[\sqrt{\frac{3}{5}} \langle 2^+,-1^+ \rangle + \sqrt{\frac{2}{5}} \langle 1^+,0^+ \rangle \right] e(V_1 + V_2)$$
$$\times \left[\sqrt{\frac{3}{5}} \langle 1^+,-2^+ \rangle + \sqrt{\frac{2}{5}} \langle 0^+,-1^+ \rangle \right] d\tau_1 d\tau_2 = 0$$

$$H_{1-2} = \iiint \left[\sqrt{\frac{3}{5}} \langle 2^+,-1^+ \rangle + \sqrt{\frac{2}{5}} \langle 1^+,0^+ \rangle \right] e(V_1 + V_2) \langle 0^+,-2^+ \rangle d\tau_1 d\tau_2 = 0$$

$$H_{1-3} == \iiint \left[\sqrt{\frac{3}{5}} \langle 2^+,-1^+ \rangle + \sqrt{\frac{2}{5}} \langle 1^+,0^+ \rangle \right] e(V_1 + V_2) \langle -1^+,-2^+ \rangle d\tau_1 d\tau_2$$
$$= -\sqrt{15} Dq$$

$$H_{03} = \iiint \left[\sqrt{\frac{1}{5}} \langle 2^+,-2^+ \rangle + \sqrt{\frac{4}{5}} \langle 1^+,-1^+ \rangle \right] e(V_1 + V_2) \langle 2^+,1^+ \rangle d\tau_1 d\tau_2 = 0$$

$$H_{02} = H_{20} = 0; \quad 4H_{01} = H_{10} = 0$$

$$H_{00} = \iiint \left[\sqrt{\frac{1}{5}} \langle 2^+,-2^+ \rangle + \sqrt{\frac{4}{5}} \langle 1^+,-1^+ \rangle \right] e(V_1 + V_2)$$
$$\times \left[\sqrt{\frac{1}{5}} \langle 2^+,-2^+ \rangle + \sqrt{\frac{4}{5}} \langle 1^+,-1^+ \rangle \right] d\tau_1 d\tau_2 = -6Dq$$

$$H_{0-1} = \iiint \left[\sqrt{\frac{1}{5}} \langle 2^+,-2^+ \rangle + \sqrt{\frac{4}{5}} \langle 1^+,-1^+ \rangle \right] e(V_1 + V_2)$$
$$\times \left[\sqrt{\frac{3}{5}} \langle 1^+,-2^+ \rangle + \sqrt{\frac{2}{5}} \langle 0^+,-1^+ \rangle \right] d\tau_1 d\tau_2 = 0$$

$$H_{0-2} = \iiint \left[\sqrt{\frac{1}{5}} \langle 2^+,-2^+ \rangle + \sqrt{\frac{4}{5}} \langle 1^+,-1^+ \rangle \right] e(V_1 + V_2) \langle 0^+,-2^+ \rangle d\tau_1 d\tau_2 = 0$$

$$H_{0-3} = \iiint \left[\sqrt{\frac{1}{5}} \langle 2^+,-2^+ \rangle + \sqrt{\frac{4}{5}} \langle 1^+,-1^+ \rangle \right] e(V_1 + V_2) \langle -1^+,-2^+ \rangle d\tau_1 d\tau_2 = 0$$

$$H_{0-3} = \iiint \left[\sqrt{\frac{1}{5}} \langle 2^+,-2^+ \rangle + \sqrt{\frac{4}{5}} \langle 1^+,-1^+ \rangle \right] e(V_1 + V_2) \langle -1^+,-2^+ \rangle d\tau_1 d\tau_2 = 0$$

$$H_{-13} = H_{3-1} = -\sqrt{15} Dq; \quad H_{-12} = H_{2-1} = 0; \quad H_{-11} = H_{1-1} = 0;$$
$$H_{-10} = H_{0-1} = 0$$

$$H_{-1-1} = \iiint \left[\sqrt{\frac{3}{5}}\left\langle 1^+,-2^+\right\rangle + \sqrt{\frac{2}{5}}\left\langle 0^+,-1^+\right\rangle\right] e(V_1 + V_2)$$
$$\times \left[\sqrt{\frac{3}{5}}\left\langle 1^+,-2^+\right\rangle + \sqrt{\frac{2}{5}}\left\langle 0^+,-1^+\right\rangle\right] d\tau_1 d\tau_2 = -Dq$$

$$H_{-1-2} = \iiint \left[\sqrt{\frac{3}{5}}\left\langle 1^+,-2^+\right\rangle + \sqrt{\frac{2}{5}}\left\langle 0^+,-1^+\right\rangle\right] e(V_1 + V_2)\left\langle 0^+,-2^+\right\rangle d\tau_1 d\tau_2 = 0$$

$$H_{-1-3} = \iiint \left[\sqrt{\frac{3}{5}}\left\langle 1^+,-2^+\right\rangle + \sqrt{\frac{2}{5}}\left\langle 0^+,-1^+\right\rangle\right] e(V_1 + V_2)\left\langle -1^+,-2^+\right\rangle d\tau_1 d\tau_2 = 0$$

$$H_{-23} = H_{3-2} = 0; \quad H_{-22} = H_{2-2} = -5Dq; \quad H_{-21} = H_{1-2} = 0;$$
$$H_{-20} = H_{0-2} = 0; \quad H_{-2-1} = H_{-1-2} = 0$$

$$H_{-2-2} == \iiint \left\langle 0^+,-2^+\right\rangle e(V_1 + V_2)\left\langle 0^+,-2^+\right\rangle d\tau_1 d\tau_2 = 7Dq$$

$$H_{-2-3} = \iiint \left\langle 0^+,-2^+\right\rangle e(V_1 + V_2)\left\langle -1^+,-2^+\right\rangle d\tau_1 d\tau_2 = 0$$

$$H_{-33} = H_{3-3} = 0; \quad H_{-32} = H_{2-3} = 0; \quad H_{-31} = H_{1-3} = -\sqrt{15}Dq$$

$$H_{-30} = H_{0-3} = 0; \quad H_{-3-1} = H_{-1-3} = 0; \quad H_{-3-2} = H_{-2-3} = 0$$

$$H_{-3-3} = \iiint \left\langle -1^+,-2^+\right\rangle e(V_1 + V_2)\left\langle -1^+,-2^+\right\rangle d\tau_1 d\tau_2 = -3Dq$$

We now have all elements to solve the secular equation $\det(\hat{H} - \varepsilon\hat{E}) = 0$ (see Equation 2.48) to find the eigenvalues of the Hamiltonian:

M_L	3	2	1	0	−1	−2	−3
3	$-3Dq - \varepsilon$	0	0	0	$-\sqrt{15}Dq$	0	0
2	0	$7Dq - \varepsilon$	0	0	0	$-5Dq$	0
1	0	0	$-Dq - \varepsilon$	0	0	0	$-\sqrt{15}Dq$
0	0	0	0	$-6Dq - \varepsilon$	0	0	0
−1	$-\sqrt{15}Dq$	0	0	0	$-Dq - \varepsilon$	0	0
−2	0	$-5Dq$	0	0	0	$7Dq - \varepsilon$	0
−3	0	0	$-\sqrt{15}Dq$	0	0	0	$-3Dq - \varepsilon$

$= 0$

We have to apply the diagonalization procedure to rearrange the determinant in smaller blocks on the diagonal:

$$
\begin{array}{c|ccccccc|c}
M_L & 3 & -1 & 2 & -2 & 1 & -3 & 0 & \\
\hline
3 & -3Dq-\varepsilon & -\sqrt{15}Dq & 0 & 0 & 0 & 0 & 0 & \\
-1 & -\sqrt{15}Dq & -Dq-\varepsilon & 0 & 0 & 0 & 0 & 0 & \\
2 & 0 & 0 & 7Dq-\varepsilon & -5Dq & 0 & 0 & 0 & \\
-2 & 0 & 0 & -5Dq & 7Dq-\varepsilon & 0 & 0 & 0 & =0 \\
1 & 0 & 0 & 0 & 0 & -Dq-\varepsilon & -\sqrt{15}Dq & 0 & \\
-3 & 0 & 0 & 0 & 0 & -\sqrt{15}Dq & -3Dq-\varepsilon & 0 & \\
0 & 0 & 0 & 0 & 0 & 0 & 0 & -6Dq-\varepsilon &
\end{array}
$$

Thus, the energy values will be given by

$$
\begin{aligned}
(-3Dq-\varepsilon)(-Dq-\varepsilon)-15(Dq)^2 &= 0 \;\Rightarrow\; \varepsilon_1 = 2Dq \;\text{ and }\; \varepsilon_2 = -6Dq \\
(7Dq-\varepsilon)^2+5(Dq)^2 &= 0 \;\Rightarrow\; \varepsilon_3 = 12Dq \;\text{ and }\; \varepsilon_4 = 2Dq \\
(-Dq-\varepsilon)(-3Dq-\varepsilon)-15(Dq)^2 &= 0 \;\Rightarrow\; \varepsilon_5 = 2Dq \;\text{ and }\; \varepsilon_6 = -6Dq \\
-6Dq-\varepsilon &= 0 \;\Rightarrow\; \varepsilon_7 = -6Dq
\end{aligned}
$$

Therefore, the solutions are $-6Dq$ (threefold degenerate), $2Dq$ (threefold degenerate), and $12Dq$ (onefold degenerate). By simple inspection, it is easy to see the correspondence between energy $12Dq$ (onefold degenerate) and $^3A_2$ representation. Both energies threefold degenerate equally probably correspond to T_1 and T_2 representations. From TS diagrams, the order of new terms in an octahedral symmetry is

$$^3T_{1g} = -6Dq = -3/5\Delta_O$$

$$^3T_{2g} = 2Dq = 1/5\Delta_O$$

$$^3A_{2g} = 12Dq = 6/5\Delta_O$$

The splitting energies that give transition energy are presented in Figure 2.55.

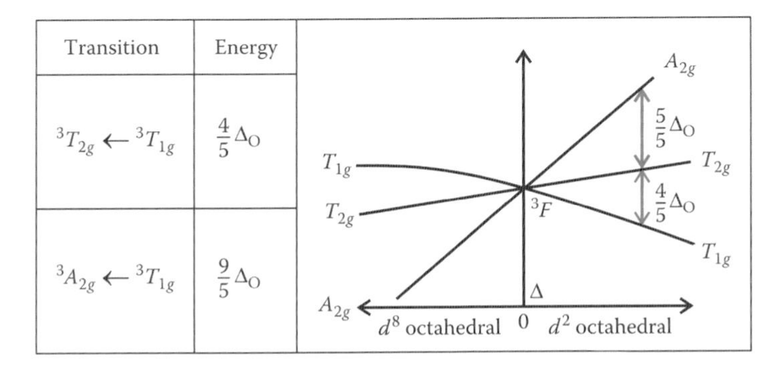

FIGURE 2.55 Splitting energies in terms of field strength and Orgel's diagram of $^3T_{2g} \leftarrow {}^3T_{1g}$ and $^3A_{2g} \leftarrow {}^3T_{1g}$ arising from splitting of the ^3F free ion term in the $\boldsymbol{O}_h$ field.

The term ^3F (of d^8 configuration) has a similar splitting but in reverse order in the $\boldsymbol{O}_h$ field; therefore, $A_{2g} + T_{2g} + T_{1g}$ because of the hole formalism, where $D < 0$.

The term ^4F (of d^3 configuration) has a similar splitting in the $\boldsymbol{O}_h$ field; therefore, $A_{2g} + T_{2g} + T_{1g}$, but the splitting energies are different. We should repeat the previous procedure in order to find them, where the wavefunction is now the product of three single-electron orbital functions.

2.10 EXAMPLE OF THE Co ION

Why cobalt?

- Cobalt compounds have been well known for a long time as blue pigments for ceramics and glasses. Cobalt is still used in modern industry for many applications.
- In compounds, it is often found in 2+ or 3+ oxidation states but it may also have others.
- In molecular complexes, the environment can be of octahedral or tetrahedral symmetry and can easily change it.

The cobalt chloride salt often presents pink color and not blue, as expected, due to the hydration. In water, the environment of the cobalt ion is octahedral $[Co(H_2O)_6]^{2+}$. The blue color means dehydrated $CoCl_2$. While preparing a solution with high concentration in chloride ions (by adding HCl), the color changes to blue and this is an indicator of the tetrahedral environment of the Co^{2+} ion. The molecular complex is $[CoCl_4]^{2-}$. The absorption spectrum shows one band located around 14,500 cm^{-1}. Also the crystal of KCl (cubic cell) doped with $CoCl_2$ also presents the same absorption band at the same energy. This band corresponds to the transition $^4T_1(P) \leftarrow {}^4A_2(F)$ (see the Orgel diagram from Figure 2.52). The first two transitions spin allowed are in the IR range due to the weaker tetrahedral field (4/9 Δ_O). The crystal growth is difficult because it is necessary to completely remove water from both the raw salts (KCl and $CoCl_2$). Table 2.45 summarizes some examples of cobalt complexes and the possible symmetries.

Co^{2+} has the d^7 electronic configuration. Figure 2.56 shows the difference between the absorption spectra of cobalt 2^+ ion in octahedral and tetrahedral symmetries, respectively. We note that the difference is mainly due to the environment symmetry and not due to the nature of ligands. The low-energy absorption band located in NIR (8000 cm^{-1} that means 1250 nm) is out of the spectral range of a standard spectrophotometer and it was not measured.

2.11 LIMITATIONS OF THE CRYSTAL FIELD THEORY

The crystal field theory is relatively simple to apply to properties of transition metal complexes using ready-to-use results and most of them are

TABLE 2.45 Examples of Divalent Cobalt Complexes and Corresponding Local Symmetry of Central Ion

	Color	Complex	Symmetry
$CoCl_2$ dehydrated powder	Blue	$[CoCl_4]^{2-}$	T_d
$CoCl_2$ hydrated powder	Pink	$[Co(H_2O)_6]^{2+}$	O_h
$CoCl_2$ single crystal	Blue	$[CoCl_6]^{4-}$	O_h
$Co:CdCl_2$ crystal		$[CoCl_6]^{4-}$	O_h
Co:KCl (NaCl) crystal (interstitial position)		$[CoCl_4]^{2-}$	T_d
Co:KCl (NaCl) crystal (substitutional position)		$[CoCl_6]^{4-}$	O_h
$CoCl_2$ water solution	Pink	$[Co(H_2O)_6]^{2+}$	O_h
$CoCl_2$ acetone solution	Blue	$[CoCl_4]^{2-}$	T_d

presented in this chapter. This theory has been well systematized over the years. The theory fails in some important aspects:

- The *order of ligands in the spectrochemical series* (Table 2.27). The spectrochemical series is the result of measurements made on different central ions and different ligands. In the crystal field theory, ligands are negative ions that interact electrostatically and change the energy of *d* orbitals of the central ion. The spectrochemical series contains either negative ions or neutral ligands. It is surprising to see only neutral ligands in the strong field range. This remark indicates how important and how strong covalency is. So, a new model of covalent bonding should be developed.

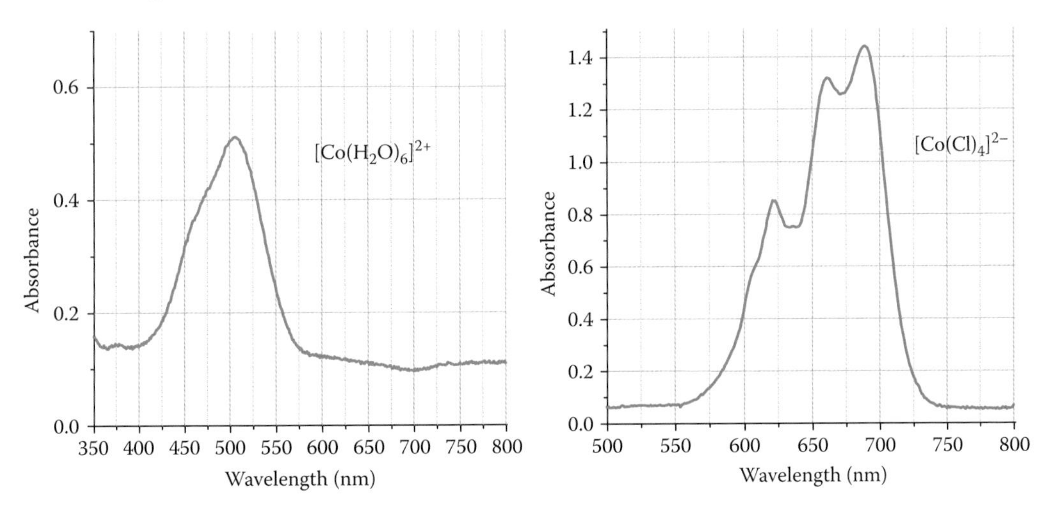

FIGURE 2.56 Absorption spectra of the Co^{2+} ion in an (a) octahedral and (b) tetrahedral environment. (The raw data of spectra were supplied by courtesy of Dr. Gabriel Stanescu.)

- The *charge transfer spectra*, that is, electron transfer from a metal to a ligand. This transition type is also very strong.
- The spectra of substances where the central ion has *full or empty d* orbitals, for example, $KMnO_4$.

STUDY QUESTIONS

2.1 How many *d*-electrons does Co^{2+} have?

2.2 What are the electron configurations of the following ions: Mn^{3+}, Fe^{3+}, Fe^{2+}, Co^{3+}, Ni^{2+}, Cu^{+}, Zn^{2+}?

2.3 Determine the configuration and the ligand field stabilization energy for each of the following complexes: $[FeCl_4]^{2-}$, $[Fe(OH_2)_6]^{2+}$, and $[Fe(CN)_6]^{3-}$.

2.4 Compare the complexes $[Co(CN)_6]^{4-}$ and $[CoCl_6]^{4}$. Are they high spin or low spin?

2.5 Construct the table of microstates for np^2 configuration. Find the free ion terms.

2.6 Fill in the missing word in Hund's rule: "The ground term has the _______ spin multiplicity."

2.7 Determine the ground term of the following ions: Mn^{3+}, Fe^{3+}, Fe^{2+}, Co^{3+}, Ni^{2+}.

2.8 Comment on the following sentence: "A pure crystal does not absorb light whatever the spectral range."

2.9 Comment on the following sentence: "A metal ion introduced in a crystal dramatically changes its absorption spectrum."

2.10 Determine the electronic configuration in an octahedral weak field of the following ions: Mn^{3+}, Fe^{3+}, Fe^{2+}, Co^{3+}, Ni^{2+}.

2.11 Comment on the following sentence: "The tetrahedral field strength is 4/6 times lower than that of octahedral field because of 4 anions present instead of 6 anions."

2.12 Would you predict the number and energy order of the transitions for the Ni^{2+} ion in octahedral and tetrahedral symmetry?

2.13 Which of the following complexes of Ti^{3+} shows the shortest wavelength absorption in the visible range: $[Ti(H_2O)_6]^{3+}$; $[Ti(CN)_6]^{3-}$; $[TiCl_6]^{3-}$?

REFERENCES

Adamson, A. W. 1969. *Photochemistry of Transition Metal Coordination Compounds*. www.iupac.org/publications/pac/20/1/0025/pdf/

Aitken, M. J. 1985. *Thermoluminescence Dating*. London: Academic Press.

Baggaley, E., J. A. Weinstein, and J. A. Gareth Williams. 2012. Lighting the way to see inside the live cell with luminescent transition metal complexes. *Coordination Chemistry Reviews* 256: 1762–1785.

Ballhausen, C. J. 1962. *Introduction to Ligand Field Theory*. New York: McGraw-Hill Book Company.

Ballhausen, C. J. 1972. The crystal field splittings in the 5f complexes. *Theoretical Chemistry Acta* (Berlin) 24: 234–240.

Bethe, H. 1929. Termaufspaltung in Kristallen. *Annalen der Physik, Fünfte Folge* 3: 133–206.

Bransden, B. H. and C. J. Joachain. 1983 or 2003. *Physics of Atoms and Molecules*. Edinburgh: Pearson Education Ltd.

Condon, E. U. and G. Shortley. 1935. *The Theory of Atomic Spectra*. Cambridge: University Press.

Cotton, F. A. 1990. *Chemical Applications of Group Theory*, 3rd edition. New York: Wiley Interscience.

Cutrubinis, M., D. Chirita, D. Savu, C. E. Secu, R. Mihai, M. Secu, and C. Ponta. 2007. Preliminary study on detection of irradiated foodstuffs from the Romanian market. *Radiation Physics and Chemistry* 76: 1450–1454.

Di Bartolo, B. 1978. *Luminescence of Inorganic Solids*. New York: Plenum Press.

FocTek, 2013. Ti:sapphire. http://www.foctek.net/products/Ti_Sapphire.htm?gclid=CMaM keXP8rQCFcNV3godpj8Acg.

Koo, C. K., K. L. Wong, C. W. Y. Man, H. L. Tam, S. W. Tsao, K. W. Cheah, and M. H. W. Lam. 2009. Two-photon plasma membrane imaging in live cells by an amphiphilic, water-soluble cytometalated platinum(II) complex. *Inorganic Chemistry* 48: 7501–7503.

KryLight, 2013. Laser crystals data. http://www.crylight.com/msg.php?id=27.

Kusuma, H. H., Z. Ibrahim, and M. K. Saidin. 2010. Optical absorption and refractive index study of Ti: Al_2O_3 single crystal. *Journal of Chemistry and Chemical Engineering* 4: 59–62.

Mc Clure, D. 1959. *Electronic Spectra of Molecules and Ions in Crystals*. London: Academic Press.

McKeever, S. W. S. 1985. *Thermoluminescence of Solids*. Cambridge Solid State Series, Cambridge: Cambridge University Press.

Secu, M., D. Vainshtein, A. A. Turkin, and H. W den Hartog. 2007. Thermoluminescence and optically stimulated luminescence of gamma-irradiated mineral zircon, *Physica Status Solidi* (c) 4: 1020–1023.

Secu, M., S. Jipa, C. E. Secu, T. Zaharescu, R. Georgescu, and M. Cutrubinis. 2008. Processes involved in the high-temperature thermoluminescence of a Mn^{2+}-doped MgF_2 phosphor. *Physica Status Solidi* (b) 245: 159–162.

Secu, C. E., M. Cherestes, M. Secu, C. Cherestes, V. Paraschiva, and C. Barca. 2011. Retrospective dosimetry assessment using the 380C thermoluminescence peak of tooth enamel. *Radiation Measurements* 46: 1109–1112.

Slater, J. C. 1960. *Quantum Theory of Atomic Structure*. Vol. I, New York: McGraw-Hill.

Strehlow, W. H. and E. L. Cook. 1973. Compilation of energy band gaps in elemental and binary compound semiconductors and insulators. *Journal of Physical and Chemical Reference Data* 2: 163–200. http://www.nist.gov/data/PDFfiles/jpcrd22.pdf

Tanabe, Y. and S. Sugano. 1954. On the absorption spectra of complex ions. I and II. *Journal of Physics Society of Japan* 9: 753–766 and 766–779.

Trutia, Ath. 2005. Optical spectra of divalent cobalt complexes. *Journal of Optoelectronics and Advanced Materials* 7: 2677–2686.

Van Vleck, J. H. and A. Sherman. 1935. The quantum theory of valence. *Reviews of Modern Physics* 7: 167–228.

Wilson, E. G., J. C. Decius, and P. C. Cross. 1955. *Molecular Vibrations—The Theory of Infrared and Raman Vibrational Spectra*. New York: McGraw-Hill (or 1980 Dover edition).

Symmetry and Molecular Orbitals Theory

3

3.1 MOLECULAR ORBITALS

3.1.1 MOLECULAR ORBITALS THEORY

There are two types of bonds:

- Ionic bonds
- Covalent bonds

Ionic bonds involve the *electrostatic attraction* of positive and negative ions. An electrostatic interaction occurs after one or more electron transfers from a metal atom to a halide atom. For example, a sodium atom will lose a single valence electron and a chlorine atom will gain an electron. A metal and a nonmetal transfer electrons to form an ionic compound. The bond between these ions is called an ionic bond.

A covalent bond consists of *pairs of electrons shared* by two bonded atoms. The valence electrons can be *shared by an entire molecule.* Covalent bonds are specific to nonmetallic atoms. For example, a hydrogen molecule is formed by sharing single electrons of both hydrogen atoms. All organic compounds are based on covalent bonds.

Crystal field theory was applied to ionic crystals and molecular complexes based on simple electrostatic interactions between point charges. The electrons of a metal ion are not allowed to mix with the electrons of the surrounding ions. The resulting energy diagram is not always correct because many molecular complexes are formed by the contribution of covalent bonds. This is the reason that VanVleck modified crystal field theory to ligand field theory, where ion orbitals partly overlap in order to take into consideration the covalency.

The nature of covalent bonds is completely different from the electrostatic interaction of ionic bonds. Covalent bonds are based on quantum mechanics and deal with delocalized electrons. From Chapter 1, we know that atomic orbitals are described by wavefunctions. Similarly, a molecular orbital is described by a molecular wavefunction. In this chapter, we will present how atomic orbitals combine to form molecular orbitals, and symmetry considerations will help show us how. For interested readers, a large collection of three-dimensional molecular orbital drawings is presented in the book by Jorgensen and Salem (1973).

The following are some basic guidelines for applying of molecular orbitals theory:

1. Start with atomic orbitals. Molecular orbitals will be constructed as linear combinations of atomic orbitals (LCAOs) from all atoms:
   - The total number of molecular orbitals equals the total number of atomic orbitals.
   - Atomic orbitals must have a suitable symmetry.
   - Atomic orbitals will interact if they overlap.
   - Atomic orbitals will interact if they have closer energies.
2. Arrange the molecular orbitals (as single-electron occupancy) in order of increasing energy using the quantum mechanics protocols.
3. Add electrons to the molecular orbitals:
   - First, fill the lowest-energy orbital.
   - Follow the Pauli exclusion principle: add only two electrons in the same molecular orbital with their spins paired.
   - Follow Hund's rule: fill a degenerate molecular orbital with spin parallel electrons before pairing them.
4. Construct the molecular states from the many electron configurations just obtained by adding electrons to the molecular orbitals.

The types of molecular orbitals are as follows:

- σ, with cylindrical symmetry with respect to the interatomic line, which means that the maximum electron density is located along the line; it appears as a result of the overlap between atomic orbitals oriented through the line (s or p).
- π, not symmetrical with respect to the interatomic line, which means that the maximum electron density is located outside the line and has two phases with different signs; it results from the overlap between atomic orbitals perpendicular to the line (p).
- δ, the maximum electron density is also located outside the interatomic line but in two perpendicular planes that intersect along the line; it has two pairs of opposite lobes, with each pair having its own sign, and can result only from the overlap of d orbitals.

We start with the simplest case of diatomic molecules with similar atoms.

3.1.2 BONDING AND ANTIBONDING σ ORBITALS

The molecular orbitals (MO) can be either *bonding* or *antibonding* orbitals.

When two atoms come together to form a molecule, their atomic orbitals overlap. The molecular wavefunctions arise from superimposing the atomic wavefunctions. Therefore, if we have two atomic orbitals coming together from two different atoms and they combine, the final form will be a molecular orbital. We may suppose that the molecular wavefunction is a linear combination of atomic orbitals (*LCAO method*). The resulting molecular wavefunctions keep the orthonormality property of atomic wavefunctions.

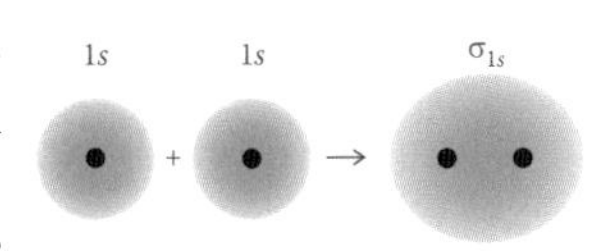

FIGURE 3.1 Two atomic orbitals 1*s* combine to form a molecular bonding orbital σ_{1s}. The sphere in the middle represents the location of the nucleus, much enlarged so that you can see it.

3.1.2.1 EXAMPLE OF H_2 MOLECULE

The hydrogen atom has the lowest orbitals of *s* type. We put two 1*s* atomic orbitals with spherical symmetry close together. They combine to form a molecular orbital σ_{1s} with cylindrical symmetry around the bond axis between the two nuclei as shown in Figure 3.1. We note that there are *no nodal planes along the bond axis*. The nodal plane is the geometrical place where the phase changes sign.

Question: Where are the electrons located in the molecule?

The answer is not obvious. We must calculate the probability density of finding an electron in a given small-volume orbital. We begin from the eigenfunction of a hydrogen atom. It is the product of a radial function and an angular function, as previously shown by Equation 1.92. The squared eigenfunction, that is, $|\Psi^*\Psi|^2$, gives a measure of the probability of finding an electron at a given position in space around the nucleus. Thus, only the radial function counts when expressing probability. The radial function of atomic orbital 1*s* is

$$1s(r) = R_{1s}(r) = \frac{1}{\sqrt{\pi}}e^{-r} \tag{3.1}$$

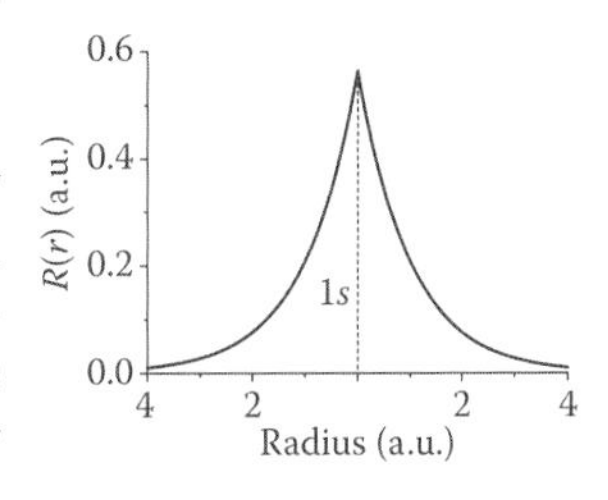

FIGURE 3.2 Dependence of the radial function of atomic orbital 1*s* with respect to the radius.

where *r* is the distance from the nucleus to the electron. We illustrate in Figure 3.2 the radial function dependence of the atomic orbital 1*s* of hydrogen with respect to its distance from the nucleus.

When we put two hydrogen atoms together to form a molecule, then the molecular wavefunction will be a linear combination of atomic wavefunctions:

$$\sigma_{1s} = \Psi_g = \frac{1}{\sqrt{2}}(\psi_{1s}(r_A) + \psi_{1s}(r_B)) \tag{3.2}$$

$$\sigma_{1s}^* = \Psi_u = \frac{1}{\sqrt{2}}(\psi_{1s}(r_A) - \psi_{1s}(r_B)) \tag{3.3}$$

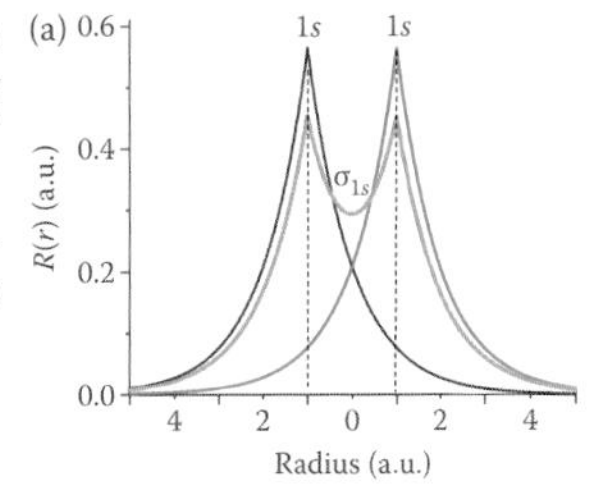

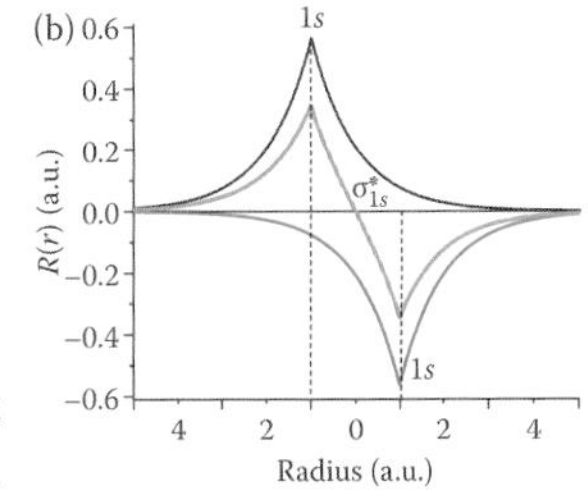

FIGURE 3.3 Shape of gerade (a) and ungerade (b) molecular wavefunctions of the hydrogen molecule.

They are even (gerade) and odd (ungerade), respectively, with respect to the inversion operation. The symmetry center is located in between the nuclei. Their dependence with respect to the radius is illustrated in Figure 3.3.

Another intuitive understanding of these phenomena is to see them as interference effects. Each electron has an associated wave. The interference occurs when the waves overlap. If both waves have the same phase, the interference is constructive and results in a bonding interaction. If the waves are in antiphase, the interference is destructive and results in an antibonding interaction.

Now, we can calculate the probability density of finding electrons in a given small-volume orbital for both orbitals. Therefore, the probability density is given by the square of the wavefunction:

$$(\sigma_{1s})^2 = |\Psi_g|^2 = \frac{1}{2}\left[(\psi_{1s}(r_A))^2 + (\psi_{1s}(r_B))^2 + 2(\psi_{1s}(r_A)\cdot\psi_{1s}(r_B))\right]$$

$$= \frac{1}{2}\left[(1s_A)^2 + (1s_B)^2 + 2(1s_A)(1s_B)\right]$$

$$(\sigma^*_{1s})^2 = |\Psi_u|^2 = \frac{1}{2}\left[(\psi_{1s}(r_A))^2 + (\psi_{1s}(r_B))^2 - 2(\psi_{1s}(r_A)\cdot\psi_{1s}(r_B))\right]$$

$$= \frac{1}{2}\left[(1s_A)^2 + (1s_B)^2 - 2(1s_A)(1s_B)\right]$$

The orbitals' radius dependence is illustrated in Figure 3.4. The electron probability density is nonzero on the line between the two nuclei. Thus, the nuclei are bonded due to the presence of the negative electric charge between the nuclei in the σ_{1s} orbital. Both nuclei are attracted to the electrons between them and compensate for the repulsive forces of the positive nuclei. If the atomic wavefunctions are in antiphase, there is a node in the center between the nuclei; therefore, the probability density is zero right in the middle as shown in Figure 3.4b.

In terms of energy, the energy of molecular orbital σ_{1s} is lower compared to atomic orbitals, whereas the energy of molecular orbital σ^*_{1s} is higher. Thus, the bonding molecular orbital is more favorable for electrons than atomic orbitals. The electrons fill the lowest energy orbital and must respect the Pauli principle. It is the same rule as for filling atomic orbitals and it is illustrated in Figure 3.5. The bonding orbital is more stable. As a consequence of the illustration, *the molecule is more stable*

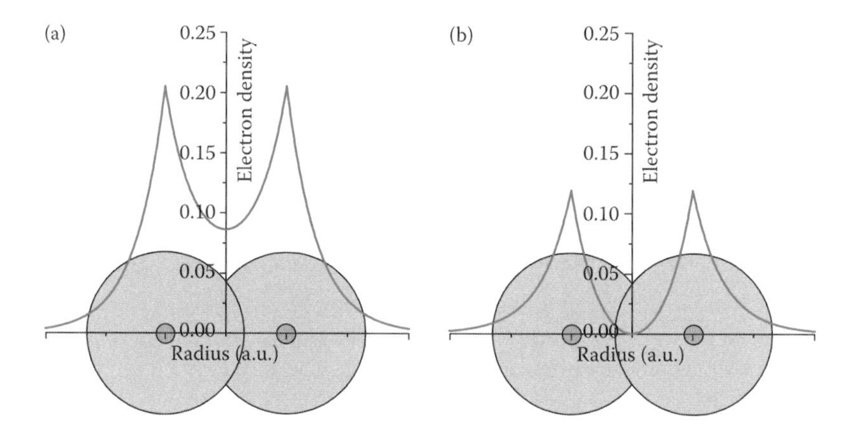

FIGURE 3.4 The electron probability density is nonzero between nuclei for the gerade molecular orbital (a) and zero right in the middle for the ungerade molecular orbital (b).

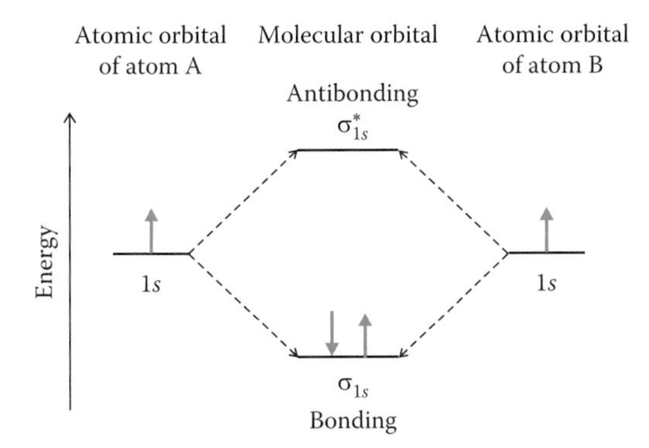

FIGURE 3.5 Energy diagram of the hydrogen molecule. The bonding orbital is filled by two electrons with an opposite spin, so the hydrogen molecule is more stable than individual atoms.

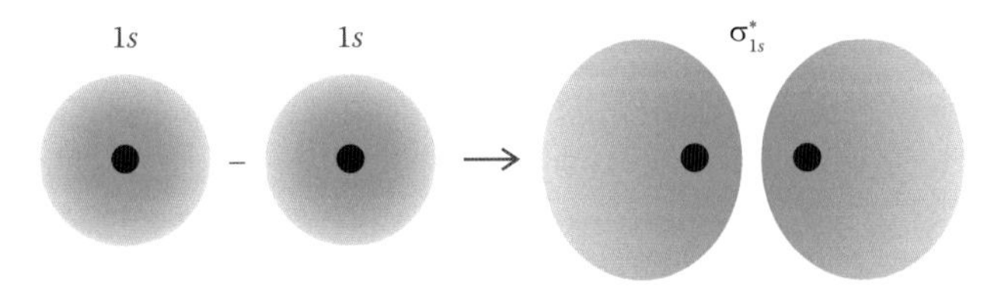

FIGURE 3.6 Antibonding σ^*_{1s} orbital.

than individual atoms. The bonding and antibonding energy levels are equally displaced compared with the atomic level.

The shape of the antibonding orbital σ^*_{1s} is illustrated in Figure 3.6. There is a nodal plane right in the middle that is perpendicular to the bonding line.

To construct a helium molecule, use the molecular energy diagram of the hydrogen molecule. The difference lies in the presence of four electrons. Two electrons fill the bonding orbital and the other two electrons fill the antibonding orbital. Keep in mind that the antibonding orbital is not a "neutral" (nonbonding) orbital. Thus, the total energy of the helium molecule equals the energy of the free atoms, and therefore it is not a stable molecule.

We note that the symbol σ denotes a single-electron molecular state, that is, a molecule with one electron. The final state of the molecule with two electrons in σ state is denoted by the symbol Σ. It follows the same procedure as that of free atoms.

The reader can make the next molecules in order of increasing number of electrons as an exercise. We now continue with *p* orbitals.

3.1.3 BONDING AND ANTIBONDING π ORBITALS

A free atom has three identical *p* orbitals related to the Cartesian axes. When two identical atoms come closer and their atomic orbitals overlap, two different cases are formed:

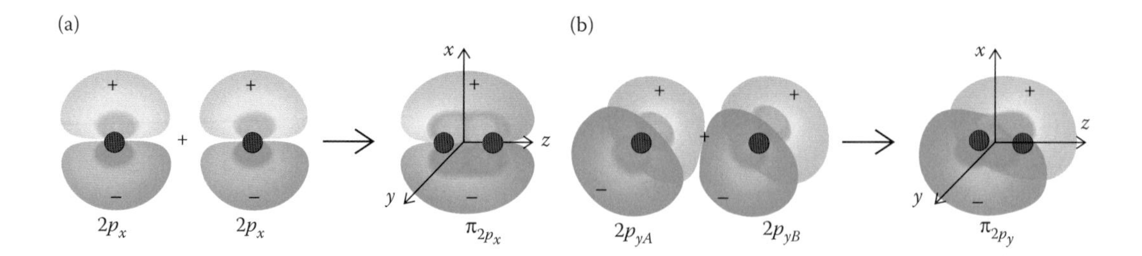

FIGURE 3.7 (a) The constructive interference of two $2p_x$ orbitals results in a π_{2p_x} bonding molecular orbital. (b) The $2p_y$ orbitals also result in a bonding molecular orbital called π_{2p_y}.

- Two *p* orbitals (it does not matter which ones) will always be perpendicular to the bonding line between atoms.
- One *p* orbital (the third) will always be along the bonding line.

As a consequence, the results of the orbitals overlapping (the interference) will be different. So we must treat the two cases separately. In what follows the p_x and p_y orbitals are perpendicular to the bonding line and p_z is along the line.

We start by constructing bonding orbitals formed as LCAOs $2p_x$ and $2p_y$. Both *p*-type orbitals have similar behaviors, but they do not mix with each other. The combination of two $2p_x$ orbitals is illustrated in Figure 3.7a. They can constructively interfere and the result is the π_{2p_x} bonding molecular orbital. The π molecular orbital has a nodal plane through the bond axis in the plane *yOz*. The symbols + and – are used in many books to indicate that the phases of the orbital lobes are *opposite*. We also note that it is customary to represent the *p* orbital by the symbol ∞ (connected). It is easier but it is not accurate. It is more accurate to represent both orbital lobes separately as shown in Figure 3.7. The rule is similar for cross sections of π molecular orbitals.

The constructive interference of the two $2p_y$ atomic orbitals that results in the π_{2p_y} bonding molecular orbital is illustrated in Figure 3.7b. The molecular orbital π_{2p_y} is similar to π_{2p_x} but its single nodal plane is located in the plane *xOz*.

Orbitals $2p_z$ with opposite orientations combine to give a bond along the internuclear axis. The constructive interference results in a σ_{2p_z} bonding molecular orbital that is illustrated in Figure 3.8. This σ molecular orbital has two nodal planes perpendicular to the bond axis.

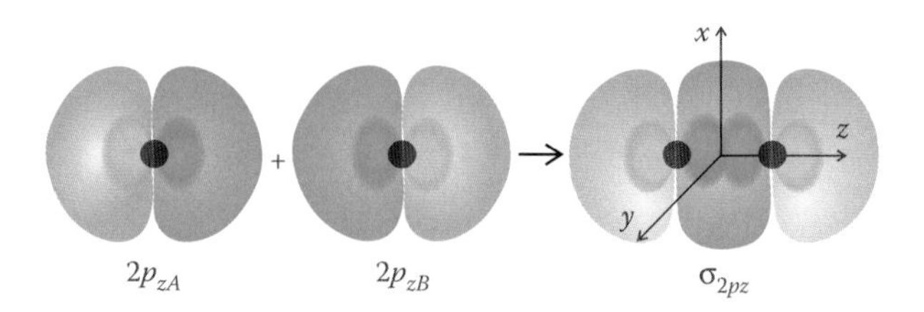

FIGURE 3.8 The σ_{2p} bonding molecular orbital appears as a result of the constructive combination of two *p* atomic orbitals directed along the internuclear axis.

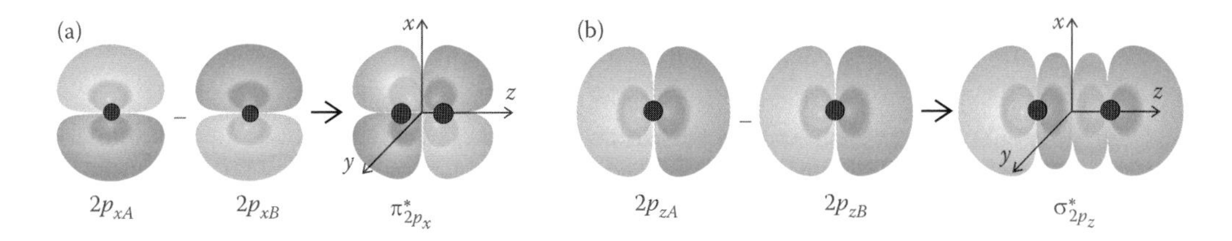

FIGURE 3.9 Antibonding orbitals $\pi^*_{2p_x}$ (a) and $\sigma^*_{2p_z}$ (b).

We illustrate in Figure 3.9 the antibonding molecular orbitals π^* (a) and σ^* (b) that appear as a result of the destructive interference of two $2p$ orbitals. The antibonding molecular orbital $\pi^*_{2p_y}$ is similar to the orbital $\pi^*_{2p_x}$ but is rotated 90° around the z axis.

An energy diagram of a molecule formed as a result of the combination of two identical atoms with s and p orbitals is shown in Figure 3.10. The energies change with atomic number, so the illustrated diagram is valuable for only some molecules. Then, the filling with electrons follows the following procedure:

1. The lowest-energy molecular orbital is filled first.
2. There are no more than two electrons in the same molecular orbital with antiparallel spins.
3. When two molecular orbitals have the same energy, they could be half filled with parallel spin electrons (see the orbitals $\pi^*_{2p_x}$ and $\pi^*_{2p_y}$ of Figure 3.10).

We note that the σ bond is stronger than the π bond due to the fact that the maximum density of electrons is on the bonding line for σ bond while in the second case it is zero (there is a nodal plane) on the bonding line.

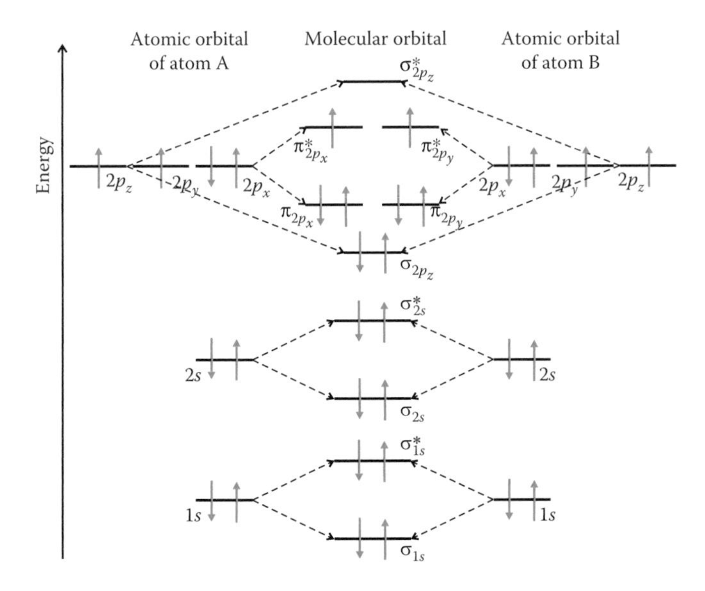

FIGURE 3.10 Energy diagram of the homonuclear molecule of O_2 and the filling with electrons of the molecular orbitals.

3.2 HYBRIDIZATION SCHEME FOR σ ORBITALS

In Section 3.1, we treated the simplest case of the homonuclear diatomic molecules. They are linear molecules. We begin to discuss about complex molecules with a spatial arrangement. Hence, all the atomic orbitals with spherical symmetry should be arranged in space in such a way as to bond to specific directions. Let us take as an example the methane molecule (CH_4), which has a carbon atom in the center and four hydrogen atoms placed in vertices of a tetrahedron. In Chapter 1, we found that the symmetry of the methane molecule is $\boldsymbol{T}_d$. The electron configuration of the carbon atom is $(1s)^2(2s)^2(2p)^2$. The energy of a 2*s* orbital is lower than that of 2*p* orbitals. However, there are only a few combinations in which the carbon atom enters as a bivalent. Instead, there are thousands of combinations in which the carbon enters as a tetravalent. It means that the carbon loses some amount of energy to promote one electron from the *s* orbital to the *p* orbital and gains more energy when it enters in a combination, where it is more stable than in the free state. Thus, its electron configuration changes to $(1s)^2(2s)^1(2p)^3$. In this configuration, with one electron *s* and three electrons *p*, the carbon atom is not yet ready to bond four atoms of hydrogen placed in equivalent positions. The atomic orbitals should combine to give four identical orbitals with $\boldsymbol{T}_d$ symmetry. This process is called hybridization. It is an *internal hybridization*.

3.2.1 sp^3 HYBRIDIZATION

The spatial representation of a methane molecule suggests that the carbon orbitals must participate in four equivalent σ orbitals with tetrahedral symmetry. σ orbitals are more suitable because they are oriented along the bond line and can accommodate in any angle. It is but natural to think that the electrons are concentrated along the bond line between the atoms, but it is not in the spirit of MO theory to localize the electrons. Hence, we should make a combination of one 2*s* orbital with three 2*p* orbitals to obtain four sp^3 hybrid orbitals.

We represent any σ orbital as a vector oriented from a carbon atom to hydrogen (Figure 3.11). The set of four vectors $\vec{r}_1$, $\vec{r}_2$, $\vec{r}_3$, and $\vec{r}_4$ is a basis that can generate a representation. In order to find the representation, we apply all operations of the group to this basis and we keep the matrix trace corresponding to each operation. All matrix traces will finally provide the characters of the representation. Therefore,

$$E\begin{bmatrix}\vec{r}_1\\ \vec{r}_2\\ \vec{r}_3\\ \vec{r}_4\end{bmatrix}=\begin{bmatrix}1\vec{r}_1+0\vec{r}_2+0\vec{r}_3+0\vec{r}_4\\ 0\vec{r}_1+1\vec{r}_2+0\vec{r}_3+0\vec{r}_4\\ 0\vec{r}_1+0\vec{r}_2+1\vec{r}_3+0\vec{r}_4\\ 0\vec{r}_1+0\vec{r}_2+0\vec{r}_3+1\vec{r}_4\end{bmatrix}\Rightarrow E=\begin{bmatrix}1&0&0&0\\0&1&0&0\\0&0&1&0\\0&0&0&1\end{bmatrix}\Rightarrow \chi(E)=4$$

The next operation is a threefold rotation around a bond. The effect of a C_3 rotation around r_1 leaves r_1 unchanged and shifts r_1 into r_4, r_4 into r_3,

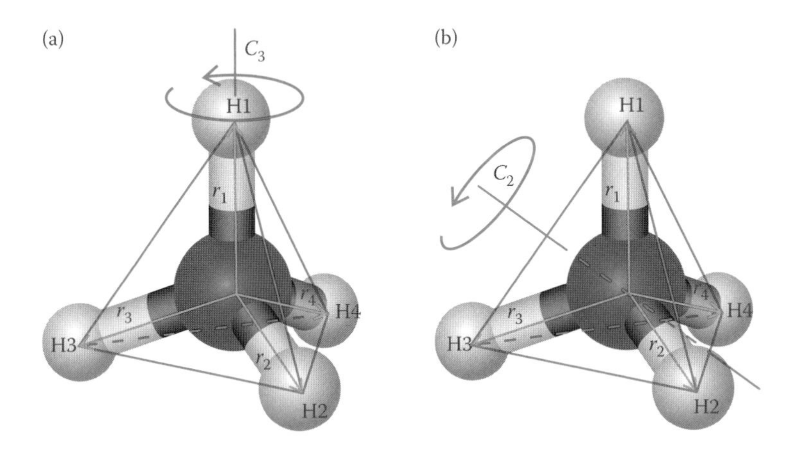

FIGURE 3.11 Each σ (CH) bond of the methane molecule is represented by a vector. They form a basis that generates a reducible representation of the $\boldsymbol{T}_d$ group. Two symmetry elements are represented: (a) threefold rotation axis c_3 that passes through the C and H1 atoms; (b) twofold rotation axis c_2 that passes through the middle of two opposite edges of the tetrahedron.

and r_3 into r_2 as shown in Figure 3.11a. Thus, the corresponding matrix of C_3 will be

$$C_3 = \begin{bmatrix} 1 & 0 & 0 & 0 \\ 0 & 0 & 1 & 0 \\ 0 & 0 & 0 & 1 \\ 0 & 1 & 0 & 0 \end{bmatrix} \Rightarrow \chi(C_3) = 1$$

The C_2 rotation around the axis that passes through the middle of two opposite edges shifts all vectors: $r_1 \to r_3$, $r_2 \to r_4$, $r_3 \to r_1$, and $r_4 \to r_2$ as shown in Figure 3.11b. Thus, the matrix character will be *zero* due to the fact that all vectors are shifted from their initial place. The improper rotation S_2 will have the same effect and then the matrix character will be zero. The dihedral plane Σ_d through the atoms H1 and H3 and the middle of the distance between atoms H2 and H4 will leave the first two atoms unchanged. So, the character of its generated matrix will be 2. The representation generated by the σ orbitals (as we supposed it to be) has the order four and it can be reduced to an irreducible representation of the $\boldsymbol{T}_d$ group by using Equation 1.77. It can be reduced to $A_1 + T_2$ representations as shown in Table 3.1. The totally symmetric representation A_1 has

TABLE 3.1 σ Orbitals of the T_d Molecule Formed by sp^3 or sd^3 Hybridization

T_d	E	$8C_3$	$3C_2$	$6S_2$	$6\Sigma_d$	Reduced to	Atomic Orbitals A_1	Atomic Orbitals T_2
Γ_σ	4	1	0	0	2	$A_1 + T_2$	$x^2 + y^2 + z^2 = s$	$(x, y, z) = (p_x, p_y, p_z)$
								$(xy, xz, yz) = (d_{xy}, d_{xz}, d_{yz})$
			σ					sp^3
								sd^3

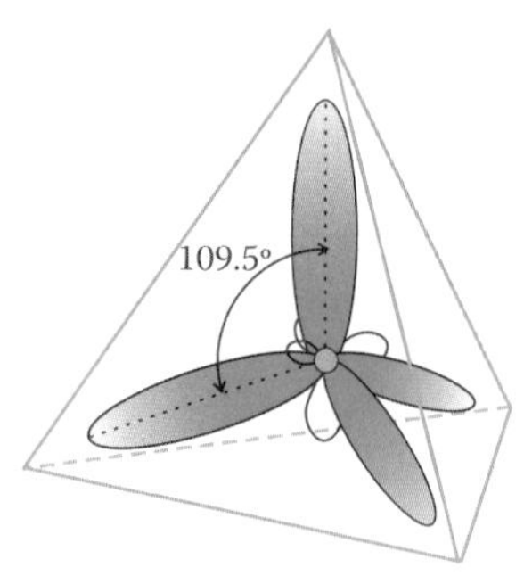

FIGURE 3.12 The atomic orbitals of the central atom in a tetrahedral molecule of AB_4 combine in four sp^3 hybrid orbitals oriented toward the vertices of the tetrahedron. They will contribute to the σ bonds between the central atom A and the outer atoms B.

a basis of second order that corresponds to the atomic orbital s. The sets of atomic orbitals (p_x, p_y, p_z) and (d_{xy}, d_{xz}, d_{yz}) have the symmetry T_2. The result is that the four molecular orbitals of σ type will be formed by the hybridization of atomic orbitals sp^3 or sd^3 or a mixture of both.

The previous results are valuable for any AB_4 molecule with tetrahedral symmetry. It is obvious that we have to make a combination from sp^3 and sd^3. For the methane molecule, it is simple to decide which combination is better because the lowest $3d$ orbitals of the carbon atom have a much higher energy than the $2p$ orbitals. Thus, we can determine that sp^3 hybrid orbital has the major contribution in forming the σ bonds. The four sp^3 orbitals are equivalent and then they are directed toward the vertices of the tetrahedron as shown in Figure 3.12. Other examples of AB_4 molecules or molecular complexes similar to the CH_4 are SiF_4, $AlCl_4^-$, or $ZnCl_4^{2-}$ (Cotton 1990). We will discuss later how to construct these hybrid orbitals.

Another example of sp^3 hybridization of atomic orbitals to form σ molecular orbitals is the ammonia molecule. Its symmetry point group is $\boldsymbol{C}_{3v}$, where the nitrogen atom is located at the top vertex of the trigonal pyramid (see Figure 1.44). It is not in the center of the molecule and this will have important consequences. We attach a vector to each N–H bond that we intend to construct. We then apply the symmetry operations of the $\boldsymbol{C}_{3v}$ group to obtain the representation generated by σ orbitals. The identity operation will have the character 3. The C_3 rotation shifts all vectors, so its character will be 0. The Σ_v reflection leaves one vector unchanged, so the character will be 1. The representation can be reduced to the irreducible representations as follow:

$$\Gamma_\sigma = A_1 + E$$

From the table of characters of the $\boldsymbol{C}_{3v}$ group, we can see that

Symmetry	**Atomic Orbitals**		
A_1	s	p_z	~~d_{z^2}~~
E	(p_x, p_y)	~~$(d_{x^2-y^2}, d_{xy})$~~	~~(d_{xz}, d_{yz})~~

Therefore, the σ orbital will be done as a result of the overlap of one orbital of A_1 symmetry and two orbitals of E symmetry. The $3d$ orbitals of nitrogen lie high above the $2s$ and $2p$ orbitals. Hence, they will not contribute to the molecular orbitals. Because the bond angles in the NH_3 molecule are nearly tetrahedral, the hybridization around nitrogen will be sp^3; therefore, the orbitals $2s$ and $2p$ will participate as shown in Figure 3.13a. When four atomic orbitals overlap, four identical molecular orbitals will be generated. They should accommodate all the space around the nitrogen atom in a tetrahedral arrangement because they originate in the set of (p_x, p_y, p_z) orbitals that are perpendicular to each other. However, the molecule has only three hydrogen atoms. It means that three sp^3 orbitals of nitrogen will make σ bonds with three s orbitals of the hydrogen atoms

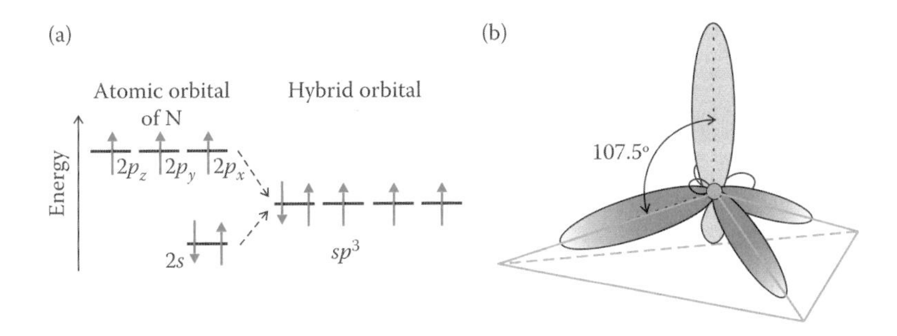

FIGURE 3.13 (a) The four valence atomic orbitals of nitrogen will combine to give four hybrid orbitals sp^3. (b) In the ammonia molecule, three of them, filled by a single electron, will contribute to the σ bonds and one hybrid orbital filled by one pair of electrons rests unbounded. As a consequence the molecule is slightly distorted from a perfect tetrahedron.

while one pair of electrons rests in the fourth sp^3 orbital located around nitrogen as illustrated in Figure 3.13b.

3.2.2 sp^2 HYBRIDIZATION

The sp^2 hybridization, that is, the combination of 2*s* and two 2*p* atomic orbitals to form three sp^2 molecular orbitals, is specific to planar molecules AB_3 because any two *p* orbitals determine a plane. The spherical symmetry of the *s* orbital does not have a preferred orientation. Examples of trigonal planar molecules are CO_3^{2-} and BF_3.

The BF_3 molecule belongs to the $\boldsymbol{D}_{3h}$ group that contains 12 symmetry operations previously illustrated in Figure 1.37. Following the above procedure, we can find the representation generated by three σ planar bonds. The representation can be reduced to $\Gamma_\sigma = A_1' + E'$. The atomic orbitals of 3*d* type are too high in the energy scale to mix with 2*s* or 2*p*. So, a hybrid orbital sp^2 will participate in σ bonding:

D_{3h}	E	$2C_3$	$3C_2$	Σ_h	$2S_3$	$3\Sigma_v$	Reduced to	Atomic Orbitals A'_1	Atomic Orbitals E'
Γ_σ	3	0	1	3	0	1	$A'_1 + E'$	s d_{z^2}	(p_x, p_y) $(d_{x^2-y^2}, d_{xy})$
			σ						sp^2
									~~sd^2, dp^2, d^3~~

All three valence electrons will fill three identical orbitals sp^2 and they will be distributed as one electron to each orbital (Figure 3.14). Thus, all sp^2 orbitals will be involved in bonds.

Another example of sp^2 hybridization of atomic orbitals of the carbon atom is the σ bonding formation in the ethylene molecule ($H_2C{=}CH_2$). Each carbon atom uses three identical hybrid orbitals sp^2 to bond with two hydrogen atoms and the other carbon atom. All six sp^2 orbitals of the molecule are arranged in the same plane.

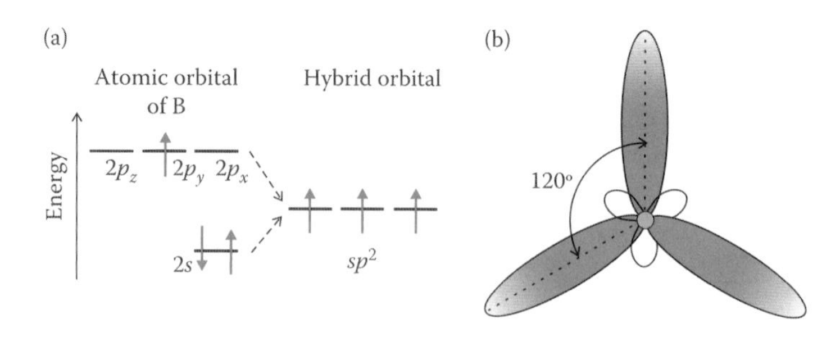

FIGURE 3.14 Hybrid orbitals sp^2 of the boron atom (a) used in the formation of σ bonds (b) of the BF_3 molecule.

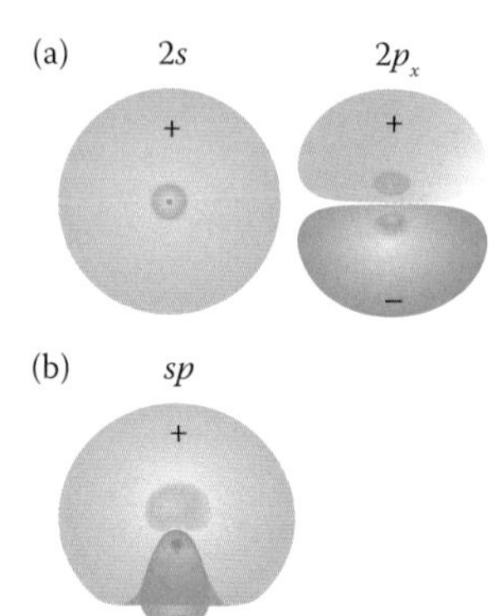

FIGURE 3.15 (a) Illustration of both the 2*s* and the $2p_x$ orbitals of the carbon atom when they are not mixed. (b) The *sp* hybrid orbital resulted from the combination of the *s* and *p* orbitals: the positive lobe of the *p* orbital adds to the *s* orbital; the negative lobe of the *p* orbital subtracts from the *s* orbital.

3.2.3 *sp* HYBRIDIZATION

The *sp* hybridization is used in linear molecules such as CO_2 and acetylene (C_2H_2). It is the simplest case of a combination between single *s* and *p* orbitals.

In the following example, we combine the 2*s* orbital with the $2p_x$ orbital of the carbon atom. The positive lobe of the *p* orbital adds to the s orbital (positive too) and the electron density increases on that side of the atom. The negative lobe of the *p* orbital subtracts from the *s* orbital and the electron density decreases on the opposite side of the atom. It results in two lobes of different size. This is illustrated in Figure 3.15.

We summarize the possible hybridizations of the atomic orbitals of a carbon atom when it is bonded to four, three, or two other atoms in Table 3.2.

3.3 HYBRIDIZATION SCHEME FOR π ORBITALS

The π bond can be formed by the overlap of *p* orbitals that are perpendicular to the bonding line as shown previously in Section 3.1.3. The π bond has two lobes above and below the nodal plane. The nodal plane passes through the bond line.

It is possible to simultaneously make two π bonds between two atoms *A* and *B* as illustrated in Figure 3.16. They must be perpendicular to each other and have in common the same axis. The orbitals of these two π bonds do not overlap.

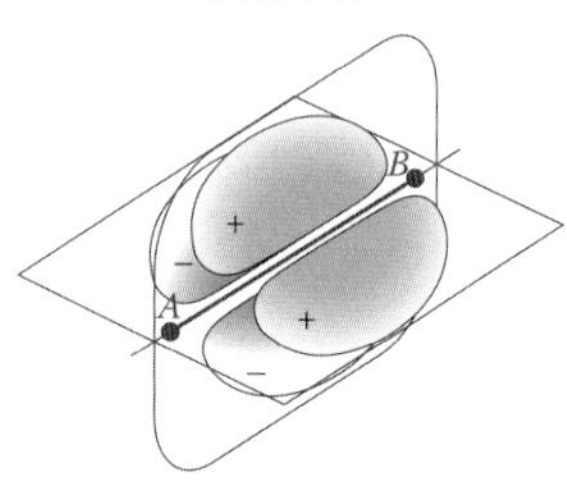

FIGURE 3.16 Two atoms can be connected by two π bonds perpendicular to each other.

TABLE 3.2 Internal Hybridization of Atomic Orbitals of the Carbon Atom for σ Bonding

Number of σ bonds	4	3	2
Hybridization type	sp^3	sp^2	*sp*
Number of combined orbitals	4	3	2
Angle between bonds	109.5°	120°	180°
Geometrical arrangement of σ bonds	Tetrahedral	Trigonal	Linear

We must determine which orbitals of the A atom should participate in π bonding with the B atom. The atomic orbitals of the A atom will combine in such a way as to form π hybrid orbitals. Then the hybrid orbitals of the A atom will connect to the π hybrid orbitals of the B atom. The maximum number of hybrid orbitals equals two times the number of the B type atoms placed around and π bonded to the A atom. The π orbitals (maximum two) of each atom can be represented by two vectors perpendicular to each other and perpendicular to the A–B line. In order to find the representation generated by these vectors, we may take the corresponding vectors attached to the B atoms because they transform equally under the action symmetry operations (Cotton 1990). Each π orbital is a pair of hybrid orbitals from the A and B atoms, respectively, and any symmetry operation will transform both in the same way.

3.3.1 AB_3 MOLECULE WITH D_{3h} SYMMETRY

We take again as an example the simple molecule of boron trifluoride already discussed at sp^2 hybridization for σ bonding. The BF_3 molecule has the advantage of being planar and easier to work with. We attach two vectors to each F atom: one is perpendicular to the plane of the molecule and the other is in the plane (Figure 3.17). Then we apply the symmetry operations of the $\boldsymbol{D_{3h}}$ group to obtain traces of the corresponding matrices. The maximum dimension will be 6, as given by the character of the identity operation. The threefold rotation C_3 shifts all vectors and its character will be 0. The character of the twofold rotation C_2 is −2 (!) because both vectors of the F2 atom do not shift but change direction (sign). The reflection through the horizontal plane leaves unchanged the horizontal vectors and changes the direction of the vertical vectors. Thus, the character of Σ_h will be 0 (+3 − 3 = 0). The improper rotation shifts all vectors and its character will be 0. The reflection through the vertical plane determined by the axis c_3 and the atom F2 leaves unchanged the vertical vector and changes the direction of the horizontal one. So, the character will be 0 (+1 − 1 = 0). The results are shown in Table 3.3.

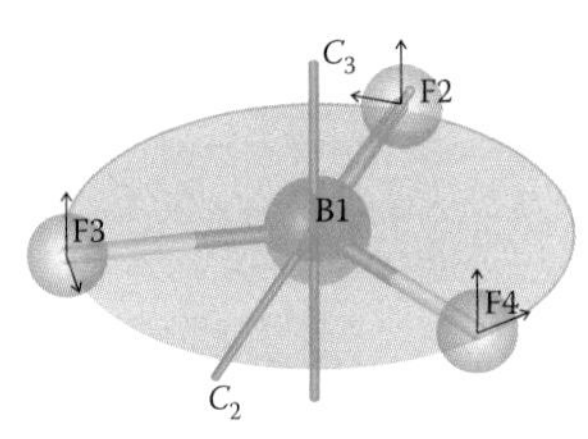

FIGURE 3.17 Two vectors perpendicular to each other will be attached to every peripheral atom in order to obtain the reducible representation generated by the π orbitals.

It is important to underline that no symmetry operation mixes horizontal vectors with vertical vectors and vice versa. It is specific to the planar molecules. The consequence is that we can split the representation into

TABLE 3.3 Characters of Reducible Representation Generated by π Hybrid Orbitals of AB_3 Molecule and Atomic Orbitals That Can Combine to Form π Bonds

D_{3h}	E	$2C_3$	$3C_2$	Σ_h	$2S_3$	$3\Sigma_v$	Reduced to	Atomic Orbitals A_2	Atomic Orbitals E
Γ_π	6	0	−2	0	0	0			
$\Gamma_{\pi vert}$	3	0	−1	−3	0	1	$A_2'' + E''$	p_z	(d_{xz}, d_{yz})
$\Gamma_{\pi horiz}$	3	0	−1	3	0	−1	$A_2' + E'$	–	$(p_x, p_y)(d_{x^2-y^2}, d_{xy})$
				π_v					pd^2
				π_h					?

two independent representations: one for vertical vectors and the other for horizontal vectors. The characters of these representations generated by either vertical vectors or horizontal vectors are shown in Table 3.3. Vertical and horizontal are attributes related to the symmetry operations of the group where the plane of the molecule is denoted as a horizontal plane and so on. From Table 3.3, one observes that the atomic orbitals p_z, d_{xz}, and d_{yz} should combine to form π hybrid orbitals out of the molecular plane (π vertical bonds). It is a conclusion from the point of view of molecular symmetry only. Remember that the overlap of orbitals should be both adapted to the symmetry and allowed by the energy of orbitals. The boron atom with three valence electrons ($2s^2p^1$) promotes one electron from 2*s* orbital into $2p_y$ orbital to construct sp^2 hybrid orbitals for σ bonds, and thus boron already used all electrons (see Figure 3.14) in the BF_3 molecule. The σ bond is lower in energy scale (more stable) than the π bond and the electrons prefer to stay there. So, boron has no other electrons available for the π bond. Regarding the horizontal orbitals, we note that there are no atomic orbitals available with A'_2 symmetry. For reasons of symmetry, we can only state that pairs of *p* orbitals or *d* orbitals or a mixture of both could partly contribute to form π hybrid orbitals in the plane of the molecule. It is not so in the case of the BF_3 molecule. We expect the p_{vertical} orbital (empty) of the boron atom to accept paired electrons in order to make covalent bonds. There are two hypotheses. The first states that the p_z orbital of the boron atom rests empty in order to keep neutrality. The second states that the p_{vertical} orbital (empty) of the boron atom overlaps with three p_{vertical} orbitals of fluorine atoms, each of them filled with a pair of electrons, as the symmetry allows. Thus, three stronger B–F π_{vertical} bonds appear and they are filled by the paired electrons from fluorine atoms. This involves a reactivity of the molecule that distorts the planar geometry to a final pyramidal structure in the complexes (Franca and Diez 2009).

3.3.2 AB_4 MOLECULE WITH D_{4h} SYMMETRY

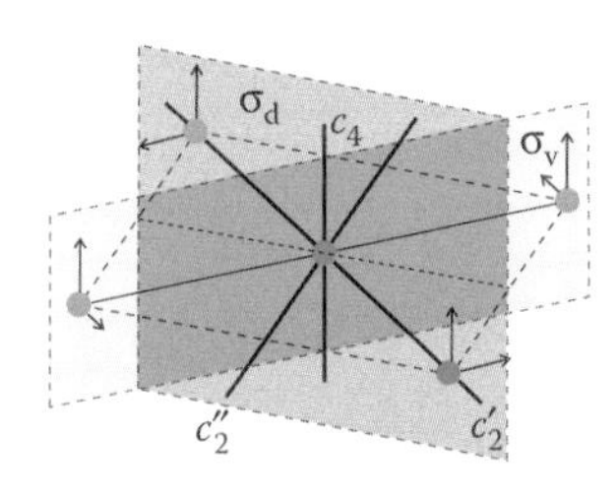

FIGURE 3.18 Illustration of some symmetry elements of the D_{4h} group that will be used to generate the reducible representations of π orbitals.

You can exercise the procedure on the AB_4 planar molecule using the illustration provided in Figure 3.18 and the character table shown in Table 3.4. We represented some symmetry elements in order to avoid confusion. The total reducible representation can be split into two representations corresponding to the vertical and horizontal vectors, respectively. Those representations must be reduced to the irreducible representations of the group. The results to be obtained are also shown in the table.

Comments:

1. Both the vertical and horizontal π bonds require four atomic orbitals and only three orbitals have the required symmetry. So, the π bonds will not be complete.
2. The vertical hybrid orbitals can use the p_z, d_{xz}, d_{yz} orbitals because they certainly will not contribute to the σ bond. They are perpendicular to the molecular plane. Thus, the molecule can have π bonds out of the molecular plane, if the energy allows them (overlapping between the *p* and *d* orbitals).

TABLE 3.4 Character Table of the D_{4h} Point Group

D_{4h}	E	$2C_4$	$C_2(C_4^2)$	$2C_2'$	$2C_2''$	I	$2S_4$	Σ_h	$2\Sigma_v$	$2\Sigma_d$	Reduced to	Atomic Orbitals
A_{1g}	1	1	1	1	1	1	1	1	1	1		x^2+y^2, z^2
A_{2g}	1	1	1	−1	−1	1	1	1	−1	−1		
B_{1g}	1	−1	1	1	−1	1	−1	1	1	−1		x^2-y^2
B_{2g}	1	−1	1	−1	1	1	−1	1	−1	1		xy
E_g	2	0	−2	0	0	2	0	−2	0	0		(xz, yz)
A_{1u}	1	1	1	1	1	−1	−1	−1	−1	−1		
A_{2u}	1	1	1	−1	−1	−1	−1	−1	1	1		z
B_{1u}	1	−1	1	1	−1	−1	1	−1	−1	1		
B_{2u}	1	−1	1	−1	1	−1	1	−1	1	−1		
E_u	2	0	−2	0	0	−2	0	2	0	0		(x, y)
$\Gamma\pi$	8	0	0	−4	0	0	0	0	0	0		Hybrid orbitals ↓
$\Gamma\pi_v$	4	0	0	−2	0	0	0	−4	2	0	$A_{2u}+B_{2u}+E_g$	$\pi_v = p_z d_{xz} d_{yz}$
$\Gamma\pi_h$	4	0	0	−2	0	0	0	4	−2	0	$A_{2g}+B_{2g}+E_u$	$\pi_h = d_{xy} p_x p_y$

3. The π bonds in the molecular plane compete with the σ bonds, where the p_x and p_y orbitals are used. Thus, the π_h bonds should be formed only by d_{xy} orbitals.

3.4 HYBRID ORBITALS AS LINEAR COMBINATIONS OF ATOMIC ORBITALS

We know to find the atomic orbitals that could combine to make hybrid orbitals. The symmetry can only suggest the possible combinations, but not if and how they are done. The energy of orbitals is the main criterion. The Schrödinger equation $\hat{H}\Psi = E\Psi$ should be solved in order to find the energy. $\hat{H}$ is the Hamiltonian operator of the molecule and Ψ is the eigenfunction of the molecular orbital. In an atom, all orbitals are centered on the nucleus, while in a molecule they are located on two or more nuclei. Thus, the electron is delocalized or, more precisely, localized on many atoms. This property of delocalizing electrons provides the stability of the molecule because the repulsion between two nuclei is compensated for by the attraction between electrons and both nuclei. We note that the last sentence is just the expression of a classic way to imagine a purely quantum phenomenon, so it is not exact. If molecular orbitals are localized on an entire molecule, then the molecular symmetry properties will show which integrals are zero. This will reduce the calculation to be done.

3.4.1 LCAO IN A TWO-CENTER BOND

The eigenfunction of the hybrid orbital can be written as a linear combination of eigenfunctions of atomic orbitals centered on atoms and involved in the construction of a molecular orbital. For a diatomic molecule, two functions are employed to describe the atomic orbitals Φ_1 and Φ_2 and we can write the wavefunction for *each electron*:

$$\Psi_i = c_{i1}\Phi_1 + c_{i2}\Phi_2 \quad (3.4)$$

The coefficients c_{i1} and c_{i2} will show the contribution of each atomic orbital to the hybrid orbital. In the homonuclear diatomic molecule, the symmetry suggests that the contributions are equal. Then, we introduce the hybrid function in the Schrödinger equation:

$$\hat{H}(c_{i1}\Phi_1 + c_{i2}\Phi_2) = E_i(c_{i1}\Phi_1 + c_{i2}\Phi_2) \quad (3.5)$$

We multiply on the left side by Φ_1^* and integrate the entire equation:

$$c_{i1}\int \Phi_1^* \hat{H} \Phi_1 \, d\tau + c_{i2}\int \Phi_1^* \hat{H} \Phi_2 \, d\tau = E_i\left(c_{i1}\int \Phi_1^*\Phi_1 \, d\tau + c_{i2}\int \Phi_1^*\Phi_2 \, d\tau \right)$$

The atomic eigenfunctions are orthonormal; thus,

$$c_{i1}\int \Phi_1^* \hat{H} \Phi_1 \, d\tau + c_{i2}\int \Phi_1^* \hat{H} \Phi_2 \, d\tau = E_i\left(c_{i1} + c_{i2}\int \Phi_1^*\Phi_2 \, d\tau \right) \quad (3.6)$$

We simplify writing the equation by denoting

$$\begin{cases} H_{11} = \int \Phi_1^* \hat{H}\Phi_1 \, d\tau \\ H_{12} = \int \Phi_1^* \hat{H}\Phi_2 d\tau \\ S_{12} = \int \Phi_1^*\Phi_2 d\tau \end{cases} \quad (3.7)$$

The matrix element H_{11} gives the energy of the atomic orbital of atom 1. The matrix element H_{12} gives the energy of interaction between the atomic orbital of atom 1 and the atomic orbital of atom 2. Be aware that the functions Φ_1 and Φ_2 could denote the same orbital type, for example, $2p_z$ and $2p_z$ in the homonuclear molecule, but they belong to different atoms 1 and 2, respectively. So, they are not orthogonal! S_{12} is called the *overlap integral* and it takes values in the range [0–1]. The method presented here is also called *LCAO approximation* because the electrons are treated distinctly while each covalent bond represents in fact a shared electron *pair*.

We may repeat the procedure to multiply Equation 3.5 by Φ_2^* and finally we obtain

$$c_{i1}(H_{11} - E_i) + c_{i2}(H_{12} - E_iS_{12}) = 0 \quad (3.8)$$

$$c_{i1}(H_{21} - E_iS_{21}) + c_{i2}(H_{22} - E_i) = 0 \quad (3.9)$$

Equations 3.8 and 3.9 form a system of linear equations in c_{i1} and c_{i2}. The system of equations can be solved using the secular equation:

$$\begin{vmatrix} H_{11} - E & H_{12} - ES_{12} \\ H_{21} - ES_{21} & H_{22} - E \end{vmatrix} = 0 \tag{3.10}$$

H_{11} and H_{22} correspond roughly to the pure electrostatic interaction in atom (1) and atom (2), respectively. They are both denoted by the symbol α. H_{12} and H_{21} correspond to the interaction energy between atomic orbitals of adjacent atoms and must be equal. They are both denoted by the symbol β. The notations are summarized as follows:

$$\begin{cases} H_{11} = \alpha_1 \quad \text{and} \quad H_{22} = \alpha_2 & \rightarrow \text{Coulomb integrals} \\ H_{12} = H_{21} = \beta & \rightarrow \text{resonance integral} \\ S_{12} = S_{21} & \rightarrow \text{overlap integral} \end{cases} \tag{3.11}$$

For diatomic molecules of

- Homonuclear type (identical atoms) – $\alpha_1 = \alpha_2$
- Heteronuclear type (different atoms) – $\alpha_1 \neq \alpha_2$

We make an approximation to replace S_{ij} by δ_{ij}, that is, the overlap is zero. Thus, the secular determinant given by Equation 3.10 reduces to

$$\begin{vmatrix} \alpha - E & \beta \\ \beta & \alpha - E \end{vmatrix} = 0 \tag{3.12}$$

Equation 3.12 has two solutions:

$$\begin{cases} E_1 = \alpha + \beta \\ E_2 = \alpha - \beta \end{cases}$$

Because two electrons are in the same state, the total energy is

$$\begin{cases} E_1 = 2(\alpha + \beta) \\ E_2 = 2(\alpha - \beta) \end{cases} \tag{3.13}$$

E_1 is the lowest energy, that is, it represents the energy of the most stable molecular orbital called *bonding MO*. E_2 is the highest energy, that is, it represents the energy of the less stable molecular orbital called *antibonding MO*. They are illustrated in Figure 3.5. The bonding MO is also more stable compared to the free atom state, where the atoms are far from each other. The splitting energy (2β) in the molecular state is given by the value of the resonance integral.

TABLE 3.5 Energies and Expressions of Molecular Orbitals of the Diatomic Homonuclear Molecule for One Electron

Energy	Function	Character
$\alpha - \beta$	$\sigma_u = \frac{1}{\sqrt{2}}(\Phi_1 - \Phi_2)$	Antibonding
$\alpha + \beta$	$\sigma_g = \frac{1}{\sqrt{2}}(\Phi_1 + \Phi_2)$	Bonding

If we take a non-zero value of the overlap integral, but small in order to improve the precision of calculus, then the solutions of the corresponding secular equation will be

$$\begin{cases} E_1 = \dfrac{\alpha + \beta}{1 + S} \\ E_2 = \dfrac{\alpha - \beta}{1 - S} \end{cases} \text{ that can be approximated as } \begin{cases} E_1 = \alpha + \beta(1 - S) \\ E_2 = \alpha - \beta(1 - S) \end{cases}$$

All the calculus and the results presented above can be used to describe the π bond between two carbon atoms in different molecules, that is, in the ethylene molecule (C_2H_4).

What about the eigenfunction describing the hybrid orbitals of the diatomic molecule? The value of E_1 given by Equation 3.13 is inserted into Equations 3.8 and 3.9, and then the system of equations may be solved for the coefficients c_{11} and c_{12}. We obtain just $c_{11} = c_{12}$. We apply the normalization condition for the molecular orbital $|\Psi^*\Psi| = 1$ that finally gives $c_{11} = c_{12} = 1/\sqrt{2}$. They will determine the expression of the bonding molecular orbital. Similarly, the value of E_2 substituted in the same equations will give the coefficients of the antibonding molecular orbital. Both results are summarized below in Table 3.5.

3.4.2 THREE-CENTER BONDING

We can continue to extend the application of the above procedure to more complicated molecules: three identical atoms, four identical atoms, and so on. The secular determinant will not only change in size (e.g., 3×3 or 4×4) but also depend on the *shape* of the molecule. For example, a molecule with three carbon atoms can be linear or triangular. For the linear arrangement C1–C2–C3, the atoms C1 and C3 do not interact and the secular determinant will be

$$\begin{vmatrix} \alpha - E & \beta & 0 \\ \beta & \alpha - E & \beta \\ 0 & \beta & \alpha - E \end{vmatrix} = 0$$

Thus, the third-order equation will be simple and have the following solutions:

$$\begin{cases} E_3 = \alpha - \sqrt{2}\beta \\ E_2 = \alpha \\ E_1 = \alpha + \sqrt{2}\beta \end{cases}$$

The energy of the bonding orbital will be $2\alpha + 2\sqrt{2}\beta$. Remember that you can fill a molecular orbital only with two electrons.

In case of a triangular arrangement, the secular determinant and its solutions are different from those of the linear arrangement:

$$\begin{vmatrix} \alpha - E & \beta & \beta \\ \beta & \alpha - E & \beta \\ \beta & \beta & \alpha - E \end{vmatrix} = 0 \quad \text{and} \quad \begin{cases} E_1 = \alpha - \beta \\ E_2 = \alpha - \beta \\ E_3 = \alpha + 2\beta \end{cases}$$

Thus, the energy of the bonding orbital will be $2\alpha + 4\beta$ and we can conclude that this arrangement will be more stable than the linear one.

To describe the corresponding molecular orbitals, we need to determine the eigenfunctions using equations similar to Equations 3.8 and 3.9. For the highest energy level $(\alpha - \sqrt{2}\beta)$ of the linear arrangement, we can write the following system, where only two equations are linearly independent:

$$\begin{cases} \sqrt{2}c_{31} + 1 \cdot c_{32} + 0 \cdot c_{33} = 0 \\ 1 \cdot c_{31} + \sqrt{2}c_{32} + 1 \cdot c_{33} = 0 \\ 0 \cdot c_{31} + 1 \cdot c_{32} + \sqrt{2}c_{33} = 0 \end{cases}$$

So, we may express the orbital function as one dependent parameter:

$$\Psi_3 = c_{31}(\Phi_1 - \sqrt{2}\Phi_2 + \Phi_3)$$

The normalization condition of the wavefunction helps us to find the coefficient c_{31}:

$$\Psi_3 = \frac{1}{2}(\Phi_1 - \sqrt{2}\Phi_2 + \Phi_3)$$

By following the above procedure, we can also find the eigenfunctions of the middle and lowest energy levels α and $\alpha + \sqrt{2}\beta$, respectively:

$$\Psi_2 = \frac{1}{\sqrt{2}}(\Phi_1 - \Phi_3) \quad \text{and} \quad \Psi_1 = \frac{1}{2}(\Phi_1 + \sqrt{2}\Phi_2 + \Phi_3)$$

An example of this molecule type is an allyl anion $(CH_2{=}CH{-}CH_2)$.

3.5 SYMMETRY ADAPTED LINEAR COMBINATIONS

3.5.1 PROJECTION OPERATOR

The construction work of molecular orbitals as LCAOs should generate combinations that must be both orthonormal and bases for irreducible representations of the point group of the molecule. Therefore, during the construction process, the symmetry restrictions will help us to reduce the possibilities of combination. The procedure is somehow reversed: we know the symmetry of molecular orbitals and we have to find which atomic orbitals will fit better in these combinations. Thus, we will generate functions that are called *symmetry adapted linear combinations* (SALCs).

The SALCs of a molecule may be constructed using a *projection operator*. Its name shows that it will reduce the dimension of an object; in our case, the set of starting functions. There are two ways to write the projection operator. You can find the complete expression of it in Cotton (1990, p. 116). In order to use it, you need to know all matrices of each representation of the group, which is a complex task. Usually, you have direct access to the characters table of the group. We therefore present here only the reduced expression of the projection operator:

$$\hat{P}^{\Gamma_i} = \frac{l_i}{h} \sum_R \chi_i(R)\hat{R} \tag{3.14}$$

where Γ_i is the irreducible representation under consideration, l_i is the dimension of the irreducible representation, h is the order of the group, and $\chi_i(R)$ is the character of the irreducible representation for the symmetry operation R of the group. The operator sums over all symmetry operations of a group, not just classes as they appear in the character table.

Note: To construct the SALCs, one refers to the atoms that are interchanged by the symmetry operations, for example, two hydrogen atoms of the water molecule, three hydrogen atoms of ammonia, and six fluorine atoms of the SF_6.

We will illustrate how it works through the following few examples.

3.5.2 SALCs FOR σ BONDS OF WATER

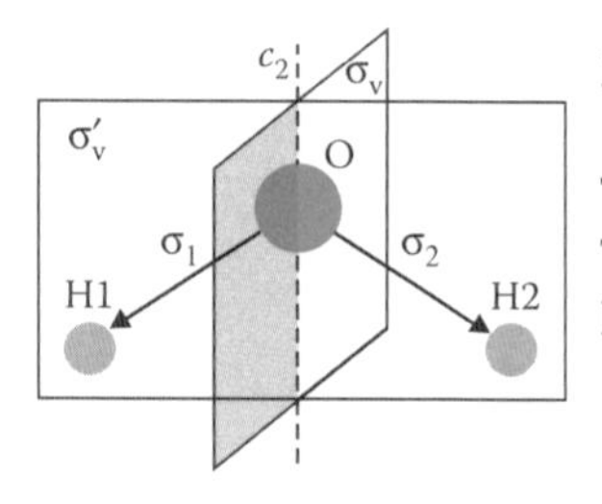

FIGURE 3.19 Bonding vectors and symmetry elements of the water molecule.

The first example is the water molecule. It belongs to the $\boldsymbol{C_{2v}}$ point group. The basis set is represented by two bond vectors σ_1 and σ_2 as shown in Figure 3.19.

1. We tabulate the characters of the irreducible representations of the group (Table 3.6). We separate the operations from the same class. It is not the case for $\boldsymbol{C_{2v}}$. We apply all symmetry operations to the σ_1 bond vector. Under the action of the rotation C_2, the vector σ_2 takes the place of σ_1. The reflection Σ_v will have the same result.

TABLE 3.6 Projection Operator Method to Obtain SALCs for the Water Molecule

C_{2v}	E	C_2	Σ_v	Σ'_v		
A_1	1	1	1	1		
A_2	1	1	−1	−1		
B_1	1	−1	1	−1		
B_2	1	−1	−1	1		
$\hat{R}\sigma_1$	σ_1	σ_2	σ_2	σ_1	Sum ↓	Normalized SALC
$\chi_{(A_1)}\hat{R}\sigma_1$	σ_1	σ_2	σ_2	σ_1	$2(\sigma_1+\sigma_2)$	$\frac{1}{\sqrt{2}}(\sigma_1+\sigma_2)$
$\chi_{(A_2)}\hat{R}\sigma_1$	σ_1	σ_2	$-\sigma_2$	$-\sigma_1$	0	
$\chi_{(B_1)}\hat{R}\sigma_1$	σ_1	$-\sigma_2$	σ_2	$-\sigma_1$	0	
$\chi_{(B_2)}\hat{R}\sigma_1$	σ_1	$-\sigma_2$	$-\sigma_2$	σ_1	$2(\sigma_1-\sigma_2)$	$\frac{1}{\sqrt{2}}(\sigma_1-\sigma_2)$

2. The results obtained after the action of all operations (row 6) will be multiplied by the characters of each representation. Then we collect the terms on the row.
3. The SALC functions will be normalized.

So, the SALC functions for the water molecule are

$$\Psi_1 = \frac{1}{\sqrt{2}}(\sigma_1+\sigma_2) \quad \text{and} \quad \Psi_2 = \frac{1}{\sqrt{2}}(\sigma_1-\sigma_2) \tag{3.15}$$

The linear combinations will always be the sum and difference functions whenever two equivalent functions are involved.

We note that the method of projection operator in order to construct SALCs is also very useful to obtain the *normal modes of vibration of the molecule.* The functions of Equations 3.15 show that the water molecule has two stretching vibrations: symmetric and antisymmetric.

Another way is to begin finding the reducible representation generated by the bond vectors σ_1 and σ_2. Then it will be decomposed into irreducible representations of the group. Finally, one generates the SALC of orbitals that arise only from the found irreducible representations. We summarize below the results:

C_{2v}	E	C_2	Σ_v	Σ'_v	**Reduced to**	
Γ_σ	2	0	0	2	A_1+B_2	
$\hat{R}\sigma_1$	σ_1	σ_2	σ_2	σ_1	Sum ↓	Normalized SALC
$\chi_{(A_1)}\hat{R}\sigma_1$	σ_1	σ_2	σ_2	σ_1	$2(\sigma_1+\sigma_2)$	$\Psi_{A_1}=\frac{1}{\sqrt{2}}(\sigma_1+\sigma_2)$
$\chi_{(B_2)}\hat{R}\sigma_1$	σ_1	$-\sigma_2$	$-\sigma_2$	σ_1	$2(\sigma_1-\sigma_2)$	$\Psi_{B_2}=\frac{1}{\sqrt{2}}(\sigma_1-\sigma_2)$

We recommend checking the orthogonality of the SALC functions obtained at the end. In this case, the orthogonality is obvious by a simple visual inspection.

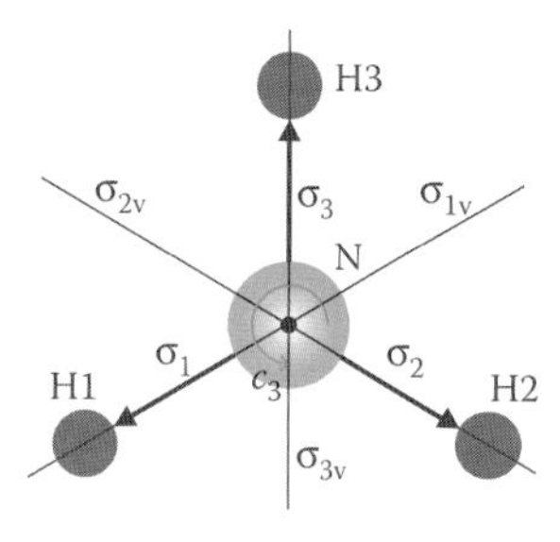

FIGURE 3.20 Bonding vectors and symmetry elements for the ammonia molecule.

3.5.3 SALCs FOR σ BONDS OF AMMONIA

The ammonia molecule (NH_3) belongs to the C_{3v} group. We attach a set of three equivalent bond vectors directed from nitrogen toward the hydrogen atoms, as shown in Figure 3.20.

The results of operations can be seen in Table 3.7. The representation Γ_σ generated by the set of three vectors (row 5) reduces to $A_1 + E$. The classes are already separated. The projection operator is applied to the σ_1 vector (row 6). Rows 2 and 6 are multiplied to project the linear combination with A_1 symmetry (row 7). Rows 4 and 6 are multiplied to project the linear combination with E symmetry (row 8). Both linear combinations are then normalized. We obtained two SALCs: Ψ_{A1} and Ψ_{1E}. They are normalized and orthogonal. Checking the orthogonality means to multiply the resulting functions taking into consideration that the starting vectors σ_1, σ_2, and σ_3 are orthonormal (i.e., $\sigma_1\ \sigma_1 = 1$, $\sigma_1\ \sigma_2 = 0$, and so on). The third SALC will be obtained by applying the same procedure to the vector σ_2 (rows 9, 10, 11). The linear combination with A_1 symmetry is the same as that obtained for σ_1. The function Ψ_{2E} is the second SALC with the required symmetry but it *is not orthogonal* to Ψ_{1E}.

We note that many ways can be used to construct an orthogonal pair of functions that transform like the E representation. We can write the function in a general way:

$$\Psi_{2E} = a\sigma_1 + b\sigma_2 + c\sigma_3$$

TABLE 3.7 Projection Operator Method to Obtain SALCs for the Ammonia Molecule

1	C_{3v}	E	C_3	C_3^2	Σ_{1v}	Σ_{2v}	Σ_{3v}		
2	A_1	1	1	1	1	1	1		
3	A_2	1	1	1	−1	−1	−1		
4	E	2	−1	−1	0	0	0	Reduced to ↓	
5	Γ_σ	3	0	0	1	1	1	$A_1 + E$	
6	$\hat{R}\sigma_1$	σ_1	σ_2	σ_3	σ_1	σ_3	σ_2	Sum ↓	Normalized SALC
7	$\chi_{(A1)}\hat{R}\sigma_1$	σ_1	σ_2	σ_3	σ_1	σ_3	σ_2	$2(\sigma_1 + \sigma_2 + \sigma_3)$	$\Psi_{A1} = \frac{1}{\sqrt{3}}(\sigma_1 + \sigma_2 + \sigma_3)$
8	$\chi_{(E)}\hat{R}\sigma_1$	$2\sigma_1$	$-\sigma_2$	$-\sigma_3$	0	0	0	$(2\sigma_1 - \sigma_2 - \sigma_3)$	$\Psi_{1E} = \frac{1}{\sqrt{6}}(2\sigma_1 - \sigma_2 - \sigma_3)$
9	$\hat{R}\sigma_2$	σ_2	σ_3	σ_1	σ_3	σ_2	σ_1		
10	$\chi_{(A1)}\hat{R}\sigma_2$	σ_2	σ_3	σ_1	σ_3	σ_2	σ_1	$2(\sigma_1 + \sigma_2 + \sigma_3)$	Ψ_{A1}
11	$\chi_{(E)}\hat{R}\sigma_2$	$2\sigma_2$	$-\sigma_3$	$-\sigma_1$	0	0	0	$(2\sigma_2 - \sigma_3 - \sigma_1)$	$\Psi_{2E} = \frac{1}{\sqrt{6}}(2\sigma_2 - \sigma_3 - \sigma_1)$

It must be orthogonal to Ψ_{A1} and Ψ_{1E} and results in two equations:

$$\begin{cases} a + b + c = 0 \\ 2a - b - c = 0 \end{cases}$$

Thus, $a = 0$ and $b = -c$. We conclude that the third SALC of the ammonia molecule is

$$\Psi_{2E} = \frac{1}{\sqrt{2}}(\sigma_2 - \sigma_3)$$

Important note: From a simple inspection of Table 3.7, it has been found that the actions of reflections are just a repeat of what is obtained under the rotation operators. So, we can reduce the work by just taking the subgroup $\boldsymbol{C_3}$. This conclusion is very useful for higher-order groups.

3.5.4 SALCS FOR σ BONDS OF THE AB_6 MOLECULE

The AB_6 molecule belongs to the $\boldsymbol{O}_h$ group. The order of the group $h = 48$! We should use the above note in order to work with the subgroup $\boldsymbol{O}$. The representation generated by the six bond vectors (Figure 3.21) can be reduced to the irreducible representations: A_1, E, and T_1. However, this result is only partly correct. It gives you information about the required symmetry and dimension of the functions but not about the function parity. The character even (gerade) or odd (ungerade) will be given only by using the character table of the $\boldsymbol{O}_h$ group. Thus, the representation generated by the set of six bond vectors can be reduced into irreducible representations of the $\boldsymbol{O}_h$ group as follows:

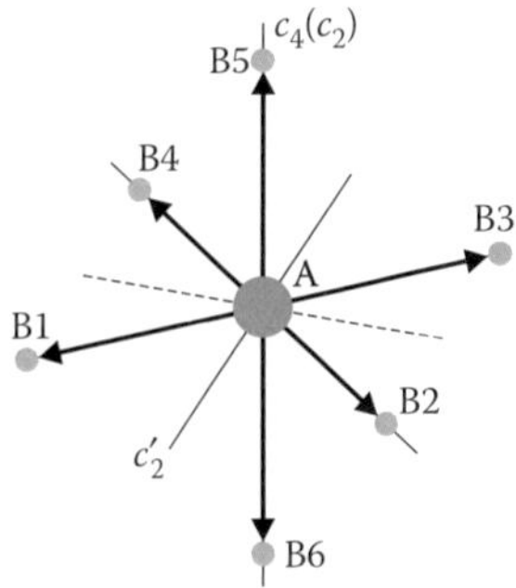

FIGURE 3.21 Bonding vectors and rotation axes for the AB_6 molecule.

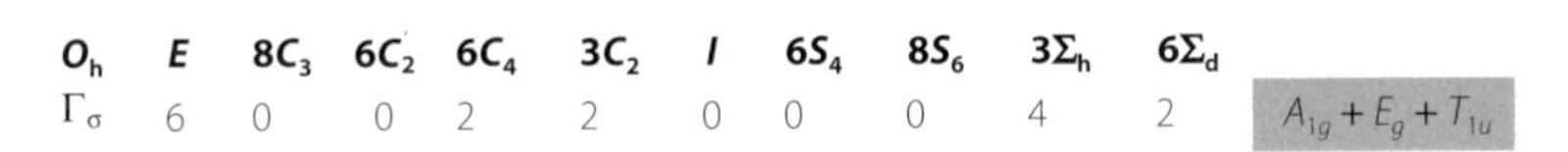

O_h	E	$8C_3$	$6C_2$	$6C_4$	$3C_2$	I	$6S_4$	$8S_6$	$3\Sigma_h$	$6\Sigma_d$	
Γ_σ	6	0	0	2	2	0	0	0	4	2	$A_{1g} + E_g + T_{1u}$

Finding the irreducible representations of σ bonds is not a difficult task, so it can be done using the larger table of the $\boldsymbol{O}_h$ group. When we apply the projection operator to σ_1 and σ_2 bonds, it will be easier to work on the table character of the $\boldsymbol{O}$ subgroup. It is important to use this shortcut from now on because the projection operator will be applied to each symmetry operation and each irreducible representation while the work of finding the irreducible representations was done only for classes of operations.

The table is too lengthy to be presented in this book, and therefore we present only the results. You should try to do the calculus and check them. Be aware that the outcome depends on how you number the atoms but the conclusion is the same. We have applied the projection operator to the set of σ_1 and σ_2 bonds for 24 symmetry operations. We present here the normalized functions:

$$\Psi_{A_1}(\sigma_1) = \frac{1}{6}(\sigma_1 + \sigma_2 + \sigma_3 + \sigma_4 + \sigma_5 + \sigma_6)$$

$$\Psi_E(\sigma_1) = \frac{1}{2\sqrt{3}}(2\sigma_1 - \sigma_2 + 2\sigma_3 - \sigma_4 - \sigma_5 - \sigma_6)$$

$$\Psi_{T1}(\sigma_1) = \frac{1}{\sqrt{2}}(\sigma_1 - \sigma_3)$$

$$\Psi_{A1}(\sigma_2) = \frac{1}{6}(\sigma_1 + \sigma_2 + \sigma_3 + \sigma_4 + \sigma_5 + \sigma_6)$$

$$\Psi_E(\sigma_2) = \frac{1}{2\sqrt{3}}(-\sigma_1 + 2\sigma_2 - \sigma_3 + 2\sigma_4 - \sigma_5 - \sigma_6)$$

$$\Psi_{T1}(\sigma_2) = \frac{1}{\sqrt{2}}(\sigma_2 - \sigma_4)$$

There is no need to continue the calculation for σ_3, σ_4, σ_5, and σ_6. We can get the conclusions:

$$\Psi_{1A1} = \frac{1}{6}(\sigma_1 + \sigma_2 + \sigma_3 + \sigma_4 + \sigma_5 + \sigma_6);$$

$$\Psi_{2E} = \frac{1}{2\sqrt{3}}(2\sigma_1 - \sigma_2 + 2\sigma_3 - \sigma_4 - \sigma_5 - \sigma_6);$$

$$\Psi_{4T1} = \frac{1}{\sqrt{2}}(\sigma_1 - \sigma_3); \quad \Psi_{5T1} = \frac{1}{\sqrt{2}}(\sigma_2 - \sigma_4); \quad \Psi_{6T1} = \frac{1}{\sqrt{2}}(\sigma_5 - \sigma_6)$$

The function $\Psi_E(\sigma_2)$ is not orthogonal to its pair function $\Psi_E(\sigma_1)$. So, we should change it to any other linear combination that is orthogonal to all other functions. We have found that the second function basis of the E representation is

$$\Psi_{3E} = \frac{1}{2}(\sigma_2 + \sigma_4 - \sigma_5 - \sigma_6)$$

We now have the complete set of SALCs of the AB_6 molecule. For further work on using these results, you may change the pair of SALCs with E symmetry with the following one:

$$\Psi_{2E} = \frac{1}{2\sqrt{3}}(-\sigma_1 - \sigma_2 - \sigma_3 - \sigma_4 + 2\sigma_5 + 2\sigma_6);$$

$$\Psi_{3E} = \frac{1}{2}(\sigma_1 - \sigma_2 + \sigma_3 - \sigma_4)$$

3.5.5 SALCs FOR CYCLIC π SYSTEMS

We can extend the application of the SALC method to the construction of π molecular orbitals. They are very important for cyclic organic molecules. As the first example, we take the simple molecule of cyclobutadiene (C_4H_4). The main bonds between carbon atoms are of σ type. Although there is a controversy as to whether the ring shape is a square or a rectangle (Allinger and Tai 1968), we have used the last one to sketch the molecule. It is just an exercise in the application of the SALC method. The molecule as shown in Figure 3.22 belongs to the $\boldsymbol{D_{2h}}$ point group. There are four p orbitals of carbon atoms perpendicular to the molecular plane. They are involved in π bonding, so four molecular combinations will result.

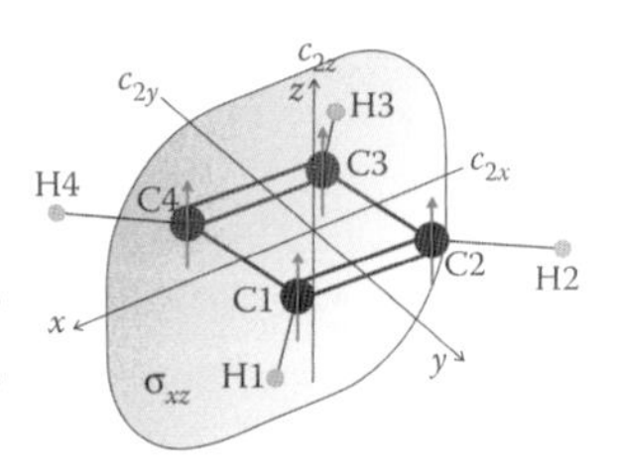

FIGURE 3.22
Representation of p_z atomic orbitals of carbon used to construct π bonds.

The representation generated by the four vectors has nonzero characters only for identity and the horizontal plane (all vectors change sign). It is a matter of simple visual inspection of the character table (Table 3.8) in order to reduce the Γ_π representation. You should look only for the presence of the character –1 on the column of the horizontal reflection Σ_{xy} [4 + (–1) (–4)]. So, the representation Γ_π can be reduced to the following irreducible representations: $B_{2g} + B_{3g} + A_u + B_{1u}$. As a result, we have to find four molecular orbitals from SALCs of the four atomic p_z orbitals.

We apply the projection operator to the vector attached to the C1 atom. The identity operation leaves it unchanged. The vertical C_{2z} rotation transforms π_1 into π_3. The rotation around a horizontal axis transforms π_1 into a neighbor reversed vector. The inversion transforms π_1 into its opposite vector with a changed sign. The reflection through a horizontal plane just reverses the vector π_1. The reflection through a vertical plane transforms π_1 into a neighbor vector. After multiplication of the values of the row 11 by the characters of the four irreducible representations, we obtain four different linear combinations with adapted symmetries to the structure of the molecule. It is important that they are orthogonal to each other and we just have to normalize the founded combinations. Thus, the cyclobutadiene has the following SALCs for π bonding:

TABLE 3.8 Character Table of the D_{2h} Group and the Representation Generated by the Out-of-Plane Vectors of Carbon Atoms in the Cyclic C_4H_4 Molecule

1	$\boldsymbol{D_{2h}}$	$\boldsymbol{E}$	$\boldsymbol{C_{2z}}$	$\boldsymbol{C_{2y}}$	$\boldsymbol{C_{2x}}$	$\boldsymbol{I}$	$\boldsymbol{\Sigma_{xy}}$	$\boldsymbol{\Sigma_{xz}}$	$\boldsymbol{\Sigma_{yz}}$	
2	A_g	1	1	1	1	1	1	1	1	
3	B_{1g}	1	1	–1	–1	1	+1	–1	–1	
4	B_{2g}	1	–1	1	–1	1	–1	1	–1	
5	B_{3g}	1	–1	–1	1	1	–1	–1	1	
6	A_u	1	1	1	+1	–1	–1	–1	–1	
7	B_{1u}	1	1	–1	–1	–1	–1	1	1	
8	B_{2u}	1	–1	1	–1	–1	1	–1	1	
9	B_{3u}	1	–1	–1	1	–1	1	1	–1	Reduced to ↓
10	Γ_π	4	0	0	0	0	–4	0	0	$B_{2g} + B_{3g} + A_u + B_{1u}$
11	$\hat{R}\pi_1$	π_1	π_3	$-\pi_2$	$-\pi_4$	$-\pi_3$	$-\pi_1$	π_4	π_2	Sum ↓
12	$\chi_{(B2g)}\hat{R}\pi_1$	π_1	$-\pi_3$	$-\pi_2$	π_4	$-\pi_3$	π_1	π_4	$-\pi_2$	$2(\pi_1 - \pi_2 - \pi_3 + \pi_4)$
13	$\chi_{(B3g)}\hat{R}\pi_1$	π_1	$-\pi_3$	π_2	$-\pi_4$	$-\pi_3$	π_1	$-\pi_4$	π_2	$2(\pi_1 + \pi_2 - \pi_3 - \pi_4)$
14	$\chi_{(Au)}\hat{R}\pi_1$	π_1	π_3	$-\pi_2$	$-\pi_4$	π_3	π_1	$-\pi_4$	$-\pi_2$	$2(\pi_1 - \pi_2 + \pi_3 - \pi_4)$
15	$\chi_{(B1u)}\hat{R}\pi_1$	π_1	π_3	π_2	π_4	π_3	π_1	π_4	π_2	$2(\pi_1 + \pi_2 + \pi_3 + \pi_4)$

$$(3.16)\quad \begin{cases} \Psi_{B2g} = \dfrac{1}{2}(\pi_1 - \pi_2 - \pi_3 + \pi_4) \\ \Psi_{B3g} = \dfrac{1}{2}(\pi_1 + \pi_2 - \pi_3 - \pi_4) \\ \Psi_{Au} = \dfrac{1}{2}(\pi_1 - \pi_2 + \pi_3 - \pi_4) \\ \Psi_{B1u} = \dfrac{1}{2}(\pi_1 + \pi_2 + \pi_3 + \pi_4) \end{cases}$$

What about the combinations generated as a result of the projecting operator action on the other vectors π_2, π_3, and π_4? Exercise the method on them. From the symmetry point of view, there is no reason to obtain a different result from that presented by Equation 3.16. It is an advantage of the chosen symmetry to have only nondegenerate (one-dimensional) irreducible representations. The square shape, which belongs to the $\boldsymbol{D}_{4h}$ point group, has a combination of two nondegenerate and one degenerate (two-dimensional) irreducible representations. This involves a longer calculus to obtain four orthogonal functions (see Table 7.1 of Cotton 1990). It is a characteristic of the higher-symmetry structures.

Important note: For cyclic molecules, we may apply only the identity and rotations to find the SALCs, that is, we may reduce the calculus to find SALCs to the operations of the rotational subgroup $\boldsymbol{C}_4$. A simple inspection of rows 12–15 of Table 3.7 results in the effect of the projection operator (cells barred) for both the inversion and reflections being the same as for the first four operations.

The SALCs given by Equation 3.16 help us to imagine the π bonding in the C_4H_4 molecule as sketched in Figure 3.23. There is an arrangement

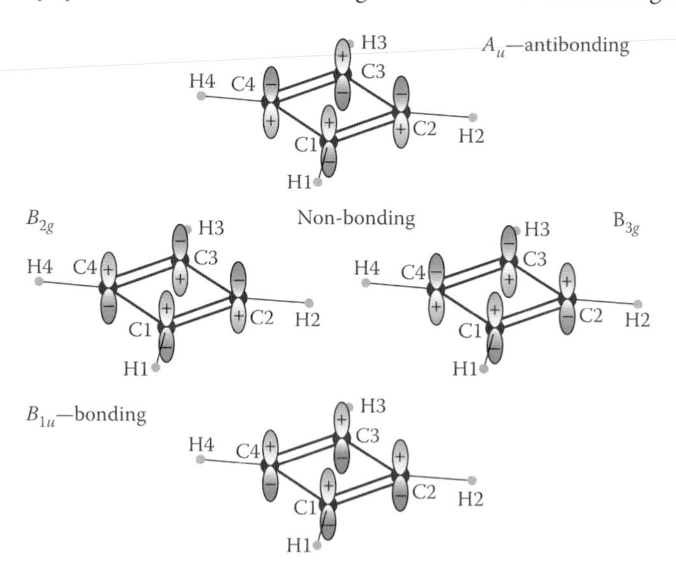

FIGURE 3.23 The atomic orbitals p_z of the carbon atoms in cyclobutadiene can combine into one bonding MO (B_{1u}), two nonbonding MOs, and one antibonding MO.

of symmetry B_{1u} where all p_z atomic orbitals have the same orientation and give a nearly continuous overlapping over the carbon ring. This arrangement is the most stable and forms a π bond with two rings positive and negative on both sides of the molecular plane. This special π bond will be filled only by two electrons that are delocalized on all carbon atoms. In the other two arrangements of symmetry, either B_{2g} or B_{3g}, the p_z orbitals form both a π bond with a positive phase above the molecular plane and a π bond with a positive phase under the plane. These are nonbonding cases. The molecular orbital of symmetry A_u has an antibonding character because any two adjacent p_z orbitals are out of phase.

3.6 EXAMPLES OF SYMMETRY APPLICATION TO MOLECULAR ORBITALS

3.6.1 MO DESCRIPTION OF THE BONDING IN FERROCENE

The organometallic compounds of the transition metals are defined as containing at least one metal–carbon bond between an organic molecule and a metal. Their history practically begins with a synthesis of ferrocene ($Fe(C_5H_5)_2$). Its complete name is bis(cyclopentadienyl)iron (IUPAC 1999). It has a sandwich structure with two cyclopentadienyl rings parallel and an iron ion between them. Each organic ring has a negative charge. The molecule in the form most known has an inversion center and belongs to the $\boldsymbol{D_{5d}}$ point group (see Figure 1.38). We should look to this molecule as an iron ion bonded to two ligands (C_5H_5 rings). We suppose that the bonding can be only between the five *p* atomic orbitals out of plane of each ring and the metal orbitals (Figure 3.24). The bonds modeling must pass through three moments:

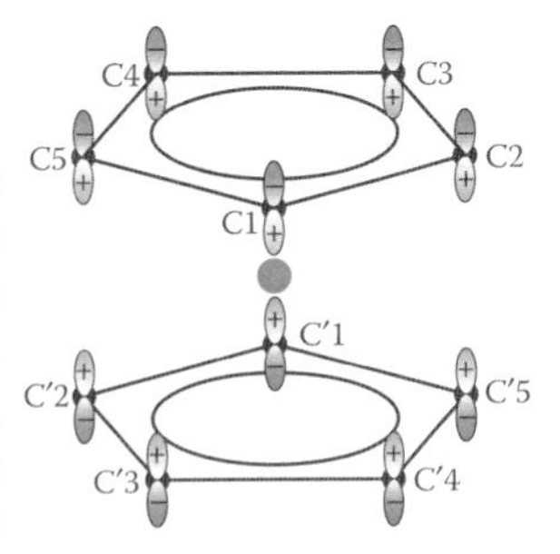

FIGURE 3.24 Ten p_z orbitals of carbon atoms participate to construct the molecular orbitals.

1. In the first moment, the atomic orbitals combine to form σ and π bonding of carbon atoms in each ring.
2. Then the π orbitals of two rings will combine.
3. The molecular orbitals of the two rings just found will be involved in the bonding to the metal ion.

The sp^2 hybridization scheme of atomic orbitals is used by each carbon to bond to adjacent carbon atoms and the hydrogen atom. So, the in-plane *p* atomic orbitals are used to construct the planar ring. This hybridization leaves the out-of-plane p_z orbital on each carbon unused. The five p_z orbitals will be used to construct the π bonds (common) of the ring. Therefore, we begin to find the SALCs of π bonding of each ring. We take the upper ring. It belongs to the $\boldsymbol{D_{5h}}$ point group. The five p_z orbitals will generate the following reducible representation Γ_π:

D_{5h}	E	$2C_5$	$2C_5^2$	$5C_2$	Σ_h	$2S_5$	$2S_5^3$	$5\Sigma_v$	
Γ_π	5	0	0	−1	−5	0	0	1	$A_2'' + E_1'' + E_2''$

We should apply the shortcut given by the note from cyclobutadiene; therefore, we should take into consideration only the effect of the projection operator for the rotational subgroup $\boldsymbol{C}_5$:

C_5	E	C_5	C_5^2	C_5^3	C_5^4	$\varepsilon = \exp(i\alpha) = \exp(2\pi i/5)$
$\hat{R}\pi_1$	π_1	π_2	π_3	π_4	π_5	
$\chi_{(A)}\hat{R}\pi_1$	π_1	π_2	π_3	π_4	π_5	$\Psi_{A''_2} = \frac{1}{\sqrt{5}}(\pi_1 + \pi_2 + \pi_3 + \pi_4 + \pi_5)$
$\chi_{(E1)}\hat{R}\pi_1$	π_1	$\varepsilon\,\pi_2$	$\varepsilon^2\,\pi_3$	$\varepsilon^{2*}\,\pi_4$	$\varepsilon^*\,\pi_5$	$\Psi_{aE_1} = (\pi_1 + \varepsilon\pi_2 + \varepsilon^2\pi_3 + \varepsilon^{2*}\pi_4 + \varepsilon^*\pi_5)$
	π_1	$\varepsilon^*\,\pi_2$	$\varepsilon^{2*}\,\pi_3$	$\varepsilon^2\,\pi_4$	$\varepsilon\,\pi_5$	$\Psi_{bE_1} = (\pi_1 + \varepsilon^*\pi_2 + \varepsilon^{2*}\pi_3 + \varepsilon^2\pi_4 + \varepsilon\pi_5)$
$\chi_{(E2)}\hat{R}\pi_1$	π_1	$\varepsilon^2\,\pi_2$	$\varepsilon^*\,\pi_3$	$\varepsilon\,\pi_4$	$\varepsilon^{2*}\,\pi_5$	$\Psi_{aE_2} = (\pi_1 + \varepsilon^2\pi_2 + \varepsilon^*\pi_3 + \varepsilon\pi_4 + \varepsilon^{2*}\pi_5)$
	π_1	$\varepsilon^{2*}\,\pi_2$	$\varepsilon\,\pi_3$	$\varepsilon^*\,\pi_4$	$\varepsilon^2\,\pi_5$	$\Psi_{bE_2} = (\pi_1 + \varepsilon^{2*}\pi_2 + \varepsilon\pi_3 + \varepsilon^*\pi_4 + \varepsilon^2\pi_5)$

We must convert the complex functions into linear combinations of real functions by using the Euler identity:

$$e^{i\alpha} = \cos\alpha + i\sin\alpha$$

to derive some trigonometric identities:

$$\cos\alpha = \frac{e^{i\alpha} + e^{-i\alpha}}{2} \quad \text{and} \quad \sin\alpha = \frac{e^{i\alpha} - e^{-i\alpha}}{2i} \tag{3.17}$$

Thus, in order to convert the paired complex functions into real functions, we have to both add and subtract them according to Equations 3.17. We will obtain the orbital functions for the E_1 representation as follows:

$$\begin{aligned}\Psi_{1E_1} &= N_{1E_1}\frac{(\pi_1 + \varepsilon\pi_2 + \varepsilon^2\pi_3 + \varepsilon^{2*}\pi_4 + \varepsilon^*\pi_5) + (\pi_1 + \varepsilon^*\pi_2 + \varepsilon^{2*}\pi_3 + \varepsilon^2\pi_4 + \varepsilon\pi_5)}{2}\\ &= N_{1E_1}[\pi_1 + (\cos\alpha)\pi_2 + (\cos 2\alpha)\pi_3 + (\cos 2\alpha)\pi_4 + (\cos\alpha)\pi_5]\\ &= \sqrt{\frac{2}{5}}[\pi_1 + (\cos\alpha)\pi_2 + (\cos 2\alpha)\pi_3 + (\cos 2\alpha)\pi_4 + (\cos\alpha)\pi_5]\end{aligned}$$

$$\begin{aligned}\Psi_{2E_1} &= N_{2E_1}\frac{(\pi_1 + \varepsilon\pi_2 + \varepsilon^2\pi_3 + \varepsilon^{2*}\pi_4 + \varepsilon^*\pi_5) - (\pi_1 + \varepsilon^*\pi_2 + \varepsilon^{2*}\pi_3 + \varepsilon^2\pi_4 + \varepsilon\pi_5)}{2i}\\ &= N_{2E_1}[(\sin\alpha)\pi_2 + (\sin 2\alpha)\pi_3 - (\sin 2\alpha)\pi_4 - (\sin\alpha)\pi_5]\\ &= \sqrt{\frac{2}{5}}[(\sin\alpha)\pi_2 + (\sin 2\alpha)\pi_3 - (\sin 2\alpha)\pi_4 - (\sin\alpha)\pi_5]\end{aligned}$$

The normalization factor was calculated by replacing α by $2\pi/5$. In a similar way, we can find the orbital linear combinations adapted to the E_2 symmetry:

$$\Psi_{1E_2} = \sqrt{\frac{2}{5}}[\pi_1 + (\cos 2\alpha)\pi_2 + (\cos\alpha)\pi_3 + (\cos\alpha)\pi_4 + (\cos 2\alpha)\pi_5]$$

and

$$\Psi_{2E_2} = \sqrt{\frac{2}{5}}[(\sin 2\alpha)\pi_2 - (\sin\alpha)\pi_3 + (\sin\alpha)\pi_4 - (\sin 2\alpha)\pi_5]$$

We summarize all previous calculus in Table 3.9 by replacing the trigonometric functions with their numeric value.

We are now able to sketch the π molecular orbitals of the cyclopentadienyl ring as shown in Figure 3.25. Except for the molecular orbital with symmetry A_2'', in all other combinations the atomic orbitals have different contributions. The MO of A_2'' symmetry has one nodal plane (the atoms plane), the highest symmetry, and probably the lowest energy (most stable). The MOs of E_1'' symmetry have two nodal planes (one horizontal and one vertical) and higher energy while the MOs of E_2'' symmetry have three nodal planes (one horizontal and two vertical planes) and play as antibonding orbitals.

The second moment consists of putting closer two rings of cyclopentadienyl. How can we arrange them? If we combine the molecular orbitals of two rings, it is reasonable to take the most stable arrangement of each ring, that is, A_2''. However, there are two possible combinations:

- In-phase, that is, the positive lobe of the upper ring interacts with the positive lobe of the lower ring. It is the combination of gerade symmetry.
- Out-of-phase, that is, the positive lobe of the upper ring interacts with the negative lobe of the lower ring. It is the combination of ungerade symmetry.

Thus, 10 p orbitals form a set of vectors that are the basis of the following reducible representation:

TABLE 3.9 π Molecular Orbital Functions of the Cyclopentadienyl Molecule (One Ring of Ferrocene)

D_{5h}	
A_2''	$\Psi_{A_2''} = \frac{1}{\sqrt{5}}(\pi_1 + \pi_2 + \pi_3 + \pi_4 + \pi_5)$
E_1''	$\Psi_{1E_1''} = \sqrt{\frac{2}{5}}(\pi_1 + 0.309\pi_2 - 0.809\pi_3 - 0.809\pi_4 + 0.309\pi_5)$
	$\Psi_{2E_1''} = \sqrt{\frac{2}{5}}(0.951\pi_2 + 0.587\pi_3 - 0.587\pi_4 - 0.951\pi_5)$
E_2''	$\Psi_{1E_2''} = \sqrt{\frac{2}{5}}(\pi_1 - 0.809\pi_2 + 0.309\pi_3 + 0.309\pi_4 - 0.809\pi_5)$
	$\Psi_{2E_2''} = \sqrt{\frac{2}{5}}(0.587\pi_2 - 0.951\pi_3 + 0.951\pi_4 - 0.587\pi_5)$

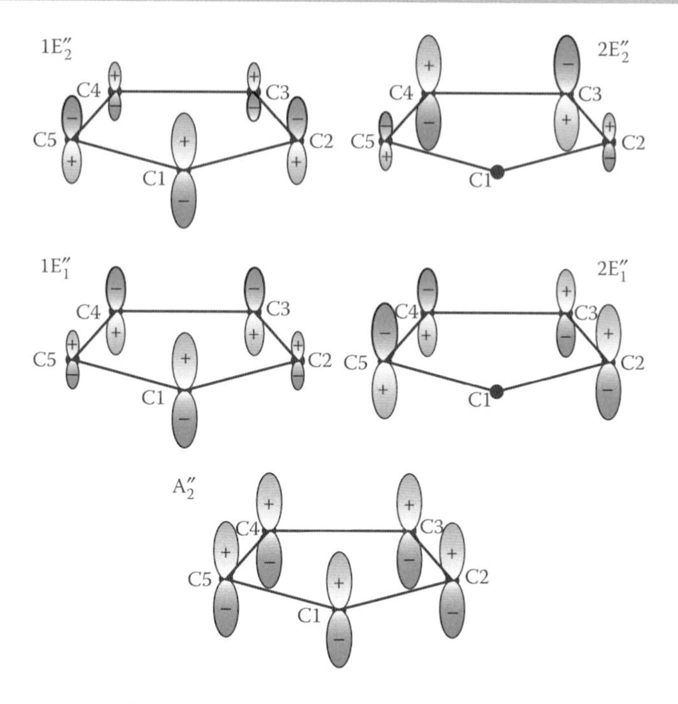

FIGURE 3.25 The illustration of the π molecular orbitals possible for the cyclopentadienyl. The size of each p_z orbital was adjusted to give an indication about its contribution to the molecular orbital.

D_{5d}	E	$2C_5$	$2C_5^2$	$5C_2$	i	$2S_{10}$	$2S_{10}^3$	$5\Sigma_d$	
Γ_π	10	0	0	0	0	0	0	2	$A_{1g}+E_{1g}+E_{2g}+A_{2u}+E_{1u}+E_{2u}$

Except the identity and reflection through the dihedral plane, all other operations shift all vectors and therefore have the character of 0. To make its reduction into irreducible representations, it is simple to inspect the first and last columns of the character table in order to find the total 20 (1·10 + 5·2). This representation of order 10 can be reduced to the irreducible representations: $A_{1g} + E_{1g}+ E_{2g}+ A_{2u} + E_{1u}+ E_{2u}$.

Any SALCs of two rings is given by in-phase or out-of-phase combinations of the similar symmetry orbitals of each ring. Therefore, the SALC of A_{1g} symmetry is the addition of $\Psi_{1(A_2'')} + \Psi_{2(A_2'')}$ as illustrated in Figure 3.26. The difference between the same functions $\Psi_{1(A_2'')} - \Psi_{2(A_2'')}$ is the ungerade combination, that is, the SALC of A_{2u} symmetry (Figure 3.26). The molecular orbitals of symmetry E_{1g} is the result of the addition of $\Psi_{1(E_1'')} + \Psi_{2(E_1'')}$ ring orbitals, and so on.

The last moment is the interaction of the molecular orbitals of two cyclopentadienyl rings with the atomic orbitals of the iron ion. The orbitals 4*s*, 4*p*, and 3*d* of the iron ion should fit to the symmetry of the molecular complex in order to make combinations. We should inspect Table 3.10 in which we have summarized the symmetries of both the metal atomic orbitals and the just obtained molecular orbitals of the two

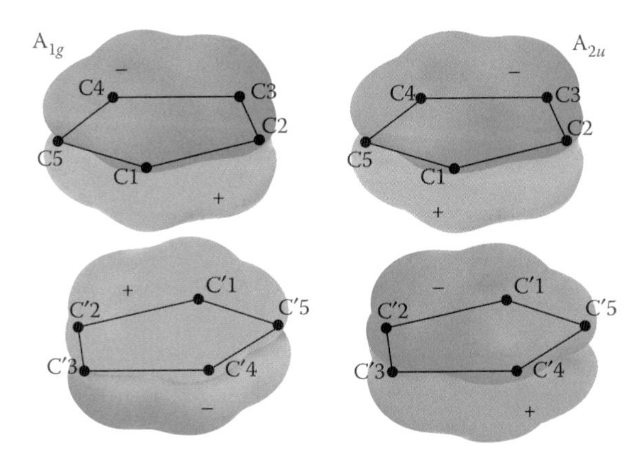

FIGURE 3.26 The molecular orbitals of two closer rings are the result of the in-phase (A_{1g}) and out-of-phase (A_{2u}) addition of the molecular orbitals of each ring, respectively.

rings. It is obvious from the symmetry reasons and by looking at Figure 3.26 that d_{2z^2} and p_z orbitals are very suitable to combine with ligand SALCs a_{1g} and a_{2u}, respectively. These combinations are illustrated in Figure 3.27.

The inspection of Table 3.10 results in all possible combinations between iron orbitals and ligands SALCs: d_{2z^2} and a_{1g}, p_z and a_{2u}, d_{xz} and e'_{1g}, d_{yz} and e''_{1g}, $d_{x^2-y^2}$ and e'_{2g}, d_{xy} and e''_{2g}, p_x and e'_{1u} p_y and e''_{1u}. The SALCs of e_{2u} symmetry constructed from the out-of-phase combination of two antibonding e''_2 SALCs rest as nonbonding ferrocene orbitals. Regarding their relative positions on the energy scale, we can observe that

- The $3d_{2z^2}$ orbital has the big positive lobe oriented toward the center of the SALC a_{1g}, where the charge density is lowest (see Figure 3.28). Thus, the ion orbital and the ring MO low overlap and combine. As a consequence, they realize a weak π bond. The π bond obtained as a result of interaction between the p_z ion orbital and the a_{1u} SALC is a weak bond too.
- The $3d_{xz}$ and $3d_{yz}$ orbitals combine with e_{1g} SALCs. They have a more suitable orientation, that of the $3d_{2z^2}$ orbital. Moreover, the

TABLE 3.10 Symmetry of Both the Atomic Orbitals of the Iron Ion and the MO of Two Rings in the Ferrocene Complex

D_{5d}	Atomic Orbitals of the Iron Ion		MO of Two Rings
A_{1g}		s, d_{2z^2}	a_{1g}
E_{1g}		(d_{xz}, d_{yz})	e_{1g}
E_{2g}		$(d_{x^2-y^2}, d_{xy})$	e_{2g}
A_{2u}	p_z		a_{2u}
E_{1u}	(p_x, p_y)		e_{1u}
E_{2u}			e_{2u}

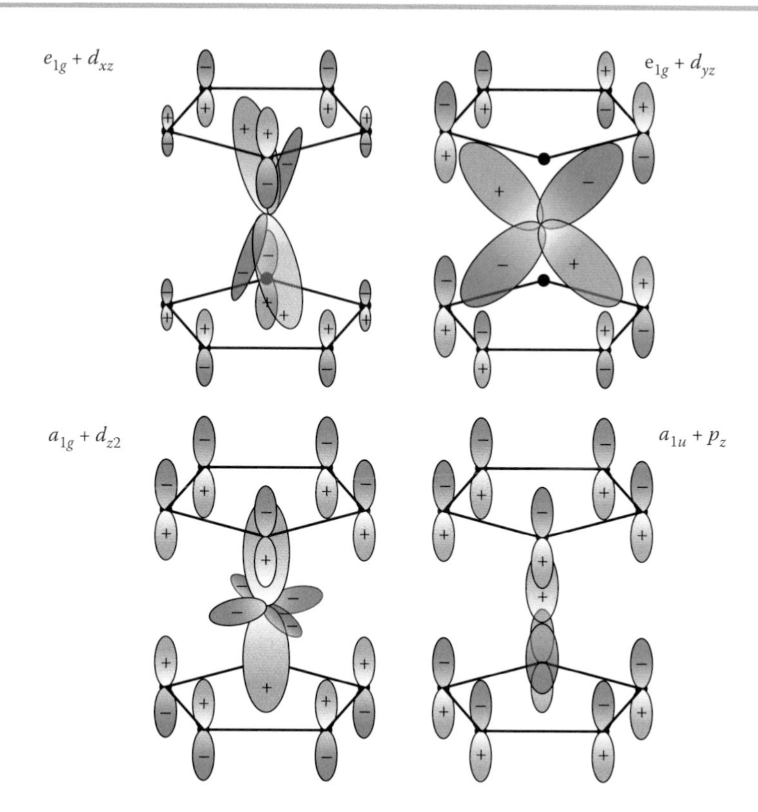

FIGURE 3.27 Few illustrations of symmetry allowed combinations of the atomic orbitals of the central iron ion with the SALCs of two cyclopentadienyl rings in the ferrocene molecule.

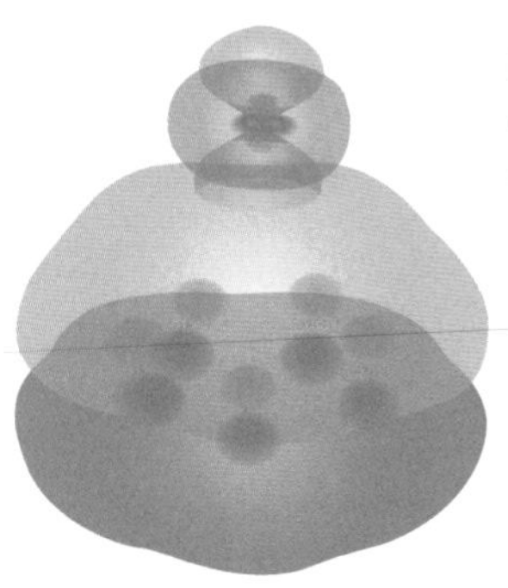

FIGURE 3.28 Graphical simulation of the overlap between the $3d_{2z^2}$ iron orbital and the π MO a_{1g} of the cyclopentadienyl ring.

two combinations ($3d_{xz} + e_{1g}$) and ($3d_{yz} + e_{1g}$) overlap and probably stabilize better together. So, we expect that they have a lower position on the energy scale.

- The $3d_{x^2-y^2}$ and $3d_{xy}$ orbitals combine with e_{2g} SALCs, which originate in two antibonding orbitals, that is, combine with higher orbitals, so they are also higher on the energy scale.

We note that only a rigorous calculation of the secular determinant can provide reliable information about their relative energy (Cotton 1990).

3.6.2 DETAILED MO TREATMENT OF THE BENZENE MOLECULE

The benzene molecule (C_6H_6) is a classic example of the application of molecular orbitals theory. The theoretical results can be experimentally verified in the laboratory, where the absorption spectra in ultraviolet, infrared, and the Raman spectra can be easily measured.

We begin talking about σ bonding, although it is not useful to describe the optical spectra because σ bonding is very strong and may give absorption in far UV. Each carbon atom has three neighbors: two carbon atoms and a hydrogen atom. Carbon has four valence electrons and four orbitals (s, p_x, p_y, and p_z) available for bonding. For three σ bonds, the orbitals s, p_x, and p_y of each atom combine to make a hybrid orbital sp^2 (see Figure 3.14). Therefore, each carbon atom makes three identical hybrid orbitals sp^2 used to bond both two carbon atoms and a hydrogen atom. Finally, a single C–C bond is constructed as a result of the interaction between two sp^2 hybrids (one of each atom). The C–H bond is the result of the interaction between one sp^2 hybrid of carbon and the orbital $1s$ of hydrogen. So, there are three σ bonds in which a carbon atom is involved. Each σ bonding orbital is filled by two electrons. There are also three antibonding σ orbitals that are empty. Since each carbon atom in the ring has the same sp^2 hybridization, which forms a triangle, it is reasonable to think that the benzene ring is planar too. The σ bonds of a benzene molecule are illustrated in Figure 3.29. For any planar and cyclic molecule, the p_z orbital or, more precisely, the $2p$ orbital perpendicular to the molecular plane cannot be involved in the construction of σ bonds, that is, bonds with maximum density along the interatomic line.

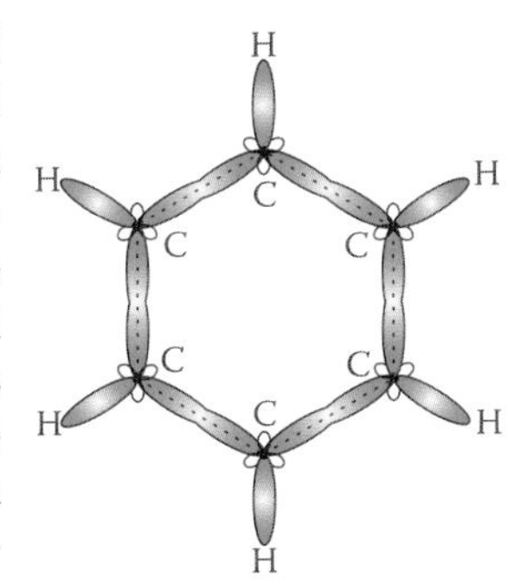

FIGURE 3.29 Top view of the σ bonds of the benzene molecule.

The p_z orbitals will be used to construct the π orbitals because they have the suitable symmetry for this. The molecule belongs to the $\boldsymbol{D_{6h}}$ point group and the p_z orbitals will generate the following reducible representation Γ_π:

D_{6h}	E	$2C_6$	$2C_3$	C_2	$3C_2'$	$3C_2''$	I	$2S_3$	$2S_6$	Σ_h	$3\Sigma_d$	$3\Sigma_v$	
Γ_π	6	0	0	0	−2	0	0	0	0	−6	2	0	$B_{1g}+A_{2u}+E_{1g}+E_{2u}$

We look to the characters of E, C_2', Σ_h, and Σ_d but mainly where the Σ_h operation gives negative characters and then we can quickly decompose Γ_π into the following irreducible representations of the $\boldsymbol{D_{6h}}$ point group: $B_{1g}+A_{2u}+E_{1g}+E_{2u}$.

In order to construct the SALCs, we will apply the projection operator but for the rotational subgroup $\boldsymbol{C_6}$:

C_6	E	C_6	C_3	C_2	C_3^2	C_6^5	$\varepsilon=\exp(i\alpha)=\exp(2\pi i/6)$
$\hat{R}\pi_1$	π_1	π_2	π_3	π_4	π_5	π_6	
$\chi_{(B)}\hat{R}\pi_1$	π_1	$-\pi_2$	π_3	$-\pi_4$	π_5	$-\pi_6$	$\Psi_{B_{1g}}=\frac{1}{\sqrt{6}}(\pi_1-\pi_2+\pi_3-\pi_4+\pi_5-\pi_6)$
$\chi_{(A)}\hat{R}\pi_1$	π_1	π_2	π_3	π_4	π_5	π_6	$\Psi_{A_{2u}}=\frac{1}{\sqrt{6}}(\pi_1+\pi_2+\pi_3+\pi_4+\pi_5+\pi_6)$
$\chi_{(E1)}\hat{R}\pi_1$	π_1	$\varepsilon\pi_2$	$-\varepsilon^*\pi_3$	$-\pi_4$	$-\varepsilon\pi_5$	$\varepsilon^*\pi_6$	$\Psi_{aE_1}=(\pi_1+\varepsilon\pi_2-\varepsilon^*\pi_3-\pi_4-\varepsilon\pi_5+\varepsilon^*\pi_6)$
	π_1	$\varepsilon^*\pi_2$	$-\varepsilon\pi_3$	$-\pi_4$	$-\varepsilon^*\pi_5$	$\varepsilon\pi_6$	$\Psi_{bE_1}=(\pi_1+\varepsilon^*\pi_2-\varepsilon\pi_3-\pi_4-\varepsilon^*\pi_5+\varepsilon\pi_6)$
$\chi_{(E2)}\hat{R}\pi_1$	π_1	$-\varepsilon^*\pi_2$	$-\varepsilon\pi_3$	π_4	$-\varepsilon^*\pi_5$	$-\varepsilon\pi_6$	$\Psi_{aE_2}=(\pi_1-\varepsilon^*\pi_2-\varepsilon\pi_3+\pi_4-\varepsilon^*\pi_5-\varepsilon\pi_6)$
	π_1	$-\varepsilon\pi_2$	$-\varepsilon^*\pi_3$	π_4	$-\varepsilon\pi_5$	$-\varepsilon^*\pi_6$	$\Psi_{bE_2}=(\pi_1-\varepsilon\pi_2-\varepsilon^*\pi_3+\pi_4-\varepsilon\pi_5-\varepsilon^*\pi_6)$

The first functions, that is, Ψ_{B1g} and Ψ_{A2u}, are orthogonal to each other. We have already written of them as normalized. In order to convert each pair of complex functions into two real functions, we have to both add and subtract them according to Equations 3.17. Thus, the SALC functions for the E_1 and E_2 representation will be as follows:

$$\Psi_{1E_1} = \frac{1}{\sqrt{12}}[2\pi_1 + \pi_2 - \pi_3 - 2\pi_4 - \pi_5 + \pi_6]$$

$$\Psi_{2E_1} = \frac{1}{2}[\pi_2 + \pi_3 - \pi_5 - \pi_6]$$

$$\Psi_{1E2} = \frac{1}{\sqrt{12}}[2\pi_1 - \pi_2 - \pi_3 + 2\pi_4 - \pi_5 - \pi_6]$$

$$\Psi_{2E2} = \frac{1}{2}[\pi_2 - \pi_3 + \pi_5 - \pi_6]$$

The arrangement of the p_z atomic orbitals to form π molecular orbitals described by these functions is sketched in Figure 3.30. The nodal lines, that is, the places where the molecular orbital changes sign, are also indicated. They can give a qualitative indication about the stability of the molecular orbital; the number of nodal lines increases as the stability decreases.

The relative ordering of the molecular orbitals presented in Figure 3.30 must be confirmed by the energy values of these orbitals. Because all the obtained functions are now orthonormal, they can be used to calculate the energy of the molecular orbitals.

In Section 3.5.1, we applied the secular equation to calculate the energy of σ bonds of very simple molecules. We now have to apply the secular equation to the benzene molecule as follows:

$$\det\left|H_{ij} - ES_{ij}\right| = 0, \quad \text{where } i \text{ and } j = 1, 2, \ldots, 6$$

The calculation of integrals $H_{ij} = \int \Psi_i \hat{H} \Psi_j d\tau$ will be simplified by using the wavefunctions expressed as SALCs because the *integrals are zero when* Ψ_i and Ψ_j belong to different irreducible representations. The overlap integrals are also approximated by 1 and 0, that is, $S_{ij} = \delta_{ij}$; therefore, *orbitals of different atoms do not overlap*! However, we have just shown in Figure 3.30 the molecular orbitals obtained as a consequence of the atomic orbitals overlapping. This is a coarse approximation but one that is extremely efficient to simplify the calculus.

We denote

$\int \Phi_i H \Phi_j d\tau = \alpha$ for $j = i = 1, 2, \ldots, 6$, because all atomic orbitals are $2p_z$ type of carbon atoms

$\int \Phi_i H \Phi_j d\tau = \beta$ for $j = i \pm 1$, that is, only adjacent atoms interact

$\int \Phi_i H \Phi_j d\tau = 0$ for all remaining combinations

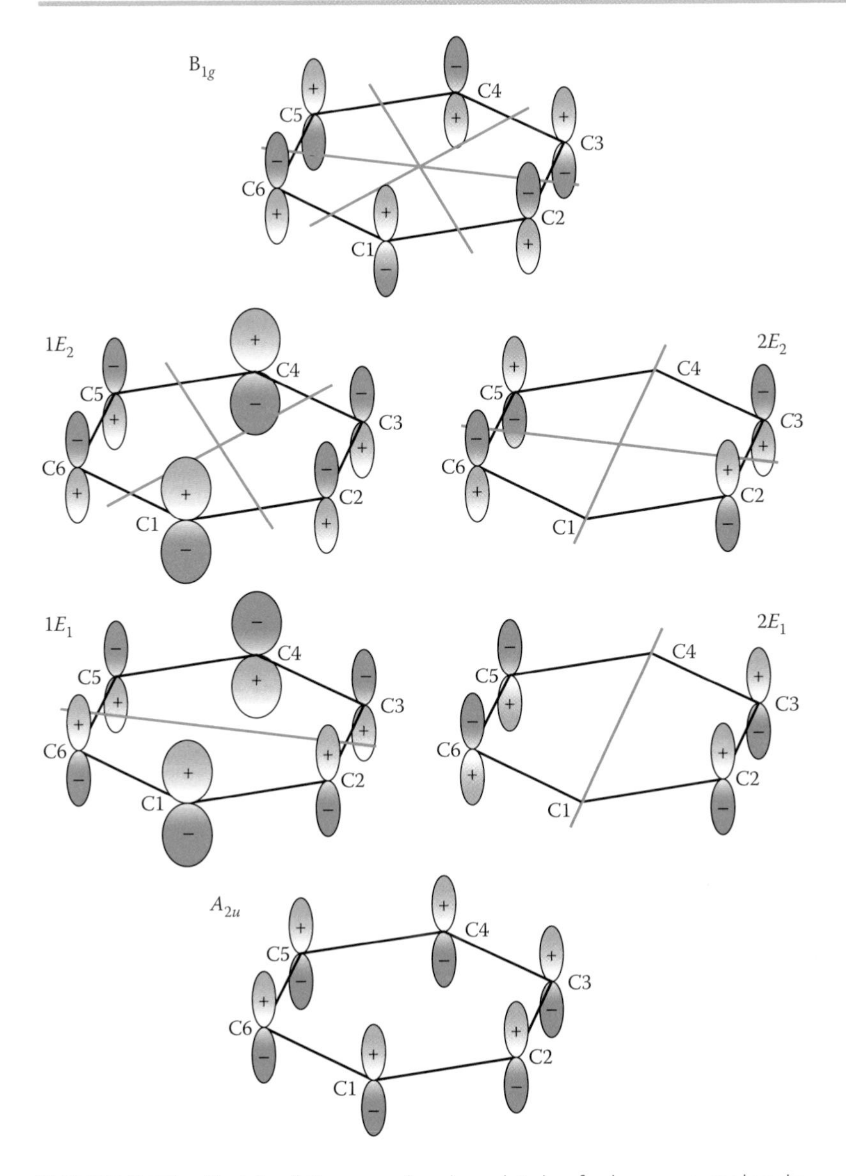

FIGURE 3.30 Sketch of the π molecular orbitals of a benzene molecule. They are grouped according to their symmetry. The supplementary lines indicate where the nodal planes intersect the molecular plane.

and all nonzero elements of the secular determinant will be calculated as follows:

$$H_{11} = \int \Psi_A H \Psi_A \, d\tau = \frac{1}{6}\left(6\int \Phi_i H \Phi_i \, d\tau + 6 \cdot 2 \int_{i=j\pm1} \Phi_i H \Phi_j \, d\tau\right) = \alpha + 2\beta$$

$$H_{22} = \int \Psi_B H \Psi_B \, d\tau = \frac{1}{6}(6\alpha - 6 \cdot 2\beta) = \alpha - 2\beta$$

$$H_{33} = \int \Psi_{1E1} H \Psi_{1E1}\, d\tau$$

$$= \frac{1}{12}\Big[(4\alpha + 2\beta + 2\beta) + (2\beta + \alpha - \beta) + (-\beta + \alpha + 2\beta) + (2\beta + 4\alpha + 2\beta)$$

$$+ (2\beta + \alpha - \beta) + (-\beta + \alpha + 2\beta)\Big] = \frac{1}{12}(12\alpha + 12\beta) = \alpha + \beta$$

$$H_{44} = \int \Psi_{2E1} H \Psi_{2E1}\, d\tau = \frac{1}{4}[(\alpha + \beta) + (\beta + \alpha) + (\alpha + \beta) + (\beta + \alpha)] = \alpha + \beta$$

$$H_{55} = \int \Psi_{1E2} H \Psi_{1E2}\, d\tau = \frac{1}{12}[(4\alpha - 2\beta - 2\beta) + (-2\beta + \alpha + \beta)$$

$$+ (\beta + \alpha - 2\beta) + (-2\beta + 4\alpha - 2\beta) + (-2\beta + \alpha + \beta)$$

$$+ (\beta + \alpha - 2\beta)\big] = \frac{1}{12}(12\alpha - 12\beta) = \alpha - \beta$$

$$H_{66} = \int \Psi_{2E2} H \Psi_{2E2}\, d\tau = \frac{1}{4}\big[(\alpha - \beta) + (-\beta + \alpha) + (\alpha - \beta) + (-\beta + \alpha)\big]$$

$$= \alpha - \beta$$

Thus, the secular determinant for π bonds of the benzene molecule in the Hückel approximation will be reduced to the diagonal elements as follows:

$$\begin{vmatrix} \alpha+2\beta-E_A & 0 & 0 & 0 & 0 & 0 \\ 0 & \alpha-2\beta-E_B & 0 & 0 & 0 & 0 \\ 0 & 0 & \alpha+\beta-E_{E1} & 0 & 0 & 0 \\ 0 & 0 & 0 & \alpha+\beta-E_{E1} & 0 & 0 \\ 0 & 0 & 0 & 0 & \alpha-\beta-E_{E2} & 0 \\ 0 & 0 & 0 & 0 & 0 & \alpha-\beta-E_{E2} \end{vmatrix} = 0$$

So, the solutions of the secular equation, that is, the orbital energies, are

$$E_A = \alpha + 2\beta$$

$$E_B = \alpha - 2\beta$$

$$E'_{E1} = E''_{E1} = \alpha + \beta$$

$$E'_{E2} = E''_{E2} = \alpha - \beta$$

We are now able to order π bonds on the energy scale, though we do not know the values of α and β. α is the electron energy in the p_z orbital of the carbon atom and can be chosen as the origin of the energy scale. β is negative because it results from the interaction of two adjacent atomic orbitals. This interaction will stabilize, more or less, the atoms in a molecule than in the free state. Figure 3.31a shows the levels of energy of π bonds in *single-electron occupancy*. The available six electrons can fill the lower levels in the ground state of the molecule. Thus, the total energy of the molecule in the ground state is $2 \cdot 2\beta + 4 \cdot \beta = 8\beta$ (where $\beta < 0$). So, the benzene molecule is more stable than the six free carbon atoms.

Looking to Figure 3.31a, we can classify the orbitals π_{A2u}, π_{1E1}, and π_{2E1} as bonding orbitals and the orbitals π_{1E2}, π_{2E2}, and π_{B1g} as antibonding orbitals (they are marked by*).

We note that the value of the β parameter (called *resonance integral*) is better determined by comparing the UV measured spectra of benzene with the computed energies. The value of β was given to be –2.6 eV by Mulliken, who was awarded the Nobel Prize for Chemistry (Roothaan and Mulliken 1948).

In the ground state, the electron configuration is written as $a_{2u}^2 e_{1g}^4$. (We have used the single-electron symmetry denoted by the lowercase letter.) The first excited state is obtained by promoting one electron from the e_{1g} degenerate orbital to the orbital e_{2u} as shown in Figure 3.31b. The excited electron configuration will be $a_{2u}^2 e_{1g}^3 e_{1u}^1$. The energy of the excited state is $2 \cdot 2\beta + 3 \cdot \beta + 1 \cdot (-\beta) = 6\beta$. Thus, the energy difference between the excited state and the ground state is 2β. Replacing the β parameter by the Mulliken value –2.6, we can find that the energy difference equals 5.2 eV. Thus, the electron can be promoted to the first state by absorption of one photon of 5.2 eV energy. The conversion of 5.2 eV energy into wavelength gives the value of 238 nm. It means that the absorption spectrum of benzene should have the first absorption electronic band in ultraviolet at 238 nm (1240 divided by 5.2). The absorption spectrum of benzene

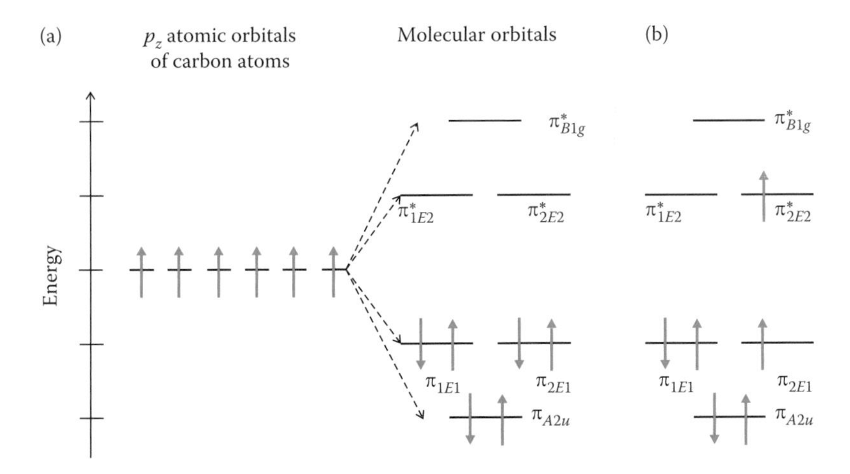

FIGURE 3.31 (a) Energy diagram of the molecular orbitals of the benzene molecule in the ground state. Filling the MO with electrons will create a more stable system than six free atoms. (b) First excited molecular configuration results from promoting one electron from E_1 bonding MO to E_2 antibonding MO.

has few bands of different intensities in ultraviolet but not exactly at this value. There are two reasons:

- The photon absorption produces as a result the electron transition between two molecular states (many-electron occupancy).
- The electron transition must obey the selection rules that we have to discuss in Chapter 4.

At the moment, we can observe by inspection of the energy diagram of the molecular orbitals (Figure 3.31) that the ground configuration has the total spin 0 (all electrons are paired) and the first excited configuration has the total spin 1 (two unpaired electrons). So, the previous expected transition is forbidden by the spin selection rule. Herzberg's (winner of the Nobel Prize in Chemistry in 1971) book is recommended for an understanding of the complete treatment of the benzene spectra (Figure 141 of Herzberg 1966).

3.6.3 ML_6 COMPLEX

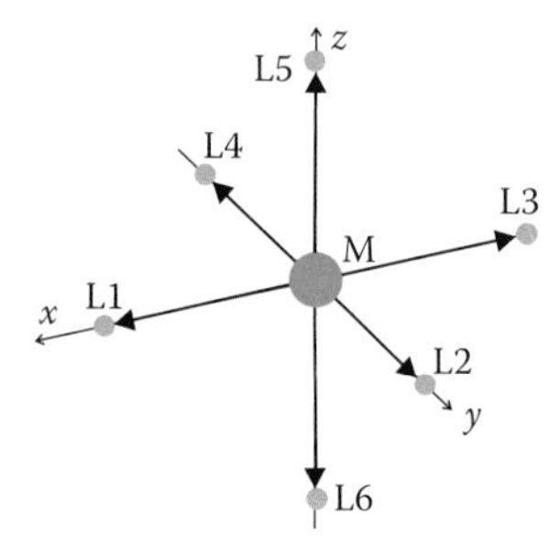

FIGURE 3.32 Illustration of the ML_6 complex and the Cartesian coordinates attached.

We have to apply the molecular orbitals theory for inorganic complexes with a transition metal ion surrounded by six halide ions, often called ligand ions. The sketch of the molecule is shown in Figure 3.32.

We previously obtained the SALCs of σ bonds for the $\boldsymbol{O}_h$ point group. The following are the results:

$$\Psi_{1A1} = \frac{1}{6}(\sigma_1 + \sigma_2 + \sigma_3 + \sigma_4 + \sigma_5 + \sigma_6);$$

$$\Psi_{2E} = \frac{1}{2\sqrt{3}}(-\sigma_1 - \sigma_2 - \sigma_3 - \sigma_4 + 2\sigma_5 + 2\sigma_6);$$

$$\Psi_{3E} = \frac{1}{2}(\sigma_1 + \sigma_3 - \sigma_2 - \sigma_4)$$

$$\Psi_{4T1} = \frac{1}{\sqrt{2}}(\sigma_1 - \sigma_3); \quad \Psi_{5T1} = \frac{1}{\sqrt{2}}(\sigma_2 - \sigma_4);$$

$$\Psi_{6T1} = \frac{1}{\sqrt{2}}(\sigma_5 - \sigma_6)$$

The atomic orbitals of metal ions, which are available to combine in molecular orbitals, are $3d$, $2s$, and $4p$. The halide ion can participate with s and p orbitals. The character table of the $\boldsymbol{O}_h$ group shows the symmetry of atomic orbitals of the metal.

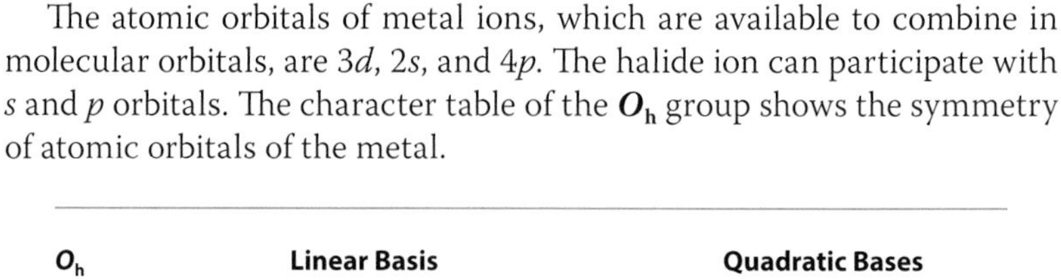

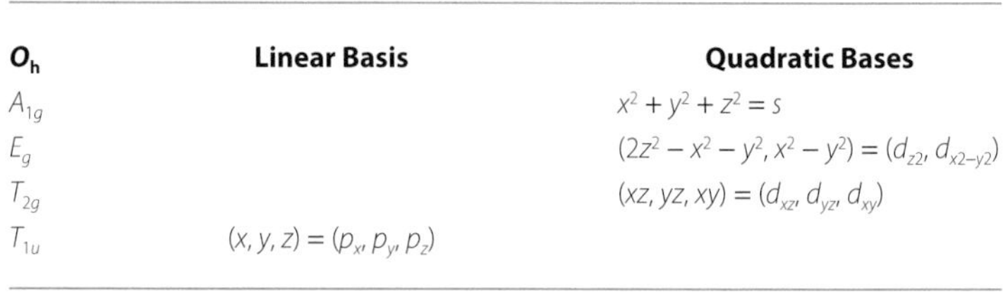

O_h	Linear Basis	Quadratic Bases
A_{1g}		$x^2 + y^2 + z^2 = s$
E_g		$(2z^2 - x^2 - y^2, x^2 - y^2) = (d_{z2}, d_{x2-y2})$
T_{2g}		$(xz, yz, xy) = (d_{xz}, d_{yz}, d_{xy})$
T_{1u}	$(x, y, z) = (p_x, p_y, p_z)$	

The s, p_x, p_y, p_z, d_{z2}, d_{x2-y2} orbitals have suitable symmetries to participate in σ bonds and antibonds with the corresponding symmetry SALCs. However, the orbitals d_{xz}, d_{yz}, d_{xy} do not have a suitable orientation relative to ligands in order to contribute to σ bonding, that is, the lobes are not directed to ligands. Thus, the molecular orbitals of σ bonds will appear as a result of combination between the atomic orbitals of the central atom and the SALCs of ligand atoms corresponding to the same irreducible representation.

AO of the Metal Ion	SALCs of Ligands	Resulting σ Molecular Orbital
s	Ψ_{1A1g}	a_{1g}
$d_{2z^2-(x^2+y^2)}$	Ψ_{2Eg}	e_g
$d_{x^2-y^2}$	Ψ_{3Eg}	
p_x	Ψ_{4T1u}	t_{1u}
p_y	Ψ_{5T1u}	
p_z	Ψ_{6T1u}	

The diagram of molecular orbitals of σ bonding and antibonding is shown in Figure 3.33. We have already used the small letters to denote the molecular states in order to underline that the new states were obtained as single-electron occupancy. Therefore, we did not take into account the

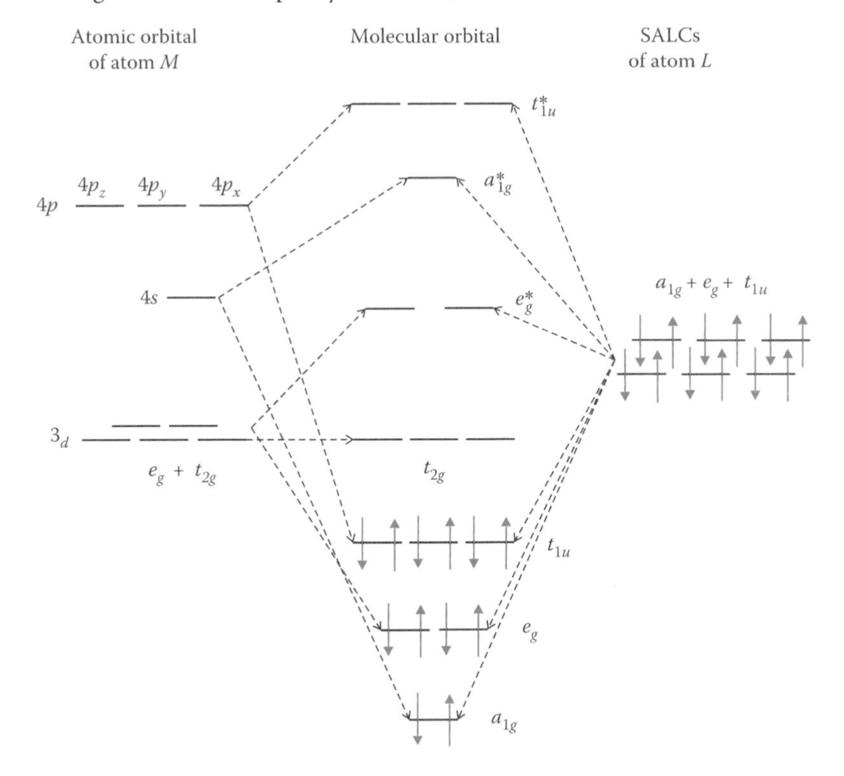

FIGURE 3.33 Diagram of molecular orbitals of σ bonds of the ML_6 complex. The energy positions of levels are arbitrary. The electrons shared by the ligands are used to fill the bonding orbitals.

electronic interaction. The energy positions of different orbitals are arbitrary. We also suggested the way they can be filled by the shared electrons provided by the ligands. The electrons provided by the metal ion can fill the next two levels: t_{2g} and e_g^*. The energy separation between t_{2g} and e_g^* levels is the parameter of crystal field Δ_O. The filling of these two levels is similar to that resulting from the crystal field theory. When you add electrons to the t_{2g} level, they are not shared with ligands. Are *d* electrons of metal ion, which fill the e_g^* level, shared or not? The molecular orbital was obtained as a combination between both metal and ligand atomic orbitals. The answer depends on the relative position of *d* orbital and SALCs of ligands, which depends on both the metal ion and the ligand species. In the case (arbitrarily taken) illustrated in Figure 3.33 where the energy of the e_g^* level is closer than that of σ SALC, the electrons of the e_g^* level are more localized on the ligands. If the energy is closer than that of *d* orbitals of the free ion, the electrons are more localized on the metal ion. This is the case most frequently encountered in the crystal field theory, which is an extreme case of molecular orbitals theory.

Figure 3.34 is a rough illustration of the spatial orientation of atomic orbitals before they combine to form σ bonds of symmetries specific to

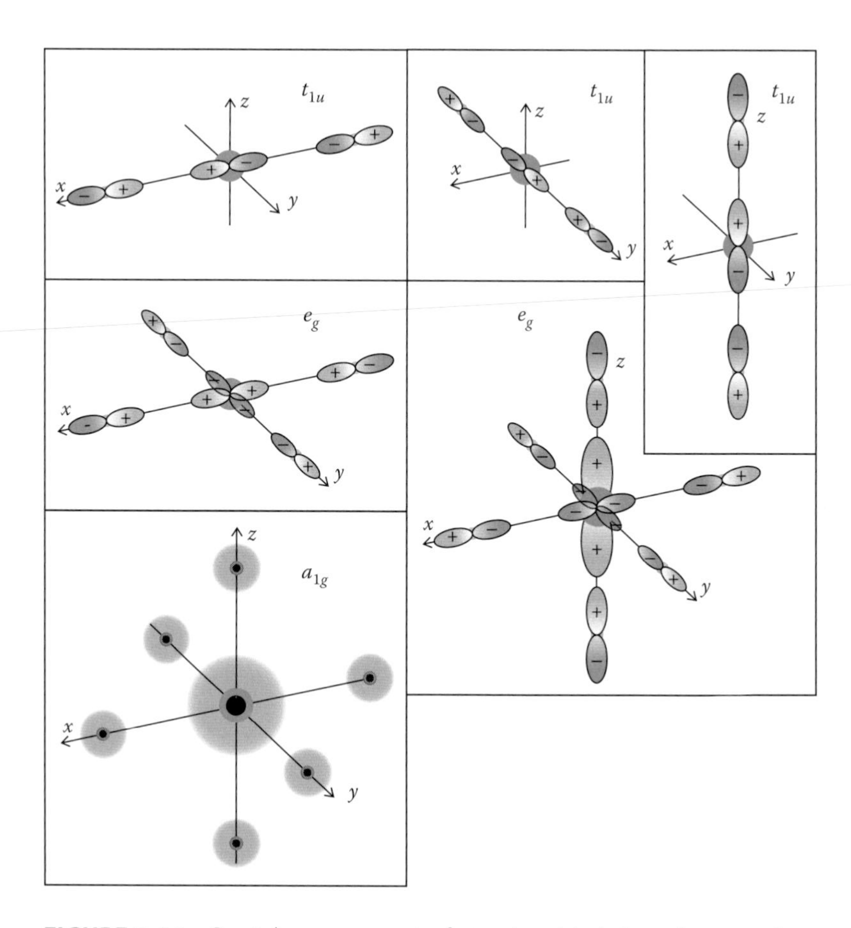

FIGURE 3.34 Spatial arrangement of atomic orbitals in order to make σ bonds of specific symmetries.

the O_h group. These sketches could be done prior to applying the projection operator, just knowing the irreducible representations generated by the σ bonds. The orientation of σ bonds along the line connecting M–L atoms and the shape of the metal orbitals are enough to have an image of what we must obtain.

In order to find the SALCs symmetry of the π bonds, we should use a new set of vectors representing the π orbitals of ligands that can combine into the π bonds M-L. We must attach to each ligand atom two vectors that are perpendicular to bond line M-L. Therefore, we will use 12 vectors as a basis to generate a representation of SALCs. They are sketched in Figure 3.35.

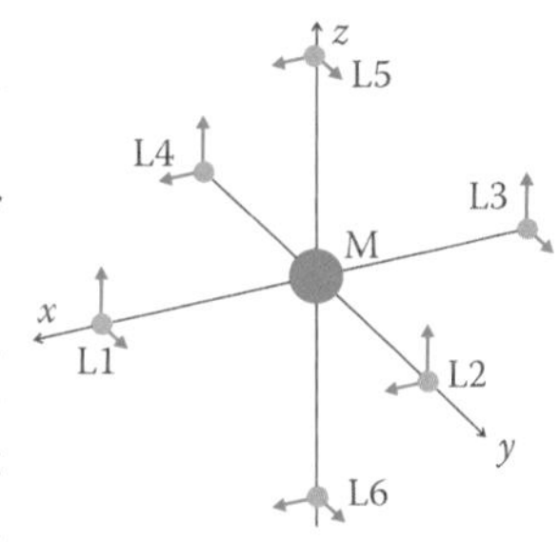

FIGURE 3.35 Set of 12 vectors perpendicular to the bond lines attached to ligands to generate the orthogonal π orbitals.

Before beginning to search for characters, remember the rules: an arrow shifted gives 0, an arrow unshifted gives 1, and an arrow reversed gives –1. There are only three operations that leave all or several vectors unshifted: *E*, C_2 (twofold rotation around the *z* axis), and reflection Σ_h through the horizontal plane *xOy*. The characters of the representation Γ_π generated by the set of 12 vectors are summarized below together with the decomposition into irreducible representations.

O_h	E	$8C_3$	$6C_2'$	$6C_4$	$3C_2$	I	$6S_4$	$8S_6$	$3\Sigma_h$	$6\Sigma_d$	
Γ_π	12	0	0	0	–4	0	0	0	0(+4 – 4)	0	$T_{1g} + T_{2g} + T_{1u} + T_{2u}$

The inspection of the character table of the O_h group results in that there are orbitals with symmetries $T_{2g} + T_{1u}$. The orbitals d_{xy}, d_{yz}, and d_{xz} have t_{2g} symmetry, so they will combine with the T_{2g} π SALCs to form π bonds. The other combination, apparently possible, is between *p* orbitals (empty) of metal and T_{1u} SALCs of ligands. However, these orbitals were already used to form σ bonds, which are more stable than π bonds. So, such a combination is unlikely and it will generate a weak π bond. We can now imagine the orientation of atomic orbitals of both metal ions and ligands (see the illustration of Figure 3.36).

What functions describe the π SALCs of T_{2g} symmetry? The illustration of Figure 3.36 will help us to write SALCs of ligands that can combine with the orbitals d_{xy}, d_{yz}, d_{xz} having t_{2g} symmetry, without using the projection operator. The small illustration from the left-bottom corner of Figure 3.36 indicates how to denote the vectors (*p* orbitals). Each atom has its own coordinate system. The local *z* axis is always oriented

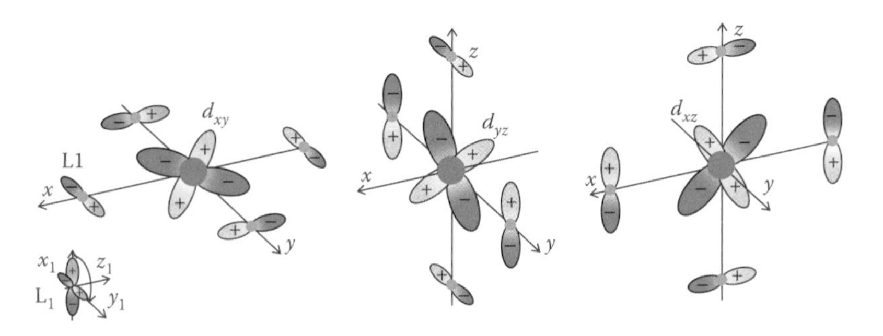

FIGURE 3.36 Arrangement of atomic orbitals of metal ions and ligands before combining into π orbitals of O_h symmetry.

toward the central atom. The x and y axes are perpendicular to the z axis. A screwdriver turned in the sense from x to y always advances in the positive direction of the z axis. Therefore, the SALCs are given by the following functions:

$$\Psi_{1T2g} = \frac{1}{2}(\pi_{1y} + \pi_{2x} + \pi_{3x} + \pi_{4y});$$

$$\Psi_{2T2g} = \frac{1}{2}(\pi_{2y} + \pi_{4x} + \pi_{5x} + \pi_{6y});$$

$$\Psi_{3T2g} = \frac{1}{2}(\pi_{1x} + \pi_{3y} + \pi_{5y} + \pi_{6x}).$$

The molecular orbitals of the π bond results from the positive overlap between SALCs and d orbitals. The corresponding π antibond results from the negative overlap of the same functions.

Thus, we can complete the diagram of the molecular orbitals of σ bonding with those corresponding to the π bonding. The final diagram of the molecular orbitals of the molecular complex ML_6 is shown in Figure 3.37. We note that it is not possible to construct a diagram of molecular orbitals valuable for any combination ML_6. The positions of levels and their filling with electrons depend on many factors but the most important are the identity of the metal ion and ligands and the energy differences between the atomic orbitals of the metal and ligands.

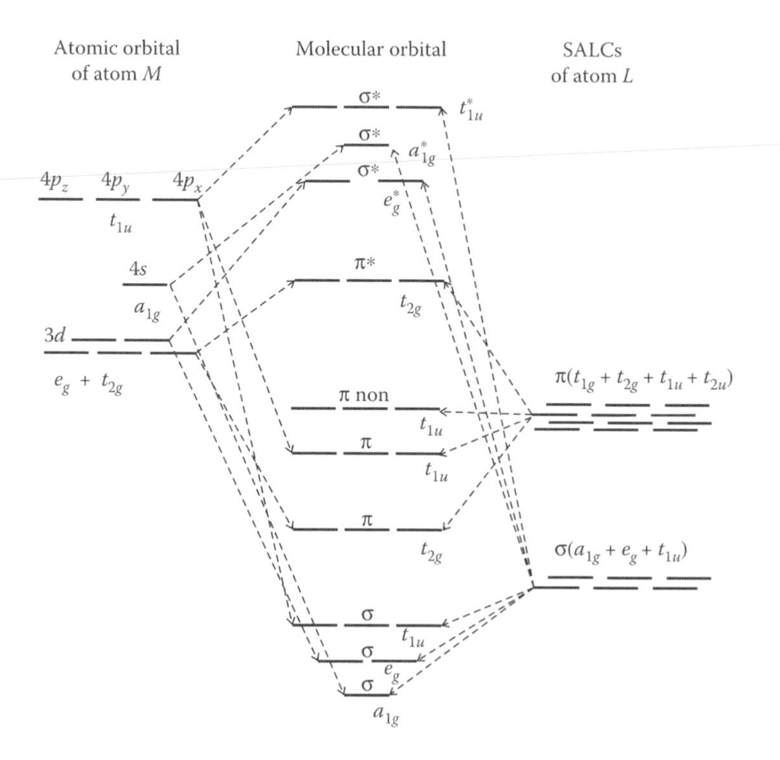

FIGURE 3.37 Approximate diagram of molecular orbitals of the ML_6 complex. The energy levels were not calculated.

STUDY QUESTIONS

3.1 Fill in the missing words in the following sentence: "Hybridization is the process by which ______ ________ combine to form a set of 'mixed' orbitals when bonding covalently."

3.2 Fill in the missing words in the following table:

Relation between hybridization and local geometry	
sp	Linear
sp^2	______ ______
___	Tetrahedral
sp^3d	Trigonal bipyramidal
sp^3d^2	_______

3.3 Comment on the following sentence: "The covalent bond consists of electrostatic interaction between electrons shared by two bonded atoms."

3.4 Comment on the following sentence: "The σ bonds are formed by *s* orbitals and π bonds are formed by *p* orbitals."

3.5 How can you explain that a carbon atom with four electrons can bond maximum four atoms while a metal with one *d* electron can bond six ligands?

3.6 Comment on the following sentence: "more nodes in a molecular orbital, the higher the position on the energy scale."

3.7 Which of the following hybridization patterns belong to the PCl_5 molecule: sp, sp^2, sp^3, sp^3d?

3.8 What is the hybridization of carbon atoms in the propane molecule shown in the picture?

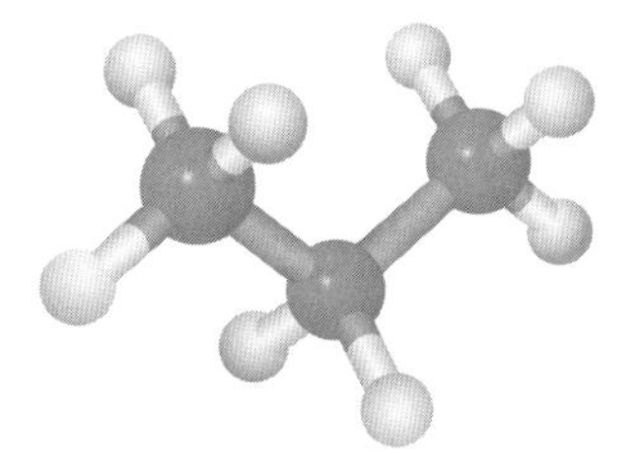

3.9 What is the hybridization of carbon atoms in the propene molecule?

3.10 Deduce an MO diagram for the N_2O molecule.

REFERENCES

Allinger, N. L. and J. C. Tai. 1968. Further theoretical studies of the structure and electronic spectrum of cyclobutadiene. *Theoretica Chimica Acta (Berlin)* 12(1): 29–33.

Cotton, A. F. 1990. *Chemical Applications of Group Theory*, 3rd edition. New York: John Wiley & Sons. Inc.

Franca, C. A. and R. P. Diez. 2009. The Lewis acidity of boron trihalides revisited. *Journal of the Argentine Chemical Society* 97(1): 119–126.

Herzberg, G. 1966. *Molecular Spectra and Molecular Structure.* Princeton: Van Nostrand Inc.

IUPAC. 1999. Nomenclature of organometallic compounds of the transition elements. *Pure and Applied Chemistry.* 71(8): 1557–1585.

Jorgensen, W. L. and L. Salem. 1973. *The Organic Chemist's Book of Orbitals.* New York: Academic Press.

Roothaan, C. C. J. and R. S. Mulliken. 1948. Molecular orbital treatment of the ultraviolet spectra of benzene and borazole. *The Journal of Chemical Physics* 16: 118.

Symmetry of Molecular Vibrations and Selection Rules

4

4.1 MOLECULAR VIBRATIONS

In the first chapter, we learned how to find the symmetry of a molecular complex: the group to which it belongs, and the representations generated by the atomic orbital functions. We then learned what happens to atomic orbitals in a given symmetry environment and the possible energies they can have. Knowing the new energy diagram of a metal ion in a symmetric environment is important to predict the transitions that may occur and therefore to explain the electronic spectra we can measure for a given ion in a solution or crystal. We have already seen that there are some selection rules (spin rule, Laporte's rule) that normally indicate which transition is possible or not. However, the reality is more complex and the *molecules do not strictly obey the selection rules*. Why? We have just discussed that the symmetry of a molecule is somewhat broken by molecular vibrations. However, pure electronic transitions are not solely responsible for the optical spectra of molecules. Molecules can *absorb light in the infrared range* too or can *produce elastic and inelastic scattering of light*. What happens to energy as a result of light absorption? Several molecules may emit light after light absorption producing the *photoluminescence* phenomenon. Molecular vibrations are involved in all new phenomena and we will try to explain them briefly in this chapter. To do so, we have to understand the different branches of optical spectroscopy:

- Vibrational spectroscopy
- Raman (scattering) spectroscopy
- Fluorescence spectroscopy

A molecule (or a molecular complex) consists of nuclei and electrons. Most electrons are bound to the atom to which they belong. Only a few electrons interact with the nearest neighbor atoms. These electrons can interact with the electric field of light in certain conditions and they can change their energy by light energy absorption. This light absorption phenomenon is so fast that nuclei (which are very heavy compared to electrons) do not change their positions during light absorption. We can consider a pure interaction between a valence electron and the electric field of light. Therefore, the nuclei do not participate in energy absorption. However, it does not mean that nuclei are fixed. They are continuously moving. We can reduce the movement of nuclei to two types: vibration and rotation. It is obvious that the movement of nuclei will influence the movement and then the energy of electrons. Remember that the pure electrostatic interaction electron–nucleus is strongly dependent on the electron–nucleus distance. How strong is the dependence on distance? Optical spectroscopy can give the answer: while the pure electronic transitions are "visible" between 10,000 and 50,000 cm^{-1}, the vibrational spectra are located in infrared range at a few thousand or hundreds of cm^{-1} and the rotation bands are located in a microwave range at 1 cm^{-1}. Thus, the vibration energy is more than one order of magnitude weaker than the electronic movement energy and the rotation is also one–two orders of magnitude weaker than the vibration. The above values are not strict. Here, we may note that microwave ovens operate at a frequency of 2.45 GHz and this is not related to any resonant frequency of a water molecule. The polar water molecule only twists and turns in the electric field of microwaves. The value of the working frequency was only chosen as a useful average frequency.

These values allow us to treat electron movement, vibration, and rotation of nuclei separately as if they were independent of each other. As a consequence, the wavefunction may be written as a simple product of

$$\Psi = \Psi_e \cdot \Psi_v \cdot \Psi_r \tag{4.1}$$

where Ψ_e is the wavefunction for electrons, Ψ_v is the wavefunction for nuclei vibration, and Ψ_r is the wavefunction for nuclei rotation. Later, we will suppose that the electron function and the vibrational function are not completely independent.

Thus, the internal energy of a molecule is composed of electronic, vibrational, and rotational energies:

$$E = E_e + E_v + E_r \tag{4.2}$$

The wavefunction separation allows us to solve three simple equations instead of one:

$$\begin{cases} H\Psi_e = E_e\Psi_e \\ H\Psi_v = E_v\Psi_v \\ H\Psi_r = E_r\Psi_r \end{cases} \tag{4.3}$$

You can find the contribution to energy of each term in any textbook dedicated to molecular spectroscopy. The objective of our book is not to solve quantum mechanics equations but to show how to use symmetry in spectra prediction.

We must often evaluate integrals on a symmetric range of integration. The *integral of an odd function calculated on a symmetric range vanishes*. (In terms of group theory, an integral overall space is nonzero if the function to be integrated belongs to the totally symmetric representation.) Moreover, we can write that

$$\int_{-\infty}^{+\infty} (-) \cdot x \cdot (-) dx = \int_{-\infty}^{+\infty} (+) \cdot x \cdot (+) dx = 0$$

$$\int_{-\infty}^{+\infty} (-) \cdot x \cdot (+) dx = \int_{-\infty}^{+\infty} (+) \cdot x \cdot (-) dx \neq 0$$

In other words, the *integral vanishes when the two functions have the same sign referred to as inversion* (or the center of symmetry).

An atomic system may absorb energy from light if the energy of light equals the energy difference between the final state and initial state:

$$h\nu = E_f - E_i \tag{4.4}$$

We name this process a *resonant* process; therefore, light is absorbed only if its energy is resonant with energy difference between the two atomic states. It is like the resonant transfer of energy between two mechanical oscillators: the transferred energy has a maximum value when the frequencies of two oscillators are equal.

This energy-resonant condition is necessary but not enough for the light absorption to produce. Any atomic system obeys the laws of quantum mechanics and the probability for the system to absorb light is given by

$$\Pi_{fi} = \int_{-\infty}^{+\infty} \Psi_f^* \hat{P} \Psi_i \, d\tau \tag{4.5}$$

where Ψ_i and Ψ_f are the initial and final wavefunctions of the atomic system and $\hat{P}$ is the operator describing the mechanism of transition. For simplicity, we consider that the atomic system is composed of a positive electric charge and many negative electric charges moving fast around the nucleus. The average distribution of the negative charges has a "weight" center (called charge center) that may or may not coincide with that of the positive charge. If the negative and positive charge centers differ then the atomic system is equivalent to an *electric dipole*, in the first approximation. The electric dipole expression can be written as $e \cdot \vec{r}$. The electric dipole will oscillate forced by the electric field of light. The free atom has a *spherical symmetry* with an inversion center. The electric dipole or each component of it is *odd (antisymmetric) with respect to inversion.* Thus, each *electric dipole type transition between two states of the same parity*, for example, $s \leftarrow s$ $(+ \leftarrow +)$, $p \leftarrow p$ $(- \leftarrow -)$, $d \leftarrow d$ $(+ \leftarrow +)$, and $f \leftarrow f$ $(- \leftarrow -)$, gives a *probability equals to 0*. So this type of transition is not allowed (Laporte's rule). The rule also applies for transitions as $s \leftarrow d$,

$p \leftarrow f$. However, transitions between states of different parity are allowed, for example, $s \leftrightarrow p$ $(+ \leftrightarrow -)$, $p \leftrightarrow d$ $(- \leftrightarrow +)$, and $d \leftrightarrow f$ $(+ \leftrightarrow -)$. This rule is known as the selection rule of angular quantum number $\Delta l = \pm 1$.

When the ion is placed in a ligand field (molecular complex or crystal) with octahedral symmetry that has an inversion center, the rule is still working. In a transition metal ion, the orbital functions are d type or are linear combinations of d type. Thus, any transition between new terms that appear by splitting in the ligand field is not allowed, too, as in the free ion case. Changing the environment to tetrahedral symmetry, where the inversion center is missing, transitions are allowed and the spectra become stronger by orders of magnitude.

We know that the octahedral compounds of metal transition ions are colored and present absorption bands but they are much weaker than those of tetrahedral compounds. It means that Laporte's rule is partly working. The vibrations of nuclei can partly destroy the inversion center of the molecular complex and the selection rule will not be very efficient.

The rotation movement and corresponding energy are not sensitive to symmetry changes. However, vibration is strongly dependent on symmetry properties. We can now suppose that the electron function and vibrational function are not completely independent and the transition probability can be written as

$$\Pi_{fi} = \int_{-\infty}^{+\infty} (\Psi_{ef}\Psi_{vf})^{*}\, \hat{P}(\Psi_{ei}\Psi_{vi})d\tau \tag{4.6}$$

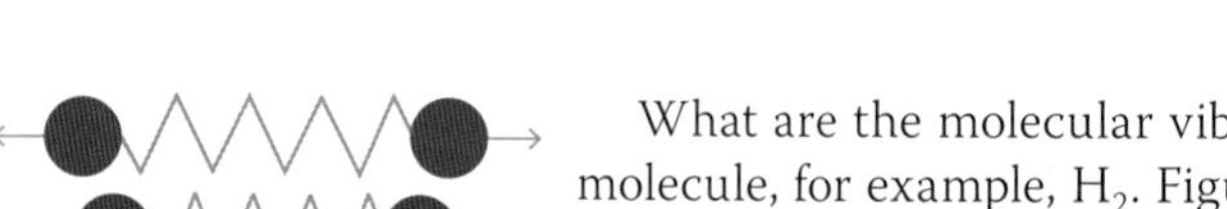

FIGURE 4.1 The simplest molecule of H_2 can be illustrated as two balls connected by a spring that is stretched or compressed.

What are the molecular vibrations? The simplest model is a diatomic molecule, for example, H_2. Figure 4.1 illustrates the oscillator model of a hydrogen molecule composed of two balls connected by a spring. The balls represent the nuclei and the spring represents the bonding force. During vibration, the atoms move toward or away from each other. Vibration has two important characteristics:

- *Frequency* (characteristic frequency): Its value depends on the square root of the ratio between spring constant (strength) and ball mass, $\sqrt{k/m}$; in the case of the molecule, the bonding strength is equivalent to the spring strength and the atomic mass is equivalent to the ball mass. Therefore, the lighter the atom, the higher the frequency. The frequency is also higher for a stronger bond.
- *Energy* (which gives the amplitude): Its value is given by the extent of the initial spring stretching or compression and the spring strength, $1/2k \cdot (\Delta x)^2$.

At the atomic level, *the frequency and the energy* are quantized; therefore, they *can have only certain values*.

All motions of the molecule (translation, vibration, and rotation) can be described as displacements of atoms in Cartesian coordinates. We can

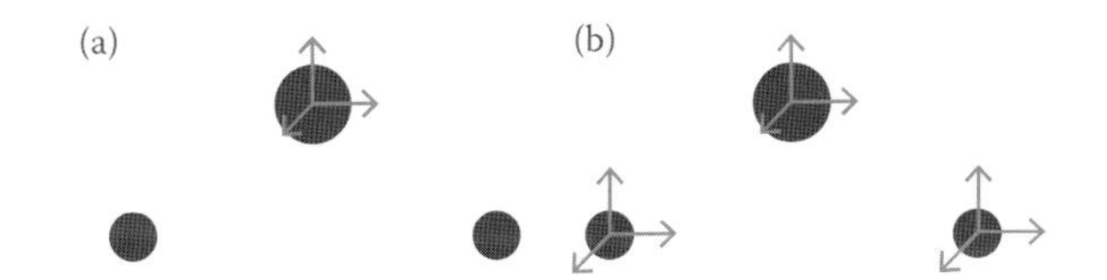

FIGURE 4.2 Each atom has attached a Cartesian coordinate system (a) and the total number of degrees of freedom is 3*N* (b). The illustration was done for the water molecule.

attach to each atom a system of coordinates as shown in Figure 4.2. It means each atom has 3 degrees of freedom. The molecule of N atoms has a total of $3N$ coordinates and *3N degrees of freedom.*

The $3N$ Cartesian coordinates can describe all atomic motions, namely:

- Three translational motions
- Three rotational motions
- $3N$–6 vibrational motions

During one vibration that is very fast, the mass center of the molecule does not move and no rotation is performed. As a consequence, it is important to note that the *three types of motion can be separately treated.*

A linear molecule composed of N atoms has two rotational motions, each one around the two axes perpendicular to the linear molecular axis (Figure 4.3a), and $3N - 5$ vibrations. The atoms are approximated to be without size, and hence the inertia momentum around the principal axis (y) vanishes. A nonlinear molecule has three rotations (each one around three mutually perpendicular principal axes of inertia as shown in Figure 4.3b), and $3N - 6$ vibrations.

So, the hydrogen molecule of Figure 4.1 has $3 \times 2 - 5 = 1$ vibration: the stretching–compression of the bond between the atoms. Carbon dioxide, a linear molecule too (Figure 4.3a), has $3 \times 3 - 5 = 4$ vibrations. The water molecule (Figure 4.3b) has $3 \times 3 - 6 = 3$ vibrations.

There are two types of molecular vibrations:

- Stretching (symmetric and asymmetric)
- Bending (scissoring, twisting, rocking, and wagging)

The vibrational modes of CO_2, shown in Figure 4.4, are responsible for the "greenhouse" effect. The CO_2 molecules trap the heat radiated from the earth in the atmosphere. The arrows indicate the directions of

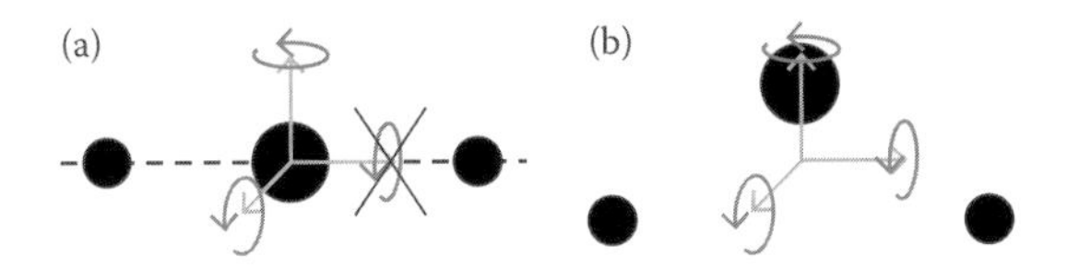

FIGURE 4.3 (a) A linear molecule has two distinct rotations and (b) a nonlinear molecule has three rotations.

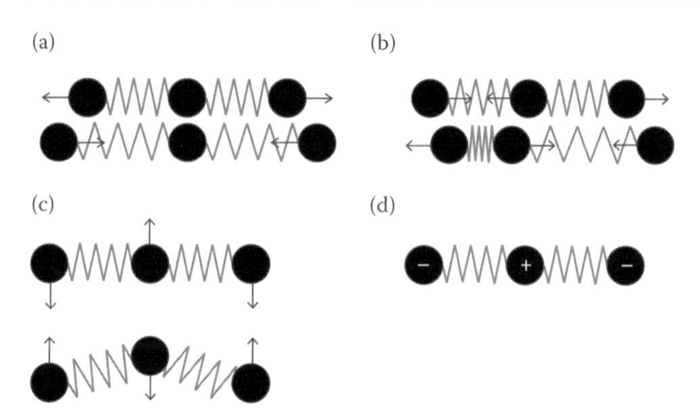

FIGURE 4.4 All vibrational modes of the CO_2 molecule: (a) The symmetric stretching mode, (b) the asymmetric stretching mode, (c) the bending mode (scissoring) in the paper plane, and (d) the bending mode out of the paper plane; the plus and minus signs represent vibration directions perpendicular to the paper plane (+above and −below).

motion. Vibrations, labeled as (a) and (b), represent the *stretching* of the chemical bonds, one in a symmetric (Figure 4.4a) way, in which both C = O bonds lengthen and contract together, and the other in an asymmetric (Figure 4.4b) way, in which one bond shortens while the other bond lengthens. The symmetric stretch is *not infrared active*, and so this vibration is not observed in the infrared spectrum of CO_2. To observe a molecular vibration, there must be a change in the dipole moment of the molecule during that vibration. The asymmetric stretch (Figure 4.4b) is infrared active because there is a *change in the molecular dipole moment* during this vibration (allowed transition). The infrared radiation at 2347 cm^{-1} (4.26 μm) excites this particular vibration. The two *bending* vibrations of CO_2 (Figure 4.4c and d) are identical except that one bending mode is in the paper plane, and the other is perpendicular to the plane. The infrared radiation at 667 cm^{-1} (14.99 μm) excites both vibrations. Thus, the bending vibration is twofold degenerate. The out-of-plane mode is illustrated in Figure 4.4d by placing + or − signs on the atoms: plus means motion above the paper plane and minus means motion below the plane. In general, stretching a chemical bond requires more energy than bending.

The three vibrations of CO_2 active in an infrared range can be seen in the absorption spectrum illustrated in Figure 4.5. Note that bending modes have absorption at the same value of energy and appear as one absorption band.

Now, let us take another example of the simple but nonlinear molecule of water. The H_2O molecule has three normal modes of vibration shown in Figure 4.6. The first mode is called the symmetric stretch, the second mode is called a bend, and the third mode is called the asymmetric stretch. The symmetric stretch and asymmetric stretch keep constant the bonding angle. The bending (scissoring) mode keeps constant the interatomic distance.

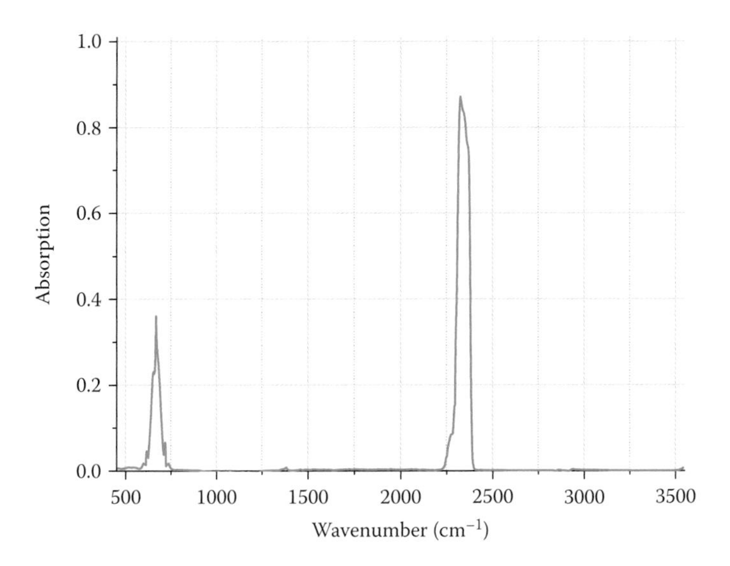

FIGURE 4.5 The infrared absorption spectrum of a gaseous sample of carbon dioxide. The sample absorbs at 2347 cm^{-1} (4.26 μm) and at 667 cm^{-1} (14.99 μm).

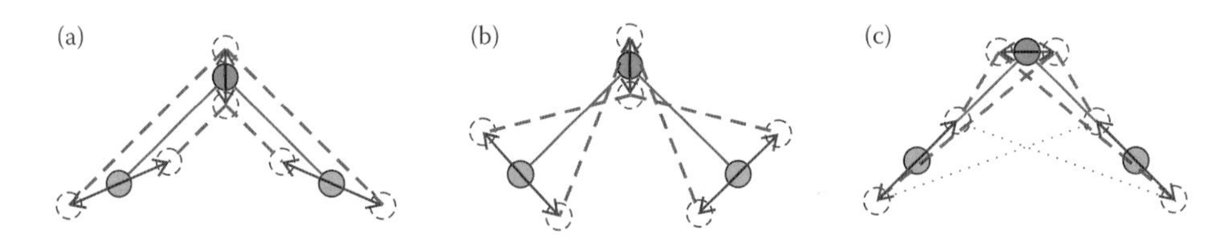

FIGURE 4.6 The normal modes of the water molecule: symmetric stretch (a), bend (scissor) (b), and asymmetric stretch (c). The dashed lines indicate extreme bonding positions. Each arrow direction indicates the motion direction. The arrow lengths are not equal to leave the mass center position unchanged and lengths are drawn excessively high compared to the interatomic distance.

Note that the direction of nuclear motion is not the same as the bonding direction. The nuclei motions occur in such directions that the mass center of molecule rests unchanged.

4.2 SYMMETRY OF NORMAL MODES

The four possible vibration modes previously presented are called *normal modes*. Each absorption band in a vibrational spectrum corresponds to a normal mode. In general:

- Linear molecules have $3N - 5$ normal modes
- Nonlinear molecules have $3N - 6$ normal modes

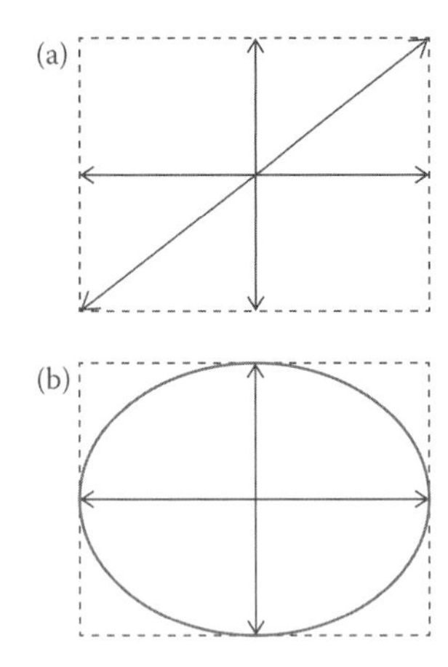

FIGURE 4.7 (a) The superposition of two perpendicular vibrations of the same frequency in phase results in a periodic motion (vibration) on a diagonal direction. (b) The superposition of two perpendicular vibrations of the same frequency with a phase difference of $\pi/2$ results in a periodic motion on an elliptical trajectory.

Note that the molecule can have a complex and aperiodic vibration at a time (apparently chaotic movement of atoms). This vibration is the result of *superposition* of two or more normal modes of vibrations. Each normal mode has its own frequency. A very simple example is the superposition of two perpendicular vibrations, in particular when they have the same frequency. If the two linear vibrations are in phase, the resulting superposition is a linear vibration, too, but in a diagonal direction as shown in Figure 4.7a. The superposition results in a periodic motion on an elliptical trajectory, as shown in Figure 4.7b, if there is a difference of phase of $\pi/2$ between the two linear vibrations.

The *characteristics of normal modes* are summarized below.

1. Each normal mode acts like a harmonic oscillator.
2. The center of mass does not move during vibration.
3. All atoms pass through their equilibrium positions at the same time.
4. Normal modes are independent; they do not interact.

What is the *symmetry of normal modes*?

Let us take a simple molecule as an example: CCl_4. The CCl_4 is a nonlinear molecule and its symmetry as a rigid molecule belongs to the $\boldsymbol{T}_d$ group. It has $3 \cdot 5 - 6 = 9$ normal modes. Figure 4.8 illustrates all normal vibrations of the carbon tetrachloride molecule.

We have to apply all symmetry operations of the $\boldsymbol{T}_d$ group $\{E, 8C_3, 3C_2, 6\Sigma_d, 6S_4\}$ to each normal mode to find its symmetry. In our illustrations of Figure 4.8, each normal mode of vibration is represented by vectors attached to atoms. The vector attached to a chlorine atom indicates the direction of motion of the chlorine atom with respect to the carbon central ion. In other words, it is like we are "located" on a central ion and we look at the chlorine atoms and how they move with respect to us. (Here, we note that the carbon atom is much lighter than the chlorine atom, and thus it moves on a longer distance than the chlorine atom, but we are interested to use the relative motion to describe the vibration of the bond C–Cl. It is similar to the description of the relative motion of the Sun as seen from the Earth. The distance between two bodies and the frequency of repetition does not depend on each other if the origin is related to the Sun or to the Earth.) The length of the vector shows how far the chlorine atom goes with respect to the central atom.

The illustration of Figure 4.8a shows that all bonds of C–Cl simultaneously stretch by the same amount. The chlorine atoms move all together toward the central atom. The identity operation E leaves the molecule unchanged. Any C_3 rotation around one C–Cl bond brings the vectors into an equivalent configuration. The C_2 rotation around the z axis changes the upper vectors one into another and the bottom vectors one into another. Each dihedral plane leaves two vectors unchanged and changes the two other vectors one into another. The improper rotation S_4 around the z axis changes each upper vector into a bottom vector and vice versa. Thus, the vibration vectors generate the totally symmetric representation A_1. So, the

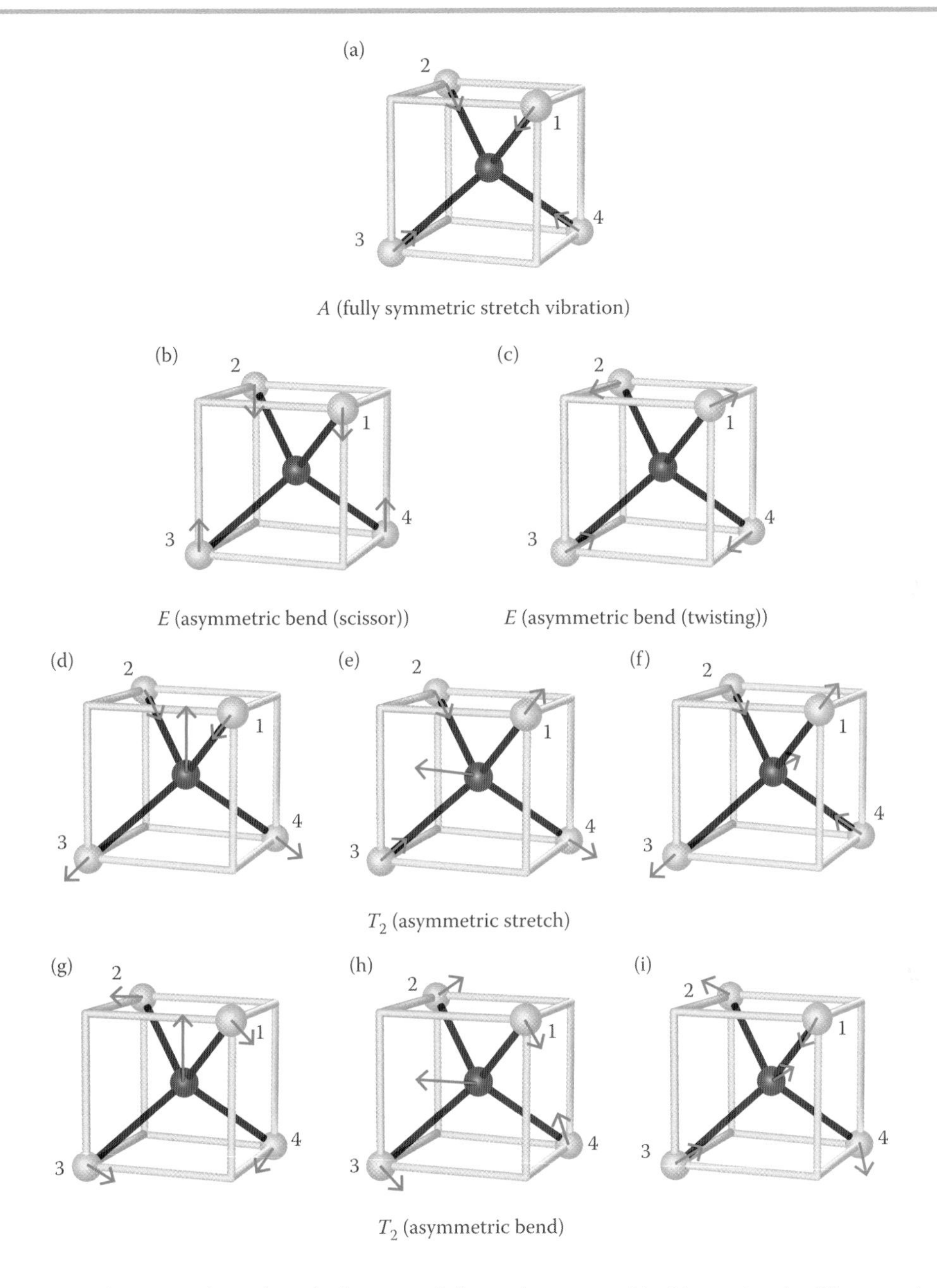

FIGURE 4.8 The normal modes of vibration of the carbon tetrachloride molecule. $2T_2$ normal modes are active in infrared and A_1, E, and $2T_2$ normal modes are active in Raman.

first illustration of the normal mode of frequency ν_1 has A_1 symmetry. This case is the simplest.

Finding symmetry in the next case is more complicated, but the general idea is that the vibration vectors must generate a representation as a result of the action of symmetry operations. The C_3 rotation transforms the vectors of Figure 4.8b into the vectors of Figure 4.8c. It means that both sets of four vectors together form a basis of twofold representation, that is, E. So, both vibrations called ν_{2a} and ν_{2b} have the same frequency. We can write the result of each operation as a linear combination of ν_{2a} and ν_{2b} using the matrix representation as follows:

$$E\begin{bmatrix}\nu_{2a}\\ \nu_{2b}\end{bmatrix}=\begin{bmatrix}\nu_{2a}\\ \nu_{2b}\end{bmatrix}\Rightarrow E=\begin{bmatrix}1 & 0\\ 0 & 1\end{bmatrix}\Rightarrow \chi(E)=2$$

$$C_3\begin{bmatrix}\nu_{2a}\\ \nu_{2b}\end{bmatrix}=\begin{bmatrix}\nu_{2b}\\ -\nu_{2b}\end{bmatrix}\Rightarrow C_3=\begin{bmatrix}0 & 1\\ 0 & -1\end{bmatrix}\Rightarrow \chi(C_3)=-1$$

$$C_2\begin{bmatrix}\nu_{2a}\\ \nu_{2b}\end{bmatrix}=\begin{bmatrix}\nu_{2a}\\ \nu_{2b}\end{bmatrix}\Rightarrow C_2=\begin{bmatrix}1 & 0\\ 0 & 1\end{bmatrix}\Rightarrow \chi(C_2)=2$$

$$\Sigma_d\begin{bmatrix}\nu_{2a}\\ \nu_{2b}\end{bmatrix}=\begin{bmatrix}\nu_{2a}\\ -\nu_{2b}\end{bmatrix}\Rightarrow \Sigma_d=\begin{bmatrix}1 & 0\\ 0 & -1\end{bmatrix}\Rightarrow \chi(\Sigma_d)=0$$

$$S_4\begin{bmatrix}\nu_{2a}\\ \nu_{2b}\end{bmatrix}=\begin{bmatrix}\nu_{2a}\\ -\nu_{2b}\end{bmatrix}\Rightarrow S_4=\begin{bmatrix}1 & 0\\ 0 & -1\end{bmatrix}\Rightarrow \chi(S_4)=0$$

Therefore, both vibrations ν_{2a} and ν_{2b} have E symmetry.

In the third case, described by illustrations of Figures 4.8d through f, two chlorine atoms move toward the cube center while the other two chlorine atoms simultaneously move away from the center. The central atom (carbon) also moves in such a way as to keep the position of the mass center unchanged. The vibrations ν_{3a}, ν_{3b}, and ν_{3c} generate a representation resulting in the next characters:

$$E\begin{bmatrix}\nu_{3a}\\ \nu_{3b}\\ \nu_{3c}\end{bmatrix}=\begin{bmatrix}\nu_{3a}\\ \nu_{3b}\\ \nu_{3c}\end{bmatrix}\Rightarrow E=\begin{bmatrix}1 & 0 & 0\\ 0 & 1 & 0\\ 0 & 0 & 1\end{bmatrix}\Rightarrow \chi(E)=3$$

$$C_3\begin{bmatrix}\nu_{3a}\\ \nu_{3b}\\ \nu_{3c}\end{bmatrix}=\begin{bmatrix}\nu_{3c}\\ \nu_{3a}\\ \nu_{3b}\end{bmatrix}\Rightarrow C_3=\begin{bmatrix}0 & 0 & 1\\ 1 & 0 & 0\\ 0 & 1 & 0\end{bmatrix}\Rightarrow \chi(C_3)=0$$

$$C_2\begin{bmatrix}\nu_{3a}\\ \nu_{3b}\\ \nu_{3c}\end{bmatrix}=\begin{bmatrix}\nu_{3a}\\ -\nu_{3b}\\ -\nu_{3c}\end{bmatrix}\Rightarrow C_2=\begin{bmatrix}1 & 0 & 0\\ 0 & -1 & 0\\ 0 & 0 & -1\end{bmatrix}\Rightarrow \chi(C_2)=-1$$

$$S_4\begin{bmatrix}\nu_{3a}\\ \nu_{3b}\\ \nu_{3c}\end{bmatrix}=\begin{bmatrix}-\nu_{3a}\\ -\nu_{3c}\\ \nu_{3b}\end{bmatrix}\Rightarrow S_4=\begin{bmatrix}-1 & 0 & 0\\ 0 & 0 & -1\\ 0 & 1 & 0\end{bmatrix}\Rightarrow \chi(S_4)=-1$$

$$\Sigma_d\begin{bmatrix}\nu_{3a}\\ \nu_{3b}\\ \nu_{3c}\end{bmatrix}=\begin{bmatrix}\nu_{3a}\\ \nu_{3c}\\ \nu_{3b}\end{bmatrix}\Rightarrow \Sigma_d=\begin{bmatrix}1 & 0 & 0\\ 0 & 0 & 1\\ 0 & 1 & 0\end{bmatrix}\Rightarrow \chi(\Sigma_d)=1$$

Therefore, the vibrations ν_{3a}, ν_{3b}, and ν_{3c} have T_2 symmetry.

In the fourth case, described by illustrations of Figures 4.8g through i, two chlorine atoms move in such a way as to increase the bending angle while the other two chlorine atoms simultaneously move to reduce the bending angle. The central atom (carbon) also moves in such a way as to keep the position of the mass center unchanged. The vibrations ν_{4a}, ν_{4b}, and ν_{4c} generate a representation resulting in the next characters:

$$E\begin{bmatrix}\nu_{4a}\\ \nu_{4b}\\ \nu_{4c}\end{bmatrix}=\begin{bmatrix}\nu_{4a}\\ \nu_{4b}\\ \nu_{4c}\end{bmatrix}\Rightarrow E=\begin{bmatrix}1 & 0 & 0\\ 0 & 1 & 0\\ 0 & 0 & 1\end{bmatrix}\Rightarrow \chi(E)=3$$

$$C_3\begin{bmatrix}\nu_{4a}\\ \nu_{4b}\\ \nu_{4c}\end{bmatrix}=\begin{bmatrix}\nu_{4c}\\ \nu_{4a}\\ \nu_{4b}\end{bmatrix}\Rightarrow C_3=\begin{bmatrix}0 & 0 & 1\\ 1 & 0 & 0\\ 0 & 1 & 0\end{bmatrix}\Rightarrow \chi(C_3)=0$$

$$C_2\begin{bmatrix}\nu_{4a}\\ \nu_{4b}\\ \nu_{4c}\end{bmatrix}=\begin{bmatrix}\nu_{4a}\\ -\nu_{4b}\\ -\nu_{4c}\end{bmatrix}\Rightarrow C_2=\begin{bmatrix}1 & 0 & 0\\ 0 & -1 & 0\\ 0 & 0 & -1\end{bmatrix}\Rightarrow \chi(C_2)=-1$$

$$S_4\begin{bmatrix}\nu_{4a}\\ \nu_{4b}\\ \nu_{4c}\end{bmatrix}=\begin{bmatrix}-\nu_{4a}\\ -\nu_{4c}\\ \nu_{4b}\end{bmatrix}\Rightarrow S_4=\begin{bmatrix}-1 & 0 & 0\\ 0 & 0 & -1\\ 0 & 1 & 0\end{bmatrix}\Rightarrow \chi(S_4)=-1$$

$$\Sigma_d \begin{bmatrix} \nu_{4a} \\ \nu_{4b} \\ \nu_{4c} \end{bmatrix} = \begin{bmatrix} \nu_{4a} \\ \nu_{4c} \\ \nu_{4b} \end{bmatrix} \Rightarrow \Sigma_d = \begin{bmatrix} 1 & 0 & 0 \\ 0 & 0 & 1 \\ 0 & 1 & 0 \end{bmatrix} \Rightarrow \chi(\Sigma_d) = 1$$

Therefore, the vibrations ν_{4a}, ν_{4b}, and ν_{4c} have T_2 symmetry, too.

The method previously described is not productive enough to find the symmetry of vibrations. For a molecule composed of only five atoms, we analyzed nine illustrations. You may imagine how much work is necessary for a bigger molecule. It is not necessary, though, to exactly know how the molecule may vibrate but only the symmetry of each normal mode to explain the spectrum molecule. So, we need a general method, that is relatively easy to be applied to a complicated molecule as well and this will be presented in Chapter 5.

4.3 DETERMINING THE SYMMETRY OF NORMAL MODES

4.3.1 METHOD OF CARTESIAN DISPLACEMENT VECTORS

Consider a molecule containing N atoms. Translation, rotation, and vibration of the molecule can be approximated by considering only nuclear motions (the Born–Oppenheimer approximation) because most of the mass resides in the nuclei. Each nucleus possesses 3 degrees of freedom in a three-dimensional space, and thus all nuclei together have $3N$ degrees of freedom. Three of these degrees describe translational motion of the mass center and three more degrees of freedom describe rotations of the molecular complex around the center of mass. The symmetry of translations, rotations, and vibrations can be determined by considering the character of the representation generated by *Cartesian vectors localized on each atom.*

For example, let us consider again the CCl_4 molecule of Figure 4.9. We attach a set of three Cartesian vectors to each atom to determine the reducible representation generated by normal modes; therefore, in total, $3N = 15$ vectors for the CCl_4 molecule. The choice of the coordinate system is arbitrary. The rigid molecule belongs to the $\boldsymbol{T}_d$ group. The operations of the $\boldsymbol{T}_d$ group transform the x, y, z vectors into themselves with a $\pm$ sign or else into one another. To compute the character of reducible representation spanned by the 15 coordinates, we need to sum only the diagonal elements of the 15-dimensional transformation matrix.

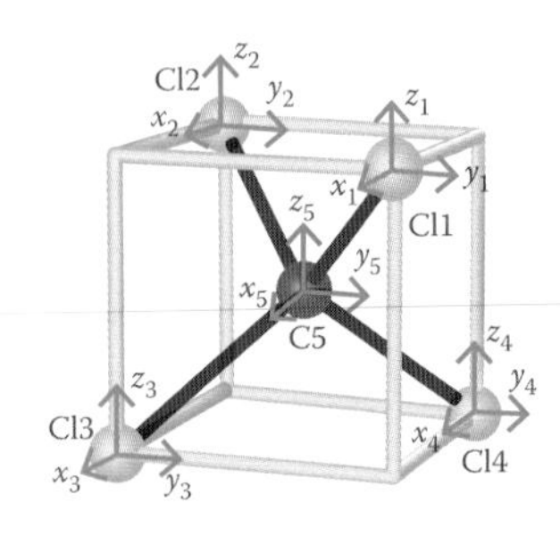

FIGURE 4.9 To determine the representation of normal modes, we attach a set of three Cartesian vectors to each atom; therefore, in total, $3N = 15$ vectors for the carbon tetrachloride molecule.

We start to apply the identity operation E and the results are shown in Table 4.1. The top row lists the original vectors and the left column lists the resulting vectors after application of symmetry operation. The identity operation leaves unchanged each vector and this will give 1. Thus, the identity operation will have the character 15.

We now apply a threefold rotation to the set of Cartesian vectors with the results illustrated in Figure 4.10. The rotation C_3 is performed around one principal diagonal of the cube or, in other words, around one bond C–Cl. We must again construct the matrix expressing these results but the procedure requires more attention. The symmetry operation C_3 rotates all the vectors associated with atom C5 in such a way that the new vectors

TABLE 4.1 Matrix Expressing the Effect of the Identity Operation on 15 Cartesian Coordinates for CCl_4

E	x_1	y_1	z_1	x_2	y_2	z_2	x_3	y_3	z_3	x_4	y_4	z_4	x_5	y_5	z_5
x_1'	1	0	0	0	0	0	0	0	0	0	0	0	0	0	0
y_1'	0	1	0	0	0	0	0	0	0	0	0	0	0	0	0
y_1'	0	0	1	0	0	0	0	0	0	0	0	0	0	0	0
x_2'	0	0	0	1	0	0	0	0	0	0	0	0	0	0	0
y_2'	0	0	0	0	1	0	0	0	0	0	0	0	0	0	0
z_2'	0	0	0	0	0	1	0	0	0	0	0	0	0	0	0
x_3'	0	0	0	0	0	0	1	0	0	0	0	0	0	0	0
y_3'	0	0	0	0	0	0	0	1	0	0	0	0	0	0	0
z_3'	0	0	0	0	0	0	0	0	1	0	0	0	0	0	0
x_4'	0	0	0	0	0	0	0	0	0	1	0	0	0	0	0
y_4'	0	0	0	0	0	0	0	0	0	0	1	0	0	0	0
z_4'	0	0	0	0	0	0	0	0	0	0	0	1	0	0	0
x_5'	0	0	0	0	0	0	0	0	0	0	0	0	1	0	0
y_5'	0	0	0	0	0	0	0	0	0	0	0	0	0	1	0
z_5'	0	0	0	0	0	0	0	0	0	0	0	0	0	0	1

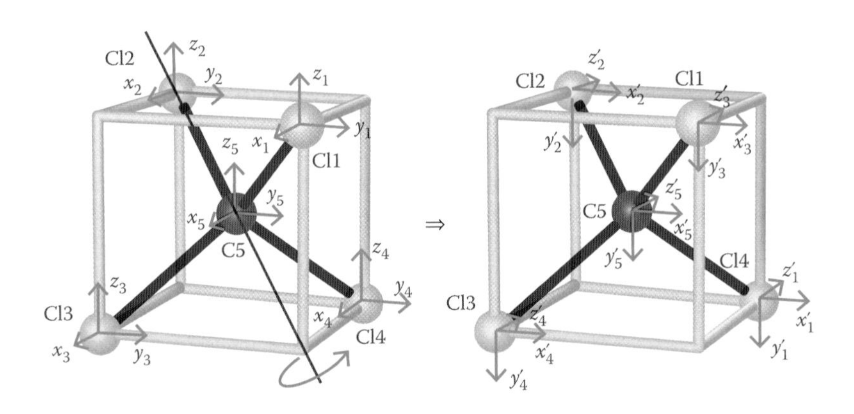

FIGURE 4.10 The effect of the C_3 rotation around the axis that passes through atoms Cl2 and C5 on Cartesian vectors attached to all atoms.

are equal to $x_5' = y_5$, $y_5' = -z_5$, and $z_5' = -x_5$. We find similar results in the transformation of vectors of the atom Cl2: $x_2' = y_2$, $y_2' = -z_2$, and $z_2' = -x_2$. The same operation C_3 moves all the vectors originally associated with atom Cl3 to atom Cl1. Thus, we may write the new vectors of the Cl1 atom as $x_3' = y_1$, $y_3' = -z_1$, and $z_3' = -x_1$. We can continue in the same way for atoms Cl1 and Cl4 and the final results are summarized by the matrix of the C_3 operation shown in Table 4.2.

Thus, the matrix generated by the C_3 operation has the character equal to 0. The result was expected because the symmetry operation moves all

TABLE 4.2 Matrix Expressing the Effect of the Rotation C_3 on 15 Cartesian Coordinates for CCl_4

C_3	x_1	y_1	z_1	x_2	y_2	z_2	x_3	y_3	z_3	x_4	y_4	z_4	x_5	y_5	z_5
x_1'	0	0	0	0	0	0	0	0	0	0	1	0	0	0	0
y_1'	0	0	0	0	0	0	0	0	0	0	0	−1	0	0	0
z_1'	0	0	0	0	0	0	0	0	0	−1	0	0	0	0	0
x_2'	0	0	0	0	1	0	0	0	0	0	0	0	0	0	0
y_2'	0	0	0	0	0	−1	0	0	0	0	0	0	0	0	0
z_2'	0	0	0	−1	0	0	0	0	0	0	0	0	0	0	0
x_3'	0	1	0	0	0	0	0	0	0	0	0	0	0	0	0
y_3'	0	0	−1	0	0	0	0	0	0	0	0	0	0	0	0
z_3'	−1	0	0	0	0	0	0	0	0	0	0	0	0	0	0
x_4'	0	0	0	0	0	0	0	1	0	0	0	0	0	0	0
y_4'	0	0	0	0	0	0	0	0	−1	0	0	0	0	0	0
z_4'	0	0	0	0	0	0	−1	0	0	0	0	0	0	0	0
x_5'	0	0	0	0	0	0	0	0	0	0	0	0	0	1	0
y_5'	0	0	0	0	0	0	0	0	0	0	0	0	0	0	−1
z_5'	0	0	0	0	0	0	0	0	0	0	0	0	−1	0	0

vectors from their original positions and it results in the fact that all diagonal elements are zero. Now, we can conclude for the next inspection that we could determine the character of the matrix by *ignoring all the vectors that are shifted from one atom to another.*

When the molecule is subjected to the next symmetry operation, rotation C_2 that passes through the atom C5 and is parallel to the z axis, the vectors of Cl1–Cl2 and Cl3–Cl4 interchange and they do not make any contribution to the character. Thus, we write zero on the diagonal. For the central atom C5, the x_5 and y_5 vectors *are transformed into their own negatives* while the z_5 vector *remains unchanged.* These results are expressed in the matrix shown in Table 4.3, in which only the elements relating to the vectors on atom C5 are given. The value of the character equals –1.

The improper rotation axis S_4 also passes through the atom C5 and is parallel to the z axis. The reflection plane is horizontal and passes through C5 too. When the molecule is subjected to the improper rotation S_4, all the vectors of the Cl1, Cl2, Cl3, and Cl4 atoms are shifted and they do not make any contribution to the matrix character. Concerning the central atom C5, x_5 becomes y_5 and y_5 becomes $-x_5$, while the z_5 vector is transformed into its negative. These results are expressed in the matrix shown in Table 4.4, in which only the z_5 vector appears on the diagonal with –1, and thus the value of the character equals –1.

The last symmetry operation is the dihedral reflection Σ_d through the plane that passes through the atoms C5, Cl1, and Cl2 as illustrated in Figure 4.11. The plane is parallel to the z axis and leaves unchanged the

TABLE 4.3 Simplified Matrix Expressing the Effect of the Rotation C_2 on 15 Cartesian Coordinates for CCl_4

C_2	x_1	...	x_5	y_5	z_5
x_1'	0		0	0	0
...		...			
x_5'	0		−1	0	0
y_5'	0		0	−1	0
z_5'	0		0	0	1

TABLE 4.4 Simplified Matrix Expressing the Effect of the Improper Rotation S_4 on 15 Cartesian Coordinates for CCl_4

S_4	x_1	...	x_5	y_5	z_5
x_1'	0				
...		...			
x_5'			0	1	0
y_5'			−1	0	0
z_5'			0	0	−1

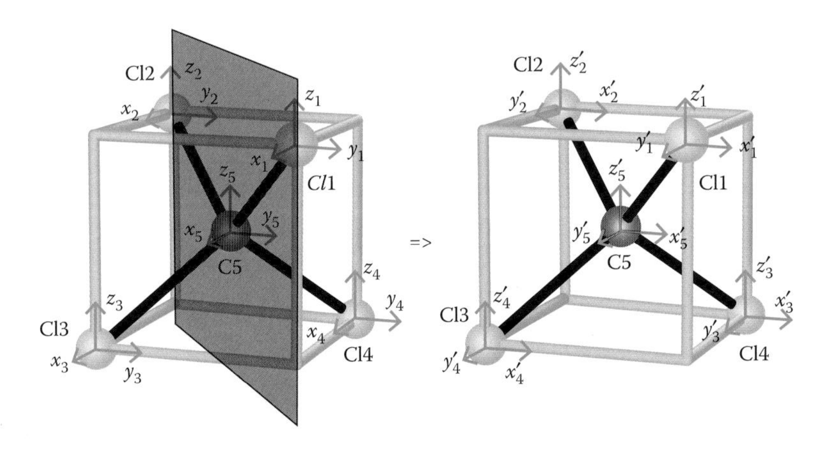

FIGURE 4.11 Illustration of the effect of the reflection through the dihedral plane.

TABLE 4.5 Simplified Matrix Expressing the Effect of the Reflection Σ_d on 15 Cartesian Coordinates for CCl_4

Σ_d	...	z_1	...	z_2	...	z_5
...	0	...	...	...	...	...
z_1'	...	1	...	...	...	...
...	...	...	0	...	...	...
z_2'	...	...	...	1	...	...
...	...	...	...	...	0	...
z_5'	...	...	...	...	...	1

z vectors of atoms 1, 2, and 5, and therefore z_1, z_2, and z_5. The x and y vectors of the same atoms are interchanged, for example, $x_5' = y_5, y_5' = x_5$, and so on. They give zero to the diagonal elements and do not contribute to the matrix character as Cl3 and Cl4 do. The results are expressed in the matrix shown in Table 4.5. The value of the character equals 3.

We may now append all the values obtained for characters of the reducible representation given by 15 Cartesian vectors to the character table of the $\boldsymbol{T}_d$ group as shown in Table 4.6.

The representation Γ_{3N} may be reduced to the irreducible representations of the groups as follows:

$$\Gamma_{3N} = A_1 + E + T_1 + 3T_2 \tag{4.7}$$

We know that the molecule has three translations, three rotations, and nine pure normal modes of vibrations. Looking at the last two columns of the character table, we can see that the translations represented by vectors (x, y, z) are the basis of T_2 representation and the rotations (R_x, R_y, R_z) are the basis of the T_1 representation. Finally, we conclude that the normal modes of vibrations generate the representation Γ_{nmv} reducible to

$$\Gamma_{nmv} = A_1 + E + 2T_2 \tag{4.8}$$

TABLE 4.6 Character Table of the T_d Group with the Characters of the Reducible Representation Generated by the Displacement Vectors of the CCl_4 Molecule

T_d	E	$8C_3$	$3C_2$	$6S_4$	$6\Sigma_d$	Linear Basis	Quadratic Basis
A_1	1	1	1	1	1		$x^2 + y^2 + z^2$
A_2	1	1	1	−1	−1		
E	2	−1	2	0	0		$(2z^2 - x^2 - y^2, x^2 - y^2)$
T_1	3	0	−1	1	−1	(R_x, R_y, R_z)	
T_2	3	0	−1	−1	1	(x, y, z)	(xy, xz, yz)
Γ_{3N}	15	0	−1	−1	3		

Therefore, we have found the symmetry of normal modes of the carbon tetrachloride molecule that can be A_1, E, and T_2.

Note that the previous discussion and results are valuable for *any AB_4 molecule or molecular complex with $\boldsymbol{T}_d$ symmetry,* for example, methane, CH_4, $SiCl_4$, CF_4, NH_4^+, and SO_4^{2-}. Thus, the infrared absorption spectra or Raman spectra of this class must have the same number of bands but with different positions (energies), because the mass ratio between the central atom and outer atoms differ and change the frequency. Be careful that not any AB_4 molecule has tetrahedral symmetry, for example, XeF_4, $PtCl_4^{2-}$ that have a *planar* quadratic symmetry ($\boldsymbol{D}_{4h}$)!

The exposed procedure to find the symmetry of normal modes of vibration is good for any type of molecule:

1. Attach a set of Cartesian displacement vectors to each atom of the molecule.
2. Apply each symmetry operation of the symmetry group of the molecule to the set of Cartesian vectors.
3. Take into consideration only vectors left to the original atom.
4. The vector left unchanged will give +1, the vector reversed will give −1, the vectors interchanged (e.g., $\pm x \leftrightarrow \pm y$) will give 0, and the vector expressed as a linear combination of the original vectors will give the coefficient of its corresponding original vector (e.g., $x' = \mathbf{0.7}x + 0.7y$, $y' = -0.7x + \mathbf{0.7}y$).
5. Write the matrix generated by each symmetry operation using only nonzero diagonal elements.
6. Calculate the character of the matrix.
7. Decompose the $3N$ representation obtained to irreducible representations of the group.
8. Eliminate the representations generated by translations and rotations.

4.3.2 METHOD OF INTERNAL COORDINATES

In solving the problem for the vibrations of a molecule, the *internal coordinates* are the most suitable. They are also called natural vibrational coordinates. They represent *changes of bond lengths, interbond angles, out-of-plane angles, and torsional angles* (Galabov and Dudev 1996). The method of using the internal coordinates has the advantage of understanding better the intuitive notions of bond stretching and bending. In other words, we know the correspondence between each type of vibrational motion of the molecule and its symmetry. If you remember the expression of the potential energy of the harmonic oscillator, where the internal coordinate is directly related to the energy $E_p = (1/2)k \cdot \Delta x^2$, you may understand why the use of internal coordinates offers considerable advantages from a physical point of view because the force constant characterizes the bonding force between atoms $F = k \cdot \Delta x$ (Galabov and Dudev 1996). The force constant is dependent on the electronic structural parameters of molecular subunits. The method of atomic Cartesian displacement coordinates gives only the number and symmetry of normal modes but no information about the atoms motion, and how they move relative to each other. Unlike the Cartesian coordinates, the second method has a disadvantage because the internal coordinates are not always orthogonal and linearly independent. For example, for a planar molecule, the sum of angles must always be 360° so that they are not all independent; therefore, the value of one angle of the XeF_4 molecule is always determined by the values of the other three angles. The angle between any two bonds of the CCl_4 molecule is around 109° and any increase in the value of five angles will determine the decrease in the value of the sixth angle. We must be careful to choose the internal coordinate motions in such a way as to ensure that the *mass center remains fixed.* If the mass center shifts, then the motion will not be a pure vibration but will contain some translational motion as well.

Let us take the CCl_4 molecule again as an example. We assume that $3 \cdot 5 - 6 = 9$ independent internal coordinates, such as a set of *changes* in valence bond lengths and interbond angles, have to be chosen. We may use the *stretching value* (not the length) *of each bond* C–Cl as the internal coordinate; so, there is a first set of four internal displacement coordinates as represented in Figure 4.12. The second set is made by the *increases in the five interbond Cl–C–Cl angles.* They can be treated separately because, under the action of each symmetry operation, the members of each set transform only among themselves. First, we must find the representation generated by the set of coordinates for C–Cl bond stretching, d_1, d_2, d_3, and d_4. (For simplicity, we denote by d_1 the stretching value Δr_1 of bonding length C–Cl1.) We use the experience gained with the method of Cartesian vectors. It means that any vector shifted from its position will not contribute to the character of the matrix and we are looking only at the vectors left on their original atom. Thus, the matrix of identity operation E has the character 4. The rotation C_3 around the C–Cl bond will have the character 1 (d_1 left on its position). The rotation C_2 shifts all vectors and will then give the character 0. The improper rotation S_4 has a similar effect and the character 0. The reflection Σ_d leaves unchanged the vectors d_1 and d_2 and will give the character 2. The representation generated by the action of symmetry operations on the set of stretching vectors is shown in Table 4.7.

Thus, the representation reduces to $\Gamma_{stretch} = A_1 + T_2$. It means that the normal mode of symmetry A_1 involves the totally symmetric C–Cl stretching. The representation T_2 is threefold degenerated and means there are three normal modes with the same symmetry that involves C–Cl stretching. For the degenerate T_2 stretching vibration, some motion of the central carbon atom must be added so that the center of the mass remains fixed.

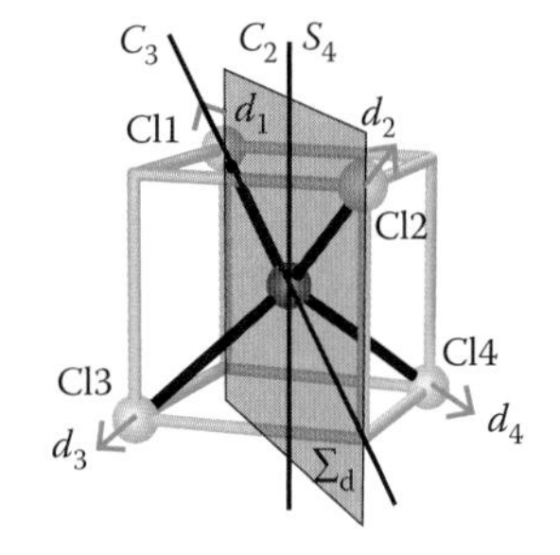

FIGURE 4.12 The choice of stretching displacement as internal coordinates for the CCl_4 molecule.

In CCl_4, the complete set of symmetrically equivalent *angular coordinates* consists of six Cl–C–Cl angle increments: δ_{12} (the increment of the interbond Cl1–C–Cl2 angle from Figure 4.12), δ_{13}, δ_{14}, δ_{23}, δ_{24}, and δ_{34}.

TABLE 4.7 Characters of the Stretching of C–Cl Bond

T_d	E	C_3	C_2	S_4	Σ_d
$\Gamma_{stretch}$	4	1	0	0	2

TABLE 4.8 Characters of Representation of the Bending of Cl–C–Cl Bonds

T_d	E	C_3	C_2	S_4	Σ_d
Γ_{bend}	6	0	2	0	2

(Again, for simplicity, we denote by δ_{12} the increment value $\Delta\alpha_{12}$ of the interbond angle Cl1–C–Cl2 and so on.) The angular coordinates of the set are not independent of geometrical considerations. It is not possible for all six angles to increase by the same amount at the same time. If the sum of angle increments always equals zero, then the set is redundant. This means that the (totally symmetric) A_1 representation that we would find by reducing the representation is spurious. We have two ways to proceed:

- It is possible to exclude one of the angles arbitrarily
- To retain the complete set and deal with the redundancy later (more convenient)

Following the same procedure, we will find the reducible representation shown in Table 4.8.

Thus, the representation reduces to $\Gamma_{bend} = A_1 + E + T_2$. It means that the normal mode of symmetry A_1 involves the totally symmetric Cl–C–Cl bending. The A_1 representation is spurious and should be discarded and there remains the twofold degenerate E representation and the threefold degenerate T_2 representation. Hence, it means that there are

- *Two normal modes* with the same symmetry that involves *symmetric Cl–C–Cl bending*
- *Three normal modes* with the same symmetry that involves *asymmetric Cl–C–Cl bending*

Note that for more complicated molecules, it is recommended that the symmetries of normal modes should be found by using the Cartesian coordinates to recognize the redundancy in advance.

The internal coordinates are used to visualize the normal modes in terms of stretching, bending, and so on (Majumder and Manogaran 2013).

4.4 NORMAL COORDINATES AND WAVE FUNCTIONS

4.4.1 SECULAR EQUATION (DETERMINATION OF FREQUENCY OF NORMAL MODES OF VIBRATION)

It is easier to understand a solution of the problem of small vibrations in classical mechanics treatment and so we present it first. In Newtonian mechanics, the motion of the linear harmonic oscillator is described by the equation

$$s_i(t) = s_i^0 \cos(2\pi\nu t + \varphi) \quad (4.9)$$

where s_i is the position of the i particle reported to the equilibrium position, s_i^0 is the amplitude, and φ is the initial phase. We can deduce the acceleration of the particle as

$$a_i = \frac{d^2 s_i}{dt^2} = -4\pi^2\nu^2 s_i \quad (4.10)$$

Then, using the second law of mechanics, we can calculate the force (elastic force):

$$F_i = -4\pi^2\nu^2 m_i s_i = -k_i s_i \quad (4.11)$$

k_i is called the *force constant*.

For a system of N particles, if a particle is moved from its equilibrium position, then an elastic force F^1 will act to bring the particle back to the equilibrium position. The elastic force is proportional to the displacement from equilibrium position. By using a system of Cartesian coordinates located on the particle, we can write the system of equations:

$$\begin{cases} F_x^1 = -k_{xx}^{11}x_1 - k_{xy}^{11}y_1 - k_{xz}^{11}z_1 \\ F_y^1 = -k_{yx}^{11}x_1 - k_{yy}^{11}y_1 - k_{yz}^{11}z_1 \\ F_z^1 = -k_{zx}^{11}x_1 - k_{zy}^{11}y_1 - k_{zz}^{11}z_1 \end{cases} \quad (4.12)$$

If all particles move from their equilibrium position when we write the system of equations, we should take into consideration the interaction between all particles as follows:

(4.13)

$$\begin{cases} F_x^1 = -k_{xx}^{11}x_1 - k_{xy}^{11}y_1 - k_{xz}^{11}z_1 - k_{xx}^{12}x_2 - k_{xy}^{12}y_2 - k_{xz}^{12}z_2 - \cdots - k_{xz}^{1N}z_N \\ F_y^1 = -k_{yx}^{11}x_1 - k_{yy}^{11}y_1 - k_{yz}^{11}z_1 - k_{yx}^{12}x_2 - k_{yy}^{12}y_2 - k_{yz}^{12}z_2 - \cdots - k_{yz}^{1N}z_N \\ F_z^1 = -k_{zx}^{11}x_1 - k_{zy}^{11}y_1 - k_{zz}^{11}z_1 - k_{zx}^{12}x_2 - k_{zy}^{12}y_2 - k_{zz}^{12}z_2 - \cdots - k_{zz}^{1N}z_N \\ F_x^2 = -k_{xx}^{21}x_1 - k_{xy}^{21}y_1 - k_{xz}^{21}z_1 - k_{xx}^{22}x_2 - k_{xy}^{22}y_2 - k_{xz}^{22}z_2 - \cdots - k_{xz}^{2N}z_N \\ F_y^2 = -k_{yx}^{21}x_1 - k_{yy}^{21}y_1 - k_{yz}^{21}z_1 - k_{yx}^{22}x_2 - k_{yy}^{22}y_2 - k_{yz}^{22}z_2 - \cdots - k_{yz}^{2N}z_N \\ F_z^2 = -k_{zx}^{21}x_1 - k_{zy}^{21}y_1 - k_{zz}^{21}z_1 - k_{zx}^{22}x_2 - k_{zy}^{22}y_2 - k_{zz}^{22}z_2 - \cdots - k_{zz}^{2N}z_N \\ \vdots \\ F_z^N = -k_{zx}^{N1}x_1 - k_{zy}^{N1}y_1 - k_{zz}^{N1}z_1 - k_{zx}^{N2}x_2 - k_{zy}^{N2}y_2 - k_{zz}^{N2}z_2 - \cdots - k_{zz}^{NN}z_N \end{cases}$$

The force constant k_{xy}^{ij} shows the dependence of the component on x direction of elastic force F_x^i, which acts on particle i, and of the component on y direction of elastic force F_y^j which acts on particle j. The force constants have a helpful relationship $k_{xy}^{ij} = k_{yx}^{ji}$.

The displacements of all particles must be synchronized in a harmonic oscillation of the entire molecule with a common frequency that is a *normal mode of vibration*. The equations system should be written taking into account Equation 4.11 as follows:

(4.14)

$$\begin{cases}
4\pi^2\nu^2 m_1 x_1 = +k_{xx}^{11}x_1 + k_{xy}^{11}y_1 + k_{xz}^{11}z_1 + k_{xx}^{12}x_2 + k_{xy}^{12}y_2 + k_{xz}^{12}z_2 + \cdots + k_{xz}^{1N}z_N \\
4\pi^2\nu^2 m_1 y_1 = +k_{yx}^{11}x_1 + k_{yy}^{11}y_1 + k_{yz}^{11}z_1 + k_{yx}^{12}x_2 + k_{yy}^{12}y_2 + k_{yz}^{12}z_2 + \cdots + k_{yz}^{1N}z_N \\
4\pi^2\nu^2 m_1 z_1 = +k_{zx}^{11}x_1 + k_{zy}^{11}y_1 + k_{zz}^{11}z_1 + k_{zx}^{12}x_2 + k_{zy}^{12}y_2 + k_{zz}^{12}z_2 + \cdots + k_{zz}^{1N}z_N \\
4\pi^2\nu^2 m_2 x_2 = +k_{xx}^{21}x_1 + k_{xy}^{21}y_1 + k_{xz}^{21}z_1 + k_{xx}^{22}x_2 + k_{xy}^{22}y_2 + k_{xz}^{22}z_2 + \cdots + k_{xz}^{2N}z_N \\
4\pi^2\nu^2 m_2 y_2 = +k_{yx}^{21}x_1 + k_{yy}^{21}y_1 + k_{yz}^{21}z_1 + k_{yx}^{22}x_2 + k_{yy}^{22}y_2 + k_{yz}^{22}z_2 + \cdots + k_{yz}^{2N}z_N \\
4\pi^2\nu^2 m_2 z_2 = +k_{zx}^{21}x_1 + k_{zy}^{21}y_1 + k_{zz}^{21}z_1 + k_{zx}^{22}x_2 + k_{zy}^{22}y_2 + k_{zz}^{22}z_2 + \cdots + k_{zz}^{2N}z_N \\
\vdots \\
4\pi^2\nu^2 m_N z_N = +k_{zx}^{N1}x_1 + k_{zy}^{N1}y_1 + k_{zz}^{N1}z_1 + k_{zx}^{N2}x_2 + k_{zy}^{N2}y_2 + k_{zz}^{N2}z_2 + \cdots + k_{zz}^{NN}z_N
\end{cases}$$

The system of linear equations has a solution only if its determinant is zero that is called as the *secular equation*; therefore:

(4.15)

$$\begin{vmatrix}
k_{xx}^{11} - 4\pi^2\nu^2 m_1 & k_{xy}^{11} & k_{xz}^{11} & k_{xx}^{12} & \cdots & k_{xz}^{1N} \\
k_{yx}^{11} & k_{yy}^{11} - 4\pi^2\nu^2 m_1 & k_{yz}^{11} & k_{yx}^{12} & \cdots & k_{yz}^{1N} \\
k_{zx}^{11} & k_{zy}^{11} & k_{zz}^{11} - 4\pi^2\nu^2 m_1 & k_{zx}^{12} & \cdots & k_{zz}^{1N} \\
k_{xx}^{21} & k_{xy}^{21} & k_{xz}^{21} & k_{xx}^{22} - 4\pi^2\nu^2 m_2 & \cdots & k_{xz}^{2N} \\
\vdots & \vdots & \vdots & \vdots & \vdots & \vdots \\
k_{zx}^{N1} & k_{zy}^{N1} & k_{zz}^{N1} & k_{zx}^{N2} & \cdots & k_{zz}^{NN} - 4\pi^2\nu^2 m_N
\end{vmatrix} = 0$$

The secular equation has $3N$ roots for the variable ν^2. The eigenvalues ν_i^2 should be zero or positive. There are five roots equal to zero for linear molecules or six roots equal to zero for nonlinear molecules. These roots correspond to three modes of translational motion and three modes of rotational motion. If two or more roots have the same value of frequency, it means they are *degenerate frequencies*. There are only two possibilities: double degeneracy and triple degeneracy (the largest value of degeneracy we can find in the molecular symmetry groups). A molecule with high symmetry will have many degenerate roots.

Using a more convenient set of coordinates $q_1, q_2, \ldots, q_{3N}$ that are related to the Cartesian displacement coordinates by the relationships:

$$\begin{cases} q_1 = \sqrt{m_1}\,x_1 \\ q_2 = \sqrt{m_1}\,y_1 \\ q_3 = \sqrt{m_1}\,z_1 \\ q_4 = \sqrt{m_2}\,x_2 \\ \vdots \\ q_{3N} = \sqrt{m_N}\,z_{3N} \end{cases} \tag{4.16}$$

and the force constants derived from the potential energy of the molecule U in equilibrium position as follows:

$$f_{ij} = \left(\frac{\partial^2 U}{\partial q_i \partial q_j}\right)_0 \tag{4.17}$$

Bright Wilson et al. (1955) proposed a simpler version of the secular equation that is written as

$$\begin{vmatrix} f_{11} - \lambda & f_{12} & f_{13} & \cdots & f_{1,3N} \\ f_{21} & f_{22} - \lambda & f_{23} & \cdots & f_{2,3N} \\ \vdots & \vdots & \vdots & \vdots & \vdots \\ f_{3N,1} & f_{3N,2} & f_{3N,3} & \cdots & f_{3N,3N} - \lambda \end{vmatrix} = 0 \tag{4.18}$$

where the symbol λ replaces the frequency ν ($\lambda = \sqrt{2\pi\nu}$).

Whatever the form of writing, the secular equation is fundamental in the study of molecular vibrations. For the given values of force constants, the secular equation (Equations 4.15 or 4.18) will give the frequencies of normal modes. Then these values substituted in the law of motion are written as the system from Equation 4.14 or compacted as $\Sigma_{i=1}^{3N}(f_{ij} - \delta_{ij}\lambda)A_i = 0$, where $j = 1, 2, \ldots 3N$, can be useful to determine the value and direction of each displacement vector, and therefore the symmetry of each normal mode of vibration. There are a few simple molecules where the force constants can be theoretically calculated to be then used in frequency calculation. Usually, we try to find the force constants from the fitting of the calculated frequencies to the measured frequencies. Imagine how complicated the secular equation will be for a bigger polyatomic molecule. The use of group theory is recommended to determine the symmetry of normal modes.

4.4.2 NORMAL COORDINATES

By solving the previous system of equations, one can obtain the ratios between displacement vectors for a given value of frequency: $x_1{:}y_1{:}z_1{:}x_2{:}y_2{:}\ \ldots z_N$. The factors of proportionality between these ratios are called the *normal coordinates.* The normal coordinates are used in the *quantum treatment of molecular vibrations.* Each normal mode of vibration has to be associated with one normal coordinate. The normal coordinates can be written as linear combinations of other types of coordinates: Cartesian or internal. The set of normal coordinates is good to describe both kinetic energy (that normally depends on velocity expressed in terms of Cartesian coordinates) and potential energy (that normally depends on displacement of internuclear bond length and interbond angles in terms of internal coordinates) (Bunker and Jensen 1998).

The normal coordinates have two important properties:

- They must respect the orthonormalization condition: $\xi_i \cdot \xi_j = \delta_{ij}$
- Each of them must form a basis of a representation of the symmetry group of the molecule

4.4.2.1 CO_2 EXAMPLE

Let us take as the first example the case of the simple molecule of CO_2. The number of normal coordinates will be $3 \cdot 3 - 5 = 4$, that is, the same as the number of normal modes. We first write the normal coordinates as linear combinations of Cartesian displacement coordinates represented in Figure 4.13. To simplify the writing of formulas, we may give up the symbol Δ that expresses the variation of the coordinate and just simply write the coordinate.

(4.19)

$$\begin{cases} \xi_1 = r_{11}x_1 + r_{12}y_1 + r_{13}z_1 + r_{14}x_2 + r_{15}y_2 + r_{16}z_2 + r_{17}x_3 + r_{18}y_3 + r_{19}z_3 \\ \xi_2 = r_{21}x_1 + r_{22}y_1 + r_{23}z_1 + r_{24}x_2 + r_{25}y_2 + r_{26}z_2 + r_{27}x_3 + r_{28}y_3 + r_{29}z_3 \\ \xi_3 = r_{31}x_1 + r_{32}y_1 + r_{33}z_1 + r_{34}x_2 + r_{35}y_2 + r_{36}z_2 + r_{37}x_3 + r_{38}y_3 + r_{39}z_3 \\ \xi_4 = r_{41}x_1 + r_{42}y_1 + r_{43}z_1 + r_{44}x_2 + r_{45}y_2 + r_{46}z_2 + r_{47}x_3 + r_{48}y_3 + r_{49}z_3 \end{cases}$$

Alternatively, it can be shortly written as

(4.20)

$$\xi_i = \sum_{j=1}^{3N} r_{ij}x_j$$

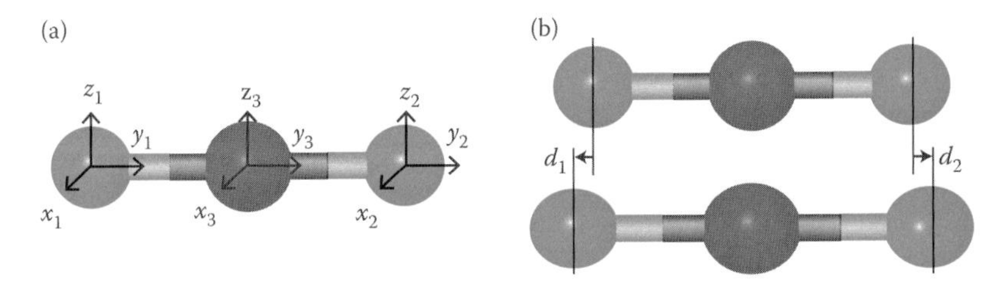

FIGURE 4.13 (a) The Cartesian displacement coordinates and (b) the internal (stretching) coordinates of the carbon dioxide molecule.

Finding so many coefficients is a heavy task that can be simplified by writing them in terms of internal coordinates; so, we can write:

$$\xi_1 = d_1 + d_2 + \delta_{xy} + \delta_{yz}$$

δ_{xy} and δ_{yz} are the increasing bending angles of the molecule in xy and yz planes, respectively. They are not illustrated in Figure 4.13. However, generally, stretching modes and bending modes are separate and we can write the first normal coordinate as a symmetric combination:

$$\xi_1 = d_1 + d_2 \tag{4.21}$$

In this simple case of CO_2, it is obvious that the second normal coordinate will be written as asymmetric:

$$\xi_1 = d_1 - d_2 \tag{4.22}$$

In terms of Cartesian displacement coordinates, we know that

$$\begin{cases} d_1 = -y_1 \\ d_2 = y_2 \end{cases}$$

Then

$$\begin{cases} \xi_1 = -y_1 + y_2 \\ \xi_2 = -y_1 - y_2 \end{cases} \tag{4.23}$$

The found signs of coordinates are not too comfortable to use because we have chosen the expanding internal coordinates instead of the compression internal coordinates. All mathematical treatments can be reinitiated starting from the compression internal coordinates, but the final results will not change. The normalization condition will give both normalization constants:

$$\begin{cases} 1 = \xi_1\xi_1 = N_1^2(-y_1 + y_2)(-y_1 + y_2) = N_1^2(y_1^2 + y_2^2 - 2y_1y_2) \\ \quad = 2N_1^2 \Rightarrow N_1 = \dfrac{1}{\sqrt{2}} \\ 1 = \xi_2\xi_2 = N_2^2(-y_1 - y_2)(-y_1 - y_2) = N_2^2(y_1^2 + y_2^2 + 2y_1y_2) \\ \quad = 2N_2^2 \Rightarrow N_2 = \dfrac{1}{\sqrt{2}} \end{cases}$$

Here, we took into consideration the orthonormalization of Cartesian coordinates, too. So, finally, the normal coordinates will be:

$$\begin{cases} \xi_1 = \dfrac{1}{\sqrt{2}}(-y_1 + y_2) \\ \xi_2 = \dfrac{1}{\sqrt{2}}(-y_1 - y_2) \end{cases} \tag{4.24}$$

In a similar way, we can write the normal coordinates as linear combinations of internal (bending) coordinates:

$$
\begin{cases}
\xi_3 = \dfrac{1}{\sqrt{2}}(\delta_{xy} + \delta_{yz}) \\
\xi_4 = \dfrac{1}{\sqrt{2}}(\delta_{xy} - \delta_{yz})
\end{cases}
\tag{4.25}
$$

The next step is to find the symmetry of each normal coordinate; therefore, what representation can generate each normal coordinate under all symmetry operations of the $\boldsymbol{D}_{\infty h}$ group? The effect of the symmetry operations on the Cartesian coordinates and then on normal coordinates is shown in Table 4.9.

TABLE 4.9 Effect of the Symmetry Operations on Components of the Normal Coordinates for the CO_2 Molecule

E	**y_1**	**y_2**	**$-y_1+y_2$**	**$-y_1-y_2$**	**δ_{xy}**	**δ_{yz}**	**$(\delta_{xy}, \delta_{yz})$**
	y_1	y_2			δ_{xy}	δ_{yz}	2
$-y_1+y_2$			1				
$-y_1-y_2$				1			
$C_\infty^\varphi(y)$							
	y_1	y_2			$\delta_{xy}\cos\varphi$	$\delta_{yz}\cos\varphi$	$\begin{vmatrix}\cos\varphi & 0 \\ 0 & \cos\varphi\end{vmatrix} = 2\cos\varphi$
$-y_1+y_2$			1				
$-y_1-y_2$				1			
$C_2(z)$							
	$-y_2$	$-y_1$			δ_{xy}	$-\delta_{yz}$	$\begin{vmatrix}1 & 0 \\ 0 & -1\end{vmatrix} = 0$
$+y_2-y_1$			1				
$+y_2+y_1$				−1			
I							
	$-y_2$	$-y_1$			$-\delta_{xy}$	$-\delta_{yz}$	$\begin{vmatrix}-1 & 0 \\ 0 & -1\end{vmatrix} = -2$
$+y_2-y_1$			1				
$+y_2+y_1$				−1			
Σ_v							
	y_1	y_2			$-\delta_{xy}$	δ_{yz}	$\begin{vmatrix}-1 & 0 \\ 0 & 1\end{vmatrix} = 0$
$-y_1+y_2$			1				
$-y_1-y_2$				1			
S_∞^φ							
	$-y_2$	$-y_1$			$\delta_{xy}\cos\varphi$	$\delta_{yz}\cos\varphi$	$\begin{vmatrix}\cos\varphi & 0 \\ 0 & \cos\varphi\end{vmatrix} = 2\cos\varphi$
$+y_2-y_1$			1				
$+y_2+y_1$				−1			

TABLE 4.10 Irreducible Representations Generated by Normal Coordinates for the CO_2 Molecule

$D_{\infty h}$	E	$2C_{\infty}^{\varphi}$	$\infty\Sigma_v$	I	$2S_{\infty}^{\varphi}$	∞C_2	
Σ_g^+	1	1	1	1	1	1	$\xi_1 = -y_1 + y_2 = d_1 + d_2$
Σ_u^+	1	1	1	−1	−1	−1	$\xi_2 = -y_1 - y_2 = d_1 - d_2$
Π_u	2	2cos φ	0	−2	2cos φ	0	$\xi_3 = \frac{1}{\sqrt{2}}(\delta_{xy} + \delta_{yz}),$
							$\xi_4 = \frac{1}{\sqrt{2}}(\delta_{xy} - \delta_{yz})$

By simple inspection of the characters, we conclude:

- The normal coordinate ξ_1 generates the totally symmetric representation Σ_g^+ (as expected)
- The normal coordinate ξ_2 generates the asymmetric representation Σ_u^+
- The pair of normal coordinates (ξ_3, ξ_4) generates the doubly degenerate representation Π_u

All results are presented in Table 4.10.

Thus, the first conclusion is: The *CO_2 molecule has four modes of vibration but only three frequencies.* How they manifest in the spectrum will be presented later.

4.4.2.2 CCl_4 EXAMPLE

As a second and more complicated example, let us continue with the carbon tetrachloride molecule. The task is to establish the expression and symmetry of $3 \cdot 5 - 6 = 9$ normal coordinates. There will be two sets of normal coordinates: one set constructed from a linear combination of stretching coordinates and the other constructed from bending coordinates. In Figure 4.14, the interbonding angles are illustrated at equilibrium instead of their increments that are the real internal coordinates, for example, α_{12} instead of its increment δ_{12}. Only five of the six increment angles are linearly independent.

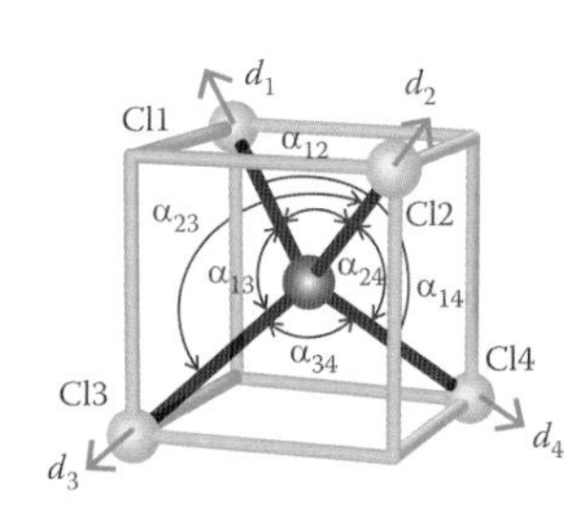

FIGURE 4.14 Internal coordinates of CCl_4 in equilibrium configuration are constructed starting from four stretching coordinates and six bending angles.

Then we build the table of transformation properties of internal coordinates under the action of symmetry operations of the $\boldsymbol{T}_d$ group shown in Figure 4.15. The results of the table are valid for the symmetry elements chosen as illustrated also on the right side of Figure 4.15. The results shown in this table depend on the choice of symmetry elements but the final results of this treatment (symmetry of normal coordinates) do not depend on it (Bunker and Jensen 1998).

The characters of representation Γ generated by the set of all coordinates are also shown in the last row of the table of Figure 4.15. Only the coordinate unshifted by a transformation contributes by 1 to the character

T_d	E	$C_{3(C5Cl4)}$	C_2	S_4	$\Sigma_{d(Cl1Cl4C5)}$
d_1	d_1	d_3	d_4	d_3	d_1
d_2	d_2	d_1	d_3	d_4	d_3
d_3	d_3	d_2	d_2	d_2	d_2
d_4	d_4	d_4	d_1	d_1	d_4
δ_{12}	δ_{12}	δ_{13}	δ_{34}	δ_{34}	δ_{13}
δ_{13}	δ_{13}	δ_{23}	δ_{24}	δ_{23}	δ_{12}
δ_{14}	δ_{14}	δ_{34}	δ_{14}	δ_{13}	δ_{14}
δ_{23}	δ_{3}	δ_{12}	δ_{23}	δ_{24}	δ_{23}
δ_{24}	δ_{24}	δ_{14}	δ_{13}	δ_{14}	δ_{34}
δ_{34}	δ_{34}	δ_{24}	δ_{12}	δ_{12}	δ_{24}
Γ	10	1	2	0	4

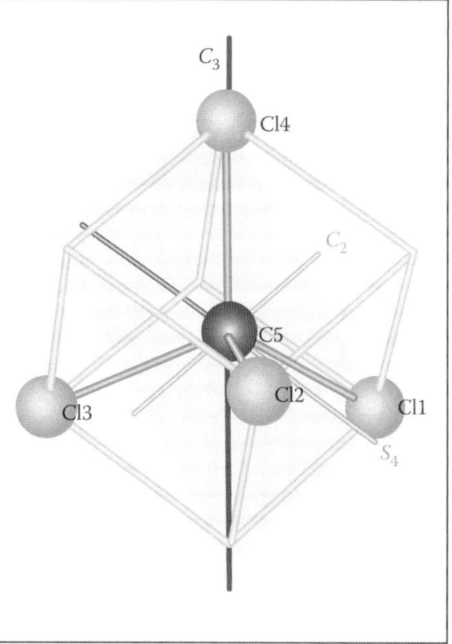

FIGURE 4.15 Transformation properties of internal coordinates of CCl_4 under symmetry operations of the $\boldsymbol{T}_d$ group and illustration of the symmetry elements used to perform the coordinates transformation.

of that transformation matrix. This representation can be expressed in terms of irreducible representations of the $\boldsymbol{T}_d$ group as

$$\Gamma = 2A_1 + E + 2T_2 \tag{4.26}$$

Owing to the redundancy of interbond angles (one of them is given by the sum of others), we subtract a totally symmetric representation and keep four symmetries (species) for normal coordinates: $\Gamma(\nu_1) = A_1$, $\Gamma(\nu_2) = E$, and $\Gamma(\nu_3) = 2T_2$.

4.4.2.3 VIBRATIONAL SYMMETRY COORDINATES

To perform the factoring of the secular equation, it is necessary to construct the symmetry coordinates, which consist of linear combinations of internal coordinates. These combinations are calculated using a simple formula (Wilson et al. 1955):

$$S^{(\Gamma)} = N\sum_R \chi_R^{(\Gamma)} RS_1 \tag{4.27}$$

where R denotes each symmetry operation of the group, RS_1 is the transformed S_1 displacement coordinate by operation R, and χ_R is the character of operation R in representation Γ. N is a normalization factor. In cases of double or triple degeneracy, pairs or triplets of symmetry coordinates of the same species and from the same internal set will be considered.

Table 4.11 contains the transformation of all internal coordinates of the CCl_4 molecule under the symmetry operations of the $\boldsymbol{T}_d$ group. The table is split into three parts because of its big size. The result of each transformation depends on the spatial orientation of the operation and the second row contains information about it. For example, C5Cl4 under the first C_3 rotation means that the rotation axis passes through the atoms carbon5 and Chlorine4. The rotation direction is always to the right (as a right screw). The notation C5(Cl1Cl4) under the first C_2 rotation means that the twofold rotation axis passes through carbon5 and the midpoint of the Cl1–Cl4 segment. The same notation is for the improper axis. Each reflection plane passes through the atoms denoted in the second row. Rows 3–12 contain the results of all symmetry operations on each internal coordinate. The rows 13–15 contain the characters of each irreducible representation (A_1, E, and T_2) of the group found to give the symmetry of normal modes. Rows 16–24 contain the product of the matrix character in the irreducible representation and the result of the internal coordinate transformation. The last column of Table 4.11 (part 3) contains the result given by Equation 4.27, which means the composition of normal modes in terms of internal coordinates.

Let us consider the totally symmetric representation. We have just found that its corresponding coordinate is a simple addition of the stretching displacement coordinates. So, the sum forms a totally symmetric normal coordinate. The normalization condition gives the normalization factor equal to $1/\sqrt{4}$. No linear combination of stretching coordinates appears as the basis of the twofold irreducible representation E.

Other linear combinations of stretching coordinates $(3d_1 - d_2 - d_3 - d_4)$ appear as the basis of the threefold irreducible representation T_2. Owing to the threefold degeneracy, it means that there must be three linear combinations that together form a basis of this representation. A practical way to find them is to use the method of the projection operator as illustrated in Figure 4.16. The atom Cl1 is placed on the point with Cartesian coordinates (1, 1, 1) and the results will correspond to this particular orientation of the molecule in space. Its projections in the x, y, and z directions are equal (d_1 multiplied by the same factor) and have the same sign +. So, the projections of d_1 are $N(d_1, d_1, d_1)$. All resulting projections of each stretching displacement vector in the x, y, and z directions are shown in Table 4.12 as well as their addition in each direction. So, we just find three linear combinations of stretching coordinates that together form a three-dimensional basis of T_2 representation. Each of them multiplied by the normalization factor of $1/\sqrt{4}$ is a normal coordinate (vibrational coordinate) of the CCl_4 molecule. To conclude, the tetrahedral molecule has three different normal modes of vibration (stretching) with the same symmetry (frequency) as illustrated in Figure 4.17.

The illustration of vibration modes of the asymmetric stretching type from Figure 4.17 is partially correct. The four vectors suggest how each bond will change starting from the equilibrium position when the symmetry of the molecule is tetrahedral. Therefore, the left figure (representation of S_{3x}) shows that the bond C–Cl1 stretches, the bond C–Cl2 compresses, the bond C–Cl3 stretches, and the bond C–Cl4 compresses.

TABLE 4.11 Transformation Properties of the Internal Coordinates of CCl_4 in Tetrahedral Symmetry and the Linear Combination Obtained from the Transformed Coordinates

(Part 1)

1	T_d	E	C_3	C_3^2	C_3	C_3^2	C_3	C_3^2	C_3	C_3^2
2	Direction		C5Cl4	C5Cl4	C5Cl1	C5Cl1	C5Cl2	C5Cl2	C5Cl3	C5Cl3
3	Rd_1	d_1	d_3	d_2	d_1	d_1	d_4	d_3	d_2	d_4
4	Rd_2	d_2	d_1	d_3	d_3	d_4	d_2	d_2	d_4	d_1
5	Rd_3	d_3	d_2	d_1	d_4	d_2	d_1	d_4	d_3	d_3
6	Rd_4	d_4	d_4	d_4	d_2	d_3	d_3	d_1	d_1	d_2
7	$R\delta_{12}$	δ_{12}	δ_{13}	δ_{23}	δ_{13}	δ_{14}	δ_{24}	δ_{23}	δ_{24}	δ_{14}
8	$R\delta_{13}$	δ_{13}	δ_{23}	δ_{12}	δ_{14}	δ_{12}	δ_{14}	δ_{34}	δ_{23}	δ_{34}
9	$R\delta_{14}$	δ_{14}	δ_{34}	δ_{24}	δ_{12}	δ_{13}	δ_{34}	δ_{13}	δ_{12}	δ_{24}
10	$R\delta_{23}$	δ_{23}	δ_{12}	δ_{13}	δ_{34}	δ_{24}	δ_{12}	δ_{24}	δ_{34}	δ_{13}
11	$R\delta_{24}$	δ_{24}	δ_{14}	δ_{34}	δ_{23}	δ_{34}	δ_{23}	δ_{12}	δ_{14}	δ_{12}
12	$R\delta_{34}$	δ_{34}	δ_{24}	δ_{14}	δ_{24}	δ_{23}	δ_{13}	δ_{14}	δ_{13}	δ_{23}
13	A_1	1	1	1	1	1	1	1	1	1
14	E	2	−1	−1	−1	−1	−1	−1	−1	−1
15	T_2	3	0	0	0	0	0	0	0	0
16	$\chi_R A_1 d_1$	d_1	d_3	d_2	d_1	d_1	d_4	d_3	d_2	d_4
17	$\chi_R A_1 d_2$	d_2	d_1	d_3	d_3	d_4	d_2	d_2	d_4	d_1
18	$\chi_R A_1 d_3$	d_3	d_2	d_1	d_4	d_2	d_1	d_4	d_3	d_3
19	$\chi_R A_1 d_4$	d_4	d_4	d_4	d_2	d_3	d_3	d_1	d_1	d_2
20	$\chi_R E d_1$	$2d_1$	$-d_3$	$-d_2$	$-d_1$	$-d_1$	$-d_4$	$-d_3$	$-d_2$	$-d_4$
21	$\chi_R T_2 d_1$	$3d_1$	0	0	0	0	0	0	0	0
22	$\chi_R T_2 d_2$	$3d_2$	0	0	0	0	0	0	0	0
23	$\chi_R E\delta_{12}$	$2\delta_{12}$	$-\delta_{13}$	$-\delta_{23}$	$-\delta_{13}$	$-\delta_{14}$	$-\delta_{24}$	$-\delta_{23}$	$-\delta_{24}$	$-\delta_{14}$
24	$\chi_R T_2\delta_{12}$	$3\delta_{12}$	0	0	0	0	0	0	0	0

(continued)

TABLE 4.11 (continued) Transformation Properties of the Internal Coordinates of CCl_4 in Tetrahedral Symmetry and the Linear Combination Obtained from the Transformed Coordinates

(Part 2)

1	T_d	C_2	C_2	C_2	S_4^1	S_4^3	S_4^1	S_4^3	S_4^1	S_4^3
2	Direction	C5(Cl1Cl4)	C5(Cl1Cl2)	C5(Cl2Cl4)	C5(Cl1Cl4)	C5(Cl1Cl4)	C5(Cl1Cl2)	C5(Cl1Cl2)	C5(Cl2Cl4)	C5(Cl2Cl4)
3	Rd_1	d_4	d_2	d_3	d_3	d_2	d_4	d_3	d_4	d_2
4	Rd_2	d_3	d_1	d_4	d_1	d_4	d_3	d_4	d_1	d_3
5	Rd_3	d_2	d_4	d_1	d_4	d_1	d_1	d_2	d_2	d_4
6	Rd_4	d_1	d_3	d_2	d_2	d_3	d_2	d_1	d_3	d_1
7	$R\delta_{12}$	δ_{34}	δ_{12}	δ_{34}	δ_{13}	δ_{24}	δ_{34}	δ_{34}	δ_{14}	δ_{23}
8	$R\delta_{13}$	δ_{24}	δ_{24}	δ_{13}	δ_{34}	δ_{12}	δ_{14}	δ_{23}	δ_{24}	δ_{24}
9	$R\delta_{14}$	δ_{14}	δ_{23}	δ_{23}	δ_{23}	δ_{23}	δ_{24}	δ_{13}	δ_{34}	δ_{12}
10	$R\delta_{23}$	δ_{23}	δ_{14}	δ_{14}	δ_{14}	δ_{14}	δ_{13}	δ_{24}	δ_{12}	δ_{34}
11	$R\delta_{24}$	δ_{13}	δ_{13}	δ_{24}	δ_{12}	δ_{34}	δ_{23}	δ_{14}	δ_{13}	δ_{13}
12	$R\delta_{34}$	δ_{12}	δ_{34}	δ_{12}	δ_{24}	δ_{13}	δ_{12}	δ_{12}	δ_{23}	δ_{14}
13	A_1	1	1	1	1	1	1	1	1	1
14	E	2	2	2	0	0	0	0	0	0
15	T_2	−1	−1	−1	−1	−1	−1	−1	−1	−1
16	$\chi_R A_1 d_1$	d_4	d_2	d_3	d_3	d_2	d_4	d_3	d_4	d_2
17	$\chi_R A_1 d_2$	d_3	d_1	d_4	d_1	d_4	d_3	d_4	d_1	d_3
18	$\chi_R A_1 d_3$	d_2	d_4	d_1	d_4	d_1	d_1	d_2	d_2	d_4
19	$\chi_R A_1 d_4$	d_1	d_3	d_2	d_2	d_3	d_2	d_1	d_3	d_1
20	$\chi_R E d_1$	$2d_4$	$2d_2$	$2d_3$	0	0	0	0	0	0
21	$\chi_R T_2 d_1$	$-d_4$	$-d_2$	$-d_3$	$-d_3$	$-d_2$	$-d_4$	$-d_3$	$-d_4$	$-d_2$
22	$\chi_R T_2 d_2$	$-d_3$	$-d_1$	$-d_4$	$-d_1$	$-d_4$	$-d_3$	$-d_4$	$-d_1$	$-d_3$
23	$\chi_R E \delta_{12}$	$2\delta_{34}$	$2\delta_{12}$	$2\delta_{34}$	0	0	0	0	0	0
24	$\chi_R T_2 \delta_{12}$	$-\delta_{34}$	$-\delta_{12}$	$-\delta_{34}$	$-\delta_{13}$	$-\delta_{24}$	$-\delta_{34}$	$-\delta_{34}$	$-\delta_{14}$	$-\delta_{23}$

(Part 3)

1	T_d	Σ_d	Σ_d	Σ_d	Σ_d	Σ_d	Σ_d	
2	*Direction*	C5Cl1Cl4	C5Cl2Cl3	C5Cl1Cl2	C5Cl3Cl4	C5Cl2Cl4	C5Cl1Cl3	
3	Rd_1	d_1	d_4	d_1	d_2	d_3	d_1	
4	Rd_2	d_3	d_2	d_2	d_1	d_2	d_4	
5	Rd_3	d_2	d_3	d_4	d_3	d_1	d_3	
6	Rd_4	d_4	d_1	d_3	d_4	d_4	d_2	
7	$R\delta_{12}$	δ_{13}	δ_{24}	δ_{12}	δ_{12}	δ_{23}	δ_{14}	
8	$R\delta_{13}$	δ_{12}	δ_{34}	δ_{14}	δ_{23}	δ_{13}	δ_{13}	
9	$R\delta_{14}$	δ_{14}	δ_{14}	δ_{13}	δ_{24}	δ_{34}	δ_{12}	
10	$R\delta_{23}$	δ_{23}	δ_{23}	δ_{24}	δ_{13}	δ_{12}	δ_{34}	
11	$R\delta_{24}$	δ_{34}	δ_{12}	δ_{23}	δ_{14}	δ_{24}	δ_{24}	
12	$R\delta_{34}$	δ_{24}	δ_{13}	δ_{34}	δ_{34}	δ_{14}	δ_{23}	
13	A_1	1	1	1	1	1	1	
14	E	0	0	0	0	0	0	
15	T_2	1	1	1	1	1	1	$\sum_R \chi_R^{(\Gamma)} RS_1$
16	$\chi_R A_1 d_1$	d_1	d_4	d_1	d_2	d_3	d_1	$6\,(d_1 + d_2 + d_3 + d_4)$
17	$\chi_R A_1 d_2$	d_3	d_2	d_2	d_1	d_2	d_4	$6\,(d_1 + d_2 + d_3 + d_4)$
18	$\chi_R A_1 d_3$	d_2	d_3	d_4	d_3	d_1	d_3	$6\,(d_1 + d_2 + d_3 + d_4)$
19	$\chi_R A_1 d_4$	d_4	d_1	d_3	d_4	d_4	d_2	$6\,(d_1 + d_2 + d_3 + d_4)$
20	$\chi_R E d_1$	0	0	0	0	0	0	0
21	$\chi_R T_2 d_1$	d_1	d_4	d_1	d_2	d_3	d_1	$2\,(3d_1 - d_2 - d_3 - d_4)$
22	$\chi_R T_2 d_2$	d_3	d_2	d_2	d_1	d_2	d_4	$2\,(-d_1 + 3d_2 - d_3 - d_4)$
23	$\chi_R E \delta_{12}$	0	0	0	0	0	0	$2\,(2\delta_{12} - \delta_{13} - \delta_{14} - \delta_{23} - \delta_{24} + 2\delta_{34})$
24	$\chi_R T_2 \delta_{12}$	δ_{13}	δ_{24}	δ_{12}	δ_{12}	δ_{23}	δ_{14}	$4\,(\delta_{12} - \delta_{34})$

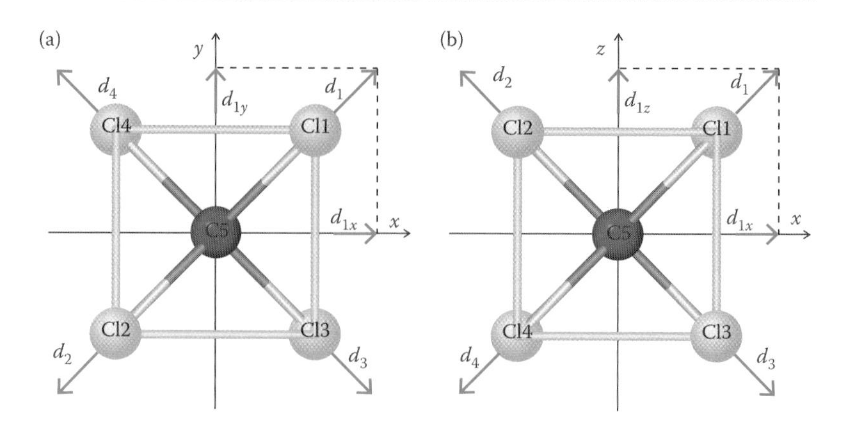

FIGURE 4.16 Illustration of the projection of stretching displacement coordinates in (a) x and y directions and (b) z direction.

TABLE 4.12 Projections of Each Stretching Displacement Vector in the x, y, and z Directions and Their Sum as They Result from Figure 4.16

	x	y	z	
d_1	d_1	d_1	d_1	
d_2	$-d_2$	$-d_2$	d_2	
d_3	d_3	$-d_3$	$-d_3$	
d_4	$-d_4$	d_4	$-d_4$	Total
Addition in each direction	$d_1 - d_2 + d_3 - d_4$	$d_1 - d_2 - d_3 + d_4$	$d_1 + d_2 - d_3 - d_4$	$3d_1 - d_2 - d_3 - d_4$

Note: The multiplication factor is the same and will be ignored because it will enter later in the normalization coefficient.

The vectors do not show the shift of atoms. To represent the shift of atoms, we must take into account the rule that the mass center does not shift as illustrated in Figure 4.18.

Exercise: Check the orthogonality property of the three normal modes! Follow the example:

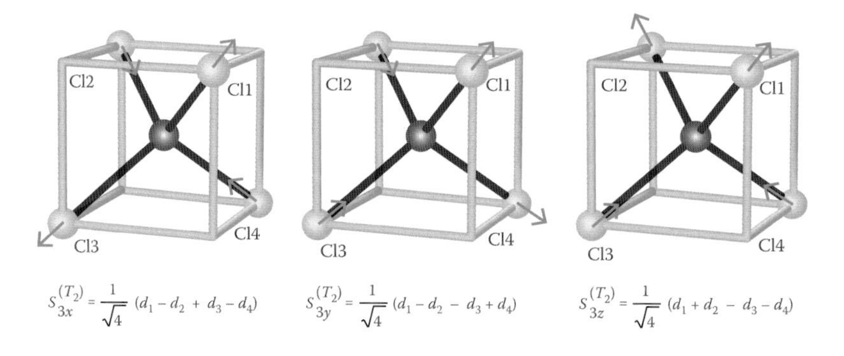

FIGURE 4.17 Spatial orientation of three normal modes of vibration (stretching) with symmetry T_2.

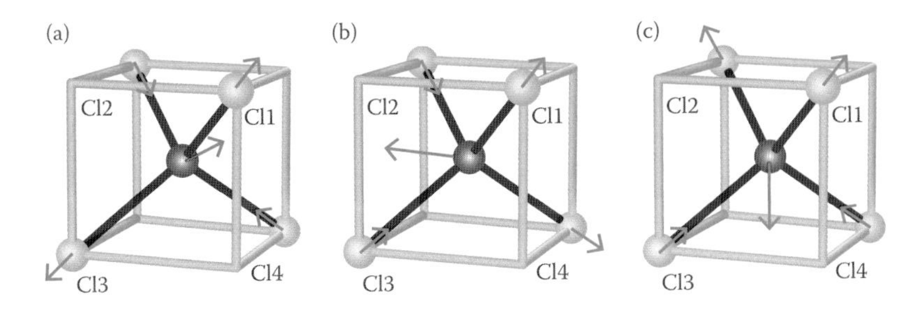

FIGURE 4.18 The shift of atoms during asymmetric vibrations: (a) the carbon atom shifts to the back, (b) the carbon atom shifts to the left, and (c) the carbon atom shifts to the bottom. These static images illustrate how the molecule vibrates asymmetrically.

$$
\begin{aligned}
S_{3x}^{(T_2)} \cdot S_{3y}^{(T_2)} &= \frac{1}{\sqrt{4}}(d_1 - d_2 + d_3 - d_4) \cdot \frac{1}{\sqrt{4}}(d_1 - d_2 - d_3 + d_4) \\
&= \frac{1}{\sqrt{16}}(d_1d_1 - d_2d_1 + d_3d_1 - d_4d_1 - d_1d_2 + d_2d_2 - d_3d_2 + d_4d_2 \\
&\quad - d_1d_3 + d_2d_3 - d_3d_3 + d_4d_3 + d_1d_4 - d_2d_4 + d_3d_4 - d_4d_4) \\
&= (1 + 1 - 1 - 1) = 0
\end{aligned}
$$

Then consider the doubly degenerate symmetry E where, according to Tables 4.8 and 4.11, only the set of Cl–C–Cl bending contributes to it:

$$
S_{2a}^{(E)} = N\sum_R \chi_R^{(E)} R\delta_{12} = \frac{1}{\sqrt{12}}(2\delta_{12} - \delta_{13} - \delta_{24} - \delta_{23} - \delta_{14} + 2\delta_{34})
$$

The twofold degenerate representation asks for a two-dimensional basis. So, the other normal coordinate will be found using the orthogonality property:

$$
\begin{aligned}
S_{2a}^{(E)} \cdot S_{2b}^{(E)} &= (2\delta_{12} - \delta_{13} - \delta_{24} - \delta_{23} - \delta_{14} + 2\delta_{34}) \\
&\quad \times (a\delta_{12} + b\delta_{13} + c\delta_{24} + d\delta_{23} + e\delta_{14} + f\delta_{34}) \\
&= 2\delta_{12}a\delta_{12} - \delta_{13}b\delta_{13} - \delta_{24}c\delta_{24} - \delta_{23}d\delta_{23} - \delta_{14}e\delta_{14} + 2\delta_{34}f\delta_{34} \\
&= 2a - b - c - d - e + 2f = 0
\end{aligned}
$$

The solutions are

1. $a = b = c = d = e = f = 1$. This means that the normal mode is totally symmetric and it is not permitted because the sum of all bending coordinates equals zero.
2. $a = f = 0$ and $b = c = 1$ and $d = e = -1$. Then $S_{2b}^{(E)} = \frac{1}{\sqrt{4}}(\delta_{13} + \delta_{24} - \delta_{23} - \delta_{14})$

The threefold degenerate coordinate (T_2) can also be constructed as a combination of interbond bending angles. The combination given by Table 4.11 (part 3) could be seen as a result of projection in the z direction. The other two combinations are quickly obtained by looking for the projections in the y and z directions. The orthogonality is also respected.

$$S_{4z}^{(T_2)} = \frac{1}{\sqrt{2}}(\delta_{12} - \delta_{34})$$

$$S_{4x}^{(T_2)} = \frac{1}{\sqrt{2}}(\delta_{13} - \delta_{24})$$

$$S_{4y}^{(T_2)} = \frac{1}{\sqrt{2}}(\delta_{14} - \delta_{23})$$

The results we have found for the carbon tetrachloride molecule (CCl_4) are also valuable for any molecule with the same spatial arrangement of atoms and, therefore, *any tetrahedral AB_4* molecule: methane CH_4 (NIST 2010), silicon tetrachloride $SiCl_4$, ammonium ion NH_4^+, sulfate ion SO_4^{2-}, and so on.

Most textbooks present the methane molecule as an example of deducing the symmetry of normal modes for tetrahedral symmetry because it is a well-known molecule. We have chosen carbon tetrachloride because it is easier to measure both the infrared absorption spectrum and the Raman spectrum for a liquid sample than for a gas sample. Carbon tetrachloride shows a stronger Raman spectrum than other substances and can be easily measured by using low sensitive (cheap) instruments. Formerly, it was widely used in industry and chemistry research. Nowadays, carbon tetrachloride is considered to be a toxic substance and must be carefully handled; you should especially *avoid inhaling its vapors.*

Final notes:

1. The results from other sources could be a little different because they depend on the spatial orientation of atoms reported to the coordinate system.
2. Another important difference is the amplitude of each atom displacement because of its mass, that is, in the case of methane, the outer atoms (hydrogen) are much lighter than the central atom (carbon) in contrast to the case of carbon tetrachloride where the outer atoms are heavier than the central atom. The lighter atom will shift further away from its equilibrium position than the heavier atom to keep the mass center unchanged because the vibration is an internal motion of the molecule.
3. *Molecular vibrations* refer to very *small and "slow" shifts of the nuclei* from their equilibrium positions. They are not vibrations of electrons that are very fast but of the outer electrons, which contribute to the molecular bond, can feel the periodic shifts of the nuclei.

4.4.3 WAVEFUNCTIONS OF NORMAL MODES

What are the normal coordinates good for?

In the quantum treatment of the linear oscillator (the simplest model), it is convenient to use the dimensionless variable ξ instead of the stretching coordinate x (Landau and Lifshitz 1965) given by the relation:

$$\xi = \sqrt{\frac{m\omega}{\hbar}}x \tag{4.28}$$

Thus, the eigenvalues E_n of the oscillator energy are

$$E_n = \left(n + \frac{1}{2}\right)\hbar\omega \tag{4.29}$$

where the vibrational quantum number $n = 0, 1, 2 \ldots$

Then the solutions of the Schrödinger equation for a *single linear oscillator* will be given by the relation:

$$\Psi_n(\xi) = Ne^{-\frac{\xi^2}{2}}H_n(\xi) \tag{4.30}$$

where N is the normalization factor and $H_n(\xi)$ is the n-th order Hermite polynomial. The Hermite polynomials are orthogonal functions given by

$$H_n(\xi) = (-1)^n e^{\xi^2}\frac{d^n}{d\xi^n}\left(e^{-\xi^2}\right)$$

The first Hermite polynomials, which are always necessary, are presented in the first column of Table 4.13.

The normal modes of the molecule can have different frequencies ω_i, where the subscript i denotes a specific vibration mode. Thus, the eigenfunctions depend on the vibration mode as follows:

TABLE 4.13 First Hermite Polynomials and the Corresponding Eigenfunctions of Vibrational Modes

$H_0(\xi) = 1$	$\Psi_{i0}(\xi_i) = N_i e^{-\frac{\xi_i^2}{2}}$	Vibrational ground state
$H_1(\xi) = 2\xi$	$\Psi_{i1}(\xi_i) = N_i e^{-\frac{\xi_i^2}{2}} 2\xi_i$	First vibrational excited state
$H_2(\xi) = 4\xi^2 - 2$	$\Psi_{i2}(\xi_i) = N_i e^{-\frac{\xi_i^2}{2}} (4\xi_i^2 - 2)$	Second vibrational excited state
$H_3(\xi) = 8\xi^3 - 12\xi$	$\Psi_{i3}(\xi_i) = N_i e^{-\frac{\xi_i^2}{2}} (8\xi_i^3 - 12\xi)$	Third vibrational excited state

(4.31) $$\Psi_{in}(\xi_i) = N_i e^{-\frac{\xi_i^2}{2}} H_n(\xi_i)$$

The expressions of the first eigenfunctions of normal modes for different values of the quantum number are presented in the second column of Table 4.13.

Note that ω_i is not exactly the oscillation frequency ν_i, but it is still used in theory to write formulas easily. In practice, it is used the wave number $\tilde{\nu}_i$ and the conversion is done using the relation:

(4.32) $$\omega_i = 2\pi\nu_i = 2\pi c\tilde{\nu}_i$$

The symmetry of each eigenfunction is given by the symmetry of its Hermitic polynomial because the exponential function is totally symmetric. If the normal modes, and therefore the normal coordinates, are the basis of irreducible representations of the symmetry group of the molecule, then their corresponding eigenfunctions must also form the basis for the group representations. For the degenerate case, if two coordinates form the basis for an irreducible representation, for example, $S_{2a}^{(E)}$ and $S_{2b}^{(E)}$, the corresponding eigenfunctions, for example, Ψ_{2a} and Ψ_{2b}, will also form the basis of the irreducible representation and will correspond to the same vibrational energy. Moreover, any set of eigenfunctions corresponding to the same vibrational energy can be written as a linear combination of the given eigenfunctions.

Let us take as an example the vibrational ground state $n = 0$. The eigenfunction Ψ_{i0} contains only the exponential term. There are two cases:

- *Nondegenerate.* The representation generated by the normal mode is one-dimensional. Thus, any symmetry operation transforms ξ_i into $\pm\xi_i$ whatever the symmetry group is. So, ξ_i^2 remains unchanged and the eigenfunction is also unchanged by any operation of the group. The resulting representation generated by the application of symmetry operations to the ground-state eigenfunction is totally symmetric.
- *Degenerate.* The representation generated by normal modes is two- or three-dimensional. Let us suppose that ξ_1 and ξ_2 are a pair of normal coordinates for a two-dimensional representation. Each symmetry operation transforms a coordinate into a linear combination of two coordinates:

$$R\xi_1 = a\xi_1 + b\xi_2$$

ξ_1 and ξ_2 are orthonormalized and the coefficients will be such that $a^2 + b^2 = 1$ such that $R\xi_1$ is also normalized $(R\xi_1)^2 = 1$. Then, the expression of the transformed eigenfunction will be

$$R\Psi_{10}(\xi_1) = \Psi_{10}(R\xi_1) = N_i e^{-\frac{(a\xi_1+b\xi_2)^2}{2}} = N_i e^{-\frac{1}{2}} = N_i e^{-\frac{\xi_1^2}{2}} = \Psi_{10}(\xi_1)$$

In conclusion, the eigenfunctions of normal modes in the ground state will always be the basis of the totally symmetric representation.

Further, the symmetry of the eigenfunction in the first excited state will be given by the product:

$$\Gamma(\Psi_{i1}) = \Gamma\left(e^{-\frac{\xi_i^2}{2}}\right)\Gamma(\xi_i) = A \cdot \Gamma(\xi_i)$$

Therefore, it will be the same as the symmetry of the normal mode. Similarly, the symmetry of the eigenfunction in the second excited state will be the same as the symmetry of ξ_i^2 and so on.

4.5 SELECTION RULES FOR FUNDAMENTAL VIBRATIONAL TRANSITIONS

4.5.1 FUNDAMENTAL VIBRATIONAL TRANSITIONS

It is time to see one of the most important utilities of group theory in spectroscopy. Group theory helps us to discover how many normal modes a molecule has and their symmetries. The number of irreducible representations found, where normal modes (normal vibrational coordinates) are the basis, shows the number of possible frequencies of the molecule oscillation. So, we have information about the molecular states. We are interested in studying the light absorbed by the matter to characterize the matter using a nondestructive procedure. The light (an electromagnetic wave) can be absorbed by the molecule, first, if the molecule has an electric dipole and, second, if the photon energy is equal to the energy difference between two molecular states. Is the fulfillment of these conditions sufficient for absorption to occur?

Let us calculate the intensity of absorbed light. In quantum mechanics, the intensity is proportional to the transition probability from the ground state to the final state. Further, the transition probability is given by the integral of Equation 4.5. To calculate the integral, we must know the wavefunctions and expression of the electric dipole. We reduce the problem to answering the question when the integral is nonzero. This occurs only if it is *invariant under all operations of the molecule's symmetry group.* The condition is fulfilled if the quantity under the integral generates a reducible representation that *contains the totally symmetric representation.*

For a given electronic state of the molecule (as initial and final state), the integral will contain only vibrational wavefunctions. We also ignore the functions of rotary motion. We may write the vibrational transition moment as

$$\Pi_{fi} = \int (\Psi_{vf})^{*}\, \hat{P}(\Psi_{vi})d\tau \qquad (4.33)$$

The vibrational wavefunction may be represented by a product of linear harmonic oscillator wavefunctions defined with respect to a set of normal (vibrational) coordinates:

$$\Psi_{v} = \Psi_{1,n1}(\xi_1) \cdot \Psi_{2,n2}(\xi_2) \cdot \cdots \Psi_{3N-6,n_{3N-6}}(\xi_{3N-6}) \tag{4.34}$$

where N is the number of atoms in the molecule and $\Psi_{i,ni}(\xi_i)$ is the wavefunction of the i-th normal mode in the n_i-th quantum state. For a linear molecule, the number of terms in the product is $3N - 5$. Note that the vibrational wavefunctions are real functions and we will give up the conjugation symbol (*) of wavefunctions in the theory below.

What is the physical interpretation we give to expression 4.34? Let us take the simple example of translational motion described in Cartesian coordinates. The straight-line motion of the molecule in whatever direction can be decomposed into three straight independent motions on each coordinate. Thus, the total wavefunction will be written as a product of all three wavefunctions about the x, y, and z coordinates. Now, imagine the same motion decomposed in a multidimensional space. Thus, the total wavefunction will be written as a product of all wavefunctions about each coordinate. The vibration of the molecule in a certain quantum state, which appears to be chaotic, can be decomposed in harmonic oscillations that are normal modes, each of them being in a certain quantum state of vibration that may be different than others. Each normal vibrational mode has its own vibrational coordinate so that the total wavefunctions will be written as a product of the individual wavefunctions defined with respect to normal coordinates. If all normal modes are in the ground state, that is, $n_i = 0$ whatever i, the molecule, is in the ground state described by the wavefunction:

$$\Psi_{v} = \Psi_{1,0}(\xi_1) \cdot \Psi_{2,0}(\xi_2) \cdot \cdots \Psi_{3N-6,0}(\xi_{3N-6}) \tag{4.35}$$

Suppose the i-th mode is excited in its first excited state $n_i = 1$; then the total wavefunction can be written as

$$\Psi_{v'} = \Psi_{1,0}(\xi_1) \cdot \Psi_{2,0}(\xi_2) \cdot \cdots \cdot \Psi_{i,1}(\xi_i) \ldots \Psi_{3N-6,0}(\xi_{3N-6}) \tag{4.36}$$

The transition of the molecule from state v to state v′ by excitation of only the i-th normal vibrational mode can be done by absorption of light ($\Delta E = \hbar\omega_i$) and it is called the *fundamental transition (tone).* If the i-th normal mode is excited in its second excited state $n_i = 2$ by absorption of light ($\Delta E = 2\hbar\omega_i$), the transition is called the *first overtone* of the fundamental. If the normal mode is excited in $n_i = 3$ state by absorption of light ($\Delta E = 3\hbar\omega_i$), the transition is called the *second overtone* of the fundamental and so on. A more complicated case is the simultaneous excitation of two different normal modes by absorption of light with energy $\Delta E = \hbar(\omega_i + \omega_j)$. The transition is called the *combination tone.* Both overtone and combination transitions appear in the spectra but we can recognize them due to they are much weaker than the fundamentals. Note that the fundamental transition can be produced as a result of the excitation of whatever mode, but its energy is different with respect to the symmetry of the excited mode. The number of fundamental transitions is usually smaller than the number of normal modes because of the degeneracy. The carbon tetrachloride molecule has nine normal modes, but only four fundamentals are permitted by the above theory.

4.5.2 SELECTION RULE OF THE QUANTUM VIBRATIONAL NUMBER

We define the electric dipole moment of a molecule as a simple vector quantity:

$$\vec{p} = e\vec{r} = e(x\vec{i} + y\vec{j} + z\vec{k}) \tag{4.37}$$

where $\vec{i},\vec{j}$, and $\vec{k}$ are the axes versors defined with respect to a Cartesian reference system fixed to the molecule. The coordinates can be transformed through the normal coordinates. In the first approximation of the linear harmonic oscillator, the molecular vibrations are very small and the electric dipole may be expressed in the Taylor series of these new coordinates:

$$\vec{p} = \vec{p}_0 + \sum_i \left(\frac{d\vec{p}}{d\xi_i}\right)_0 \xi_i + \cdots \tag{4.38}$$

The electric dipole moment is a linear function of vibrational coordinates and we keep only the first-order term.

The electric dipole of fundamental transition between the vibrational states v′ and v″ by the excitation of the i-th normal mode may be written as

$$\begin{aligned}
\Pi_{v'v''} &= \int (\Psi_{v''}) p(\Psi_{v'}) d\tau \\
&= \int [\Psi_{v''}(\xi_1,\ldots\xi_k)] \left(p_0 + \sum_j \frac{\partial p}{\partial \xi_j}\xi_j \right) [\Psi_{v'}(\xi_1,\ldots\xi_k)] d\xi_1 \ldots d\xi_k \\
&= \int [\Psi_{v''}(\xi_1,\ldots\xi_k)](p_0)[\Psi_{v'}(\xi_1,\ldots\xi_k)] d\xi_1 \ldots d\xi_k \\
&\quad + \int [\Psi_{v''}(\xi_1,\ldots\xi_k)] \left(\sum_j \frac{\partial p}{\partial \xi_j}\xi_j \right) [\Psi_{v'}(\xi_1,\ldots\xi_k)] d\xi_1 \ldots d\xi_k \\
&= p_0 \int [\Psi_{1,0}(\xi_1)\ldots\Psi_{i,v''}(\xi_i)\ldots\Psi_{k,0}(\xi_k)][\Psi_{1,0}(\xi_1)\ldots\Psi_{i,v'}(\xi_i)\ldots\Psi_{k,0}(\xi_k)] \\
&\quad \times d\xi_1 \ldots d\xi_k \\
&\quad + \int [\Psi_{1,0}(\xi_1)\ldots\Psi_{i,v''}(\xi_i)\ldots\Psi_{k,0}(\xi_k)] \left(\frac{\partial p}{\partial \xi_1}\xi_1 \cdots + \frac{\partial p}{\partial \xi_k}\xi_k \right) \\
&\quad \times [\Psi_{1,0}(\xi_1)\ldots\Psi_{i,v'}(\xi_i)\ldots\Psi_{k,0}(\xi_k)] d\xi_1 \ldots d\xi_k
\end{aligned}$$

We will use the orthonormality of wavefunctions to reduce the above expression to

$$\begin{aligned}
\Pi_{v'v''} &= p_0 \int [\Psi_{i,v''}(\xi_i)][\Psi_{i,v'}(\xi_i)]d\xi_i \\
&\quad + \sum_j \frac{\partial p}{\partial \xi_j} \int [\Psi_{1,0}(\xi_1)\ldots\Psi_{i,v''}(\xi_i)\ldots\Psi_{k,0}(\xi_k)]\xi_j \\
&\quad \times [\Psi_{1,0}(\xi_1)\ldots\Psi_{i,v'}(\xi_i)\ldots\Psi_{k,0}(\xi_k)]d\xi_1\ldots d\xi_k \\
&= p_0 \int [\Psi_{i,v''}(\xi_i)][\Psi_{i,v'}(\xi_i)]d\xi_i \\
&\quad + \frac{\partial p}{\partial \xi_i} \int [\Psi_{1,0}(\xi_1)][\Psi_{1,0}(\xi_1)]d\xi_1 \int [\Psi_{2,0}(\xi_2)][\Psi_{2,0}(\xi_2)]d\xi_2 \\
&\quad \times \int [\Psi_{i,v''}(\xi_i)]\xi_i[\Psi_{i,v'}(\xi_i)]d\xi_i + \cdots \\
&= p_0 \int [\Psi_{i,v''}(\xi_i)][\Psi_{i,v'}(\xi_i)]d\xi_i + \frac{\partial p}{\partial \xi_i} \int [\Psi_{i,v''}(\xi_i)]\xi_i[\Psi_{i,v'}(\xi_i)]d\xi_i
\end{aligned}$$

The first term is zero except for $v' = v''$, which is not a transition. For the second term, we will use the recurrence relation between Hermitic polynomials:

$$\xi H_v = \frac{1}{2} H_{v+1} + v H_{v-1}$$

Hence, the integral of the second term of the above expression of electric dipole transition becomes

$$\begin{aligned}
\int e^{-\frac{\xi_i^2}{2}} H_{v''}(\xi_i) e^{-\frac{\xi_i^2}{2}} H_{v'}(\xi_i)\xi_i d\xi_i &= \int e^{-\xi_i^2} H_{v''}(\xi_i)\left(\frac{1}{2} H_{v'+1}(\xi_i) + v' H_{v'-1}(\xi_i)\right) d\xi_i \\
&= \frac{1}{2}\int e^{-\xi_i^2} H_{v''}(\xi_i) H_{v'+1}(\xi_i) d\xi_i \\
&\quad + v' \int e^{-\xi_i^2} H_{v''}(\xi_i) H_{v'-1}(\xi_i) d\xi_i
\end{aligned}$$

Because of the orthogonality of Hermitic polynomials:

- The first term is not zero if $v' + 1 = v''$
- The second term is not zero if $v' - 1 = v''$

In conclusion, the infrared transitions can occur if they obey the *selection rule of the vibrational quantum number*

$$\Delta v = \pm 1 \quad (4.39)$$

The selection rule of the vibrational quantum number arises from the *properties of the linear harmonic oscillator wave function.*

Note that the selection rule was obtained as a result of an approximation but the spectra may also contain weaker bands forbidden by the rule (overtones and combination bands). These features exist mainly because the vibration of the molecule is not perfectly harmonic. The first overtone transitions are allowed when the vibrational quantum number changes by 2 (or more for a higher-order overtone) while the combination transitions respect the rule of changing by 1.

4.5.3 SYMMETRY SELECTION RULE FOR FUNDAMENTAL TRANSITIONS

The second selection rule arises from the *symmetry properties of vibrations.*

We return to the expression of the vibrational transition moment given by Equation 4.33 where we replace the electric dipole by its vector expression 4.37:

$$\begin{aligned} \Pi &= \int (\Psi_{v,1}) e(x\vec{i} + y\vec{j} + z\vec{k})(\Psi_{v,0}) d\tau \\ &= e\vec{i}\int \Psi_{v,1} x \Psi_{v,0}\, d\tau + e\vec{j}\int \Psi_{v,1} y \Psi_{v,0}\, d\tau + e\vec{k}\int \Psi_{v,1} z \Psi_{v,0}\, d\tau \\ &= \vec{i}\Pi_x + \vec{j}\Pi_y + \vec{k}\Pi_z \end{aligned} \tag{4.40}$$

Therefore, the transition moment has three independent components about the x, y, and z directions:

$$\begin{aligned} \Pi_x &= e\int \Psi_{v,1} x \Psi_{v,0}\, d\tau \\ \Pi_y &= e\int \Psi_{v,1} y \Psi_{v,0}\, d\tau \\ \Pi_z &= e\int \Psi_{v,1} z \Psi_{v,0}\, d\tau \end{aligned} \tag{4.41}$$

Π_x is the transition moment related to the electric dipole in the x direction and allows electric oscillation to occur only parallel to the x direction. In other words, the interaction with the electromagnetic field of radiation and the transition occur only if the electric field of radiation oscillates parallel to the x direction. The other components are similar, that is, Π_y is active only in the y direction and Π_z is active only in the z direction. For the fundamental transition to occur, at least one component of the transition moment must be nonzero.

Remember the rule: The direct product of a symmetry species by itself contains the totally symmetric representation.

The wavefunction of the ground state $\Psi_{v,0}$ always belongs to the totally symmetric representation (except free radicals). Then the integral is not zero if the wavefunction of the excited vibrational state and the corresponding Cartesian coordinate belongs to the same representation:

$$A = \Gamma(\Psi_{v,1}x\Psi_{v,0}) = \Gamma(\Psi_{v,1}x) \times \Gamma(\Psi_{v,0})$$
$$= [\Gamma(\Psi_{v,1}) \times \Gamma(x)] \times A \Rightarrow \Gamma(\Psi_{v,1}) \equiv \Gamma(x)$$

The representation of the Cartesian coordinate is found in the table of characters of the symmetry group. The representation of the molecular vibrational function is given by

$$\Gamma(\Psi_{v,1}) = \Gamma[\Psi_{1,0}(\xi_1)] \times \Gamma[\Psi_{2,0}(\xi_2)] \times \cdots \Gamma[\Psi_{i,1}(\xi_i)] \times \cdots \Gamma[\Psi_{3N-6,0}(\xi_{3N-6})]$$
$$= A \times A \times \cdots \Gamma[\Psi_{i,1}(\xi_i)] \times \cdots A = \Gamma[\Psi_{i,1}(\xi_i)] = \Gamma[(\xi_i)]$$

that is, the representation of the excited normal mode.

The selection rule of symmetry for the activity of the fundamental infrared absorption is

A normal mode that belongs to the same symmetry as any of the Cartesian coordinates will be active in the infrared spectrum.

Note: Working with polarized light, we could find the direction where the absorption occurs, that is, the polarization direction for each vibrational mode. For example, if the molecule is active only in the x direction, the absorption occurs when the light is linearly polarized in the x direction. In real systems, the light wave interacts with many molecules (thousands, millions) founded in its optical path and the property of polarized vibration can be observed if all the analyzed molecules have the same space orientation. Many samples are naturally aligned, although this alignment is not perfect, for example biological samples and polymer sheets.

For the carbon tetrachloride molecule, we can predict the components that will be active in the spectrum by using the character table and adding to it the normal modes (Table 4.14).

TABLE 4.14 Character Table, Normal Modes, IR Absorption, and Raman Activity of the CCl_4 Molecule

T_d	E	$8C_3$	$3C_2$	$6S_4$	$6\Sigma_d$	Linear Rotations	Quadratic	Normal Modes	Activity
A_1	1	1	1	1	1		$x^2+y^2+z^2$	$S_1^{(A_1)}$	Raman
A_2	1	1	1	−1	−1				
E	2	−1	2	0	0		$(2z^2-x^2-y^2, x^2-y^2)$	$S_{2a}^{(E)}S_{2b}^{(E)}$	Raman bend active
T_1	3	0	−1	1	−1	(R_x, R_y, R_z)			
T_2	3	0	−1	−1	1	(x, y, z)	(xy, xz, yz)	$S_{3x}^{(T_2)}S_{3y}^{(T_2)}S_{3z}^{(T_2)}$	IR and Raman stretch active
								$S_{4x}^{(T_2)}S_{4y}^{(T_2)}S_{4z}^{(T_2)}$	IR and Raman bend active

4.5.4 SELECTION RULES FOR RAMAN ACTIVITY

Raman spectroscopy is a powerful tool of structural analysis. The method is based on Raman scattering and will be described later in detail in Chapter 5. It is mentioned here to establish the elements of the molecule that contribute to Raman scattering and the symmetry issues.

Briefly, in the classical description when the light beam is incident on a molecule, its electric field acts to polarize the electric charges of the molecule. The incident light is in visible range, so only the electron cloud suffers a small deformation. In this way, an electric dipole appears by induction. The induced momentum P depends on the electric field strength E of the light wave:

$$\vec{P} = \alpha \vec{E} \tag{4.42}$$

where α is a constant called the *polarizability* of the molecule. For a rigid molecule, the polarizability is constant, but the real molecule oscillates and polarizability changes. The polarizability value depends on direction and it is expressed as a tensor:

$$\begin{bmatrix} P_x \\ P_y \\ P_z \end{bmatrix} = \begin{bmatrix} \alpha_{xx} & \alpha_{xy} & \alpha_{xz} \\ \alpha_{yx} & \alpha_{yy} & \alpha_{yz} \\ \alpha_{zx} & \alpha_{zy} & \alpha_{zz} \end{bmatrix} \begin{bmatrix} E_x \\ E_y \\ E_z \end{bmatrix} \tag{4.43}$$

If the polarizability changes with the normal coordinate at equilibrium configuration, then the corresponding normal mode is Raman active:

$$\left(\frac{\partial \alpha_{ij}}{\partial \xi} \right)_0 \neq 0$$

The transition momentum for Raman activity that implies a fundamental transition can be written as

$$P_{0,1} = \int \Psi_{v,1} \alpha E \Psi_{v,0}\, d\tau = E \int \Psi_{v,1} \alpha \Psi_{v,0}\, d\tau \tag{4.44}$$

An integral must be written for each component of polarizability. The transition will be Raman active if at least one component of the transition momentum is not zero. It means that the integral function must be totally symmetric. If the wavefunction of the ground state $\Psi_{v,0}$ is totally symmetric, then the wavefunction $\Psi_{v,1}$ and the polarizability component α_{ij} must have the same symmetry. The symmetry of the wavefunction is the same as the symmetry of its excited normal mode. The symmetry of polarizability is the same as the symmetry of binary products of coordinates or a combination of them. Examples of binary products are x^2, y^2, z^2, xy, xz, yz, $x^2 - y^2$, and so on.

To conclude, the selection rule for Raman activity is

> A normal mode that belongs to the same symmetry as any binary product of the Cartesian coordinates will be Raman active.

In Raman spectroscopy, the polarization of scattered light contains useful information. The linear polarized light can be used as incident light. The scattered light intensity is measured by passing it through a polarizing filter, which is arranged parallel and then perpendicular to the polarization direction of the incident light. The ratio between the intensity of the perpendicular component and the intensity of the parallel component of scattered light is called the *depolarization ratio*. The value of this ratio strongly depends on the symmetry of normal mode involved in the corresponding transition.

Only Raman transitions that involve totally symmetric vibrations are polarized.

The Raman-active modes of the carbon tetrachloride molecule are also shown in Table 4.14.

We can conclude for the CCl_4 molecule: the carbon tetrachloride molecule, CCl_4, belongs to the point group $\boldsymbol{T}_d$. It has nine normal modes of vibration with four different frequencies, that is, symmetries:

- A_1 is a *Raman*-active symmetric stretch of the CH bond
- E is a *Raman*-active ClCCl bend
- T_2 is an asymmetric stretch active in both *IR* and *Raman*
- T_2 is a bend active in both *IR* and *Raman*

Note: The *Raman spectrum* gives information about the vibrational modes in the system similar but *complementary* to that given by the *infrared absorption* spectroscopy.

As a general conclusion related to the infrared absorption and Raman spectroscopy:

To predict which normal modes will be IR and/or Raman active as fundamental transitions, follow the steps:

- A molecule will be *IR active* if the *dipole moment has to change* during vibration;
- A molecule will be *Raman active* if the *polarizability* of the molecule *has to change* during vibration.

- Find the point group of the molecule.
- Find the symmetry representations for the normal modes.
- Then look at the character table for the point group of the molecule.
- If the symmetry of a normal mode corresponds to x, y, or z, then the fundamental transition for this normal mode will be IR active.
- If the symmetry of a normal mode corresponds to the binary products and a combination of x, y, or z, then the fundamental transition for this normal mode will be Raman active.

Important note: *In molecules having an inversion center, the normal modes cannot be simultaneously IR and Raman active.* It is often called the *mutual exclusion rule*.

4.6 EXAMPLES OF IR AND RAMAN ACTIVITY OF MOLECULAR VIBRATIONS

4.6.1 WATER MOLECULE

The *water molecule* belongs to the $\boldsymbol{C}_{2v}$ group. It has 3 * 3 – 6 = 3 normal modes: symmetric stretch, bend (scissor), and asymmetric stretch (Figure 4.6). Table 4.15 shows that the molecule should have three infrared-active fundamental transitions and three Raman-active transitions at the same frequency as in infrared absorption. We expect to observe three strong features both in infrared and Raman because all are allowed by the selection rule. Two Raman lines are totally symmetric and will be polarized.

Question: All three transitions allowed in infrared absorption are sensitive to the polarization of incident light. Thus, two vibrations (symmetric stretch and bend with A_1 symmetry) will change only the z component of the water dipole momentum and one vibration (asymmetric stretch with B_2 symmetry) will change only the y component. Is it possible to measure the sensitivity of water IR absorption to the light polarization? (See Harris and Bertolucci 1978.)

4.6.2 OCTAHEDRAL MOLECULE AB_6

Examples of molecules and molecular complexes: SF_6, SeF_6, TeF_6, and $[Cr(H_2O)_6]^{3+}$. The AB_6-type molecule belongs to the $\boldsymbol{O}_h$ group. It has 3*7 – 6 = 15 normal modes: six A–B stretching modes and nine bending independent modes. Actually, the molecule has 12 B–A–B interbond angles and their corresponding increasing bending angles, but only nine bending angles are independent because they are interrelated by three equations, for example, $\delta_{F1SF2} + \delta_{F2SF3} + \delta_{F3SF4} + \delta_{F4SF1} = 0$ (their meanings are shown in Figure 4.19). The Cartesian displacement coordinates (Figure 4.19) generate the total representation Γ_{total} given in Table 4.16. Its characters can be easily obtained by visual inspection of Figure 4.19. E leaves all atoms at their positions ($\chi(E) = 21$). Rotation C_3 moves all vectors. Rotation C_2 leaves 1 vector and reverses 2 vectors ($\chi(C_2) = 1 - 2 = -1$). Rotation C_4 leaves 3 vectors unmoved ($\chi(C_4) = 3$). Rotation C_2 (from C_4^2) leaves 3 vectors and reverses 6 vectors ($\chi(C_4^2) = 3 - 6 = -3$). Inversion I reverses 3 vectors ($\chi(I) = -3$). Improper

TABLE 4.15 Point Group C_{2v} of the Water Molecule and Activity of Vibrational Modes

C_{2v}	E	$C_2(z)$	$\Sigma_v(xz)$	$\Sigma_v(yz)$	Linear Rotations	Quadratic	Normal Modes	Spectral Activity
A_1	1	1	1	1	z	x^2, y^2, z^2	Symmetric stretch Bend	IR and Raman IR and Raman
A_2	1	1	−1	−1	R_z	xy		
B_1	1	−1	1	−1	x, R_y	xz		
B_2	1	−1	−1	1	y, R_x	yz	Asymmetric stretch	IR and Raman
Γ_{total}	9	−1	1	3	$3A_1 + A_2 + 2B_1 + 3B_2$		$2A_1 + B_2$	

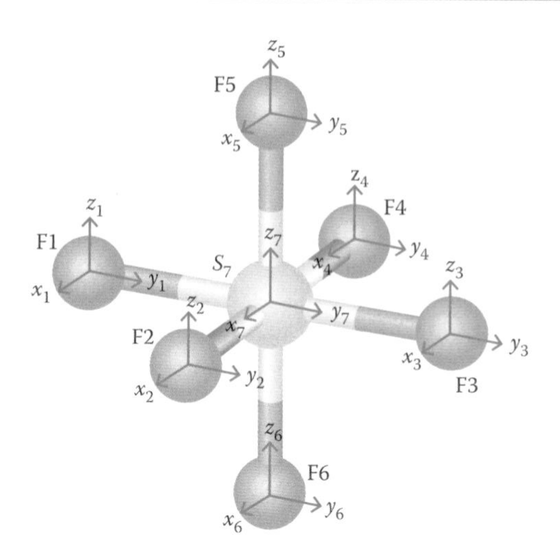

FIGURE 4.19 The set of Cartesian coordinates used to find the total representation of vibrations for SF_6 molecule.

rotation S_4 reverses 1 vector ($\chi(S_4) = -1$). Improper rotation S_6 moves all vectors ($\chi(S_6) = 0$). Reflection through the horizontal plane Σ_h leaves 2*5 vectors and reverses 1*5 vectors ($\chi(\Sigma_h) = 10 - 5 = 5$). Reflection through the dihedral plane Σ_d leaves 3 vectors unmoved ($\chi(\Sigma_d) = +3$). The total representation is then decomposed into reducible representations (see Table 4.16, last row) from where the representations corresponding to translation and rotation motions must be eliminated, that is, T_{1u} and T_{1g}, respectively. You must simply look for linear coordinates and rotations in the table. So, the total representation of normal vibrations includes $A_{1g} + E_g + T_{2g} + 2T_{1u} + T_{2u}$. The last column shows the spectral activity of the normal modes of the AB_6 molecule.

Comments

We are looking for 15 modes of vibration. They are six stretching modes and nine bending modes (You can find details on them on the Internet from the nice communication dated 1934! and recommended by Sir C.V. Raman, the result of the work of Nagendra Nath 1934.) The first stretching mode, which is totally symmetric (A_{1g}), is Raman active. The double degenerate stretching mode (E_g) is also Raman active. The triple degenerate bending mode T_{2g} is Raman active. The triple degenerate T_{1u} is a combination of stretching and bending modes and is active in infrared at two *different frequencies*. The triple degenerate T_{2u} symmetric mode (bending) rests unused (inactive). In octahedral symmetry, the Raman-active modes are completely separated from IR-active modes.

4.7 VIBRONIC COUPLING

Why do *tetrahedral cobalt* complexes $[CoCl_4]^{2-}$ absorb in the visible range more strongly than *octahedral cobalt* complexes $[CoF_6]^{3-}$? The

TABLE 4.16 Character Table of the O_h Group and Activity of Vibrational Modes

O_h	E	$8C_3$	$6C_2$	$6C_4$	$3C_2$	I	$6S_4$	$8S_6$	$3\Sigma_h$	$6\Sigma_d$	Linear Rotations	Quadratic	Normal Modes	Spectral Activity
A_{1g}	1	1	1	1	1	1	1	1	1	1		$x^2+y^2+z^2$	Stretch	Raman
A_{2g}	1	1	−1	−1	1	1	−1	1	1	−1				
E_g	2	−1	0	0	2	2	0	−1	2	0		$(2z^2-x^2-y^2, x^2-y^2)$	2 × stretch	Raman
T_{1g}	3	0	−1	1	−1	3	1	0	−1	−1	(R_x, R_y, R_z)			
T_{2g}	3	0	1	−1	−1	3	−1	0	−1	1		(xz, yz, xy)	3 × bend	Raman
A_{1u}	1	1	1	1	1	−1	−1	−1	−1	−1				
A_{2u}	1	1	−1	−1	1	−1	1	−1	−1	1				
E_u	2	−1	0	0	2	−2	0	1	−2	0				
T_{1u}	3	0	−1	1	−1	−3	−1	0	1	1	(x, y, z)		Stretch bend	IR
T_{2u}	3	0	1	−1	−1	−3	1	0	1	−1				
Γ_{total}	21	0	−1	3	−3	−3	−1	0	5	3	$A_{1g}+E_g+T_{1g}+T_{2g}+3T_{1u}+T_{2u}$		$A_{1g}+E_g+T_{2g}+2T_{1u}+T_{2u}$	

difference (dark blue versus light blue) is of an order of magnitude concerning light absorption due to electronic transitions. We have already discussed at the beginning of this chapter that the electronic transitions between states described by wavefunctions of d type are forbidden by Laporte's rule for centrosymmetric complexes. It is a consequence of the symmetry that makes the transition momentum zero. However, we can actually measure the absorption that is weaker but still exists. The answer could be either that the rule or the symmetry is not well. The physical phenomenon is very simple: certain vibrations of the molecule can remove the inversion center. It is called *vibronic coupling* because the electronic wavefunction is not completely separated from the vibrational wavefunction as shown by Equation 4.6. Here, it is another implication of molecular vibrations in spectroscopy. Note that the energy of vibrations is very low compared to the energy of visible light photons (hundreds of cm^{-1} versus tens of thousands of cm^{-1}) and *no direct interaction between vibration and light occurs.*

We assume in the initial state that the molecule is in its ground vibrational state $\Psi_{ei}\Psi_{v,0}$. The ground vibrational state of any molecule is always totally symmetric and can be ignored. In the final state, the i-th normal mode is excited; so, the transition momentum can be written in its components as follows:

$$\Pi_x = e\int \Psi_{ef}\Psi_{v,1}x\Psi_{ei}\,d\tau$$
$$\Pi_y = e\int \Psi_{ef}\Psi_{v,1}y\Psi_{ei}\,d\tau \quad (4.45)$$
$$\Pi_z = e\int \Psi_{ef}\Psi_{v,1}z\Psi_{ei}\,d\tau$$

Any integral is not zero if the representation generated by $\Psi_{ef}\Psi_{v,1}x\Psi_{ei}$ contains the totally symmetric representation. Specifically,

> if the vibrational mode involved in transition belongs to one of the representations generated by the product $\Psi_{ef}x\Psi_{ei}$, then the electronic transition is vibronically allowed (in the x direction).

Let us answer the question why the molecular complex $[CoF_6]^{3-}$ has an absorption band around 700 nm. The metal ion is Co(III), which means that the electron configuration is d^6. The fluoride ion F^- is on the weak-field side of the spectrochemical series; thus, the Co(III) will be a high-spin ion in the complex. It means that a small d splitting will be produced. The Tanabe–Sugano diagram for d^6 electron configuration (see Figure 2.53 reversed or Figure 2.54) results in only one spin-allowed transition ${}^5E_g \leftarrow {}^5T_{2g}$. In the octahedral group, x, y, and z coordinates belong to T_{1u} representation. We have enough data to calculate the direct product of representations for the transition ${}^5E_g \leftarrow {}^5T_{2g}$:

$$\Gamma[\Psi({}^5E_g)(x)\Psi({}^5T_{2g})] = E_g \times T_{1u} \times T_{2g} \quad (4.46)$$

TABLE 4.17 Reducible Representation of the Electronic Transition $^5E_g \leftarrow {}^5T_{2g}$ in Octahedral Cobalt Complex

O_h	E	$8C_3$	$6C_2$	$6C_4$	$3C_2$	i	$6S_4$	$8S_6$	$3\Sigma_h$	$6\Sigma_d$	Linear Rotations	Quadratic	Normal Modes
A_{1g}	1	1	1	1	1	1	1	1	1	1		$x^2+y^2+z^2$	Stretch
E_g	2	−1	0	0	2	2	0	−1	2	0		$(2z^2-x^2-y^2, x^2-y^2)$	2 × stretch
T_{2g}	3	0	1	−1	−1	3	−1	0	−1	1		(xz, yz, xy)	3 × bend
T_{1u}	3	0	−1	1	−1	−3	−1	0	1	1	$(\mathbf{x}, y, z)$		**Stretch, bend**
Γ_{EgxT2g}	18	0	0	0	2	−18	0	0	−2	0		$A_{1u}+A_{2u}+2E_u+\mathbf{2T_{1u}}+2T_{2u}$	

The characters of transition representation are shown in the last row of Table 4.17. It can be decomposed into irreducible representations as shown in the last row too. The result is the same for y and z coordinates, therefore: $A_{1u}+A_{2u}+2E_u+2T_{1u}+2T_{2u}$. T_{1u} is the only common irreducible representation between the electron and vibrational motions.

The above rule, stating that the integral is not zero if there is one normal vibration whose first excited state belongs to any of these representations, results in the transition $^5E_g \leftarrow {}^5T_{2g}$ that can occur by simultaneous excitation of any T_{1u} vibration (stretch or bend). So, the transition $^5E_g \leftarrow {}^5T_{2g}$ is vibronically allowed.

Note that the vibronic coupling does not depend on the aggregation state (liquid or solid) but strongly depends on temperature. When the temperature decreases, the absorption band becomes lower because the amplitude of vibration also decreases and Laporte's rule is stronger. However, working at lower temperature still has a positive effect because the band becomes thinner and we can see new features of the spectrum otherwise obscured by stronger bands.

4.8 VIBRONIC POLARIZATION

The metal ion in octahedral symmetry from the above example has the electron transition vibronically allowed whatever the direction of oscillation of electric field of light. It means that the intensity of the resulting absorption band does not depend on the polarization state of incident light.

The environment of the metal ion can be changed in such a way that the symmetry of the molecular complex becomes $\boldsymbol{D}_{4h}$. An example is the cation of *trans*-dichlorotetraammine cobalt (III) chloride *trans*-$[CoCl_2(NH_3)_4]^+ Cl^-$ whose structure is sketched in Figure 4.20 (Dunlop and Gillard 1964; Orvis et al. 2003). The presence of the tetraammine (NH_3) environment, which is located on the stronger side of the spectrochemical series, increases the ligand field strength so much that Co (III) enters as a low-spin ion. From the Tanabe–Sugano diagram for the d^6 electron configuration in $\boldsymbol{O}_h$ symmetry results that the ground state has $^1A_{1g}$ symmetry and two spin-allowed transitions are going to the $^1T_{1g}$ and $^1T_{2g}$ excited states. The complex rests centrosymmetric in the lower symmetry of $\boldsymbol{D}_{4h}$ and all spin-allowed transitions are forbidden by Laporte's

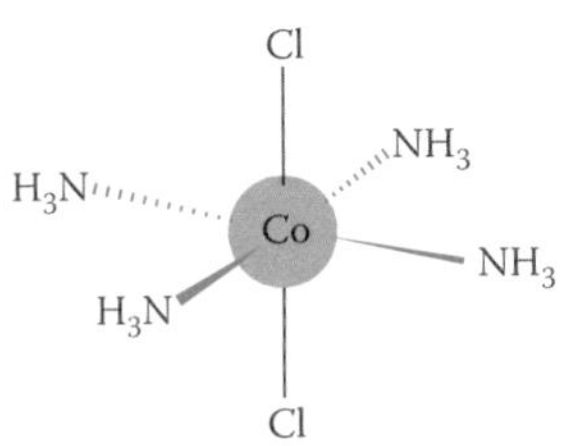

Cl
H_3N
NH_3
Co
NH_3
H_3N
Cl

FIGURE 4.20 The sketched configuration of the *trans*-dichlorotetraammine cobalt (III) cation.

rule, too. The ground state from $\boldsymbol{O}_h$ symmetry will keep the symmetry ${}^1A_{1g}$ in the new environment. The excited states split according to the correlation in Table 2.17 as follows:

- ${}^1T_{1g} \rightarrow {}^1A_{2g} + {}^1E_g$.
- ${}^1T_{2g} \rightarrow {}^1B_{2g} + {}^1E_g$.

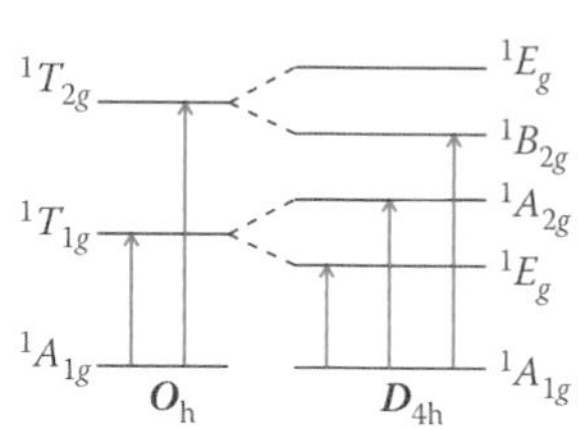

FIGURE 4.21 The splitting of energy levels of the low-spin Co (III) ion when symmetry decreases from $\boldsymbol{O}_h$ to $\boldsymbol{D}_{4h}$. Transitions $B_{1g} \leftarrow A_{1g}$ and $E_g \leftarrow A_{1g}$ give a single large band by superposition.

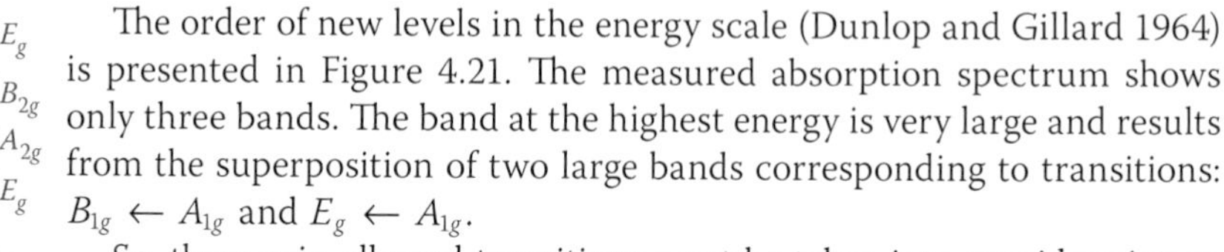

The order of new levels in the energy scale (Dunlop and Gillard 1964) is presented in Figure 4.21. The measured absorption spectrum shows only three bands. The band at the highest energy is very large and results from the superposition of two large bands corresponding to transitions: $B_{1g} \leftarrow A_{1g}$ and $E_g \leftarrow A_{1g}$.

So, three spin-allowed transitions must be taken into consideration to calculate the transition momentum. The resulted direct products for each transition in the (x,y) and z directions, respectively, are shown in Table 4.18.

The number of normal modes will be deduced as $3 \cdot 7 - 6 = 15$. The molecular complex has more than seven atoms, but we take into consideration only one central ion and the nearest two Cl ions and four nitrogen ions. We ignore the contribution of very light hydrogen atoms to vibrations. The Cartesian displacement coordinates (similar to that of the $\boldsymbol{O}_h$ illustration from Figure 4.19) generate the total representation Γ_{total} shown in Table 4.18. Its characters can be easily obtained by visual inspection of the molecule illustration and then it can be reduced to the irreducible representations (see Table 4.18). After elimination of translation- and rotation-irreducible representations, the total representation of normal modes can be written as

$$\Gamma_{nm} = 2A_{1g} + B_{1g} + B_{2g} + E_g + 2A_{2u} + B_{2u} + 3E_u \tag{4.47}$$

Note that another simple way to find the symmetries of normal modes is to begin with normal modes of the AB_6 molecule in $\boldsymbol{O}_h$ symmetry and then to reduce the symmetry of the same molecule to $\boldsymbol{D}_{4h}$ according to the correlation in Table 2.17:

$$\Gamma_{nm}(\boldsymbol{O}_h) = A_{1g} + E_g + T_{2g} + 2T_{1u} + T_{2u} \tag{4.48}$$

$$\downarrow$$

$$\Gamma_{nm}(\boldsymbol{D}_{4h}) = A_{1g} + (A_{1g} + B_{1g}) + (B_{2g} + E_g) + 2(A_{2u} + E_u) + (B_{2u} + E_u)$$

$$= 2A_{1g} + B_{1g} + B_{2g} + E_g + 2A_{2u} + B_{2u} + 3E_u$$

By simple visual inspection of the last column of Table 4.18, we can complete Table 4.19 that summarizes the electron transitions vibronically allowed and their polarization. To conclude, the bands corresponding to the transitions $A_{2g} \leftarrow A_{1g}$ and $B_{2g} \leftarrow A_{1g}$ should disappear from the spectrum when using polarized light in the z direction.

Note that to measure the absorption spectrum in the UV–Vis range using polarized light is not an simple task because the crystal does not "know" the orientation of the Cartesian system of the laboratory. So, the

TABLE 4.18 Character Table of the D_{4h} Group and Symmetry of Vibrational Modes of the AB_2C_4 Molecule

	E	$2C_4(z)$	C_2	$2C_2'$	$2C_2''$	I	$2S_4$	Σ_h	$2\Sigma_v$	$2\Sigma_d$	Linear Rotations	Quadratic	Normal Modes
A_{1g}	1	1	1	1	1	1	1	1	1	1		x^2+y^2, z^2	2 *nm*
A_{2g}	1	1	1	−1	−1	1	1	1	−1	−1	R_z		
B_{1g}	1	−1	1	1	−1	1	−1	1	1	−1		x^2-y^2	*nm*
B_{2g}	1	−1	1	−1	1	1	−1	1	−1	1		xy	*nm*
E_g	2	0	−2	0	0	2	0	−2	0	0	(R_x, R_y)	(xz, yz)	*nm*
A_{1u}	1	1	1	1	1	−1	−1	−1	−1	−1			
A_{2u}	1	1	1	−1	−1	−1	−1	−1	1	1	z		2 *nm*
B_{1u}	1	−1	1	1	−1	−1	1	−1	−1	1			
B_{2u}	1	−1	1	−1	1	−1	1	−1	1	−1			*nm*
E	2	0	−2	0	0	−2	0	2	0	0	(x, y)		3 *nm*
Γ_{total}	21	3	−3	−3	−1	−3	−1	5	5	3	$2A_{1g}+A_{2g}+B_{1g}+B_{2g}+2E_g+3A_{2u}+B_{2u}+4E_u$		
Γ_{nm}											$2A_{1g}+B_{1g}+B_{2g}+E_g+2A_{2u}+B_{2u}+3E_u$		
Γ_{EgxA1g}	4	0	4	0	0	−4	0	−4	0	0	$A_{1u}+A_{2u}+B_{1u}+B_{2u}$		
$\Gamma_{A2gxA1g}$	2	0	−2	0	0	−2	0	2	0	0	E_u		
$\Gamma_{B2gxA1g}$	2	0	−2	0	0	−2	0	2	0	0	E_u		
Γ_{EgzA1g}	2	0	−2	0	0	−2	0	2	0	0	E_u		
$\Gamma_{A2gzA1g}$	1	1	1	1	1	−1	−1	−1	−1	−1	A_{1u}		
$\Gamma_{B2gzA1g}$	1	−1	1	1	−1	−1	1	−1	−1	1	B_{1u}		

TABLE 4.19 Allowed Electronic Transitions by Vibronic Coupling on Each Coordinate Direction for *Trans*-$[Co(NH_3)_4Cl_2]Cl$

D_{4h}	(x, y)	z	Normal Modes
A_{2u}			2 *nm*
B_{2u}			1 *nm*
E_u			3 *nm*
$E_g \leftarrow A_{1g}$	$A_{1u} + A_{2u} + B_{1u} + B_{2u}$	E_u	**Allowed (x, y, z)**
$A_{2g} \leftarrow A_{1g}$	E_u	A_{1u}	**Allowed (x, y)**
$B_{2g} \leftarrow A_{1g}$	E_u	B_{1u}	**Allowed (x, y)**

first step of the work is the "orientation" of the crystal by X-ray diffraction. Then the crystal will be cut using these crystallographic directions and is now ready for measuring its behavior on polarized light. (Often, the crystal faces require to be polished to obtain two parallel faces of optical quality; otherwise, the measurement will give wrong results.)

4.9 PHOTOLUMINESCENCE AND SOLID-STATE LASERS

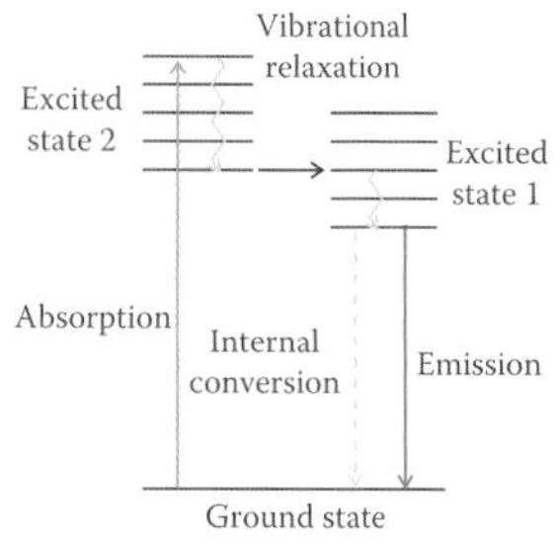

FIGURE 4.22 A simplified energy diagram (Jablonski diagram) of the molecular photoluminescence process that involves the molecular vibrations. The only radiative process is the photon emission.

Photoluminescence means absorption of light followed by emission of light in all directions, usually at lower energy (longer wavelength). Here, we mention this phenomenon because it is strongly related to vibrations. In the condensed phase, the absorption of light is followed by three competitive processes: vibrational relaxation, which involves transfer of energy to the surrounding molecules; vibrational relaxation and internal conversion; and vibrational relaxation and luminescence relaxation, which is the most likely event. A simplified illustration of the last processes concerning the involvement of vibrations is shown in Figure 4.22, currently called the Jablonski diagram. The excitation of the molecule from the ground state to a certain excited state (called 2) is produced by absorption of one photon. Absorption is followed by vibrational relaxation to the lowest vibrational level of the electronic excited state 2. Then the energy can be transferred to a certain vibrational level of the electronic excited state 1. Another vibrational relaxation is produced. The energy of the electron can now decrease from the lowest vibrational level of the electronic state 1 to the ground state in two competitive ways described in Figure 4.22 such as internal conversion (nonradiative process) or photon emission (radiative process of lower energy than that of the absorbed photon).

The problem of the impurity (dopant) ion in a dielectric solid can often be reduced to the molecular problem of the MX_n molecule. In the solid state, photoluminescence also involves vibrations. They are collective vibrations of the entire lattice because of the stronger interaction between ions and are called *phonons*. The number of optical phonons (involved in optical transitions) equals $3 \cdot N - 3$, where N is the number of atoms per primitive unit cell. The odd parity optical lattice modes are active in infrared spectroscopy. The even parity modes are active in Raman spectroscopy. In the case of noncentrosymmetric crystals, the optical modes are both infrared active and Raman active. In infrared absorption, the crystal must have an electric dipole. In an alkali halide such as a pure crystal, the vibrating electric dipole is created by the vibrations of neighboring ions and this dipole can interact with the electromagnetic field of light. The infrared absorption is a resonant process. What about the photoluminescence of crystals?

The photoluminescence of doped crystals had and still has an important place in laser developments. The first laser ever created by Theodor Maiman in 1960 had a small ruby crystal as an active medium. Solid-state lasers are lasers based on active media such as crystals or glasses doped with rare-earth or transition metal ions. Semiconductor lasers are considered separately because the lasing mechanism is different. Since photoluminescence is the process involved in their function, most of the solid-state lasers are optically pumped with flash lamps. For higher-efficiency pumping, the laser diodes are often used for pumping of new solid-state lasers in the so-called diode-pumped solid-state lasers (DPSS lasers). Flash lamps are still considered for high-energy laser systems in use. We discussed in detail the effects of a crystal field on energy levels of transition metal ions in Chapter 2 and then how Laporte's rule is relaxed by vibrational coupling. In this chapter, we will give only some examples. We insist on transition metal ions as dopants because they are strongly coupled to the neighboring ligand ions and, as a consequence, they are more strongly affected by the lattice vibrations than rare-earth ions.

4.9.1 RUBY LASER

The first laser ever made had an Cr^{3+}:Al_2O_3 active medium. The $Cr^{3+}(3d^3)$ ion has strong absorption bands and the luminescence emissions can be sharp (Henderson and Imbusch 1989). Introduced in the Al_2O_3 crystal, the Cr^{3+} ion takes the place of an Al^{3+} ion and will have an octahedral symmetry. The spin-allowed transitions are illustrated in Figure 4.23 by thick arrows. Remember that they are all forbidden by Laporte's rule. The transition ${}^4T_{1g}({}^4P) \leftarrow {}^4A_{2g}$ has a very large energy and corresponds to an absorption in far ultraviolet. It is not useful for optical pumping. Other spin-allowed transitions ${}^4T_{1g}({}^4F) \leftarrow {}^4A_{2g}$ and ${}^4T_{2g}({}^4F) \leftarrow {}^4A_{2g}$ give very broad and strong absorption bands in violet and green due to the strong coupling between the vibrational energy states of the host sapphire crystal and the electronic energy states of the active Cr^{3+} ions. Both allow a high-energy pumping (strong absorption) but a low efficiency of conversion into emitted light. However, at this value of crystal field strength for a thick piece of ruby, we can also measure a weak and thin absorption band ${}^2T_{1g}({}^2G)$ and ${}^2E_g({}^2G) \leftarrow {}^4A_{2g}$ in red range. This absorption is normally forbidden by both Laporte's rule and the spin rule. For a better view of the red lines in the absorption spectrum, it is recommended to reduce the temperature of the sample to around 80 K by using liquid nitrogen. When the temperature decreases, the lattice vibrations decrease in amplitude and the broad green absorption band becomes thinner and weaker (it is very sensitive to temperature). As a consequence, the red lines become visible from the tail of the green band.

The photoluminescence process in a ruby laser consists of green and blue light that is absorbed and they result in the emission of red light as the doublet (mainly from ${}^2E_g \rightarrow {}^4A_{2g}$). This red light is then amplified in

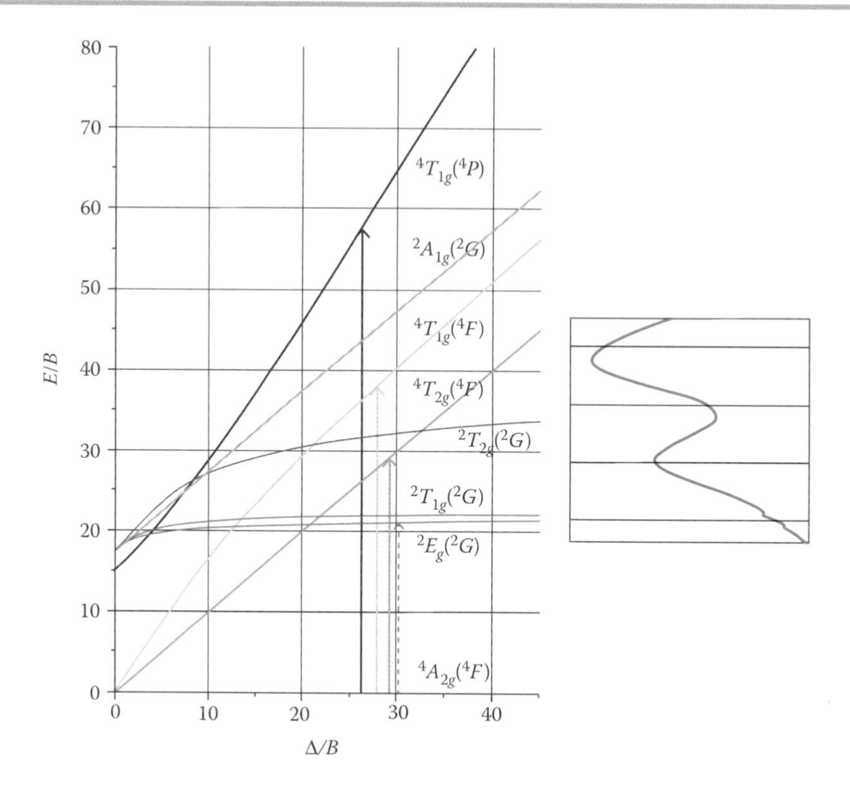

FIGURE 4.23 Spin-allowed transitions of Cr^{3+} (d^3 configuration) in Al_2O_3 with $\mathbf{O}_h$ symmetry are illustrated by thick arrows ($Dq/B = 2.8$ as given in Henderson and Imbusch). The spin-forbidden transitions that can also be experimentally measured are represented by a dashed arrow. (The representation was made using the available data by courtesy of Professor Robert Lancashire.) On the right side is shown the absorption spectrum of the crystal.

the resonant cavity of the laser and lased as a short pulse at 694.3 nm. The ruby laser is representative of the three-level laser medium. What is the basic phenomenon on which the laser is based? The first phenomenon is the achievement of *population inversion*. This means that there is a moment when more atoms are in the excited state than in the ground state. The inversion of population by optical pumping can only be achieved when using a three- or four-level system (Figure 4.24). This is not a book of lasers. However, we mention this principle of solid-state laser functioning because it is strongly related to vibrational coupling, the main subject of the chapter. The atom is initially raised into the excited state. Then it immediately tends to decay to a lower state. If the active atom is not isolated but is even strongly bonded to other atoms, as in a solid-state medium, it tends to transfer its energy to the neighbors. We denote by γ the probability of radiationless decay. If the upper state 2, for example, $^4T_{2g}$ of Cr^{3+}, is strongly coupled (γ_{21} high) through the lattice vibrations, then the active atom loses energy without radiation and quickly decays to a lower level 1, for example, ^2E_g of Cr^{3+}. The properties of this state are very important to qualify a material as a laser medium. If this intermediate state is also strongly coupled to the lattice vibrations (γ_{10} high), the next decay will be radiationless, too. If the intermediate state is low coupled to

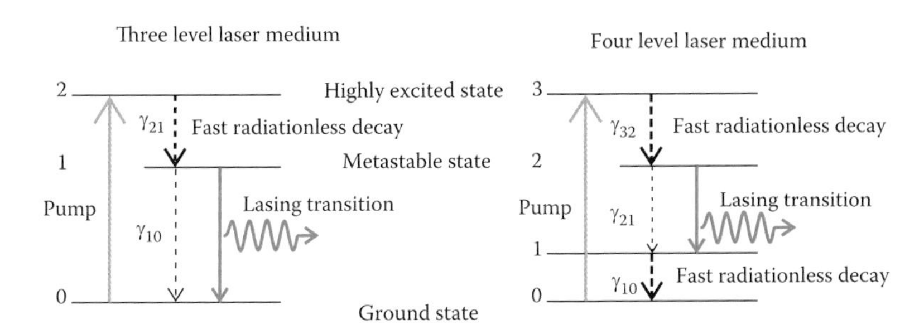

FIGURE 4.24 Diagram of a three-level and four-level laser medium, respectively. γ is the probability of radiationless decay.

the lattice vibrations (γ_{10} low), the radiation decay will be the most probable. Therefore, the active atom rests for a relatively long time (few milliseconds) in the intermediate state that is called the *metastable state.* If *at least half of the atoms* are raised to the same state by a similar process, then the inversion of population is achieved in the medium. The active medium is now ready for stimulated emission when all excited atoms will simultaneously decay to the ground state. The three-level active medium is not an efficient laser medium because it is difficult to raise half of the population at the same time. The four-level active medium offers the great advantage that population inversion can be produced only with a few atoms, if the intermediate state 1 is strongly vibrationally coupled (γ_{10} high) and, as a consequence, it will decay fast as radiationless.

In the last few years, the ruby laser was replaced by a more efficient alexandrite laser (Cr^{3+}:$BeAl_2O_4$) and four-level neodymium laser (Nd^{3+}:YAG). We note that the emission wavelength of the Nd^{3+} slightly depends on the host crystal as it is normal because of small perturbation suffered by a rare-earth ion in crystal field. So, for rare-earth ions, the lattice phonons (vibrations) are little involved in transitions.

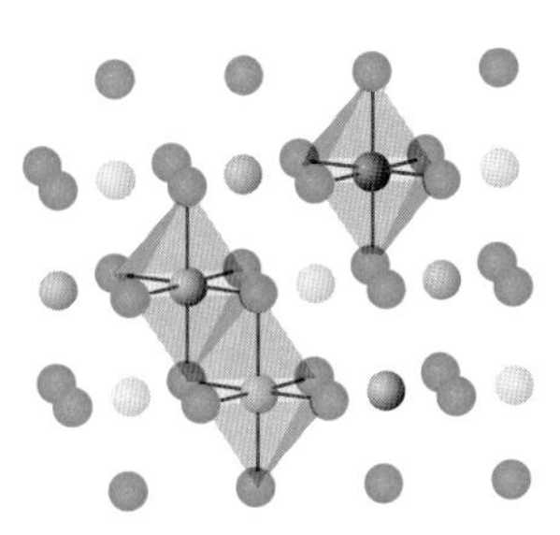

FIGURE 4.25 The Ti^{3+} ions occupy substitutional Al^{3+} sites of octahedral symmetry in sapphire (Al_2O_3).

4.9.2 TITANIUM: SAPPHIRE LASER

Is it possible to obtain more power with a laser than with all power plants over the world? How much energy does a 100 W lightbulb use in a second? 100 J, that is, $2.7 \cdot 10^{-5}$ kWh! Now, imagine you deliver the same amount of energy in a short laser pulse of 10 fs width (10^{-14} s). The result is huge and hard to imagine a power of 10^{16} W (10 PW)! This laser system is not yet ready but its construction has begun (ELI—The Extreme Light Infrastructure). In a few places throughout the world, 1 PW power has already been obtained (Mourou et al. 2006, 2011). The idea came from the exceptional properties of the Ti:sapphire crystal. The Ti^{3+} ions take the place of certain Al^{3+} ions in sapphire crystal (Al_2O_3) in an octahedral environment (Figure 4.25). The outer single $3d^1$ electron is strongly coupled to lattice vibrations. These lattice vibrations lead to both very broad absorption (from 400 to 600 nm) and emission (650–1100 nm) bands as shown in Figure 4.26 (Mourou et al. 2011). The lattice vibrations of the crystal act in the same way on all atoms in the crystal volume and they lead to a homogeneous broadening of the band.

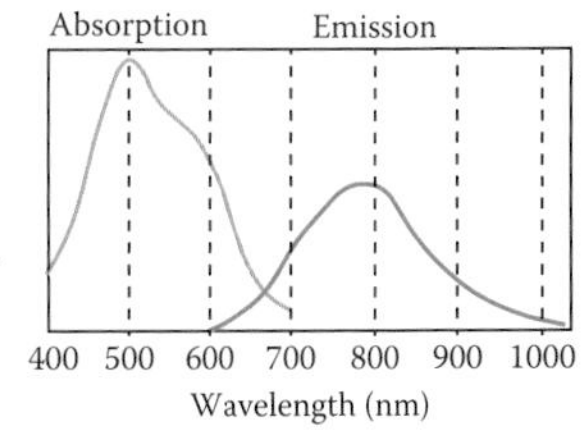

FIGURE 4.26 Both the absorption and emission bands of the Ti^{3+} ion in sapphire are exceptionally large due to the strong coupling between the vibrational states of the host crystal and the electronic states of the active ions.

There are only two energy levels of d^1 configuration: $^2T_{2g}$ and ^2E_g This can explain the presence of one large absorption band and one emission band, respectively. Two questions arise:

1. If there are only two levels, why do the two bands not coincide?
2. How does a two-level laser function?

1. Taking into account vibrational coupling, the electronic energy level can be represented by a parabola (denoted by G) in harmonic oscillator approximation instead of a line, as shown in Figure 4.27. Then the electronic level is divided into many equidistant vibrational energy levels according to the vibrational quantum number v. The diagram illustrated in Figure 4.27 is called the *configurational coordinate diagram*. The quantity Q (configurational coordinate) represents the distance between the central ion and its surrounding ligand ions. The lowest point of the parabola corresponds to the equilibrium position where nuclei pass through but never stop. Even at a temperature of 0 K, the system has the lowest vibrational energy that is represented by the first horizontal line. At room temperature, its energy is located higher. The excited state is similarly represented (denoted by E). The difference is that the lowest points of parabolas are not located on the same vertical due to the fact that the distance between nuclei in the excited state is larger than that of the ground state. The difference is denoted by Δ. This displacement between the equilibrium positions of the ground and excited states, respectively, is responsible for the difference between the absorbed photon energy and the emitted photon energy. The energy difference is called the *Stokes shift*. The Stokes shift is positive in most of the cases and is very large for the Ti^{3+} ion.

2. Is it a two-level laser? No! The illustration of Figure 4.27 shows that the vibrations are involved in two moments to complete the sequence of excitation–decay. Steps 2–3 and 4–1 are fast vibronic relaxations. Looking

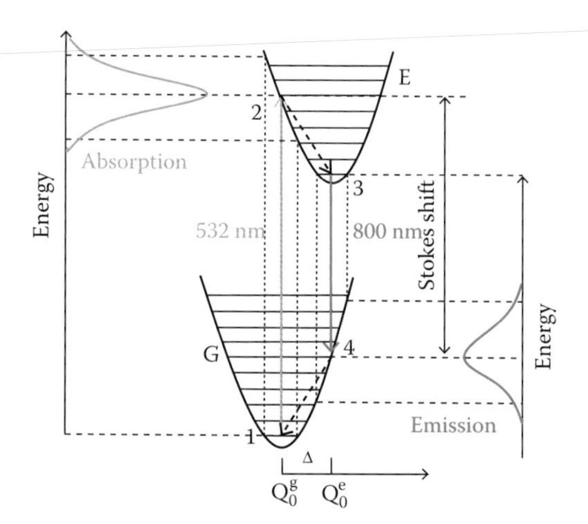

FIGURE 4.27 Illustration of light absorption (1–2), nonradiative decay (2–3), light emission (3–4), and nonradiative decay (4–1) in the configurational coordinate diagram. It explains the Stokes shift of emitted light with respect to the absorbed light due to the fact that the lattice vibration is involved.

carefully at the absorption spectrum of Figure 4.26, one can observe the asymmetry of the band. It is obvious that the band is the result of superposition of two bands. The absorption measurements at very low temperature can reveal the real structure of the spectrum (Albers et al. 1986). The substitutional replacement of the Al^{3+} ion by the Ti^{3+} ion induces a stress in the crystal lattice and lowers the local symmetry of the crystal field from $\boldsymbol{O}_h$ to trigonal symmetry. The orbital degeneracy of the *d* levels T_{2g} and E_g is partly removed by the Jahn–Teller effect. The lower-level ${}^2T_{2g}$ splits into ${}^2E$ and ${}^2A_1$. The upper-level ${}^2E_g$ also splits into two levels (Trager 2007). So, the lasing mechanism of Ti^{3+}:sapphire is more complicated than was described here and makes this active medium the only one that can give pulses as short as 5 fs on a very large tunable range (Trager 2007).

4.10 REMEMBER FOR GOOD PRACTICE

1. Group theory is a powerful method for determining the allowed transitions in optical spectra.
2. The dipole momentum (represented by x, y, and z) must change in an IR transition.
3. The polarizability of the molecule (represented by $x^2, y^2, z^2, xy, yz,$ and xz) must change in a Raman transition.
4. The transition moment is nonzero if the direct product of the symmetries of Ψ_i, H, and Ψ_f gives/contains a totally symmetric irreducible representation (such as A_1).
5. The number of normal modes is $3N - 6$ (or $3N - 5$ for linear molecules).
6. Associate with each atom a Cartesian coordinate system and apply all symmetry operations of the character table of the molecule group to calculate the character of each symmetry operation. The contribution to the character will be as follows:
   - Each coordinate left unchanged adds 1 to the character
   - Each coordinate reversed adds –1 to the character
   - Each coordinate changed adds 0

   Decompose the total representation into irreducible representations and eliminate representations of pure translations (represented by x, y, and z) and pure rotations (represented by R_x, R_y, and R_z).

   The remaining irreducible representations show the symmetries of the normal modes but do not show how they oscillate.
7. The vibrational ground state is totally symmetric and we count only the product of the excited state and electric dipole. If the direct product of a symmetry species by itself contains the totally symmetric representation, then *a normal mode that belongs to the same symmetry as any of the Cartesian coordinates* (x, y, and z) will be active in the infrared spectrum.
8. A supplementary selection rule is $\Delta v = \pm 1$ for fundamental transitions in infrared.
9. A normal mode that belongs to the same symmetry as any binary product of the Cartesian coordinates ($x^2, y^2, z^2, xy, yz,$ and xz) will be Raman active.

10. The electronic transitions between states described by the wavefunctions of d type are forbidden by Laporte's rule for *centrosymmetric* complexes. Certain vibrations of the molecule can remove the inversion center and relax Laporte's rule. It is called vibronic coupling. If the *vibrational mode involved in transition belongs to one of the representations generated by the product* $\Psi_{ef} x \Psi_{ei}$, *then the electronic transition is vibronically allowed* (in the x direction). To apply this rule, calculate the direct product of $\Psi_{ef} x \Psi_{ei}$ and decompose it into irreducible representations. If there is one normal vibration whose first excited state belongs to any of these representations, the transition is vibronically allowed in the x direction. The procedure is similar for the y and z directions.
11. If transition is vibronically allowed only in the x direction, it means that the intensity of the resulting absorption band depends on the polarization state of incident light. This type of measurement can supply supplementary information on local symmetry.

STUDY QUESTIONS

4.1 The dipole moment is an expression of the charge separation in a molecule. Could you predict from symmetry considerations which of the following molecules exhibit a permanent dipole moment: CO_2, H_2O, H_2S, CH_4, NH_3, and BF_3?

4.2 Can the energy of vibration (or frequency) be influenced by other vibrations in the molecule?

4.3 Comment on the following sentences: "A transition will be forbidden if the direct product of the symmetries of the involved electronic states and the dipole moment is odd. In other words, the transition is forbidden if the direct product contains the totally symmetric representation."

4.4 The benzene molecule C_6H_6 does not have a dipole moment because of its symmetry. Explain why benzene has, however, a very strong absorption in ultraviolet?

4.5 The hydrogen chloride stretching vibration determines a single absorption peak at 2886 cm^{-1}. Using a high-resolution IR spectrophotometer, one can observe that there is also a small peak nearby, which corresponds to the same vibration in HCl^{37}. Calculate the separation between the two peaks (in cm^{-1}) that is expected to occur.

Note that isotopic substitution is an interesting method to determine the origin of some peaks. Deuteration (substitution of hydrogen by deuterium) is a good method of tracing due to the drastically changing frequency of the molecular vibrations. Deuterium is not radioactive!

4.6 The Raman spectrum of CCl_4 shows four line shifts at 220, 312, 454, and 770 cm^{-1}. The IR spectrum of CCl_4 measured by a medium performance FTIR spectrophotometer is illustrated in the figure below. Assign the peaks in the IR spectrum using the table of normal modes of CCl_4.

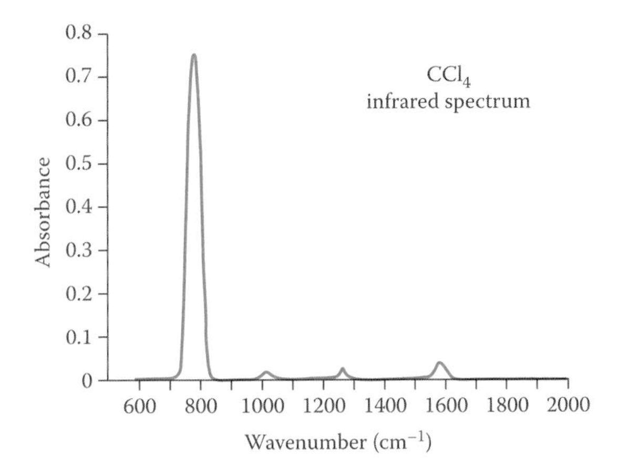

T_d	Normal modes	Activity
A_1	$\tilde{\upsilon}_1$	Raman, symmetric stretch
E	$\tilde{\upsilon}_2$	Raman, bend
T_2	$\tilde{\upsilon}_3$	IR and Raman antisymmetric stretch
	$\tilde{\upsilon}_4$	IR and Raman bend

4.7 Arrange in increasing order of frequency the stretching vibration of the following groups: C–H, C–Br, C–C, C–Cl, C–O.

4.8 The carbon dioxide molecule does not have a permanent electric dipole due to its symmetry. How can one explain, however, the presence of two peaks at 667 cm^{-1} and 2347 cm^{-1}, respectively, in the infrared absorption spectrum of CO_2?

4.9 One measures the fluorescence of a solution with a very sensitive spectrofluorometer. The emission spectrum of the original solution has a maximum at 387 nm (see figure below). After diluting the sample 100 times, the maximum of fluorescence emission shifts to 335 nm. Could you explain why? The solvent is pure water and the excitation wavelength is 300 nm.

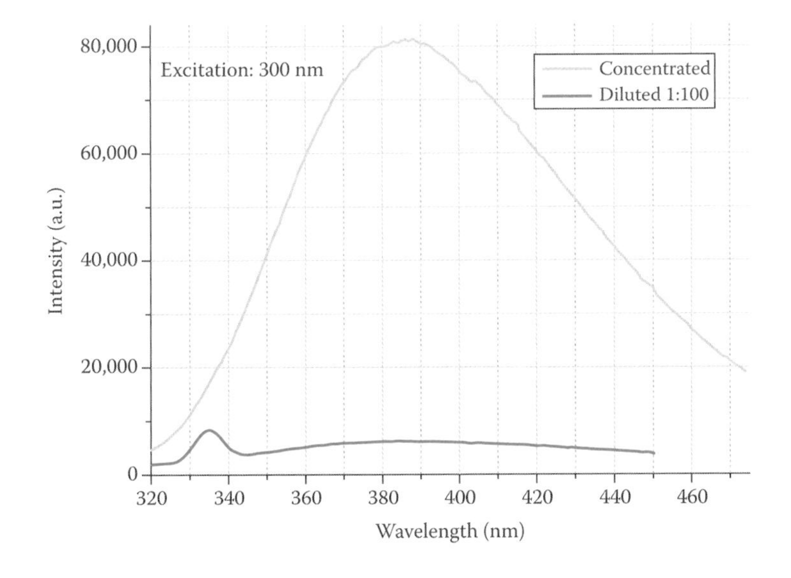

4.10 The functional group is a small group of atoms within molecules that has similar chemical properties whenever it occurs in different compounds (IUPAC Golden Book definition). The functional groups are responsible for the characteristic chemical and physical properties of those molecules. Examples of functional groups are hydrogen (–H), hydroxyl (–OH), carboxyl (–COOH), carbonyl (–CO), amino ($-NH_2$), phosphate ($-PO_4$), and methyl ($-CH_3$). How do you expect them to behave concerning the infrared absorption?

REFERENCES

Albers, P., E. Stark, and G. Huber. 1986. Continuous wave operation and quantum efficiency of titanium-doped sapphire. *Journal of Optical Society of America* B3: 134.

Bunker, P. R., and P. Jensen. 1998. *Molecular Symmetry and Spectroscopy,* 2nd edition. Ottawa: NRC Research Press (This book presents the symmetry of molecular structure and molecular spectroscopy at high level).

Dunlop, J. H., and R. D. Gillard. 1964. Tetragonal splittings in cobalt III complexes. *Molecular Physics: An International Journal of Interface between Chemistry and Physics* 7: 493–496, published online, 2006 http://www.tandfonline.com/doi/abs/10.1080/00268976300101281#.UdcJLjtmjTo.

ELI—The Extreme Light Infrastructure. http://www.extreme-light-infrastructure.eu/

Galabov, B. S., and T. Dudev. 1996. *Vibrational Spectra and Structure: Vibrational Intensities.* Amsterdam: Elsevier.

Harris, D. C., and M. D. Bertolucci. 1978. *Symmetry and Spectroscopy. An Introduction to Vibrational and Electronic Spectroscopy.* New York: Dover Publications.

Henderson, B., and G. F. Imbusch. 1989. *Optical Spectroscopy of Inorganic Solids.* Oxford: Oxford University Press.

Landau, L. D., and E. M. Lifshitz. 1965. *Quantum Mechanics Volume 3 of a Course of Theoretical Physics.* Oxford: Pergamon Press.

Majumder, M., and S. Manogaran. 2013. Redundant internal coordinates, compliance constants and non-bonded interactions—Some new insights. *Journal of Chemistry Science* 125: 9–15.

Mourou, G. A., M. Sergeev, A.V. Korzhimanov, A. A. Gonoskov, and E. A. Khazanov. 2011. Extreme light field and their fundamental applications. *Herald of the Russian Academy of Sciences* 81: 211–217.

Mourou, G. A., T. Tajima, and S. V. Bulanov. 2006. Optics in the relativistic regime. *Review of Modular Physics* 78: 309–371.

Nagendra Nath, N. S. 1934. The normal vibrations of molecules having octahedral symmetry. *Proceedings of the Indian Academy of Sciences,* Section A, 1(4): 250–259.

NIST. 2010. Methane Symmetry Operations. http://physics.nist.gov/Pubs/Methane/chap06.html

Orvis, J. A., B. Dimetry, J. Winge, and T. C. Mullis. 2003. Studying a ligand substitution reaction with variable temperature ^{1}H NMR spectroscopy: An experiment for undergraduate inorganic chemistry students. *Journal of Chemistry Education* 80(7): 803–805.

Trager, F. (ed.). 2007. *Handbook of Lasers and Optics.* New York: Springer.

Wilson, E. B., J. C. Decius, and P. C. Cross. 1955. *Molecular Vibrations.* New York: McGraw-Hill or 1980. *Molecular Vibrations: The Theory of Infrared and Raman Vibrational Spectra.* Dover Books on Chemistry.

Basic Optical Spectroscopic Techniques

5

5.1 WHAT IS SPECTROSCOPY?

Spectroscopy is now accepted as a common method to investigate matter by measuring the light absorbed or emitted, that is, studying the *interaction of light and matter.*

In this book, the term spectroscopy refers to *optical spectroscopy.* We refer to a spectral range of electromagnetic radiation larger than the visible range but extremely narrow within the entire electromagnetic spectrum (Figure 5.1). The optical range, limited to the visible range (400–750 nm) for a long time, was extended by using instruments both to longer wavelengths in infrared (~between 750 nm and 1 mm) and to shorter wavelengths in ultraviolet (~between 400 nm and 10 nm). Commercial instruments are currently available from 180 nm of ultraviolet up to 50 μm of infrared. No instrument can work on the entire optical range because of the physical phenomena involved in its operation (transmission, reflection, photoelectric effect, etc.).

Measured radiation is characterized by the following parameters:

- *Intensity*
- *Wavelength*
- *Polarization* state

The measurement unit for wavelength is the *nanometer,* as recommended by the International System of Units (SI). However, in the infrared range, it is more practical to use micrometer rather than the nanometer. In using the information given by a spectrum, the photon energy is often more practical; thus, other physical quantities are used: *wavenumber* (expressed in cm^{-1}) and even energy (expressed in electron-volt eV). The conversion from wavelength to energetic quantities is given by the following equations:

$$\begin{cases} \tilde{\nu}(cm^{-1}) = \dfrac{10^7}{\lambda(nm)} \\ E(eV) = h\dfrac{c}{\lambda} = \dfrac{1240}{\lambda(nm)} \end{cases} \tag{5.1}$$

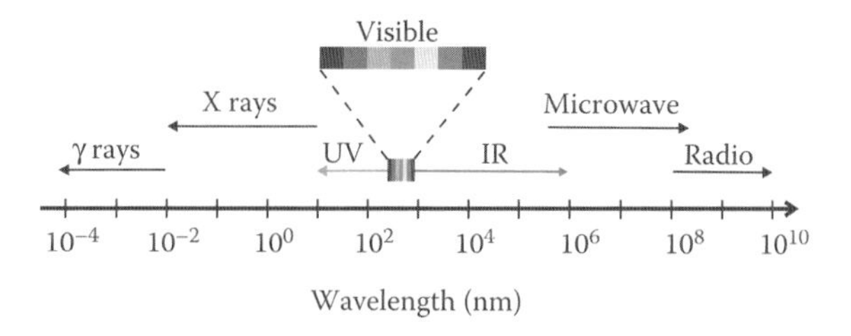

FIGURE 5.1 The optical range is a narrow part of the electromagnetic spectrum (the scale is logarithmic!). The unit is the nanometer, which is usual for the visible and ultraviolet range but not for either the shorter or longer wavelengths.

Let us now imagine that you see two apples as shown in Figure 5.2: one is green and the other is red, but you do not know whether they are good or not (it is the first time that you have seen such a fruit). Your action will be divided into two moments:

- The 1st moment: You take the green apple and bite it and you feel that it is acidic (you do not like it). Then you take the red apple and bite it and you feel that it tastes sweet (you like it).
- The 2nd moment: After some time, you see beyond the two apples: one is green and the other is red. You recall from the 1st moment that green color means that the fruit is not good and red color means that the fruit is good. As a consequence, you can decide *from a distance* without any direct interaction whether you will take one and which one. (The method is already used in agriculture to decide when to harvest.)

Conclusions:

1. Sight provides us with *information about matter*
2. It may be used from a *longer distance* than any other sense
3. To have correct information, you need to *calibrate* the analyzer!

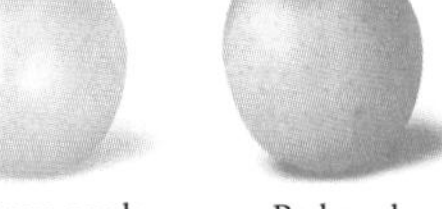

FIGURE 5.2 The information given by the scattered light (color green or red) without touching the sample is called the qualitative spectral analyzer and needs a previous calibration of the analyzer using a known sample.

Your action, divided into two moments, is called the *qualitative spectral analyses* of matter using reflected light: the 1st moment is the *calibration* of the analyzer and the 2nd moment is the *measurement.* A fundamental characteristic of spectral analysis is that you can *obtain data about the sample from a distance.* The calibration of the instrument is not required before any measurements, but you must perform it and you need *standard* samples with optical characteristics that are very well known. The calibration of the instrument can be made both as a wavelength and as the relation intensity quantity. The *wavelength calibration* is done by the producer of the instrument and you only have to check from time to time if it still meets the requirements of the parameters. It is a necessary condition to make the correct distinction between the green apple and the red apple (color is related to the wavelength).

The *spectrum* that you obtain from the spectral instrument is a 2-D representation between measured *light intensity* on the ordinate axis and *wavelength* (or energy) always on the abscissa axis. The spectrum features are called bands (if they are thick) or lines (if they are thin). The *band position* in the spectrum, that is, the wavelength of the band, gives *qualitative information* about the composition of the sample because *each atom or molecule has its characteristic spectrum*. The *band amplitude* gives *quantitative information* about the absorber component of the sample; the higher the amplitude, the higher the concentration. The quantitative calibration (relationship between signal amplitude and specimen quantity) is necessary for each type of sample and is mandatory in order to make a correct quantitative analysis.

Although Isaac Newton obtained the first spectrum of light using a glass prism, the first qualitative spectral analysis was performed by Gustav Kirchhoff and Robert Bunsen (who also were first to utilize the spectroscope, which can be seen today in any student's laboratory, for chemical analysis). Using highly purified salts of sodium, potassium, and lithium, they proved the central idea of qualitative analysis that *each element gives a unique spectrum*, and this principle can be used to identify the composition of the sample.

Today, we know the elemental composition of the Sun and many other stars only by studying the spectrum of the emitted light at a huge distance from the source. The robot Curiosity, which is stationed on Mars, has a spectral analyzer that operates from a distance using a laser as the excitation source. It can perform the spectral analysis of rocks from a distance of 3 m (NASA 2013).

Spectroscopy has two main branches (for more detailed information, see the diagram on page 13 of Bauman 1962, which is still accurate):

- Absorption spectroscopy
- Emission spectroscopy

Since this book is dedicated to symmetry applications in optical spectrum analysis, we will not refer to the spectroscopy of atoms that is mostly emission spectroscopy and only part of absorption spectroscopy. Free atoms have spherical symmetry and their spectra are not the subject of group theory presented in this book. Their spectra are composed of spectral lines and need special equipment to be resolved. We deal with molecules and molecular complexes in a condensed state whose spectra are mainly composed of bands. Most of them may be analyzed by absorption spectroscopy. Hence, we will refer mainly to instruments that are dedicated to measuring these kind of spectra. Exceptions will be highlighted.

The principle of absorption spectroscopy is to measure the light intensity I_0 incident to the sample and the light intensity I emerging from the sample (Figure 5.3). The dependence between intensities is given by a former relationship known as the *law of Beer–Bouguer–Lambert* (or absorption law): the emerging intensity depends on the incident intensity, the thickness of sample ℓ, the concentration of absorbing species c, and the extinction coefficient ε.

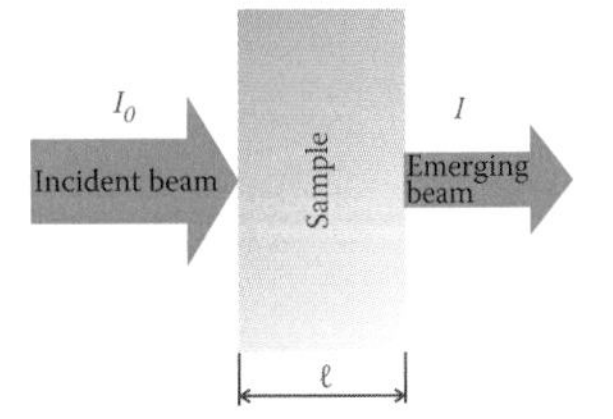

FIGURE 5.3 A beam of light incident on a sample is partly transmitted. The emerging light intensity is given by the absorption law (known as the Beer–Bouguer–Lambert law).

$$I_{(\lambda)} = I_0 e^{-\varepsilon(\lambda)c\ell} \tag{5.2}$$

By using both intensities, we can calculate either *transmittance* T or *absorbance* A of the sample:

$$\begin{cases} T = \frac{I}{I_0}\% \\ A = \ln\left(\frac{I_0}{I}\right) = \varepsilon_{(\lambda)} c\ell \end{cases} \quad (5.3)$$

Since the extinction coefficient depends on the light wavelength, the emerging light intensity will also vary with wavelength. Assuming constant incident light intensity, either transmittance or absorbance of the sample will vary as a function of the light wavelength, which results in the transmittance spectrum or absorbance spectrum. Absorbance has the advantage of direct dependence on sample characteristics, mainly on the concentration.

Note the difference between transmittance as a physical quantity and transmission as a phenomenon and, on the other hand, between the physical quantity absorbance and the phenomenon of absorption. Many articles or books call the A symbol as absorption. In the field of solid state research, it is common to use of optical density (OD) instead of absorbance. Absorbance is a dimensionless quantity. Remember that it is also a logarithmic quantity and $A = 2$ means that the emerging light intensity is very low; therefore, the sample absorbs strongly.

Comments:

1. Relation 5.2 is the classical form of the absorption law. In modern instruments, photoelectric detectors detect light and measure the *radiant power* (also called *radiant flux*) of the electromagnetic beam, which is measured in W and denoted by the symbol Φ_e.
2. In nature, the attenuation of any radiation or particle beam through matter is expressed by an exponential function, as given in relation 5.2. Because of practical reasons for simplifying the mathematical operations, one prefers to write the absorption law using decimals instead of an exponential function as below

$$\Phi_{e(\lambda)} = \Phi_{e0} 10^{-k(\lambda)c\ell} \quad (5.4)$$

   where k is the decimal extinction coefficient. In such conditions, absorbance is given by the decimal logarithm (log).
3. The absorption law is valuable at normal incidence of a parallel beam (collimated beam), at relatively low intensities of the incident light, and for homogeneous samples with a very low scattering of light and low reflection at surfaces.
4. The method is used on a large scale for solution measurements and, as a consequence, ε (or k) is called the molar (decimal) extinction coefficient (or molar absorptivity) and it will be expressed in units of $liter \cdot mol^{-1} \cdot cm^{-1}$.

5.2 OPTICS OF A SPECTROMETER

The absorption spectroscopy of molecules, liquids, and solid samples is one of the main applications of the group theory. So we will call attention on the necessary equipment for this method. An instrument dedicated to measuring the absorption spectrum must measure the light intensity before and after the sample at different wavelengths. To perform this task, the spectral instrument should be composed of three main parts:

1. Light source
2. Monochromator
3. Photodetector

We refer here only to optical components. The processing of the electrical signal given by the photodetector is another task of the instrument and it is continuously changing and fast improving.

5.3 LIGHT SOURCE

Its purpose is to supply the light used to probe the sample. What are the requirements of the light source?

- Time stability
- *Broad* emission spectrum
- *Continuum* emission spectrum
- *Flat* emission spectrum, that is, the intensity keeps a constant value for all wavelengths

There is no source that meets all these requirements, especially the last one. The most suitable is a black-body heated to hundreds or even a thousand degrees that emits light with respect to Planck's law. Figure 5.4 illustrates two important properties of black-body radiation: the emission spectrum is broad and continuum. One example is the incandescent lamp that has another advantage of low price. Any commercial instrument has an incandescent lamp, specially made for this task with a linear filament, and with a proper power source that can meet the first three requirements:

- Good time stability for almost 1000 h.
- Broad emission range—it can cover *visible and near infrared.*
- Continuum spectrum. The halogen incandescent lamps have a higher intensity in the blue part of the spectrum compared to simple lamps. In time (after several hundred hours), their spectra present a narrow band superimposed on the main band that can influence the precision of measurements.

The spectral sources of infrared radiation work on the same principle of black-body radiation. They have a bar made of either doped ceramic

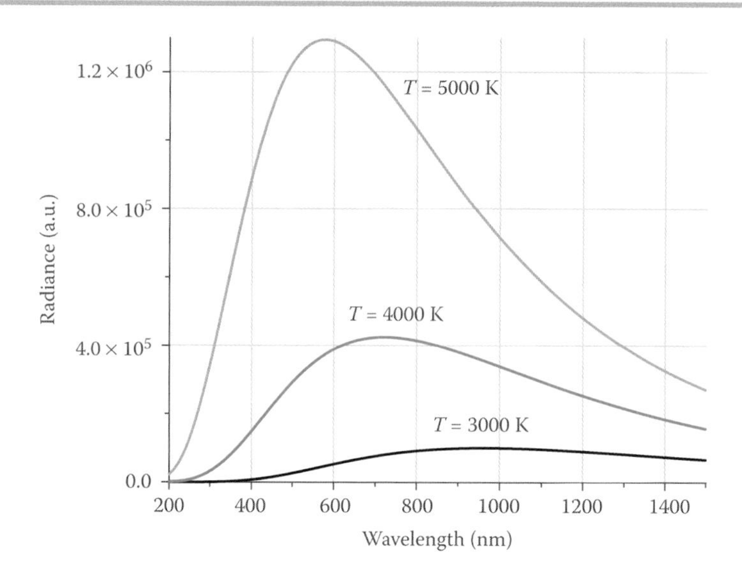

FIGURE 5.4 The light emitted by a heated black-body has a continuum and broad spectrum according to Planck's law.

(Nernst lamp) or silicon carbide (Globar lamp). The bar is heated to a well-controlled temperature but lower than the one seen in the filament lamp case. They ensure a very large spectrum in the infrared range. After decades of use, Nernst lamps have been replaced by Globar lamps have replaced in most of the new instruments because of shorter starting time and higher brightness.

The common sources with a continuum spectrum for ultraviolet are available as deuterium lamps. The deuterium lamp is filled with molecular deuterium at low pressure and an arc-type discharge is produced between two electrodes. The emitted light has a broad spectrum from 180 to 900 nm with strong atomic lines superimposed on a continuum spectrum (Figure 5.5). Fortunately, the spectrum is free of lines between 180 and 400 nm. The deuterium lamp radiance decreases so much to a longer wavelength that an instrument, which uses two lamps in order to measure both in UV and visible light, will automatically interchange the lamps around 350 nm. Thus, the measurement of a UV–vis absorption spectrum will be performed with a D_2 lamp from 180 to ~350 nm and then with a halogen incandescent lamp until 800 nm or further. Deuterium lamps emit only in one direction (of a small quartz window), and the light collection is therefore more efficient. For special cases, you can order deuterium lamps with the spectrum extended down to 160 nm. The difference comes mainly from the material used for the window to replace the quartz.

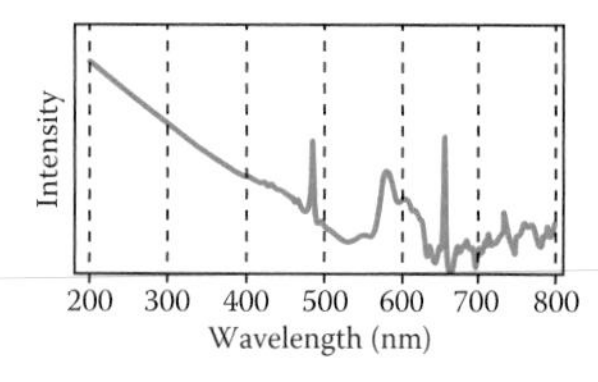

FIGURE 5.5 The emission spectrum of the deuterium lamp.

In special cases, when you need a higher intensity, you can use a xenon lamp. It is a high-pressure arc lamp that can provide a thousand times more light than all the aforementioned lamps. Its spectrum is very broad ranging from UV (180 nm) to NIR (2000 nm). It needs a big power source with a high-voltage igniter and a cooled box because of its high power ranging from 50 to 1000 W. Caution is required when using such lamps:

- The spectrum shown in Figure 5.6 has several very strong lines in the visible and NIR regions that can influence the quality of spectral measurement around their position or at double wavelength (see also hereunder the diffraction grating).
- The light beam can be harmful to the eyes, even if reflected by a metallic or glass surface. Remember that the light will be finally focused on the retina.
- When you concentrate the beam (reduce the diameter) on a relatively small surface of an absorption filter, this may break because of local heating. Any light-absorbing material can even burn!

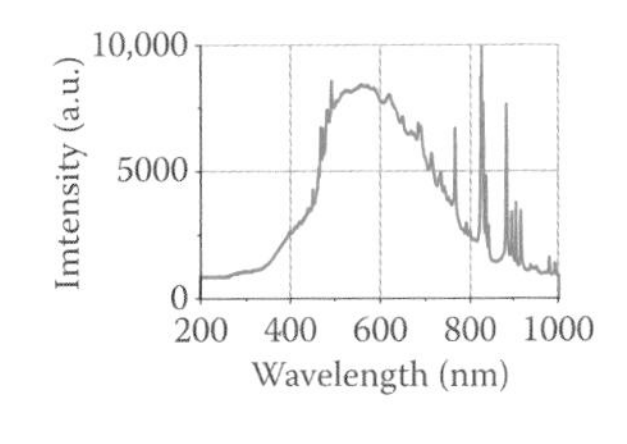

FIGURE 5.6 The emission spectrum of the xenon arc lamp.

When you need high intensity in an infrared region, a carbon arc lamp could be useful. It is very simple: a low-voltage arc discharge produced in a small space between two carbon electrodes placed in air. The spectrum is a continuum and very broad. The light is emitted by a small volume, so it can be well collimated or focused. The disadvantages are important: spatial and temporal stability because the carbon material of the electrodes is slowly burnt away. It needs a system of automatic adjustment of the distance between electrodes.

Other light sources occasionally used for the absorption measurement are lasers (most as laser diode [LD]) and LEDs. Such sources can be used only for an application at a specific wavelength that you can obtain from lasers or LEDs. Using a small laser diode with all power concentrated in a very narrow spectrum (a few nanometers), you can obtain a high sensitive and small size measurement instrument. The work at a specific wavelength eliminates the use of a monochromator, which is quite large and expensive. A simple laser diode with a proper power source is very cheap (recall a laser pointer). To change the emission wavelength without much power loss in a range of 1 nm, you can use a special power source where you can vary either the diode current (the simplest way) or the diode temperature (the better but more expensive way). The relationship between the diode temperature and the center wavelength is linear (Bartl et al. 2002). The wavelength of the maximum emission increases as the diode temperature increases. This characteristic is useful in certain spectroscopy applications of atomic and molecular spectroscopy, and process and pollution monitoring where the absorber line coincides or is around the diode emission line.

5.4 MONOCHROMATOR

5.4.1 COMPONENTS OF MONOCHROMATORS

The *monochromator* is a device "that transmits a selectable narrow band of wavelengths of electromagnetic radiation chosen from a wider range of wavelengths available at the input" (IUPAC 2012). In other words, it is a "black" box (the inner walls are black painted) where the white light enters and a monochromatic light exits. The wavelength of the exit color can be selected and changed.

Isaac Newton discovered that a prism of glass refracts the white light of the Sun into many colors and, when projected on a wall, looked like a rainbow. In his *Opticks* (Newton 1721), he described many experiments on light decomposition by prisms:

Proposition 1: "Lights which differ in colour, differ also in degrees of refrangibility (blue more than red light)."
Proposition 2: "The light of the Sun consists of rays differently refrangible."

For each experiment, Newton used a *small hole* in the window shutter to allow sunlight in before it passes through a prism. When Newton drew many incident rays on the prism, they formed a *parallel beam*. He also decomposed the light emitted from a candle and obtained approximately the same spectrum.

Nowadays, the components of the monochromator are (as illustrated by the block diagram in Figure 5.7):

- Entrance slit
- Collimator
- Dispersive element
- Projection objective
- Exit slit

The monochromator without exit slit is called a *spectrograph*. It means that you can see at once the whole spectrum of white light like the rainbow and not just one color after the exit slit as in the case of the monochromator. The Newton device was nearly close to a spectrograph as a final result. We will treat both types at the same time because the components and their functions are similar. The differences will be underlined where they appear. For an extensive treatment of spectral instrumentation reader is urged to read the excellent tutorial of Lerner and Thevenon (1988) or the up-dated short version (Lerner 2006).

The main part is the dispersive element whose task is to decompose the incident radiation into its components, giving to each component a different angular deviation according to its wavelength. Newton for the first time conducted this by using a glass prism and astronomer David Rittenhouse made a grating of wires to decompose light in 1785. A machine to make gratings was invented by H. A. Rowland (1882) but good gratings were available for commercial instruments during the later part of the twentieth century.

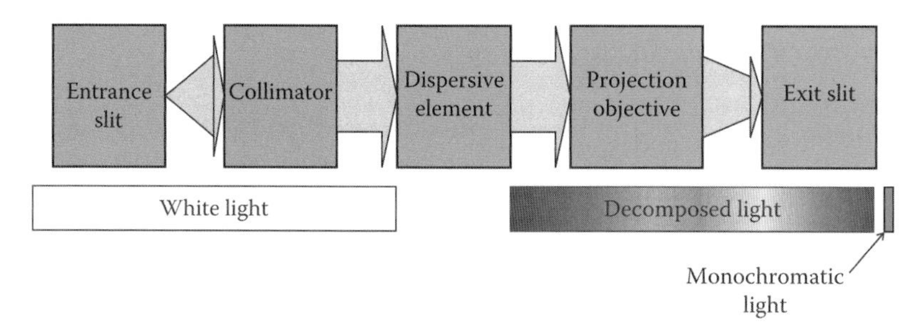

FIGURE 5.7 The block diagram of the monochromator.

5.4.2 OPTICAL PRISM

It is no longer used in most of the spectral instruments except in special cases. From a didactical point of view, it remains as a relatively easier example to understand the ray tracing.

In a transparent material, the refractive index depends on the wavelength according to the relationship of Cauchy:

$$n \approx C_0 + \frac{C_1}{\lambda^2} \tag{5.5}$$

where C_0 and C_1 are constants. The change of refraction index with respect to the wavelength change is called *dispersion* of light.

As a consequence, when light enters a transparent material, its velocity changes as shown by the definition of the refractive index:

$$n = \frac{c}{\upsilon} \Rightarrow \upsilon = \frac{c}{n} = \frac{c}{C_0 + \frac{C_1}{\lambda^2}} \tag{5.6}$$

Because $\lambda_{red} > \lambda_{violet}$, it results in $n_R < n_V$ and the phase velocity $v_R > v_V$. When the wavefront of white light is at normal incidence on the surface of a transparent material, a difference appears between the wavefront of red radiation and the wavefront of violet radiation inside the material, which will increase for higher thickness of the material (Figure 5.8). This type of separation of light components in the direction of propagation is not practically useful in spectroscopy because it is not stationary. (There are recent interesting applications in the case of ultra-fast laser beams where it is used to compensate the dispersion produced by some optical components, for example, microscope objectives, lenses, or in light amplification technology of high-power beams, for example, stretcher–compressor devices.)

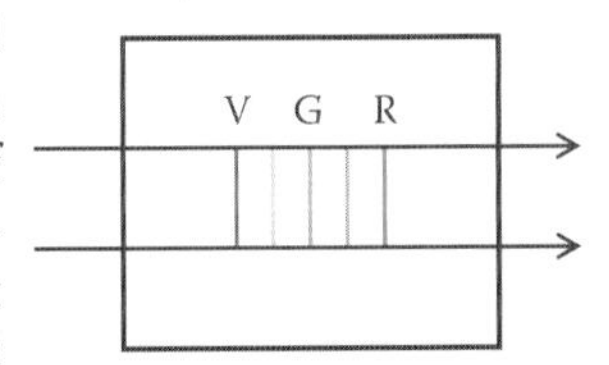

FIGURE 5.8 The red light (longer wavelength) is moving faster than the violet light (shorter wavelength) as a consequence of the dispersion.

The stationary separation of light in colors is produced by the optical prism in air, which is composed of two flat diopters, making an angle called prism angle A or the refracting angle. The relationships of optical prisms are as follows:

$$\begin{cases} \sin i = n \sin r \\ n \sin r' = \sin i' \\ A = r + r' \\ \delta = i + i' - A \end{cases} \tag{5.7}$$

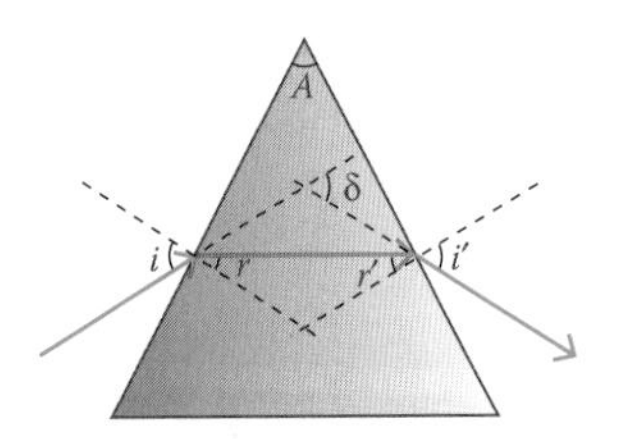

FIGURE 5.9 The ray tracing of the monochromatic light through a prism at minimum deviation.

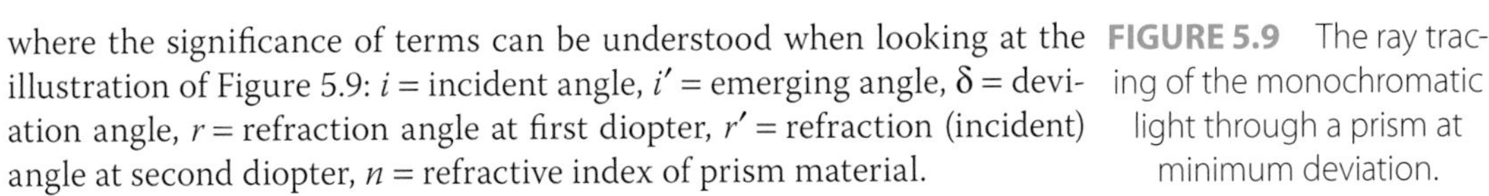

where the significance of terms can be understood when looking at the illustration of Figure 5.9: i = incident angle, i' = emerging angle, δ = deviation angle, r = refraction angle at first diopter, r' = refraction (incident) angle at second diopter, n = refractive index of prism material.

The illustration was already made in a particular case called *minimum deviation* (when δ has the smallest value), which is very important because the *optical aberrations are minimum* in that case.

The prism is in the condition of minimum deviation when either:

- *The light ray tracing has a symmetry* ($i = i'$, $r = r'$) regarding the prism axes, or
- *The light ray propagates parallel to the base prism*, or in other words, perpendicular to the prism angle bisector.

At minimum deviation, the refraction relationship is

$$n \sin \frac{A}{2} = \sin \frac{A + \delta_m}{2} \tag{5.8}$$

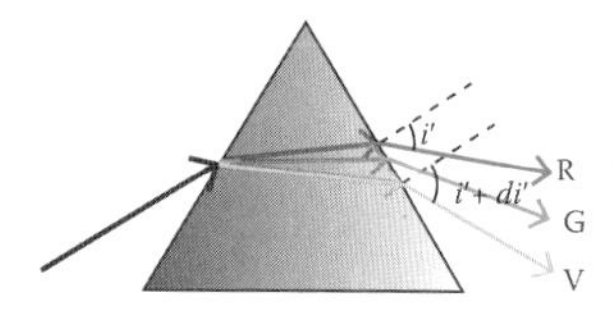

FIGURE 5.10 The light rays with different wavelengths will be deviated at different angles by the prism, with the shorter wavelength radiation (violet) at a higher angle than the longer wavelength radiation (red).

Thus, the *angle of minimum deviation depends on* the refractive index and therefore on the *radiation wavelength*. If several light rays with different wavelengths are incident on the prism at the same angle, then they will be deviated at different angles. Therefore, they will be *angular separated* by the prism as illustrated in Figure 5.10. The shorter the wavelength, the higher the deviation! This is a stationary deviation that can be observed by the naked eyes and measured by an ordinary photodetector.

The prism has two important parameters that are discussed as follows:

1. The *angular dispersion* is the variation of deviation angle with respect to the radiation wavelength.

$$D_a \equiv \frac{d\delta}{d\lambda} = \frac{di'}{d\lambda} \tag{5.9}$$

   Taking into consideration that in the fourth relation of the prism (Equation 5.7) $\delta = i + i' - A$, the incident angle i is the same for all rays and the prism angle A is constant; thus, the variation of deviation angle equals the variation of the exit angle.
2. The *power of chromatic resolution* (or shorter *resolving power*) is a measure of the prism capacity to produce fine spectral lines. The relation of its definition is

$$R \equiv \frac{\lambda}{d\lambda} \tag{5.10}$$

   where $d\lambda$ is the smallest resolvable wavelength difference between two wavelengths whose average is λ.

To understand its significance and to relate it to prism characteristics, we should recall the light diffraction phenomenon. Whenever light passes through a hole, the image of the hole has no sharp edges. Let us discuss an example of the slit that must be very thin in order to obtain a naked eye visible diffraction image. The slit illumination with a monochromatic and parallel light beam, for example, a red laser beam, results in an intensity distribution (or intensity contour) on the screen as in Figure 5.11b instead of the illustration in Figure 5.11a, which corresponds to geometrical optics. It is the result of the light diffraction. (The reader is recommended to read the book by Pedrotti (1993) for a detailed explanation on the diffraction phenomenon.)

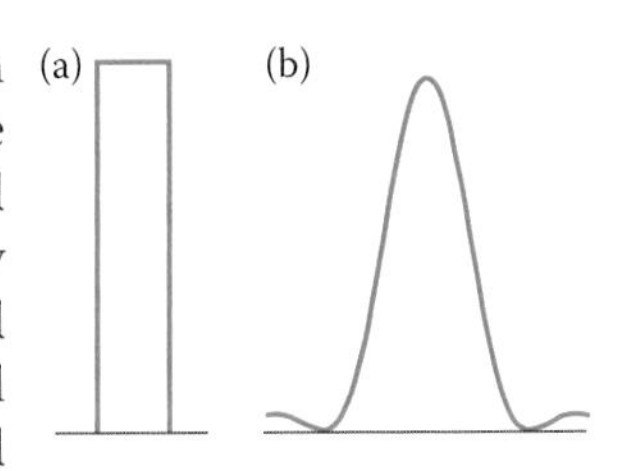

FIGURE 5.11 The intensity distribution of the light after passage through a rectangular aperture in (a) the geometrical optics approximation versus (b) the diffraction theory of the wave optics.

Now, imagine that the green beam from another laser also passes at normal incidence through the same aperture. The result will be similar to that of red light. When the green beam, simultaneous to the red beam, meets at an angle θ with the slit normal, then the green image will be superimposed on the red image as shown in Figure 5.12a. Their central maxima are well separated. The question is: How small may be the angle θ between any two beams of different wavelengths to be distinguished? The answer is given by the *Rayleigh criterion*: the fringes formed by two radiations λ and $\lambda + d\lambda$ are just resolved when the central maximum (of order zero) of diffraction image of $\lambda + d\lambda$ radiation is to occur in the same position as the first minimum of λ radiation (Figure 5.12b). In terms of intensity, two radiations λ and $\lambda + d\lambda$ are just resolved when the contour of the superimposed image has a difference between maximum and minimum intensities of about 19%. Such a contrast between "peak" and "valley" may be seen by a medium eye. So there are people who can easily see a smaller contrast, that is, two closer images, but we need a quantitative criterion generally valid, even if it does not express the best situation.

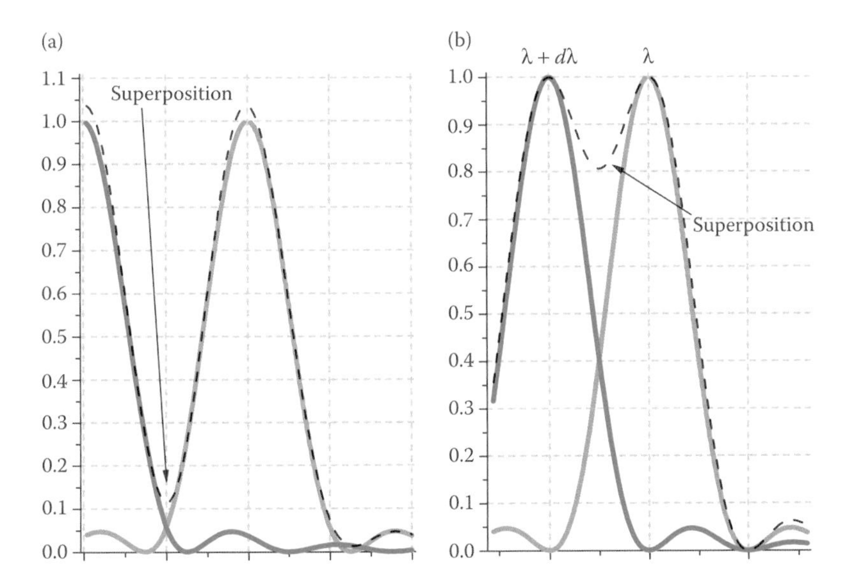

FIGURE 5.12 (a) The angle between two beams of different wavelengths is large enough in order to see two different diffraction images. (b) The Rayleigh criterion states that the fringes are just resolved if a maximum of one radiation occurs in the same position as the first minimum of the other radiation.

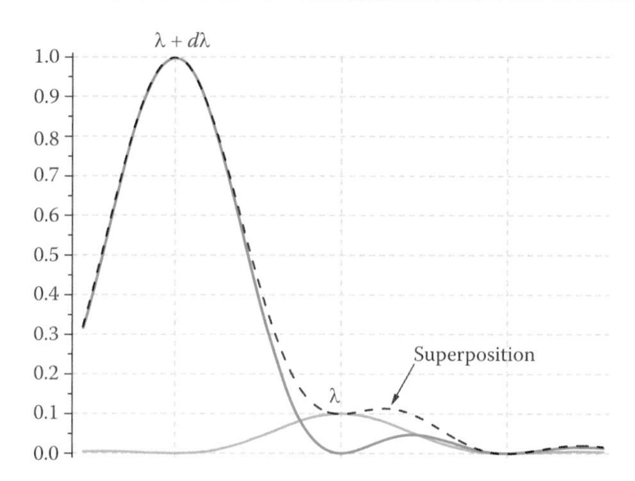

FIGURE 5.13 The superposition of two diffraction images of different intensities (here 1:10) does not obey the Rayleigh criterion.

Important note: the Rayleigh criterion applies to equal intensity radiations! Figure 5.13 illustrates that the superposition of two radiations with the ratio intensities 1:10 does not have enough contrast between "peak and valley."

Treating the diffraction by an open aperture of d width as a multiple interference, we can deduce that the optical path difference between the extreme rays passing through the slit for a monochromatic beam equals λ in order to obtain a minimum. So we may write the below relation to find the angular direction of the first minimum:

$$\sin\theta = \frac{\Delta}{d} = \frac{\lambda}{d} \tag{5.11}$$

Then, using the symbols of Figure 5.14, the angle θ is practically the angular difference di' between the two radiations that we want to resolve, which is very small. So we may rewrite the relation given by Equation 5.11 as follows:

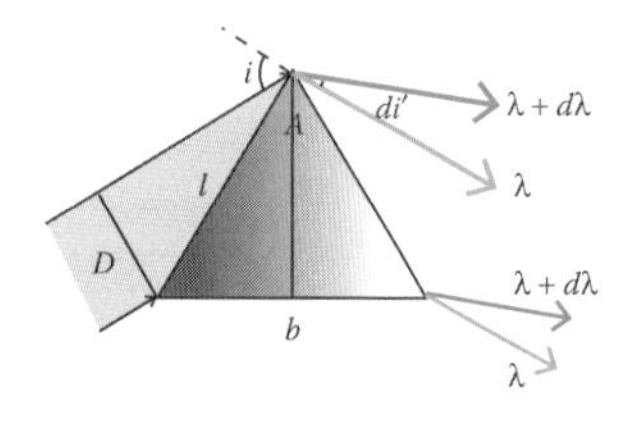

FIGURE 5.14 A light beam of width *D* at incidence angle *i* is refracted through a prism with edges *l* and base length *b* and is decomposed into two monochromatic components λ and $\lambda + d\lambda$, which are resolved by the angle *di'*.

$$di' = \frac{\lambda}{D} \Rightarrow D = \frac{\lambda}{di'} \tag{5.12}$$

We calculate the derivative of Equation 5.8 and we find the expression of the angular dispersion:

$$\frac{dn}{d\lambda}\sin\frac{A}{2} = \frac{1}{2}\left(\cos\frac{A+\delta}{2}\right)\frac{d\delta}{d\lambda} \Rightarrow D_a = \frac{b}{D}\cdot\frac{dn}{d\lambda} \tag{5.13}$$

Finally,

$$R = \frac{\lambda}{di'}\frac{di'}{d\lambda} = D\cdot D_a \Rightarrow \boxed{R = b\frac{dn}{d\lambda}} \tag{5.14}$$

Conclusions:

- The resolving power depends on the dimension of the prism (base length) and the dispersion of the material.
- Most of the prisms used in spectral applications are *equilateral*. So in order to double the resolving power buying a new one you should pay several times more because the volume of optical glass increases by eight times. Increasing the refracting angle of the prism is not a solution because it is limited by the emergence condition ($A < 2C$, critical angle) and reduces the input diameter of the beam.
- The prism material should have both a higher dispersion and a higher transmittance in the working range. Material dispersion is very high in ultraviolet range but very low in visible for a quartz prism. It is therefore recommended that a glass prism be used for visible measurements.

Nowadays, the optical prism is not usual as a dispersive element, but it is better to take it into consideration and also the above recommendations for the research experiments.

5.4.3 DIFFRACTION GRATING

Diffraction grating is another device used to obtain the stationary dispersion of light for angular separation of rays of different wavelengths. Most of the progress in grating fabrication is due to research work in astronomy. You may imagine how difficult it is to collect enough light coming from a very small bright point, that is, a star in the sky, and to separate emitted radiation by different isotopes, for example, hydrogen and deuterium (Vogt and Penrod 1988). Even Joseph von Fraunhofer, who realized the first quality diffraction grating in 1821 (Palmer and Loewen 2005), used it to measure the absorption lines of the solar spectrum.

The simplest model of grating is a glass plate with many (N = tens of thousands) parallel and equidistant scratches. In this way, we have a system with thousands of slits of the same width. A parallel light beam passes through all slits. Thus, the theory of diffraction grating is treated as N-slits interference.

When a parallel beam of monochromatic light normal impinges on a screen which has three, five, and seven narrow slits, respectively, it results in a distribution of intensity of diffraction pattern as illustrated in Figure 5.15. The phenomenon is called Fraunhofer diffraction and the final image consists of thin equidistant interference maxima (the same number whatever the number of slits) whose intensity decreases because the single-slit diffraction pattern acts as an "envelope" for the interference pattern.

Important remarks drawn from the illustration:

When the number of slits increases:	1. The main interference maxima become narrower!
	2. The intensities of secondary maxima between two main maxima become lower.

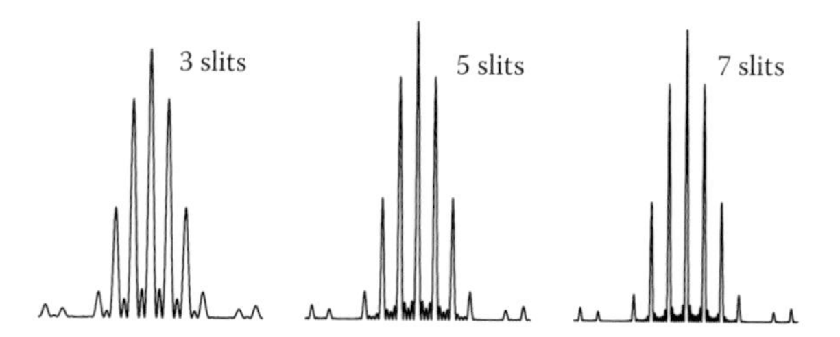

FIGURE 5.15 The calculated distribution of intensity in the diffraction pattern produced by 3, 5, and 7 slits for monochromatic light. The intensity of main interference maxima is modulated by the diffraction on each narrow slit.

Both remarks are very important for grating use in spectroscopy as follows:

As the *total number of grating slits N is greater*:	1. The *spectral lines will be sharper* (increased resolution). 2. In order to keep constant the energy when the diffraction maximum becomes sharper, it *also becomes much more intense.* 3. The *contrast* between spectral lines and the background *will be greater* (signal-to-noise ratio increases).

While the intensity distribution depends weakly on the wavelength, the angular position of each main maximum depends linearly on wavelength as shown by the *fundamental equation of diffraction grating* of optics:

$$d(\sin\alpha \pm \sin\beta) = k\lambda \quad (k = 0, \pm1, \pm2, \pm3,\ldots) \tag{5.15}$$

where d is the periodicity of the groove (distance between the middle of two adjacent grooves), α is the incidence angle, β is the diffraction angle, k is the diffraction order, and λ is the radiation wavelength. In applied spectroscopy, it is common to know the grooves density n (given in grooves per mm or grooves per inch) that is often written on the grating frame. Thus, it is more comfortable to write the fundamental equation as follows:

$$\boxed{\sin\alpha \pm \sin\beta = nk\lambda} \tag{5.16}$$

This equation immediately results in spectral use of it: when a light beam is composed of radiations of many wavelengths, which is incident on a grating (α is the same for all), then each component will be deviated at an angle dependent on radiation wavelength (β is different). This behavior will happen for each order of main maxima (except for zero order).

Let us take the example of a transmission grating of 600 gr/mm. Consider the white light beam composed of only three radiations: violet ($\lambda_V = 400$ nm), green ($\lambda_G = 550$ nm), and red ($\lambda_R = 700$ nm). The light beam impinges the grating at a normal incidence ($\alpha = 0$). What is the number of maxima for each color that we can see? The results are shown

TABLE 5.1 Angular Deviation of Three Radiations by a Diffraction Grating of 600 gr/mm

Radiation Color	Wavelength (nm)	sin β $k = 1$	 $k = 2$	 $k = 3$	 $k = 4$	 $k = 5$
Violet	400	0.24	0.48	0.72	0.96	1.2
Green	550	0.33	0.66	0.99	1.32	1.65
Red	700	0.42	0.84	1.26	1.68	2.1

Note: The results in shaded cells are forbidden.

in Table 5.1. The results of the shaded cells are not possible because $\sin \beta \leq 1$.

The answer to the above question is: eight violet maxima (four maxima symmetrical on each side), six green maxima, and four red maxima could be visible. The maximum of zero order is white (the white light is not decomposed). The phenomenon is illustrated in Figure 5.16. It was not possible to represent the 4th maximum of the violet radiation (V_4) in its correct position and the 3rd maximum of the green radiation (G_3) on the linear screen.

Remarks:

1. If there are multiple monochromatic radiations λ_1, λ_2, λ_3… they will appear in different places on the screen, because the maximum direction depends on λ according to the fundamental equation of the diffraction grating. Therefore, for white light, a rainbow spectrum of first order will be obtained between violet and red of the spectrum (V_1–R_1), another one of second order between V_2 and R_2, and so on. It is the main property for the spectral use of the diffraction grating.

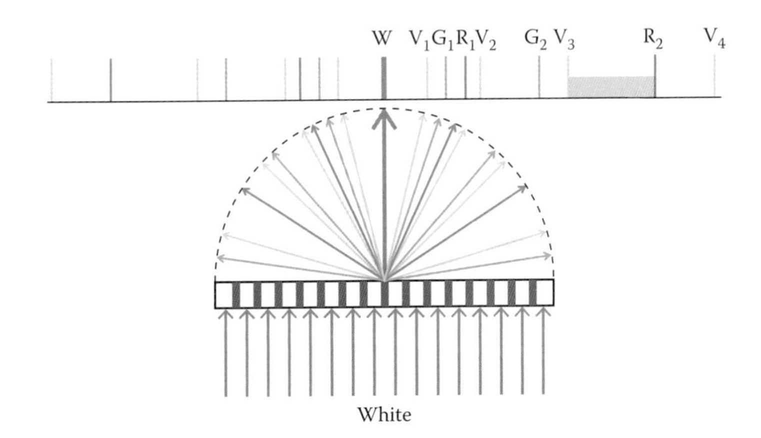

FIGURE 5.16 The white light is decomposed in its components because they are deviated at different angles depending on their wavelengths. We have used the values of Table 5.1.

2. The grating can form multiple spectra of different orders (1, 2, 3...) of the same incident light beam while the prism can form only one spectrum. At the same time this property has both positive and negative implications that will be discussed below.
3. It is negative because the same amount of light will be distributed in several spectra and luminosity will be much lower than that of the prism spectrum. This was for a very long time the most important reason to keep the (normal) gratings out of the spectroscopy field.
4. It is positive because the spectral dispersion will be almost double in the second order compared to the first order spectrum (it is correct for small angles). Working in a higher order, the grating could give an impressive dispersion impossible to be obtained by any prism or multi-prism.
5. In any optics book you can see the pattern produced by the Fraunhofer diffraction where the main maxima are equidistant (different from that of Figure 5.16). It is also mentioned that for lower angles, cos β will not differ much from unity, and so the angular dispersion (defined hereunder) is almost constant. In conclusion, any two different spectral lines in a spectrum will have an angular difference that is directly proportional to their difference in wavelength. Therefore, this linear scale of wavelengths is the main advantage of the gratings over the prism! It is quite right for gratings with a lower density of grooves. In the case of the 600 gr/mm grating presented in Table 5.1, this conclusion is not quite correct even for the first-order spectrum. In practice, one can work with 1200 or 2000 gr/mm gratings where the first order is obtained at higher angles where the above conclusion is not correct.
6. When the spectra become larger, the overlapping of neighboring spectra occurs. It is illustrated in Figure 5.16 where the red part of the second order overlaps on the blue part of the third-order spectra, that is, between V_3 and R_2 (shaded range). It means that the second-order spectrum is useful approximately in the range 400–500 nm and the third-order spectrum is useful starting at

$$\sin\beta = n2\lambda_R = n3\lambda_X \Rightarrow \lambda_X = \frac{2}{3}\lambda_R = \frac{2}{3}700 = 466 \text{ nm}$$

The range of the useful part of the spectrum is called the *free spectral range* and can be calculated as follows (according to Figure 5.17):

$$\sin\beta = k(\lambda + \Delta\lambda) = (k+1)\lambda \Rightarrow \Delta\lambda = \frac{\lambda}{k} \tag{5.17}$$

Equation 5.17 is useful for the gratings called *echelles* that work in higher orders. For example, the free spectral range for the order 100 is only 1% of

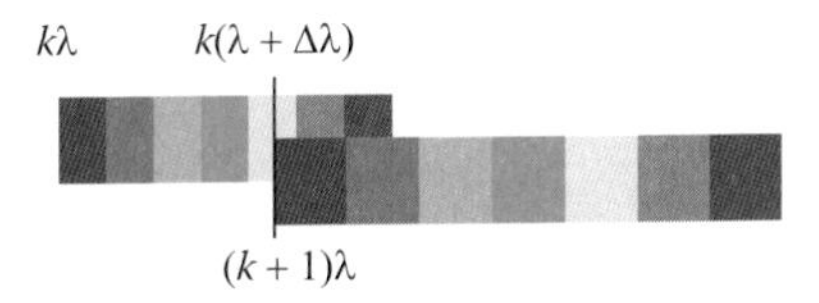

FIGURE 5.17 The overlapping of two adjacent spectra of k and $k+1$ orders, respectively.

the observation wavelength (5000 Å, $\Delta\lambda = 50$ Å). It is an important reason for that grating type to be used in conjunction with another dispersive device, usually a prism. The overlapping occurs even in the first order when you need to work with the same grating in the UV–Vis range (200–800 nm). The UV range of the second order spectra begins in the same position as the Vis range of the first order (2×200 nm $= 1 \times 400$ nm!). This is troubleshot by introducing a blocking filter in the optical path corresponding to the visible range to cut the UV range. Another solution is the crossed dispersion geometry: the grating dispersion is followed by prism dispersion in a perpendicular direction. The final image consists of points instead of lines and is used in several instruments without monochromators. It is called the polychromator.

The diffraction grating also has several main parameters as the prism.

1. The *angular dispersion* is simple to deduce starting from the differential form of the grating equation:

$$D_a \equiv \frac{d\beta}{d\lambda} = \frac{nk}{\cos\beta} \tag{5.18a}$$

It is obvious that the angular dispersion can be simply increased by changing the grating with a higher density grating (it is the expensive way) or by measuring in the second-order spectra taking into consideration the previously mentioned difficulty. The nk product can be substituted by using the fundamental equation and angular dispersion will be

$$D_a = \frac{\sin\alpha \pm \sin\beta}{\lambda \cdot \cos\beta}$$

that reduces under the particular condition $\alpha = \beta$ (called the Littrow configuration) to

$$D_a = \frac{2 \cdot \tan\beta}{\lambda} \tag{5.18b}$$

For a given value of wavelength, the angular dispersion increases several times when β increases.

2. The *power of chromatic resolution* (*resolving power*) is a measure of the grating to separate adjacent spectral lines of average wavelength λ. Starting from the Rayleigh criterion, two radiations λ and $\lambda + d\lambda$ with the same intensity are resolved when the main

maximum of diffraction image of $\lambda + d\lambda$ radiation occurs in the same position as the first minimum of λ radiation. In terms of optical path difference we can write:

$$\left.\begin{aligned} \Delta_{\lambda+d\lambda} &= k(\lambda + d\lambda) \\ \Delta_{\lambda} &= k\lambda + \frac{\lambda}{N} \end{aligned}\right\} \Rightarrow k(\lambda + d\lambda) = k\lambda + \frac{\lambda}{N}$$

We made the approximation $N - 2 \approx N$ because of the great value of N. Thus, the resolving power of grating is given as

$$\boxed{R = kN = knL} \tag{5.19a}$$

where k is the diffraction order, N is the total number of illuminated grooves of the grating, n is the grooves density, and L is the grating length (only of the scratched surface).

To increase the resolving power, it is better to review the discussion about angular dispersion. k order and n density changing are available with their consequences. The changing of the grating size L is not common because of problems in illumination.

We underline here how important the *illumination of all grating surfaces* is in order to use the resolving power of grating. The most common mistake in an experimental setup is to forget the rule of all grooves illumination that reduces a lot of the resolving power. This happens especially when using a laser beam that is sometimes thinner than the entrance slit. When you change the grating with a larger one, you must be careful because its correct illumination is given by other optical components described below.

We can rewrite the resolving power formula:

$$R = \frac{\sin\alpha + \sin\beta}{\lambda} L$$

The maximum resolving power will be obtained for $\sin\alpha + \sin\beta = 2$ and so

$$R_{max} = \frac{2L}{\lambda} \tag{5.19b}$$

This value is given at grazing incidence when $\alpha = \beta = 90°$. It means that the maximum resolving power is the ratio between the maximum optical path difference (interference is the main optical phenomenon involved) between diffracted rays and the wavelength.

The value of resolving power given by either Equations 5.19a or 5.19b is theoretical. Neither of them have included the quality of grating. This depends on the surface flatness, the uniformity of groove distance, the uniformity of the reflecting layer, and so on. All are issues depending on the producer technology.

Remember that the diffraction grating is only a component of the spectral instrument. Usually the resolving power of grating is great enough so that the spectral resolution of the instrument will be limited by other components.

3. *Grating efficiency* means how much of the incident light can be found in the area where the measurement is done. For comparison purposes, the glass prism looks perfectly transparent in the visible range, but when the light passes through it, it loses a few percent at every face and also by scattering inside the prism. Per total the prism should have more than 80% of incident light in the output spectrum. Only in the far-infrared range can the efficiency of prism of special materials be less than 50%. Concerning the diffraction grating, imagine that a perfect grating distributes all the light energy in several maxima and that you measure only one, for example, right first-order maximum, which is not the brightest. This major difficulty was surpassed by changing the shape of scratch to asymmetric as discussed below.

The diffraction grating types can be divided into transmission and reflection gratings. Transmission gratings pose a major inconvenience for spectral use: the material (glass or quartz) has a limited spectral range with higher transmittance. When the scratched glass slide is covered by a thin metal layer with high reflectivity on a very broad spectral range, this is a reflection grating. The layer can be of aluminum, silver, or gold. Aluminum has the advantage of a flat high reflectivity spectrum. It must be protected against oxidation by a thin silica layer. Gold is the best mirror layer in infrared range and is almost chemically inert. Both transmission and reflection gratings have a similar low efficiency. The efficiency was greatly improved by making asymmetric scratches as shown in Figure 5.18. The tilt angle of each scratch is calculated to concentrate light into a particular order direction. When the normal to the groove face bisects the angle between the incident ray and diffracted ray, then most of the light is directed to the −1 order because of the reflection (see the callout on the left side of Figure 5.18), namely, $\alpha + \theta = \beta - \theta$ in our example. Thus, the tilt angle θ is called the *blaze angle*. In this way, the efficiency can increase above 50%, but it is still lower than the prism efficiency.

A particular case is where the diffracted light returns in the direction of the incident light. It happens when $\alpha = \beta$, which is *called the Littrow configuration*, and plays an important role in monochromators. For a blaze grating (where $\alpha = \beta = \theta$), the diffraction equation then simplifies to

$$2\sin\theta = nk\lambda \quad (5.20)$$

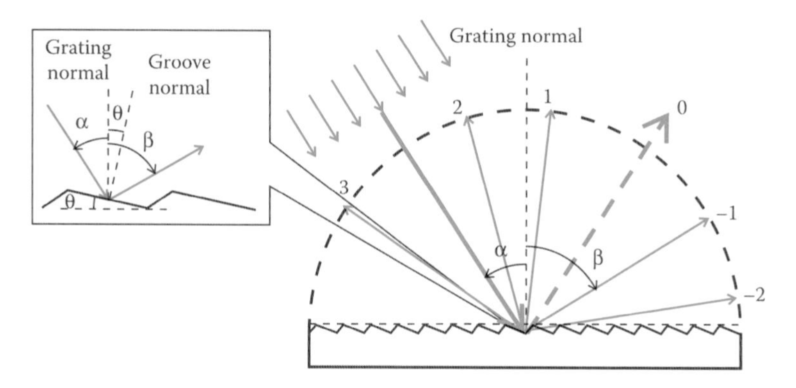

FIGURE 5.18 The ray tracing in the reflection grating with tilted grooves. The blaze condition illustrated in the left callout is that the groove normal bisects the angle between the incident and diffracted rays.

Remember that the grating equation is written for a given wavelength, and so the blazed grating is characterized by the *blazed wavelength*, that is, the wavelength where the grating has maximum efficiency. The producer catalogs typically specify the main characteristics of diffraction gratings in terms of groove density, size, angular dispersion, blaze angle, and either blaze wavelength or, better, the efficiency curve.

There are two types of gratings depending on the technology used to manufacture them (see a detailed description in Palmer and Loewen 2005):

- Ruled gratings
- Holographic gratings

Ruled gratings were earlier manufactured using a machine with a diamond tool to scratch tens of thousands of lines on a thin film of metal coating applied to a glass plate. In fact, the grating obtained in this way is the master grating that is not for sale. It is used only to make replicas of the same quality as the master grating, but cheaper. Ruled gratings generally *have a high efficiency* especially at the blaze wavelength (the efficiency curve is sharp). These gratings are often used in systems requiring a high resolution. They can also produce "ghost" spectra because of the periodic error in the ruling process of very high-density gratings, but this issue was much reduced in modern machines.

Holographic gratings have no ghost spectra and lower scattered light than ruled gratings due to the optical technology instead of mechanical displacement-based technology. The groove profile is sinusoidal, and thus they generally have a lower efficiency. When they are supplementary processed with ion etching, to give a flat face and sharp angle to the grooves, the efficiency increases. By using the ion etching technique, it is possible to obtain any groove shape for either a plane surface or curved surface. They are expensive even as replicas.

We have a dispersive element: what more do we need to have a spectral instrument?

5.4.4 COLLIMATOR

The ray-tracing diagram of prism and diffraction grating results in the incident light beam must be composed of parallel rays called *collimated beam*. The light comes from the source, even from a laser device, as a divergent beam, so an optical system of collimation should be included. The simplest collimator is either a convex lens or a concave mirror. The mirror is preferred due to the lack of chromatic aberration and for the broad spectral range of high reflectivity. A parabolic mirror can produce a parallel beam if the source is a bright point placed in the focus. A spherical mirror can approximately do a parallel beam at a lower price. The real issue for both the mirror and lens is that the light source is not a point. Another issue of using mirrors is the off-axis position of the source so that the beam path of the collimated beam is not blocked by the source.

5.4.5 PROJECTION OBJECTIVE

The emerging rays from the prism are deviated according to their wavelength, but they are still parallel for a given wavelength as the incident beam does; similarly, only the parallel rays must be considered in the case of Fraunhofer diffraction by a grating. Imagine a 50 mm diameter outgoing beam and a light detector of a few millimeters aperture that must capture all the light. So the beam should be concentrated on it and this is the task of the projection objective. It could be either a lens or a mirror. The objective does the inverse process as the collimator does and it is often the same as the first, with the same characteristics.

5.4.6 ENTRANCE SLIT

A simple question arises: since we have to collect as much light from the source, why must it place a narrow slit between the source and the collimator?

The spectral instrument we have built until now is made of: entrance slit, first lens, prism or grating, second lens (objective). If we place a photographic film (or a CCD) in the focal plane of the second lens, we have a device like a photographic camera that will record a real image of the object. The object is the slit. The image at magnification 1:1 has exactly the shape and size of the entrance slit, that is, a bright narrow line on the black background. However, the difference from a photographic camera is the presence of several lines, one for each wavelength of the radiation passing through the slit. So the lines have different colors due to the presence of the dispersive element (prism or grating). To separate two wavelengths with very close values, we have to obtain images of the slit as narrow as possible and this is done by reducing the slit width. The image lines have to be not just very narrow but also very clear from an optical point of view and this is achieved by *controlling the ray angle* entering the device. Without the entrance slit, the light would enter the spectroscope from a range of angles. As a consequence, the position of a spectral line would be spread out over a range. The entrance slit must be placed in the focus plane of the collimator to obtain the parallel beam.

The second function of the entrance slit is to *adjust the intensity* of the light that enters the instrument in such a way as to be measurable.

There are instruments where the slit width is fixed by the manufacturer. In this case, you need to know exactly what is to be measured with the instrument before you order it: you need either high selectivity or high sensitivity. We will discuss later how important the entrance slit width is for the performance of the instrument.

5.4.7 EXIT SLIT

The exit slit is twin to the entrance slit. Its presence is not obligatory. It depends on the type of detection you have to use—either single detector or multiple detectors. It is placed in the focal plane of the projection objective, where all rays are focused but in different positions transversal to the optical axis. Its purpose is to select a narrow part of the spectrum projected by the objective onto the slit wall and therefore to select a line to be measured by the detector. It is simultaneously adjustable with the entrance slit.

5.4.8 BASIC DESIGNS

Depending on the presence of the exit slit, there are two main classes of spectral instruments:

- Spectrograph, where the out product is a broad spectrum
- Monochromator, where the out product is a very narrow part of the spectrum

The optical diagram of a *spectrograph* is illustrated in Figure 5.19. We have used as an example a classic spectrograph with transmission optics because it is easier to follow the ray paths. The light source is not part of the spectrograph, but it is important to draw it in order to understand how light must enter into a spectral instrument. The illumination of the entrance slit is detailed on the left-bottom side of Figure 5.19: (a) the supplementary lens projects the collected light from the source onto the slit width in order both to obtain the proper angle to illuminate the entire cross section of the prism and to collect more light than when it is missing (b). When the prism is not completely illuminated, the resolving power decreases (the illuminated prism base is shorter). So the white light passes through the entrance slit, is collimated by an achromatic lens (or a lens system), and enters the prism that starts to deviate each ray corresponding to its wavelength. The emergent light is formed by many collimated beams that partly overlap each other. The objective lens projects and focuses all beams in its focal plane that is not perpendicular to the lens main axis due to the chromatic aberration. The detector is placed at an angle to the optical axis so that the spectral lines (images of the input slit) are clear on its surface.

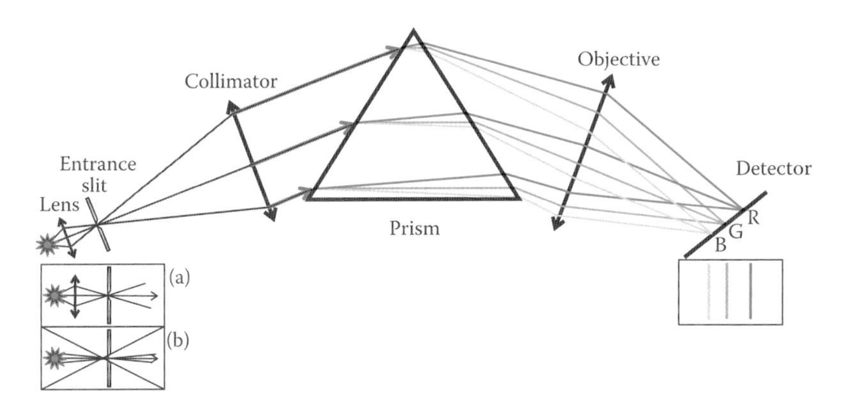

FIGURE 5.19 The optical diagram of a prism spectrograph. (a) The supplementary lens task is to collect more light and to adjust the entrance angle so that the prism may be completely illuminated (correct). (b) The direct illumination of the slit is not recommended (wrong).

The monochromator may have a supplementary output slit to deliver a narrow part of the spectrum to a single detector. In this case to measure the entire spectrum, we must scan the spectrum in front of the exit slit. This is done by the prism rotation due to a complicated mechanical system. The grating monochromator has a similar mechanism to rotate the grating. We illustrate the optical diagram of the grating monochromator by reflection in two arrangements:

- The Czerny–Turner (CT) mount consists of two concave mirrors and one plane diffraction grating as shown in Figure 5.20a

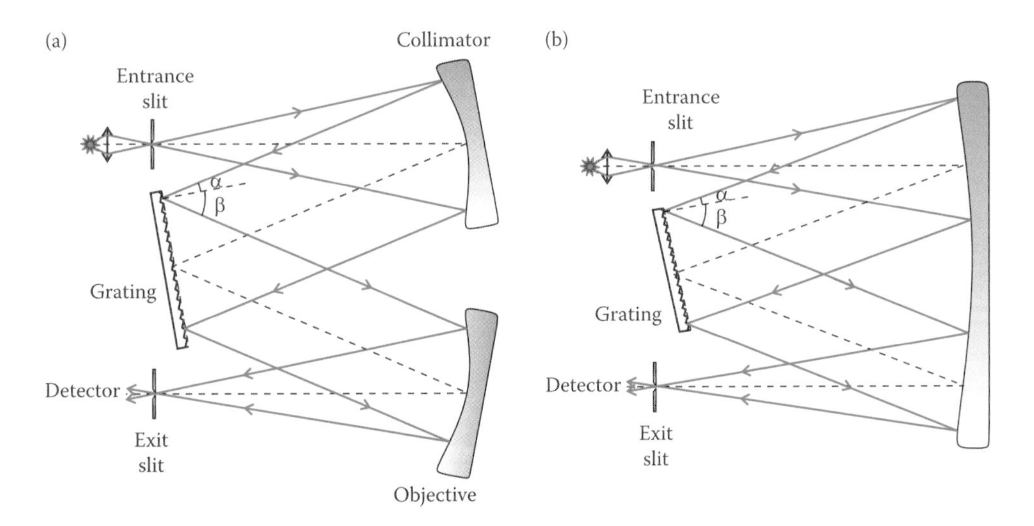

FIGURE 5.20 The optical diagrams of two arrangements for grating monochromators: (a) The Czerny–Turner mount and (b) the Ebert mount. The light source and photodetector are not components of the monochromator.

- The Ebert mount consists of a single large mirror in order to both collimate the white light and focus the dispersed light and a plane diffraction grating (Figure 5.20b). We mention that two different parts of the mirror are used to accomplish the two tasks, so the mirror must be larger than in the CT mount

Each component of the monochromator/spectrograph has an important role in the proper functioning of the instrument. However, at the same, time it contributes to increase the instrument optical aberrations or width of the spectral range by its properties of transmission/reflection. The correction of aberrations is possible with much effort but often impossible. Astronomy was (and still is) the field that poses very difficult problems: extremely low intensities, very high resolution, extreme spectral ranges. Henry Rowland, engaged in a labor of studying the Sun's spectrum, had the brilliant idea to include the collimator, grating, and objective in a single device: a concave grating. He engineered a ruling machine which he realized had to have a large number of grooves on a concave mirror. No primary optics is necessary in the Rowland mount as shown in Figure 5.21. The concave grating, entrance slit, and multiple detector (or exit slit) are placed on a circle (called the Rowland circle) the diameter of which equals the radius of the mirror curvature. This system, most frequently used as a spectrograph, is attractive for use in the extreme and far ultraviolet region where low reflectivity of common optical coatings can severely limit instrument sensitivity. Another particular advantage of the concave grating spectrograph is that it is lighter than a plane grating spectrograph and, as a consequence, it is suitable for space missions (NASA-JPL 2013). It is also commonly used in X-ray optics.

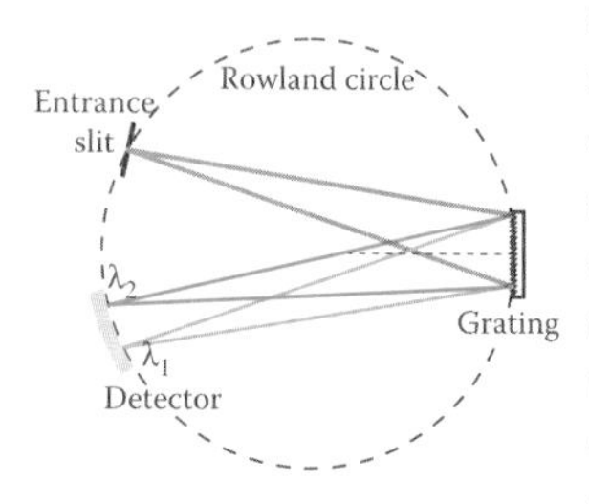

FIGURE 5.21 The Rowland spectrograph with concave grating.

5.4.9 MAIN PARAMETERS OF INSTRUMENT PERFORMANCE

5.4.9.1 LINEAR DISPERSION

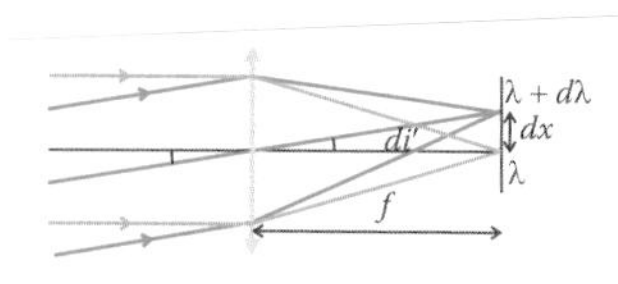

FIGURE 5.22 Two radiations just resolved by the prism/grating are focused in two lines separated by a distance *dx*.

Imagine two monochromatic components λ and $\lambda + d\lambda$, which are resolved by the angle di' by prism/grating, and focused on the objective in its focal plane as shown in Figure 5.22. It is the real image given by the instrument and seen by the detector. We do not have direct access to the angular dispersion of the dispersive element that reflects its resolving capacity, but it is more convenient to express the dispersion in terms of a linear scale at the detector rather than an angle. This is the *linear dispersion* given by

$$D_l \equiv \frac{dx}{d\lambda} \quad (5.21)$$

where dx is the minimum distance between two lines that can be seen as resolved. It can be directly measured. According to Figure 5.22, we can rewrite Equation 5.21:

$$D_l \equiv \frac{dx}{d\lambda} = \frac{dx}{di'}\frac{di'}{d\lambda} = f \cdot D_a \quad (5.22)$$

Hence, the linear dispersion is the product between the objective focal length and the angular dispersion. Greater linear dispersion means either longer focal length or greater angular dispersion.

In practice, Equation 5.21 is often inverted to give the *reciprocal linear dispersion dλ/dx*, that is, the maximum wavelength range for a given length of the detector. It is expressed in units of nm/mm. By using the reciprocal linear dispersion, we can simply calculate the spectral range that can be simultaneously measured by a multiple detector. For example, an ICCD camera with a sensor of 1024×255 pixels and a 26×26 μm pixel size will have an active detection length of approximately $1024 \times 26 = 26{,}624$ μm. The camera can be attached to a spectrograph with the following specifications: 163 mm focal length; the reciprocal linear dispersion is 9.25 nm/mm for a grating of 600 lines/mm (blazed at 500 nm). So the spectral range you can measure in a single exposure is 9.25 nm/mm × 26.624 mm ≈ 246 nm. To cover the entire visible range (400–800 nm), you should choose a grating of 300 lines/mm, which can provide a reciprocal dispersion of 19 nm/mm for the same spectrograph, but you will lose the resolving power. The grating price is much cheaper than that of the spectrograph or camera.

Remember that if two radiations are not angularly resolved by the prism/grating, they will not be resolved by the spectral instrument no matter how long the focal length of the optics is.

5.4.9.2 LUMINOSITY

The luminosity is a measure of how much light enters in the monochromator/spectrograph. The amount of light entering the instrument depends on the slit width and the solid angle subtended by the first lens/mirror, that is, by the collimator. We ignore the influence of the quality of optical components. The slit width is directly related to the capacity of the instrument to form as narrow lines as possible. Because the spectrograph is like a camera where each spectral line is the image of the entrance slit, we should keep a narrow slit in order to obtain very thin lines. Hence, the luminosity of the instrument will mainly depend on the characteristics of the collimator. From Figure 5.23, one can observe that the luminosity of the image given by a lens depends on being directly proportional to its aperture and inversely to its focal length:

$$\Lambda \propto \left(\frac{D}{f}\right)^2 \quad (5.23)$$

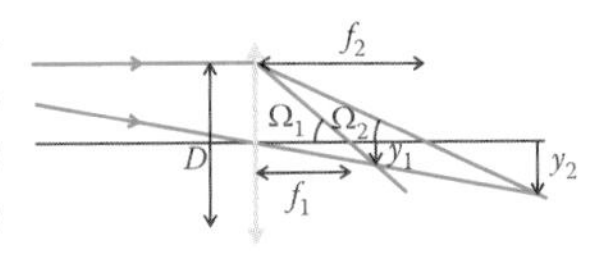

FIGURE 5.23 The luminosity of the lens decreases proportional to the square of the ratio between its diameter and its focal length, because the same light flux is distributed over a larger area. We represented two lenses of the same aperture but of different focal lengths. *D*—linear aperture of the lens, Ω—one-half the angular aperture, *f*—focal length, *y*—image size.

The square is due to the energy distributed on the image area. The value of lens diameter is given so that the collimated beam uniformly illuminates the entire active area of grating it is fixed by. So the luminosity can be adjusted by choosing the focal length of optical components. We note that a greater focal length results in both greater linear dispersion and lower luminosity! This balance results in the spectrograph being selected according to the type of measurements to be made. Therefore, you must decide what performance is more important—either linear dispersion or luminosity. In fluorescence measurements, the signal is low and the spectrum often has few broad bands. The result is that luminosity is more important and you will select a short focal length spectrograph, for example, 163 mm as in the previous example. While Raman spectroscopy means both a low signal and narrow lines overlapping, the Raman spectrograph should mainly separate lines so a long focal length spectrograph is needed, for example, 500 mm or longer. Low luminosity should be overpassed by selecting a higher intensity source (laser) for excitation.

In catalogs, the information about the luminosity of the instrument is given as: aperture ratio, *f*/value, *F* number, or numerical aperture (NA), depending on the producer. The numerical aperture NA was defined in microscopy as given by the following equation:

$$NA = n \cdot \sin \Omega \tag{5.24}$$

where n is the refractive index of the medium in front of the microscope objective and Ω is half of the angular aperture (angle of the cone made by the light entering into the objective). In spectroscopy, $n = 1$ usually. Higher NA means more light captured by the objective.

Because the spectrograph is an "imaging" instrument, its luminosity is often described as that of any optical system by *F*-number or *f*/value, most known from photographic techniques. It is the ratio between the focal length and diameter of the lens:

$$f/\# = \frac{f}{D} \tag{5.25}$$

For example, if the spectrograph has a focal length equal to 163 mm and its pupil diameter is 45 mm, the *F*-number is 3.6 and it would be expressed as *f*/3.6. If you have the option to select a double focal length spectrograph, the *F*-number will become two times higher for the same pupil diameter. However, remember that you will measure a double image, where the energy is distributed on the detector area, and the illuminance of each detector pixel decreases four times (!) according to Equation 5.23. In such a case, you need to increase by four times the recording (acquisition) time.

5.4.9.3 STRAY LIGHT

When you intend to use a monochromator to measure at specific wavelengths, the radiation from all emitted spectrum of the light source enters the instrument and only the radiation selected exits. What happens with the remaining radiation that is dominant? It is absorbed by the inner surface of the monochromator walls. A very small quantity of all spectral ranges escapes through the exit slit and it is called a "stray light." There are many causes that can lead to the formation of larger or smaller amounts of stray light (Bauman 1962), but it is important to know that it can influence the precision of quantitative measurements. There are methods of measuring the value of stray light, but these are difficult for end users (ASTM 2009). The stray light issue is the same in spectrographs.

In particular applications, where the light source is several orders of magnitude stronger compared to the signal to be measured, it is required that the second monochromator or the third monochromator be used (as in the most performing instruments of Raman spectroscopy). There are two or three identical monochromators connected in series, i.e., in a so-called additive geometry, the exit slit of the first monochromator simultaneously acts as the entrance slit of the second monochromator, and so on. The mechanical systems are coupled in such a way that they select the same wavelength. If a single monochromator achieves 10^3 suppression of the stray light, then a double monochromator can reach a factor of 10^6. It is recommended whenever either the light source is a high power broad band as the xenon lamp or measurements must be made near the laser line. It is the most expensive method and the most efficient in order to reduce the stray light. A much cheaper method is to use rejection filters.

5.4.9.4 RESOLUTION AND SPECTRAL BANDWIDTH

A consequence of the uncertainty principle of Heisenberg is that purely monochromatic radiation does not exist. In spectroscopy, the term "line" is defined by its *width and shape* (profile of intensity). If we measure a radiation emitted from a low-pressure gas discharge lamp using a high-performance spectral instrument, we obtain the natural line profile called the Lorentz profile (Figure 5.24). The displacement of atoms during emission produces another type of broadening (Doppler) that results in the so-called Gaussian profile (Figure 5.24). The Gaussian profile strongly depends on temperature and is often used while the Lorentz profile is rarely used. The combination of a thermal and natural broadening of line is called the Voigt profile.

The question is how large is a line? There is no limit either to the right or to the left to start/end measuring. So we must find a criterion to define the line width. The profile always has a maximum value. For a symmetric line, the line width is defined as the spectral range measured at half of the peak maximum and it is called *full-width at half-maximum* (FWHM) that is shown for the Gaussian profile in Figure 5.24.

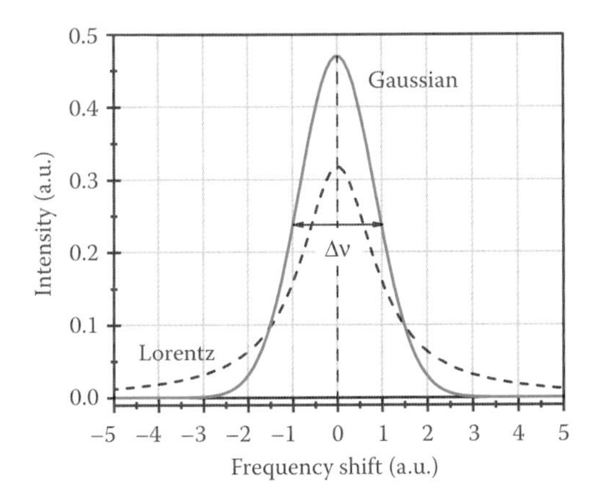

FIGURE 5.24 The Lorentz and Gaussian profiles of natural lines.

We note that the natural line broadening is of the order of 10^{-4}–10^{-3} nm. The broadening affects both emission and absorption lines, although the causes are not the same. In absorption, the line broadening can be asymmetric and is larger than that of emission lines.

A new question arises: Can we measure such a line profile using a regular monochromator? Another question: can we get such spectral width at the exit of a monochromator? The monochromator can produce a monochromatic radiation unless the entrance/exit slit is infinitely small, in which case it would be impossible to measure the energy.

From the definition previously mentioned results that a monochromator "transmits a narrow band of wavelengths." It is called *resolution* $\Delta\lambda$ (or *bandpass*) and can be calculated by the following relationship:

$$D_l = \frac{dx}{d\lambda} = fkn \Rightarrow d\lambda = \frac{dx}{fkn} \tag{5.26}$$

For example, a monochromator with the focal length $f = 500$ mm and a grating of $n = 1200$ lines/mm blazed in the first order has the bandpass through the exit slit (width = 30 μm):

$$d\lambda = \frac{dx}{fkn} = \frac{30 \cdot 10^{-3}\ \text{mm}}{500\ \text{mm} \cdot 1 \cdot 1200\ \text{mm}^{-1}} = 5 \cdot 10^{-8}\ \text{mm} = 0.05\ \text{nm}$$

Although it is a very good spectral instrument, it cannot be used to measure the natural line profile. Using a high-quality grating, the focal length becomes critical, but a longer focus means loss of stability and the mechanical issues increase the price. The above resolution is good enough for most of the molecular, liquid, and solid sample measurements, but it is not good for atomic spectroscopy. The spectrograph has no exit slit, but the problem is similar according to the detector pixel size. A high sensitivity detector needs a larger pixel size, for example, 26 μm against 15 μm for high resolution (ANDOR 2013). Such a pixel size acts almost like the abovementioned exit slit of 30 μm.

Several of the world's most performing spectral instruments are at the National Institute of Standards and Technology (NIST 2010). Using the value 0.078 nm/mm of reciprocal dispersion of the Eagle spectrograph with a focal length 10.7 m (!), one can obtain for a virtual exit slit of 30 μm the extraordinary value of 0.078 nm/mm × 30 μm = 0.00234 nm. The spectrograph is still using a photo-plate recording, where the photosensitive grain size is much smaller and, as a consequence, the resolution should even be higher than that calculated. The highest resolution can be obtained with a Fabry–Perot interferometer.

5.5 PHOTODETECTOR

The photodetector converts the optical signal into an electrical signal in order to be processed and displayed. It is somehow similar to what happens in sight of the animal body. The spectral photodetector is the fastest changing component during the last few decades. For over 100 years, the spectral photographic plates or films were almost the only photodetectors used. They have good characteristics, but in order to be more efficient we need fast optical–electrical conversion and then analog-to-digital ones. This is possible only by using photoelectric detectors. They can be classified as follows:

Quantum detectors:	External photoelectric effect	Photomultiplier tubes
	Internal photoelectric effect	Photodiodes and phototransistors
		Photoconductors
		Charge-coupled detectors (CCDs)
Thermal detectors:	Thermocouples, thermistors, and bolometers	
	Pyroelectric	
	Golay cell	

We can also make another classification as

- Single channel detectors (all mentioned detectors except CCD)
- Multichannel detectors (CCD and photographic plate)

Quantum detectors directly convert the incident photons into an electrical signal based on either an external photoelectric effect or an internal photoelectric effect, namely by creating free electrons or electron–hole pairs after photon absorption.

Before describing a few of the most used spectral detectors, we mention the existence of the PIN photodiodes known as avalanche photodiodes (APD) that can be a strong competitor for the photomultiplier tube. Its advantages: broad spectral range (200–1100 nm), very small size (few mm), very fast, extremely sensitive (comparable to the photomultiplier), internal amplification, low-voltage operation, easy to be coupled to an optical fiber, resistant to shocks or vibrations, can be easily thermoelectrically cooled, often integrated in small size single photon counting systems. Its use raises the issue of focus stability on the small size of photo-sensitive areas.

Warning! All very sensitive detectors are easily damaged, sometimes permanently, by overexposure to light.

5.5.1 PHOTOMULTIPLIER TUBE

The *photomultiplier tube* (PMT) is based on external photoelectric effect and it is still probably the most used detector in spectroscopy after 70 years from the first Beckman DU spectrophotometer that operated using a PMT detector. It reduced the time of analysis from hours to minutes. It is a single channel detector meaning that it will measure only the "monochromatic" radiation. Therefore, it is mounted after the exit slit of the monochromator. When an incoming photon strikes the metal plate of the photocathode, few electrons can be ejected with respect to photoelectric effect laws, only if the photon energy is greater than the extracting work specific of the metal. The extracted electrons are accelerated toward the first supplementary

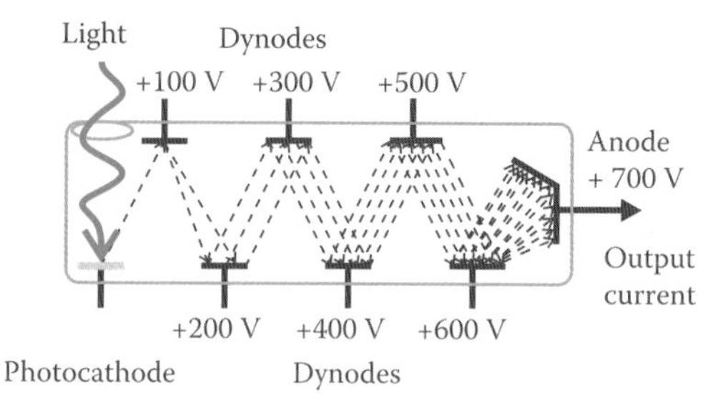

FIGURE 5.25 The simplified diagram of a photomultiplier tube. The photoelectrons ejected from the cathode are accelerated and multiplied step by step by dynodes and finally collected by the anode.

electrode, called dynode, and positive polarized compared to the cathode (Figure 5.25). Secondary electron will be ejected from the dynode surface as a result of electron bombardment. Each dynode can generate more secondary electrons than incident primary electrons. Thus, the anode will collect a hundred thousand electrons for each electron ejected from the cathode. The dynode chain provides internal amplification that can be in the range 10^6–10^7. More dynodes mean a greater amplification factor. The nature of cathode material is responsible for the spectral range of light that can be detected. They are used on a large scale for UV–vis light detection. The main issue of PMT use is to know the long-wavelength cutoff given by the extracting work of cathode material. It is usual for the producer to cover the cathode with a thin layer of low extraction work materials such as Cs_2O, bialkali, or GaAs in order to make PMT sensitive for red and NIR detection until 900 nm. Special combinations for the cathode covering layer can extend the spectral range to both sides, that is, to 115 or 1500 nm (Hamamatsu 2013). The nature of window material is also important for lower wavelengths, that is, a quartz window becomes mandatory in the UV range.

When you need to select a PMT from a producer, ask for

- The *spectral response* to match to your spectral range.
- The *sensitivity* to be high enough and without great variation on your spectral range.
- The *response time* (rise time and fall time) if you intend to measure short signals; the PMT is a very fast detector (several nanoseconds is usual).
- The *linearity* is important for quantitative analysis; the PMT exhibits a large linearity of orders of magnitude of incident light flux.
- To reduce the dark current for very low intensity light measurements, it is better to select a PMT with *cathode cooling*.
- For a very low light level, it is recommended to choose a PMT working in the *photon counting* mode. The photon counting is a digital working mode used to measure very low optical signals. It is also called single photon detection. The number of electrons emitted by photocathode per photon is one or zero and the detector works as a digital device that may count photons reaching the detector into a period of time.

A disadvantage of PMT is the HV supply that must work at 1000–2000 V within 10^{-5} stability. Because of its size and power supply, it is not suitable for miniaturization.

5.5.2 CCD

The *charge-coupled device* (CCD) is the most important competitor of the PMT. Actually it is the most used detector in spectrographs and especially in all mini-spectrographs. It is a multi-channel detector, that is, it detects the "whole" spectrum at once. It is actually a diode array made on the same chip as shown in Figure 5.26. A diode is a pixel, that is, an active element of square shape. When a photon is absorbed by the semiconductor active layer, a pair electron–hole is produced in respect of internal photoelectric effect, only if the photon energy is greater than the energy gap of the semiconductor. The local electric field separates the charge carriers and a small integrated capacitor accumulates the electric charge produced by the light incident on the surface of that pixel. So after illumination of this diode array, each pixel will accumulate a specific amount of electric charge proportional to the light flux incident to that pixel. There is a pattern of electric charge distributed on the sensitive area of the detector. Now you must imagine that the CCD is a more complex electronic device integrated on a chip with many circuits. After conversion of the light into electric charge for a given time (exposure time), the device starts to measure the charge of each pixel. The charge is shifted row by row toward a serial shift register, that is, a separated row that allows reading of each charge accumulated. Then each cell charge is converted into voltage that is transferred outside. Each outgoing voltage pulse is proportional to the charge of each pixel. When the reading sequence (shifting–conversion–delivery) ends, the device is ready to record a new frame. The frame transfer time is a dead time when the detector can lose interesting moments in evolution of a fast process that you measure.

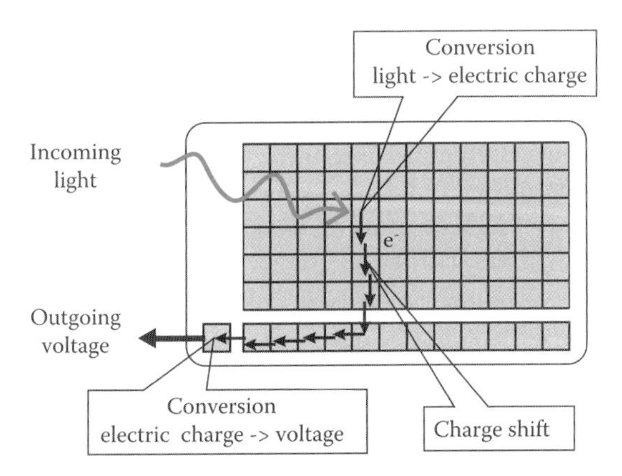

FIGURE 5.26 The simplified CCD diagram and internal processes of light-charge conversion, parallel charge shift, serial charge shift, and charge–voltage conversion.

We note that the CCDs of spectral use are different from those of photographic use. While the commercial cameras compete in having CCD sensors with an increasing number of pixels, the sensors for spectral use have a low number of large area pixels distributed in rectangular shape, for example, 1024 × 256 similar to the photographic plate. To be competitive with PMT, the CCD producers invented ICCD (intensified CCD) and EMCCD (electron multiplied CCD) with thermoelectric cooling to −100°C integrated in spectral cameras (in order to reduce electronic thermal noise). They are fast, extremely sensitive for different spectral ranges, but much more expensive than a good PMT system.

When selecting a spectral CCD detector, you should be aware of

- The *spectral sensitivity* (manufacturers give the quantum efficiency) that shows what spectral range it will best cover
- The *pixel size* that indicates where it will perform: either higher sensitivity (large pixel size of 26 μm) or higher resolution (small pixel size of 13 μm)
- The thermoelectric *cooling* availability (best is −100°C) or even liquid nitrogen cooling (−196°C)

The CCD, ICCD, and EMCCD are well suited for low light applications. For higher resolution, they will be integrated into a long focal length spectrograph (e.g., 500 mm) supporting different diffraction gratings.

5.5.3 INFRARED DETECTORS

We start the new topic from the applications: the current application is vibrational absorption spectroscopy. This involves measurements in the spectral range 2.5–25 μm, that is, 4000–400 cm^{-1}. From this point of view, there are two regions:

- The functional group region from 4000 cm^{-1} to approximately 1500 cm^{-1}
- The fingerprint region of molecular vibrations characteristic of the molecule from 1500 to 450 cm^{-1}

From the first commercial instrument in 1944 (Perkin Elmer 2005), the detection of infrared radiation was based on the use of thermal detectors because they are heat-sensitive devices. They produce an electrical response based on a temperature change in the sensor. Thermal detectors were mainly thermocouple devices and pyroelectric devices. They are sensitive but relatively slow devices.

During the last decade, the fast evolution of ternary semiconductors technology imposed a new IR detector type based on internal photoelectric effect: Hg_{1-x}–Cd_x–Te (called MCT). The main quality of this compound is that it accepts a large variation of composition formula leading to the variation of the energy gap. It results in a semiconductor detector with a variable gap that is proper to detect IR radiation from 8000 to 600 cm^{-1} or even 450 cm^{-1}. The high-energy value is already in the near infrared

(1250 nm) region, which is an exceptional feature. Other important features are both high-speed detection and the greatest IR sensitivity at the moment. One difficulty in its operation is the need for permanent cooling of the nitrogen liquid temperature (–196°C) during operation and you have to wait until the detector reaches the thermal equilibrium.

The deuterated doped triglycine sulfate (DTGS) is a pyroelectric crystal. The electric polarization of the crystal strongly depends on the temperature even at room temperature. The crystal is placed between two electrodes and its temperature increases when IR radiation is absorbed by a black coating of one electrode. The DTGS or DLaTGS (deuterated L-alanine-doped triglycine sulfate) detectors have significantly lower sensitivity than MCT but they are proper for IR microscopy, where high spatial resolution is required. They are miniaturized and it is easy to cool them by a Peltier element (thermoelectric cooling). As thermal detectors, they have wide and *flat* spectral sensitivity, which is a great advantage (unique!) in spectroscopy. The DTGS detectors are sensitive from 6000 to 350 cm^{-1}, which represents a broad enough spectral range for molecular spectroscopy.

5.6 FIBERS AND FIBER COUPLING ACCESSORIES

An optical fiber is a transparent fiber used to transport light. Optical fibers are well known for their use in telecommunications. They are very cheap, but they are not useful in spectroscopy. The optical fibers of spectral use are different and specially made for spectroscopy applications:

- The material from which they are made must have a high transmittance on a broad spectral range. Fused quartz is for UV, glass and plastic is for visible, and silica and chalcogenides are for IR.
- While telecom fibers are a few microns in diameter, spectral fibers are a hundred microns in diameter in order to transport more light. A higher diameter is required to realize an optimum connection with light sources, photodetectors, monochromators/spectrographs, and optics devices. Sometimes, bundles of several optical fibers are used. In such way, a contact area of more than 1–2 mm diameter is possible in order to still keep the flexibility of the cable.
- They are short of a few meters because of the applications where they are used.
- They often need connectors and adapters in order to realize a fast connection to a device that is dedicated to fiber use. There are several types of connectors, but the SMA and FC types are often used. You can choose the connector type when you order the cable. If you need to couple two fibers using connectors in order to obtain a longer fiber, beware that you lose power compared to the single fiber of the same length.

When are optical fibers useful? Anytime the sample cannot be placed in the sample compartment of the spectrometer. The adapter accessories dedicated to the instrument you use are required to couple the fiber to the entrance/exit slit of the monochromator and/or to the photodetector. Optical fibers are always used in mini-spectrographs in order to collect light from the sample and to transport the light to be analyzed to the entrance of the spectrograph. In case you intend to use a laser diode as the light source for an experiment, we recommend ordering the laser diode already connected to the fiber. It is better to proceed in a similar way when using a photodiode as a detector. It is called a pig-tail diode. A bundle of many fibers is very useful in fluorescence measurements. The bundle is an assembly of several fibers to transport the excitation light to the sample and other fibers to transport the light emitted by the sample to the emission spectrograph.

Notes:

- For high-power light transportation as in the case of pulsed lasers or xenon lamps, the standard fibers are not resistant. The damage threshold depends on both the light power to be transmitted and the operating wavelength.
- When working in the UV, the solarization can produce and the transmission of fiber decreases due to the color centers formation. Use solarization-resistant fibers when working for a long time under 300 nm.
- Avoid bending too much over the fiber because it can block the propagation of light.

5.7 DISPERSIVE SPECTROMETERS

A dispersive spectrometer is an instrument dedicated to measuring the intensity of light in interaction with matter at different wavelengths using either a prism or diffraction grating or both (cross dispersion). The purpose is to determine the composition of a sample using two sources of information:

- Line/band positions
- Line/band intensities

We can measure the light emitted by the sample. This can happen at high temperature where the sample is decomposed into atoms. The light emitted by excited atoms can be analyzed by a spectrograph as shown previously in Figure 5.19. The method is called atomic emission spectrometry. The emitted radiations are narrow and very close to each other. So we need a high resolution instrument. The dispersive element is always a diffraction grating with high resolving power or better a prism-grating pair (cross dispersion). The optics has a long focal length (1 m or more) in order to increase the linear dispersion. For radiation detection a multichannel detector, that is, a (two dimensional) CCD or ICCD, is recommended. It is important to achieve the simultaneous detection of different lines because often the spectral source does not have a high stability even in plasma excitation.

In atomic absorption spectrometry, spectral analysis is often made by a monochromator that also has a long focal length and the detection is done by a PMT.

Symmetry and group theory has applications in molecular spectra and specific methods are shown in the following.

The *absorption spectrometers for ultraviolet and visible* have a classic design, though their components have much evolved. Their common name is the *UV–Vis spectrophotometer*. What must it do? To measure the amount of light of specific wavelength, the sample absorbs in the range 190–800 nm. In other words, it must measure the absorption spectrum of a sample. So it has a simple structure: light source, monochromator, detector, electronics, and mechanics.

A light source must have a broad and flat spectrum and a very stable emission. It is usually both a halogen tungsten filament lamp for part of UV and visible (350–1000 nm) and a deuterium lamp for UV (180–350 nm).

The typical monochromator, usually based on diffraction grating, has a resolution around 1 nm because the absorption spectra tend to be featureless—most measurements are related to quantitation rather than identification. The focal length is short. In older instruments the entrance and exit slits must be adjustable in order to compensate for the variation of lamp emissivity at different wavelengths. In the new and simple instruments the slit width is fixed and the compensation is performed by electronics. Most research grade instruments have a slit width adjustable by user in order to adapt spectral bandwidth of the monochromator to that required by the sample. This will improve the capacity of instrument to discriminate between sharp features.

Detection by PMT is common for desktop instruments. The position of the sample is usually after the monochromator in order to expose the sample as less as possible to the UV radiation emitted by the deuterium lamp (Figure 5.27a). This position will reduce the possible fluorescence of the sample. There are on the market many mini spectrographs that can be used for low-precision absorption measurements. In this case, you must place the sample before the entrance slit of the spectrograph and you also need a light source (Figure 5.27b). They are very useful on the field for in-situ measurements where you can even use the sunlight! You measure first the spectrum without the sample (reference spectrum) and then the spectrum with the sample (sample spectrum). Then the software (installed on your laptop which also provides the power supply of the

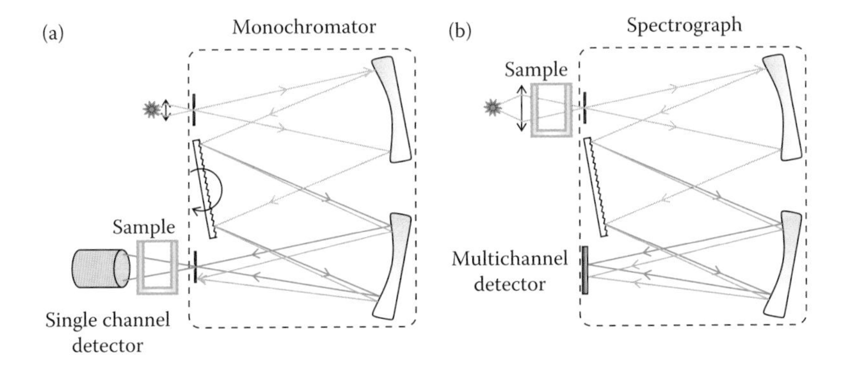

FIGURE 5.27 The single beam spectrophotometer diagram. The sample can be placed either between the monochromator and detector (a) or between the light source and spectrograph (b). To measure the absorption spectrum of the sample, two measurements must be performed (with the sample and without the sample) and then subtracted from each other.

spectrograph) will compare both spectra to calculate by subtraction the real spectrum of the sample. The procedure of the two spectra measurement and subtraction is also valuable for the first type. We just described a *single beam spectrophotometer.*

The single beam spectrophotometer is often used as a low-cost version for specialized measurements. Therefore, when you must determine the concentration of a specific component by absorption measurements made at a given wavelength, following a standard method, daily on tens or hundreds of similar samples, you can use a single beam spectrophotometer, as shown in Figure 5.27a, or even much simpler, a color filter instead of a monochromator.

When you have to measure different samples using different wavelengths, you should have a monochromator in order to select the wavelength, and better still is to use a *double-beam spectrophotometer* as shown in Figure 5.28. The fundamental difference is that the monochromatic light beam is divided into two identical beams called the sample beam and reference beam that pass the first through the sample and the second through the reference. The detector will measure alternatively the sample beam and reference beam intensities. Thus two different electric signals arise and they are proportional to I and I_0, respectively. The I and I_0 quantities correspond to those appearing in the Beer–Bouguer–Lambert law as given by Equation 5.2. The main advantage is that the fluctuations caused by light sources, electronic devices, or detectors that would otherwise appear between measurements are not issues for a double-beam spectrophotometer. Therefore, the measurements are more reliable, not just more comfortable, because you do not need to put/take sample/reference. In order to cover a broad

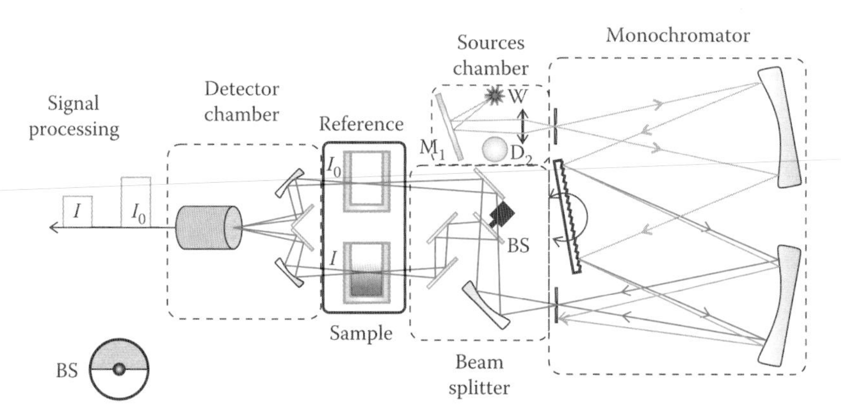

FIGURE 5.28 The optical diagram of a double beam spectrophotometer. The primary beam is divided by the beam splitter BS, which is often a rotating half-disc mirror (as shown in the left-bottom corner). The beam is either reflected by the mirror toward the sample during a half-turn or passes through the empty half-disc going toward the reference. The detector sends to the signal processor a current containing high pulses corresponding to the reference exit (I_0) and low pulses corresponding to the sample exit (I) light. The spectral range is changed when the mirror M_1 changes its position in such a way as to reflect toward the entrance slit the light emitted by either the tungsten or the deuterium lamp.

spectrum, such an instrument has two light sources, one for UV and the other for visible, that are automatically interchanged. The primary beam can be divided into twin beams by a fixed beam splitter 50/50 (filter) or by a rotating half-disc mirror. In the first case, the twin beams pass simultaneously through the sample and reference and should be detected by two identical detectors. In the half-disc rotating mirror case, the same beam is directed alternatively through either the sample or reference and their detection will be done by the same detector as shown in Figure 5.28.

Why is the reference important? The light ray is both refracted and reflected at any interface between two different media. The reflection coefficient depends on the refractive index n_1 and n_2 of both media:

$$R = \frac{I_{reflected}}{I_{incident}} = \left(\frac{n_1 - n_2}{n_1 + n_2}\right)^2 \quad (5.27)$$

Equation 5.27 is the simplest form of the Fresnel equations (see optics book) where the light ray is perpendicular to the interface. In most spectrophotometers, the light beam is collimated (ideal case) or focalized to the middle of the sample compartment (the convergence of the beam can be neglected) so that we can consider the previous formula as a good approximation. For light passing through the common air-glass interface (where $n_1 = 1$ and $n_2 = 1.5$), the reflection coefficient is around 4%. For a perfectly transparent glass sample, there are two air-glass interfaces, so the emergent energy from the sample toward the detector is only 92% ($I = 0.96 \times 0.96 \times I_0$) of the incident energy. It looks like a low absorbent sample while it is a perfectly transparent glass sample. Additionally, the glass refractive index dramatically changes with respect to the wavelength toward the near ultraviolet range. Another approximation was that the interface is perfectly flat and does not diffuse the light.

When the sample is solid, for example, a slice of doped crystal or a doped (color) glass, the reference is often just air. For a precision measurement, one should use another slice of the same matrix but from pure (undoped) material with the same thickness. It is difficult to accomplish this requirement. Preparing a solid sample for absorption measurement also involves polished surfaces in order to avoid light scattering. Use of very thin samples with parallel surfaces also can cause interference fringes. This is a specific issue mainly in infrared measurements where samples must be very thin to reduce the absorbance value in the range of the instrument. The absorption of powders and small or irregular shaped samples cannot be determined by transmission measurement. The only way is to measure the reflected light, but this is scattered in all directions and one must collect it by using an integrating sphere or cylinder (Figure 5.29). It is a small empty sphere with the inner side wall painted with a thin white diffuse layer of magnesium oxide, barium sulfate or PTFE small balls. A white standard is mandatory as the reference. The measurement is performed by placing the sample in front of the sample window and concentrating the light scattered from the sample on the detector using the sphere. The obtained value of the reflectance is reported to the reflectance of the reference standard white board, which is taken to be

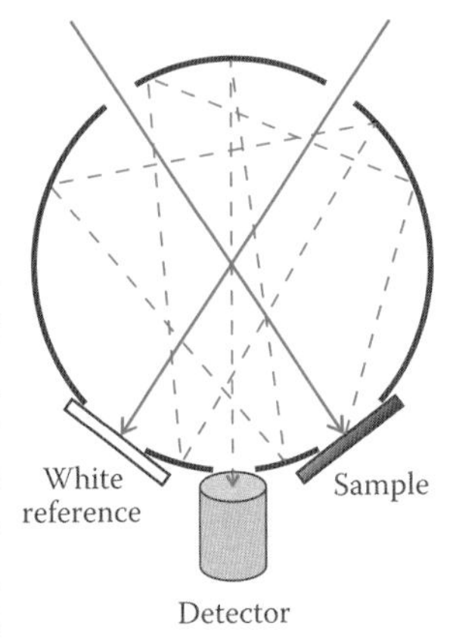

FIGURE 5.29 The integrating sphere ray paths in a double-beam spectrophotometer.

100%. This accessory must be adapted exactly to the sample compartment of your instrument and one must take into consideration this possible future application when one has to buy the instrument.

When measuring on a liquid, the sample must be placed in a glass cell (for visible), a plastic cell (for single use) or fused quartz cell (for ultraviolet). The cell has usually a rectangle cross section and two faces (windows) of optical quality. One can select the cell size in a wide range depending on the sample absorbance (low absorbance—long cell). A cell has four interfaces and the total transmission is lower than that in the above example and use of a reference cell use is mandatory. The solvent in which the absorbing species are dissolved also has an effect on the spectrum of the species and it is another main reason to use a cell filled with it as a reference. The quantitative analysis is more accurate when using liquid samples and it is often better to make solutions from solid samples and then to measure the solution. There are standard procedures of analytical chemistry, also known as sample preparation, which are fundamentals for a precise determination of a compound concentration in a given sample (Bauman 1962).

The measurement of gaseous samples is more difficult because the absorption is low and it depends also on the gas pressure. There are special closed cells for gases and for low absorption you need to use a cell with a long optical path (1–10 m), which results from multiple reflections on two opposite mirrors. Such a cell must be adapted to the sample compartment of the spectrophotometer. The composition of the gaseous sample must be well controlled to be sure that inside it is exactly what you intend to measure.

In the research work of rare earths or transition metal doped crystals, where many bands overlap at room temperature and also their intensities strongly depend on temperature (by means of vibrations), a *cryostat* allows measurements to be performed at very low temperatures where the vibrations amplitude decreases. It is the most expensive accessory of the spectrophotometer because it is a complicated device that can cool the sample at liquid nitrogen temperature of about 77 K or even at liquid helium temperature (4 K or lower!). It must have special optical windows, temperature sensors, and controllers (heating), and windows must be adapted in order to avoid water vapor condensation. With an appropriate design, the cryostat can also be used in other measurement types of crystals, for example, fluorescence and Raman spectroscopy.

It is better to have the option to select the values of resolution, either scanning speed or acquisition time, before measuring a sample. Let us remember the importance of the slit width:

- A larger slit means more light, therefore accurate measurement of absorbance (technically called signal-to-noise ratio increase).
- A narrow slit means higher precision in measuring the band (or line) shape (technically called higher resolution).

You will never have both! How should one decide which is the correct way? It is the decision of the instrument operator and it depends on both the sample characteristics and the purpose of analysis. If the sample spectrum has a few non-overlapping bands, a we do not need high resolution; so, a large width will increase the signal-to-noise ratio and therefore better accuracy in concentration determination. It is also important when the absorption of reference is high and I_0 is low (uncommon case in the UV–vis range but often in the IR range). Again, the accuracy should be increased by a larger slit in order to allow more light to pass through both the sample and the reference.

If the spectrum has overlapping bands and if one has to determine their correct positions by spectrum deconvolution then each feature of the asymmetric absorption band ("shoulder band") is important. A narrow slit width will be selected even if the signal to noise ratio will decrease. The best way to remove the overlapping bands is by decreasing the temperature if the sample allows this procedure and a cryostat is available. Also, whenever one has to determine the concentration from the band area measurement, one must use a narrow slit to improve the accuracy of the band shape profile.

5.8 FTIR SPECTROSCOPY

For several decades, the infrared absorption spectrometry was based on double-beam dispersive spectrophotometers similar to that for ultraviolet and visible. The major differences were the light source, prism material, and photodetector. Nowadays, all commercial spectrophotometers for IR absorption measurement work on a different measurement principle and the technique is called Fourier transform infrared spectroscopy (FTIR). This technique offers a number of advantages over conventional infrared systems including sensitivity, resolution, and speed but requires fast computers for data processing.

Let us imagine a pure monochromatic oscillation that is described by the trigonometric function

$$y(t) = a\cos(2\pi\nu t + \varphi_0)$$

as illustrated in Figure 5.30a for $\nu = 5$ Hz and $\varphi_0 = 0$. This ideal oscillation is infinite in the time domain, but it was illustrated on a limited range. FT relates the time domain to the frequency domain by the relation:

$$F(\nu) = \int_{-\infty}^{\infty} y(t)e^{-2\pi i\nu t}\, dt$$

Figure 5.30b illustrates the FT of the graph (a) of pure monochromatic oscillation on a limited time range. The graph shows two peaks at −5 and +5. Therefore, by FT applied to the sinusoidal signal, we obtained the correct value of oscillation frequency. In our illustration, the peaks are narrow, but normally they are infinitely narrow for an infinite time range. In Figure 5.30c, we represented a real oscillation whose amplitude increases and then decreases as an emission process that is time limited. The FT of the real oscillation also shows two peaks that are larger than that of infinite oscillation (Figure 5.30d).

In Figure 5.31a, we illustrate the case of two simultaneous oscillations of 5 and 8 Hz, respectively, limited in time. The FT can easily resolve the two oscillations as shown in Figure 5.31b. The resolution of the method depends on the number of data points of the primary graph.

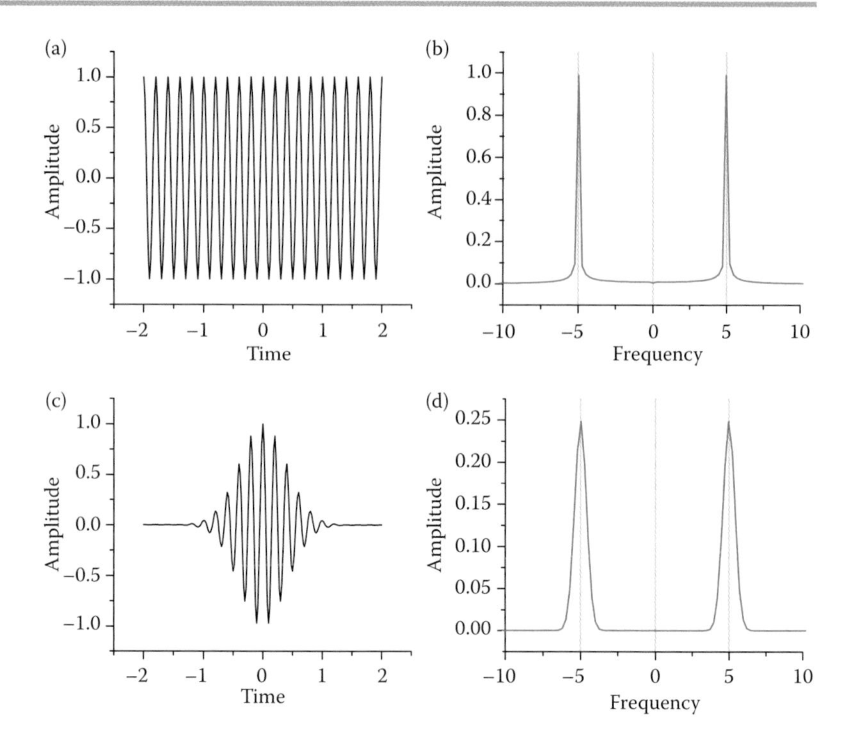

FIGURE 5.30 The Fourier transform of one infinite monochromatic oscillation of 5 Hz frequency (a) has two narrow peaks located at −5 and +5 Hz (b). In case of the increasing–decreasing monochromatic oscillation (c) the FT has two larger peaks (d) located in the same positions as earlier.

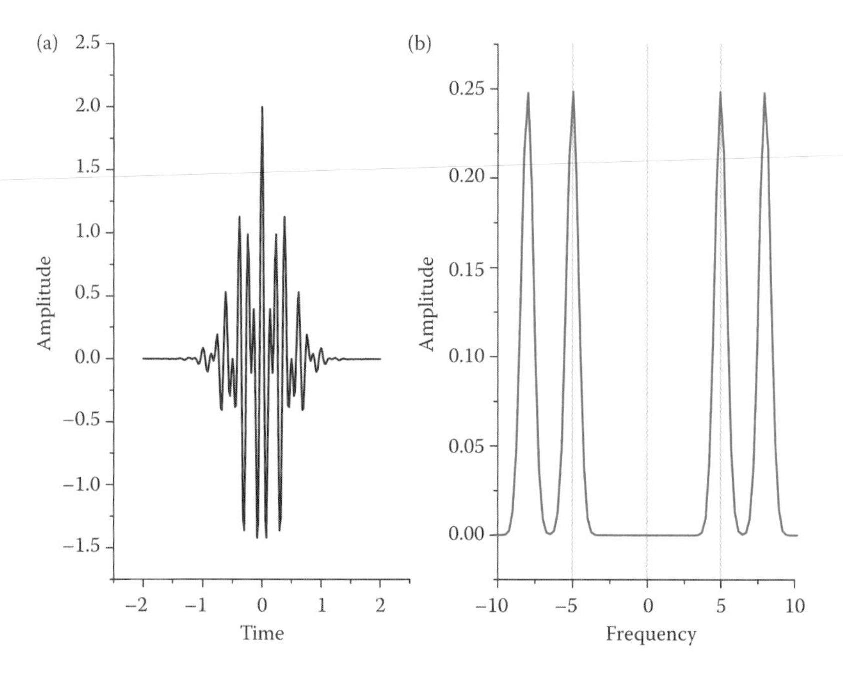

FIGURE 5.31 The Fourier transform (b) of two simultaneous oscillations (a) of 5 and 8 Hz, respectively.

Time and frequency are reciprocal physical quantities. However, FT can also relate other reciprocal physical quantities: length and wavenumber. Let us remember the Michelson interferometer from the optics course. It is a very simple instrument that made important contributions to the development of modern physics, both theoretical and experimental. It has three main optical components as shown in Figure 5.32:

- The beam splitter 50/50
- The moving mirror M_1
- The fixed mirror M_2

We use a monochromatic light source (such as a low pressure sodium vapor lamp or a laser device). (In fact the yellow emission of sodium vapor lamp contains two radiations slightly different as wavelength but one can be considered as monochromatic source.) The main light beam is divided by the beam splitter into two identical beams with reduced amplitudes, but the two corresponding waves oscillate in phase. The two beams obtained by amplitude division are sent in perpendicular directions against plane mirrors M_1 and M_2, where they are brought together again to form an interference pattern. To obtain concentric circular fringes, the mirrors M_1 and M_2 are adjusted exactly perpendicular to each other. Looking directly at the beam splitter, your eye can see the interference pattern composed by several yellow concentric rings, called Haidinger rings (fringes). Moving the mirror M_1 by translation, the interference image also changes. New rings appear in the center of the image while other rings grow from the center. When replacing the eye by the photodetector and its objective lens as shown in Figure 5.32, it will give a signal proportional to the intensity of focused light. When the mirror M_1 is in such a position that the optical path difference between the two beams is zero (called the zero path difference (ZPD)—in Figure 5.32), the interference is constructive and the

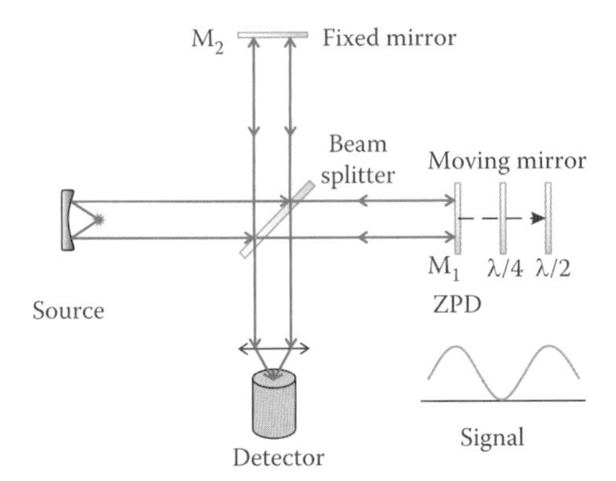

FIGURE 5.32 Schematic diagram of the Michelson interferometer which is the main component of an FTIR spectrometer. When the M_1 mirror continuously moves from the position, where the optical path difference is zero (ZPD), toward outer positions, the detector signal has a sinusoidal shape, where the interval between maximum and minimum corresponds to a mirror displacement of $\lambda/4$.

detector gives a maximum signal. When the mirror M_1 is displaced by $\lambda/4$, the optical path difference between the two beams is two times the displacement of the moving mirror, namely $\lambda/2$. It is a case of destructive interference when the detector signal is zero. Moving the mirror again through $\lambda/4$, the optical path difference between the two beams is λ and the interference is constructive again, and so on. If the moving mirror is scanned over a long distance, the signal given by the detector has a sinusoidal shape. This sinusoidal signal is called the interferogram and it represents the detector signal against the optical path difference. The signal processing using the FT will give a single peak corresponding to the wavenumber of the monochromatic radiation emitted by the lamp.

If the light source emits two radiations of different wavelengths, each radiation will produce its own sinusoidal interferogram. The maximum corresponding to ZPD will have the same position on the interferogram, but the other maxima will have different positions, because the path difference depends on wavelength $\delta = 2\,d = 2\,k\lambda$, where d is the displacement of the moving mirror reported to the ZPD position. In Figure 5.33, we illustrated how the interferogram looks for 2, 3, 4, and 5 monochromatic waves when frequencies are submultiples of the first: ν_1, $\nu_1/2$, $\nu_1/3$, $\nu_1/4$, and $\nu_1/5$. Then we apply the FT to these interferograms in order to obtain the corresponding spectrum. Figure 5.34 illustrates only the FT result of the interferogram with 5 waves. We see from Figure 5.34a that the FT does not always succeed in giving the correct answer. When the frequencies are too close to each other, they cannot be separated. Therefore, the method does not have

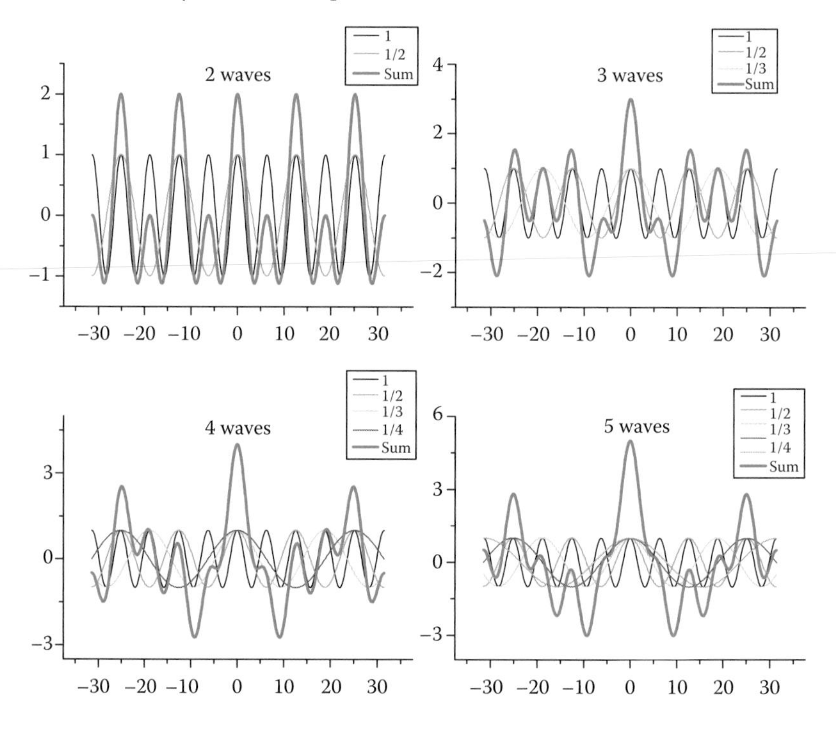

FIGURE 5.33 The interferograms (thick line) of 2, 3, 4, and 5 monochromatic waves, with frequencies as submultiples of the first one: ν_1, $\nu_1/2$, $\nu_1/3$, $\nu_1/4$, and $\nu_1/5$.

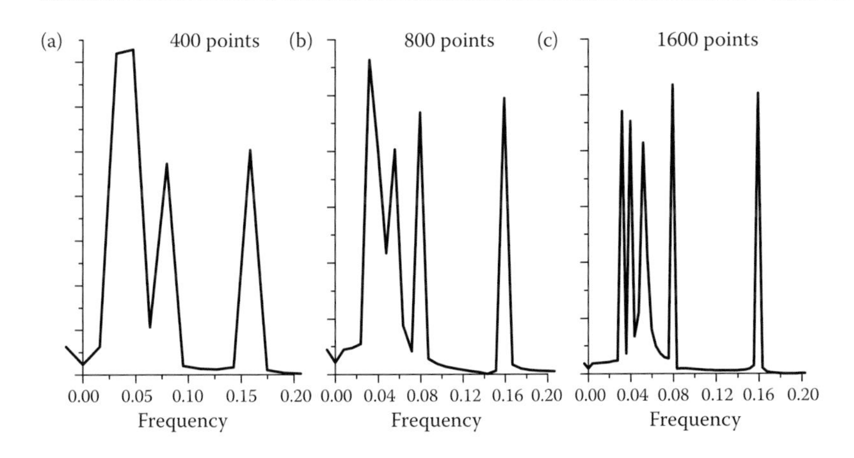

FIGURE 5.34 The result of Fourier transform of the five waves interferogram for increased displacement. A longer displacement of the moving mirror can provide for a better resolution of waves.

enough resolution. The scanning distance of the moving mirror has to be increased in order to increase the wavelength resolution of the method. This conclusion is supported by the illustrations of Figures 5.34b and c, where the scanning distance was increased by two and four times, respectively. The scanning step is the same while the number of steps is different. The fifth radiations are resolved only in the last case. We used the term of scanning instead of displacement because in practice the mirror is moved back and forth several times in the same way as in a scanning movement. Therefore, a performing instrument must have long optical paths and this involves extremely stable mechanics (a great issue!). There are technical methods to increase the optical paths without increasing the arms of the interferometer.

The basic diagram of an FTIR spectrophotometer is shown in Figure 5.35. The infrared source made of heated SiC emits a broad range of

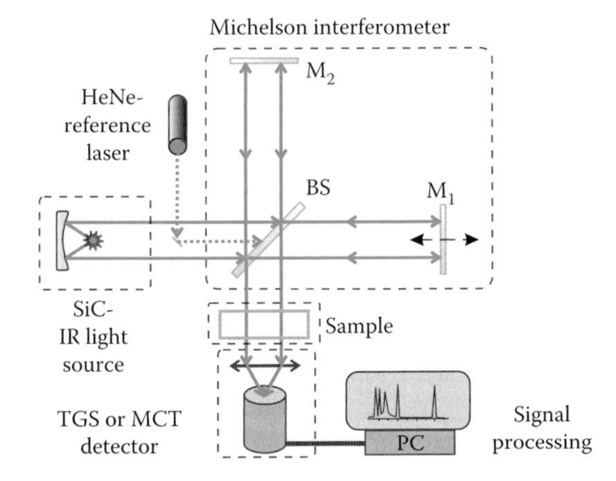

FIGURE 5.35 The basic diagram of the FTIR spectrophotometer. The IR beam is modulated by the interferometer, and then it passes through the sample and strikes the detector. The interferogram is converted by FT into the sample spectrum.

infrared radiation. The IR radiation goes through the Michelson interferometer that modulates the infrared radiation when the mirror M_1 moves. The modulated IR beam passes through the sample where it is absorbed at different wavelengths to various extents. Finally, the transmitted IR beam strikes the detector. The photodetector is often made of DTGS or, for higher sensitivity, of cooled MCT. The measured signal is the interferogram. Using the FT by computer processing, it is converted into the IR spectrum, and to the IR spectrum of the sample after normalization by the IR spectrum of the source. A small He–Ne laser is also included in the instrument in order to monitor the displacement of the moving mirror.

Features of the FTIR spectrometer:

- It is a single beam spectrometer. The background should be measured before each sample measurement.
- The detector measures simultaneous by the whole wavenumber spectrum. It results in a high speed of measurement. The total time of analysis finally depends only on the computer processing capacity.
- Almost all the light emitted by the source strikes the sample and determines a high sensitivity.
- The use of the interferometer provides higher accuracy in the wavenumber determination (increasing with the range of M_1 displacement).
- The resolution of the FTIR spectrometer may be as high as 0.001 cm^{-1}, but it is not common for all instruments because it should have an exceptional mechanical system. It requires a long displacement of the mirror and many scans in order to improve the signal-to-noise ratio.
- It allows the use of accessories such as diffuse reflectance (ATR) and IR microscope due to the high brightness.

5.9 RAMAN SPECTROSCOPY

The Raman spectroscopy is another technique for measurements of the vibrational properties of materials. It was named after C. V. Raman who won the Nobel Prize in Physics in 1930 for the experimental discovery of inelastic light scattering. It is a powerful non-destructive method of characterization (structural, qualitative, and quantitative). The instrumentation for Raman spectroscopy is commercially available from decades and it has evolved a lot. It is often treated in connection with IR absorption spectroscopy (see Chapter 4).

Any photon may interact with matter either *resonant* or *non-resonant*. The resonant interaction implies that the photon energy is equal to the energy difference between either two molecular levels (electronic, vibrational, or rotational) or two vibrational states (phonons) of a crystal as shown in Figure 5.36. The resonant energy results in the *absorption* of one photon. If the photon energy is higher or lower than the energy difference of two molecular states, absorption does not occur. It is a nonresonant interaction that can be produced at any energy of the photon using the

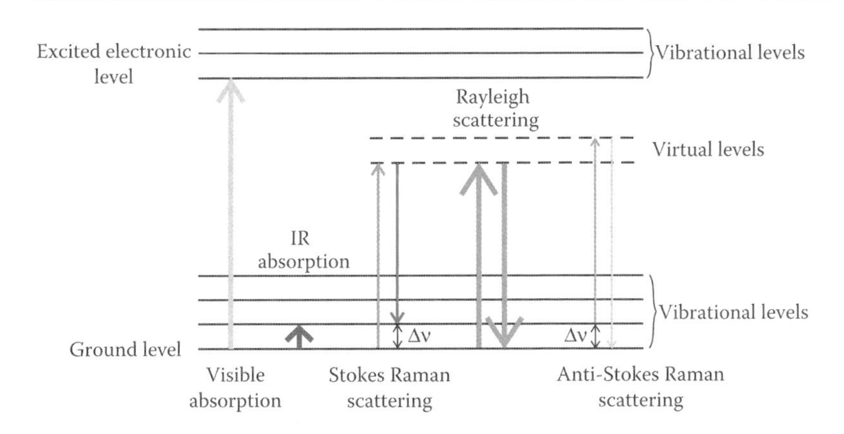

FIGURE 5.36 The diagram of resonant (absorption) and nonresonant (scattering) processes involved when the photons strike a molecule. The main data extracted from both Stokes and anti-Stokes Raman scattering measurements is the energy difference $\Delta\tilde{\nu}$ between the vibrational levels that is called the Raman shift.

so-called virtual energy levels. In that case, interaction may still exist so that the photon does not disappear. It keeps all energy, but its travel direction changes. The phenomenon is called *elastic scattering* (Rayleigh scattering). An intuitive picture of the phenomenon is the collision of billiard balls. If a ball hits a wall, it bounces back in an elastic collision with the same energy. During an elastic collision, the ball loses an amount of its kinetic energy and finally gets the same amount. In the Raman process, which is an *inelastic scattering*, when the photon hits the molecule, it loses all energy (the photon disappears); then the molecule releases part of the absorbed energy as a new photon with lower energy than that of the incident photon. The process is called *Stokes emission* and it is always measured in order to know the small amount of energy absorbed by the molecular vibrations (Figure 5.36). When the molecule releases a new photon with higher energy than that of the incident photon, the process is called *anti-Stokes emission*. The probability of anti-Stokes emission is much lower, but it still exists, due to the fact that the population of excited molecules is lower at room temperature. The Stokes and anti-Stokes emissions are symmetric with respect to the energy of the elastic scattering (Rayleigh). The explanation of the Raman effect through either electrodynamics (classical) theory or quantum theory has a hard mathematical support, but for an introduction in the field we can use this simplified image of particles collision.

Most of the measured scattered light results from elastic scattering (Rayleigh peak) and a very small amount originates from the inelastic scattering, which gives the Stokes and anti-Stokes Raman peaks. The Rayleigh peak intensity typically is 10^6 times higher than Raman peaks and it is never measured by the Raman spectrometers. We note that you must be careful not to ever let the light originating in the Rayleigh scattering to strike a high sensitivity detector, such as a photomultiplier.

The main advantage of the Raman effect lies in its capacity to detect several specific molecular vibrations as shown in Figure 5.36, and usually is complementary to IR spectroscopy. In Raman spectroscopy, the sample should be irradiated with high intensity and monochromatic light. The laser invention, which has both properties, was a great step forward in the evolution of Raman spectroscopy. The laser irradiates the sample and produces a short excitation of molecules to a virtual state from where they return either to the initial state (most of them, e.g., ≈10^6 molecules) or to the first vibrational state (e.g., 1 molecule) as in Stokes Raman scattering. The single molecule remains in an excited vibrational state and the energy difference between the Rayleigh photons and the Raman photons is caused by the excitation of a specific molecular vibration. This energy shift, called the Raman shift, is characteristic of the molecules involved in the scattering process. In general, the Raman spectrum is composed of relatively narrow bands called Raman lines (from the first decades of the photographic plate record). Qualitative and quantitative information about the composition of materials can be achieved from the position and intensity of Raman lines.

How does a Raman spectrometer work? The main components of the Raman spectrometer shown in Figure 5.37 are

- Laser source for excitation
- The monochromator to disperse the radiation emitted by the sample
- The photodetector

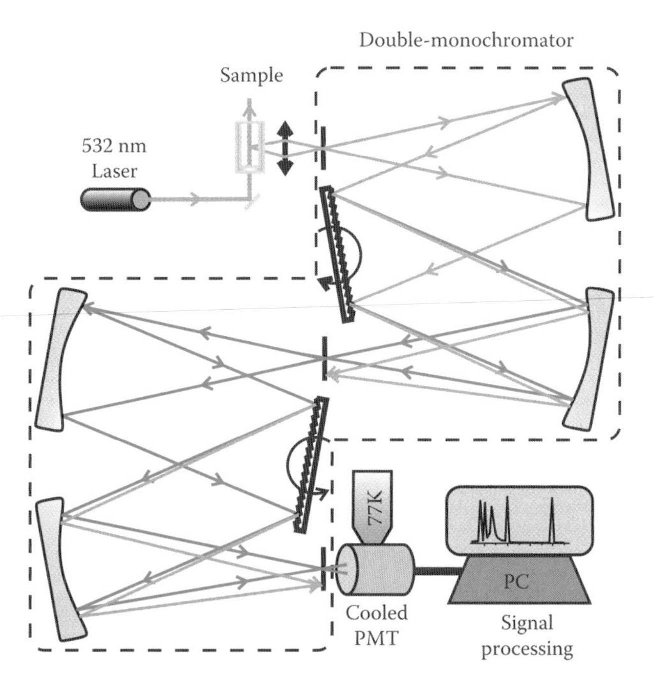

FIGURE 5.37 The basic diagram of a Raman spectrometer. The laser beam passes through the sample. The scattered light (both Rayleigh and Raman) is collected (at 90° in this example) by a lens and focused on the entrance slit of a high-resolution double monochromator ("additive geometry" in the present case). After the exit from the monochromator, the Raman radiation is measured by the detector that is cooled to increase the signal-to-noise ratio.

The *laser sources* for continuous-wave Raman spectroscopy include laser diodes, diode-pumped lasers, and ion lasers. Examples of gas lasers: argon laser with several wavelengths from 514 (most intense) to 454 nm; krypton laser with main emissions at 530.9, 568.2, 647.1, and 752.5 nm (IUPAC 2012); He–Ne laser (632.8 nm); He–Cd laser (325 and 442 nm). Examples of solid-state lasers are DPSS at 532 nm and laser diode at 785 nm. It is recommended that several laser wavelengths be available in order to achieve the optimal detection of the Raman signal (see below). There are on the market mixed gas lasers in one discharge tube that can generate several laser lines from the UV to the near IR only by adjusting the position of a prism/mirror. Why do we change the wavelength? The first reason is that the Rayleigh scattering intensity depends on λ^{-4}. Therefore, if we change the excitation wavelength from 400 to 800 nm, the Rayleigh intensity should decrease by 16 times (theoretically); the Rayleigh peak tail can even overlap some Raman peaks at low wavenumbers. The second reason is to avoid the excitation of fluorescence often encountered in organic and biological samples. The low energy photons (longer wavelengths) would be in principle better to reduce electronic excitation. However, at longer wavelengths (red or NIR), both the sensitivity of the detector and Raman scattering efficiency decrease. So shorter wavelengths will provide a better Raman signal. Moreover in a particular case of Raman microscopy (called micro-Raman), the spatial resolution is better at shorter wavelengths. The final conclusion is: you should have at least three wavelengths available and you will select the best excitation for the sample. At least we mention that the laser source must have a high stability of emission for a long time if a broad spectral range is to be collected since data acquisition with several accumulations can take several minutes or more.

The *monochromator* must have a high linear dispersion in order to resolve the Raman lines that are very close. Thus, the monochromator has a long focal length and needs to be very stable. The dispersive element is a blaze grating of good quality with high brightness. It is recommended to have a triple turret with different gratings that can be interchanged by a simple command on the computer. With regard to the weakness of the Raman signals with respect to the Rayleigh one, it is necessary to reduce as much as possible the stray light in Raman spectrometer. One efficient way is to use multiple dispersion stages. The best solution for Raman spectroscopy employing multichannel detection systems is to use a triple monochromator (Deckert et al. 1995). The cheapest method is the use of notch filters (rejecting the Rayleigh line) or edge filters (rejecting the Rayleigh line and higher energy lines, including the anti-Stokes signals). This method also requires a set of such filters for the various laser lines used.

The *photodetector* must be a very sensitive one: either a PMT with higher internal amplification for a monochromator or a large ICCD device for spectrograph mounting. Both detector types should be cooled at low temperatures to reduce the electronic noises that can be intense, especially in case of long time accumulations. The cooling at liquid nitrogen temperature gives the lowest detector noise and the best signal-to-noise ratio.

We did not mention several optical components as lenses, filters, and polarizing filters required for proper operation, and that bring key informations about the symmetries of the vibrations, especially when studying single crystals. A part of the scattered light is collected at either 90° or 180° (especially in micro-Raman set-up) to be focused on the entrance slit of the monochromator.

Low-cost Raman mini-spectrometers with optical fibers are available on the market. However, their performance is worse and they should be used only for either *in situ* industrial applications or where the sampling operation is not permitted, for example, with a painting. Even when a painting has to be analyzed, we can get very small samples (down to a few micrometers) with a high-performance Raman microscope.

5.10 FLUORESCENCE SPECTROSCOPY

Fluorescence (or, more precisely, photo-luminescence) is a two-step process of light absorption followed by light emission. The energy of emitted light can be either lower (Stokes emission) or higher (anti-Stokes emission) than the energy of absorbed light. The fluorescence phenomenon was presented in Chapter 4. To discuss the instrumentation (the spectrofluorometer), we mention the main requirements for it:

- Multi-wavelength light source
- The excitation wavelength must be tunable
- High luminosity monochromator
- Low stray light monochromator
- Very sensitive detector

As a consequence, the spectrofluorometer must have a broadband source from ultraviolet to visible (the xenon arc lamp is the best), a first double monochromator to select the excitation radiation, a sample compartment to allow the use of accessories for special samples, with a holder that permits the measurement of the emitted light at different angles, a second double monochromator to select the emitted radiation to be measured, and a high sensitivity PMT (photon counting is recommended). Usually the excitation and emission monochromators are similar, with short focal length and large (of few millimeters) but adjustable slits. The gratings of excitation and emission monochromators are blazed to provide maximum throughput in the UV and visible region, respectively. An example of a spectrofluorometer is illustrated in Figure 5.38. The laser source can be very efficient in the fluorescence excitation due to its high intensity, but the wavelength of the laser must match the energy levels of the sample. Otherwise, a tunable laser (e.g., a dye-laser) could be very useful. Another advantage of using the laser excitation is that it does not need the excitation monochromator. Because the broad range tunable laser is very expensive, it is recommended only for specific applications in research laboratories.

Unlike other techniques of analysis, fluorometry (the measurement of fluorescence) requires the measurement of two spectra: the spectrum of excitation and the spectrum of emission. The first spectrum is required

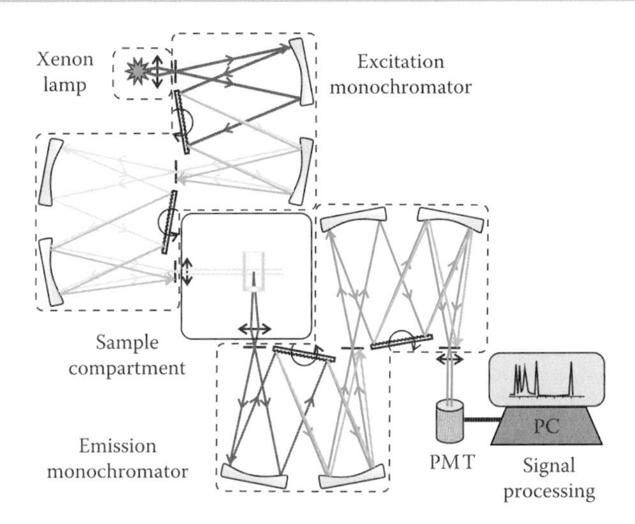

FIGURE 5.38 The basic diagram of a spectrofluorometer. The light from a xenon lamp is dispersed in the first monochromator; then it is cleaned from the stray light by the second monochromator. The selected excitation radiation (often UV) is sent to the sample. The emitted light (fluorescence) is collected (at 90° in the present case) by a lens and focused on the entrance slit of the emission double monochromator. After the exit from the emission monochromator, the selected radiation is measured by the detector.

in order to determine the optimum wavelength for excitation of the sample emission. During the scanning of excitation, the emission monochromator does not scan. Then the excitation monochromator transmits the specific wavelength for that sample. The excitation light passes through and excites the sample. Part of the emitted light is collected usually at 90° by a lens and focused on the entrance slit of the emission double monochromator. The excitation monochromator is now fixed and the emission monochromator scans to record the emission spectrum.

We note that a long exposure of sample to excitation radiation can produce so-called photobleaching that can be irreversible. The fluorometry has higher sensitivity than the absorbance measurements. There are also low-cost instruments with filters instead of monochromators dedicated to a specific sample in a wide variety of laboratories. They show only the value of the integrated emission band.

STUDY QUESTIONS

5.1 Explain the advantages of low-temperature measurement and using crystalline samples in order to study the metal complexes.

5.2 Every bond can be approximately modeled as a harmonic spring. Does this mean that we can put springs between all atoms of a molecule and, finally, the infrared spectrum of a complex organic molecule will show hundreds of features?

5.3 Fill in the missing values in the following conversion table of wavelength to wavenumber:

Wavelength	200 nm	? nm	750 nm	? nm	3 μm	? μm
Wavenumber	? cm^{-1}	25,000 cm^{-1}	? cm^{-1}	5000 cm^{-1}	? cm^{-1}	400 cm^{-1}

5.4 Fill in the missing values of the half width (indicated units) in the following table:

Band Position	Wavelength	200 nm		3 μm		
	Wavenumber (cm^{-1})		25,000		5000	400
Half-width	$\Delta\lambda$*	0.24 nm	?	1 nm	?	?
	$\Delta\tilde{\upsilon}$ (cm^{-1})*	?	15	?	15	1

* You should deduce a conversion formula and use on your computer!

5.5. You have to measure the IR absorption spectrum of an aqueous solution. Will you use a cell with NaCl or a ZnSe optical window?

5.6. The IR spectrum of the water molecule has a strong absorption around 3400 cm^{-1} arising from the overlap of two absorption bands corresponding to stretching frequencies (symmetric and antisymmetric). The sensitivity of a spectrofluorometer is usually proved by measurement of the Raman (!) peak of pure water corresponding to that stretching mode. Using an excitation radiation of 350 nm, where do you expect to get the small Raman peak? Test your spectrofluorometer at different wavelengths.

5.7. In the figure below, the transmittance of two samples is plotted. Assuming validity of Beer's law, the concentration of the sample corresponding to the dashed line compared to the concentration of the sample corresponding to the solid line is: (a) 2 times lower; (b) 1.23 times lower; (c) 2 times higher; (d) 1.23 times higher.

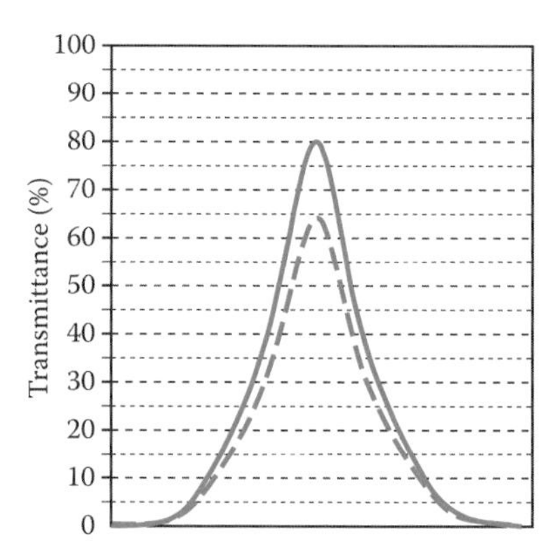

5.8. Describe what should be done when a sample absorbance in UV–Vis falls roughly at A = 4: (a) Keep the data because you trust in measurement beyond 6 absorbance, that is, the maximum value given by the instrument specifications; (b) dilute the solution 2–3 times; (c) use a shorter cell (5 mm optical path instead of 10 mm of a standard cell).

5.9. Raman spectrum of CCl_4 illustrated below was obtained using an He–Ne laser for excitation (λ_{ex} = 632.8 nm). Identify the components of the Raman spectrum.

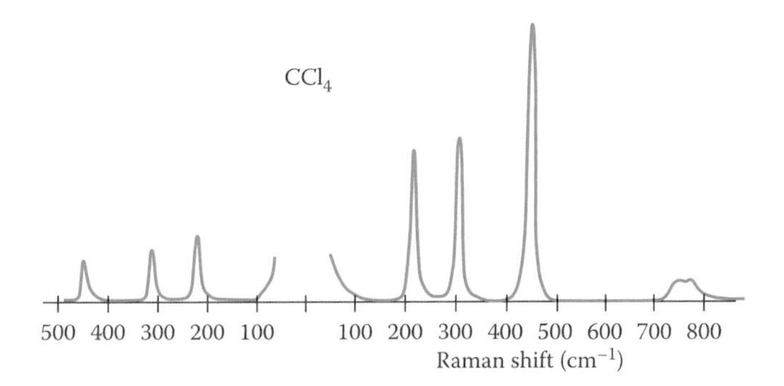

5.10. You have to reproduce a spectral measurement found in a journal as shown in the figure below. Your instrument is a small and compact spectrograph with the following specifications: the spectral range is 350–1050 nm; the number of CCD detector elements is 2048; pixel resolution is 3.2 pixels for a 10 μm slit width. Could you reproduce the measurement? Please justify.

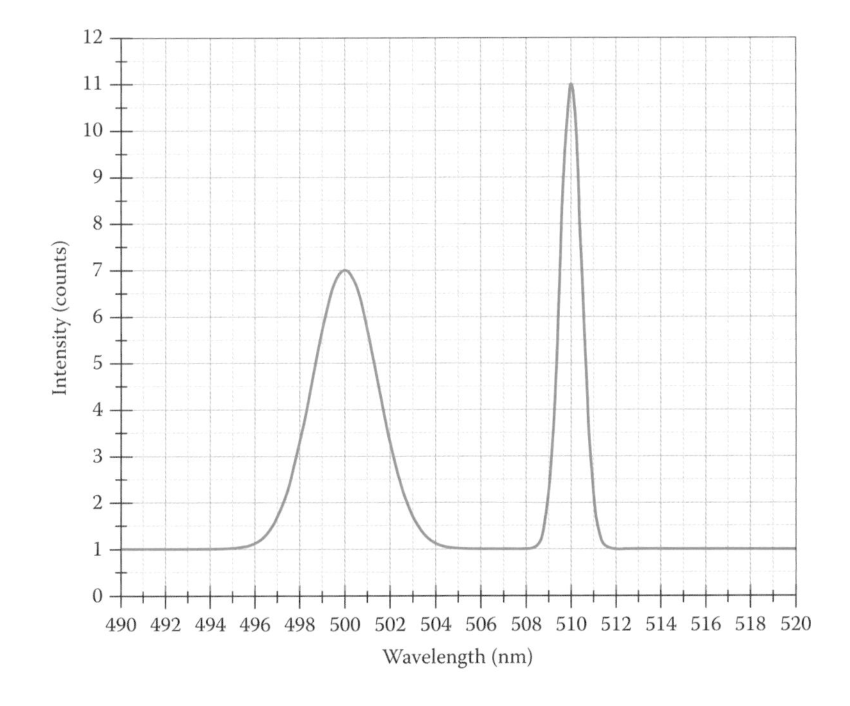

REFERENCES

ANDOR 2013. iDus spectroscopy cameras. http://www.andor.com/scientific-cameras/idus-spectroscopy-cameras

ASTM E387 - 04. 2009. Standard test method for estimating stray radiant power ratio of dispersive spectrophotometers by the opaque filter method. http://www.astm.org/Standards/E387.htm

Ball, D. W. 2006. *Field Guide to Spectroscopy*. Bellingham: SPIE.

Bartl, J., R. Fira, V. Jacko. 2002. Tuning of the laser diode. *Measurement Science Review*, *2*: 9–15.

Bauman, R. P. 1962. *Absorption Spectroscopy*. New York: Wiley.

Deckert, V., C. Fickert, D. Gernet, P. Vogt, T. Michelis, and W. Kiefer. 1995. Stray light rejection in double monochromators with multichannel detection. *Applied Spectroscopy*, *49*(2): 253–255.

Hamamatsu. 2013. *PMT-Handbook*. Chapter 4—Characteristics of photomultiplier tubes. http://www.hamamatsu.com/resources/pdf/etd/PMT_handbook_v3aE-Chapter4.pdf

IUPAC. 2012. *Compendium of Chemical Terminology*, 2nd ed. (the "Gold Book"). Compiled by A. D. McNaught and A. Wilkinson. Blackwell Scientific Publications: Oxford (1997). XML on-line corrected version: http://goldbook.iupac.org (2006–2012) created by M. Nic, J. Jirat, B. Kosata; updates compiled by A. Jenkins.

Lerner, J. M. and A. Thevenon. 1988. *The Optics of Spectroscopy*. Edison: J-Y Optical Systems/Instruments SA, Inc. *Horiba-Jobin-Yvon tutorial*: http://www.horiba.com/us/en/scientific/products/optics-tutorial/.

Lerner, J. M. 2006. Imaging spectrometer fundamentals for researchers in the biosciences—A tutorial. *Cytometry Part A*, *69A*: 712–734.

NASA. 2013. *Curiosity*. http://mars.jpl.nasa.gov/msl/

NASA-JPL 2013. Cassini Mission. http://saturn.jpl.nasa.gov/spacecraft/cassiniorbiter instruments/instrumentscassiniuvis/

Newton, I. 1721. *Opticks: Or, A Treatise of the Reflections, Refractions, Inflections and Colours of the Light*. 3rd edition. http://books.google.com

NIST 2010. Atomic spectroscopy group: Classical spectroscopy. http://www.nist.gov/pml/div684/grp01/classical.cfm

Palmer, C. and E. Loewen. 2005. *Diffraction Grating Handbook*, 6th edition. Rochester: Newport Corporation.

Pedrotti, F. and L. Pedrotti. 1993. *Introduction to Optics*, 2nd edition. New Jersey: Prentice-Hall International.

Perkin Elmer. 2005. 60 Years of PerkinElmer innovation in infrared spectroscopy. http://www.perkinelmer.com/CMSResources/Images/44-74388BRO_60 YearsInfrared Spectroscopy.pdf

Vogt, S. S. and G. D. Penrod. 1988. HiRES: A high resolution Echelle spectrometer for the Keck 10-meter telescope. In *Instrumentation for Ground-Based Astronomy*, L. B. Robinson, ed. New York: Springer-Verlag, pp. 68–103.

Appendix: Selected Character Tables*

C_2	E	C_2	Linear Functions, Rotations	Quadratic Functions
A	$+1$	$+1$	z, R_z	x^2, y^2, z^2, xy
B	$+1$	-1	x, y, R_x, R_y	xz, yz

C_4	E	C_4^1	C_4^2	C_4^3	Linear Functions, Rotations	Quadratic Functions
A	1	1	1	1	z, R_z	$x^2 + y^2, z^2$
B	1	-1	1	-1		$(x^2 - y^2, xy)$
E	1	i	-1	$-i$	$x + iy; R_x + iR_y$	(xz, yz)
	1	$-i$	-1	i	$x - iy; R_x - iR_y$	

C_5	E	C_5	C_5^2	C_5^3	C_5^4	Linear Functions, Rotations	Quadratic Functions	
A	1	1	1	1	1	z, R_z	$x^2 + y^2, z^2$	$\varepsilon = \exp\left(\frac{2\pi i}{5}\right)$
E_1	1	ε	ε^2	ε^{2*}	ε^*	$x + iy; R_x + iR_y$	(xz, yz)	
	1	ε^*	ε^{2*}	ε^2	ε	$x - iy; R_x - iR_y$		
E_2	1	ε^2	ε^*	ε	ε^{2*}		$(x^2 - y^2, xy)$	
	1	ε^{2*}	ε	ε^*	ε^2			

C_6	E	C_6	C_3	C_2	C_3^2	C_6^5	Linear Functions, Rotations	Quadratic Functions	
A	1	1	1	1	1	1	z, R_z	$x^2 + y^2, z^2$	$\varepsilon = \exp\left(\frac{2\pi i}{6}\right)$
B	1	-1	1	-1	1	-1			
E_1	1	ε	$-\varepsilon^*$	-1	$-\varepsilon$	ε^*	$x + iy; R_x + iR_y$	(xz, yz)	
	1	ε^*	$-\varepsilon$	-1	$-\varepsilon^*$	ε	$x - iy; R_x - iR_y$		
E_2	1	$-\varepsilon^*$	$-\varepsilon$	1	$-\varepsilon^*$	$-\varepsilon$		$(x^2 - y^2, xy)$	
	1	$-\varepsilon$	$-\varepsilon^*$	1	$-\varepsilon$	$-\varepsilon^*$			

* A complete list of character tables can be seen in the book by Cotton mentioned in Chapter 1.

C_{2v}	E	$C_2(z)$	$\Sigma_v(xz)$	$\Sigma'_v(yz)$	Linear Functions, Rotations	Quadratic Functions
A_1	1	1	1	1	z	x^2, y^2, z^2
A_2	1	1	−1	−1	R_z	xy
B_1	1	−1	1	−1	x, R_y	xz
B_2	1	−1	−1	1	y, R_x	yz

C_{3v}	E	$2C_3(z)$	$3\Sigma_v$	Linear Functions, Rotations	Quadratic Functions
A_1	1	1	1	z	$x^2 + y^2, z^2$
A_2	1	1	−1	R_z	
E	2	−1	0	(x, y) (R_x, R_y)	$(x^2 - y^2, xy)$ (xz, yz)

C_{5v}	E	$2C_5$	$2C_5^2$	$5\Sigma_v$	Linear Functions, Rotations	Quadratic Functions
A_1	1	1	1	1	z	$x^2 + y^2, z^2$
A_2	1	1	1	−1	R_z	
E_1	2	2 cos 72°	2 cos 144°	0	(x, y) (R_x, R_y)	(xz, yz)
E_2	2	2 cos 144°	2 cos 72°	0		$(x^2 - y^2, 2xy)$

C_{3h}	E	$C_3(z)$	C_3^2	Σ_h	S_3	S_3^5	Linear Functions, Rotations	Quadratic Functions	
A'	1	1	1	1	1	1	R_z	$x^2 + y^2, z^2$	
E'	1	ε	ε^*	1	ε	ε^*	$x + iy$	$(x^2 - y^2, xy)$	$\varepsilon = \exp\left(\frac{2\pi i}{3}\right)$
	1	ε^*	ε	1	ε^*	ε	$x - iy$		
A''	1	1	1	−1	−1	−1	z		
E''	1	ε	ε^*	−1	$-\varepsilon$	$-\varepsilon^*$	$R_x + iR_y$	(xz, yz)	
	1	ε^*	ε	−1	$-\varepsilon^*$	$-\varepsilon$	$R_x - iR_y$		

D_{3h}	E	$2C_3(z)$	$3C'_2$	$\Sigma_h(xy)$	$2S_3$	$3\Sigma_v$	Linear Functions, Rotations	Quadratic Functions
A'_1	1	1	1	1	1	1		$x^2 + y^2, z^2$
A'_2	1	1	−1	1	1	−1	R_z	
E'	2	−1	0	2	−1	0	(x, y)	$(x^2 - y^2, xy)$
A''_1	1	1	1	−1	−1	−1		
A''_2	1	1	−1	−1	−1	1	z	
E''	2	−1	0	−2	1	0	(R_x, R_y)	(xz, yz)

D_{4h}	E	$2C_4$	$C_2(C_4^2)$	$2C_2'$	$2C_2''$	I	$2S_4$	Σ_h	$2\Sigma_v$	$2\Sigma_d$	Linear Functions, Rotations	Quadratic Functions
A_{1g}	1	1	1	1	1	1	1	1	1	1		x^2+y^2, z^2
A_{2g}	1	1	1	−1	−1	1	1	1	−1	−1	R_z	
B_{1g}	1	−1	1	1	−1	1	−1	1	1	−1		x^2-y^2
B_{2g}	1	−1	1	−1	1	1	−1	1	−1	1		xy
E_g	2	0	−2	0	0	2	0	−2	0	0	(R_x, R_y)	(xz, yz)
A_{1u}	1	1	1	1	1	−1	−1	−1	−1	−1		
A_{2u}	1	1	1	−1	−1	−1	−1	−1	1	1		
B_{1u}	1	−1	1	1	−1	−1	1	−1	−1	1		
B_{2u}	1	−1	1	−1	1	−1	1	−1	1	−1		
E_u	2	0	−2	0	0	−2	0	2	0	0	(x, y)	

D_{5h}	E	$2C_5$	$2C_5^2$	$5C_2$	Σ_h	$2S_5$	$2S_5^3$	$5\Sigma_v$	Linear Functions, Rotations	Quadratic Functions
A_1'	1	1	1	1	1	1	1	1		x^2+y^2, z^2
A_2'	1	1	1	−1	1	1	1	−1	R_z	
E_1'	2	2 cos 72°	2 cos 144°	0	2	2 cos 72°	2 cos 144°	0	(x, y)	(x^2-y^2, xy)
E_2'	2	2 cos 144°	2 cos 72°	0	2	2 cos 144°	2 cos 72°	0		
A_1''	1	1	1	1	−1	−1	−1	−1		
A_2''	1	1	1	−1	−1	−1	−1	1	z	
E_1''	2	2 cos 72°	2 cos 144°	0	−2	−2 cos 72°	−2 cos 144°	0	(R_x, R_y)	(xz, yz)
E_2''	2	2 cos 144°	2 cos 72°	0	−2	−2 cos 144°	−2 cos 72°	0		

D_{6h}	E	$2C_6$ (z)	$2C_3$	C_2	$3C_2'$	$3C_2''$	I	$2S_3$	$2S_6$	Σ_h (xy)	$3\Sigma_d$	$3\Sigma_v$	Linear Functions, Rotations	Quadratic Functions
A_{1g}	1	1	1	1	1	1	1	1	1	1	1	1		x^2+y^2, z^2
A_{2g}	1	1	1	1	−1	−1	1	1	1	1	−1	−1	R_z	
B_{1g}	1	−1	1	−1	1	−1	1	−1	1	−1	1	−1		
B_{2g}	1	−1	1	−1	−1	1	1	−1	1	−1	−1	1		
E_{1g}	2	1	−1	−2	0	0	2	1	−1	−2	0	0	(R_x, R_y)	(xz, yz)
E_{2g}	2	−1	−1	2	0	0	2	−1	−1	2	0	0		(x^2-y^2, xy)
A_{1u}	1	1	1	1	1	1	−1	−1	−1	−1	−1	−1		
A_{2u}	1	1	1	1	−1	−1	−1	−1	−1	−1	1	1	z	
B_{1u}	1	−1	1	−1	1	−1	−1	1	−1	1	−1	1		
B_{2u}	1	−1	1	−1	−1	1	−1	1	−1	1	1	−1		
E_{1u}	2	1	−1	−2	0	0	−2	−1	1	2	0	0	(x, y)	
E_{2u}	2	−1	−1	2	0	0	−2	1	1	−2	0	0		

D_{2d}	E	$2S_4$	$C_2(z)$	$2C_2'$	$2\Sigma_d$	Linear Functions, Rotations	Quadratic Functions
A_1	1	1	1	1	1		x^2+y^2, z^2
A_2	1	1	1	−1	−1	R_z	
B_1	1	−1	1	1	−1		x^2-y^2
B_2	1	−1	1	−1	1	z	xy
E	2	0	−2	0	0	(x, y) (R_x, R_y)	(xz, yz)

D_{3d}	E	$2C_3$	$3C_2$	I	$2S_6$	$3\Sigma_d$	Linear Functions, Rotations	Quadratic Functions
A_{1g}	1	1	1	1	1	1		x^2+y^2, z^2
A_{2g}	1	1	−1	1	1	−1	R_z	
E_g	2	−1	0	2	−1	0	(R_x, R_y)	(xz, yz) (x^2-y^2, xy)
A_{1u}	1	1	1	−1	−1	−1		
A_{2u}	1	1	−1	−1	−1	1	z	
E_u	2	−1	0	−2	1	0	(x, y)	

D_{5d}	E	$2C_5$	$2C_5^2$	$5C_2'$	I	$2S_{10}^3$	$2S_{10}$	$5\Sigma_d$	Linear Functions, Rotations	Quadratic Functions
A_{1g}	1	1	1	1	1	1	1	1		x^2+y^2, z^2
A_{2g}	1	1	1	−1	1	1	1	−1	R_z	
E_{1g}	2	2 cos 72°	2 cos 144°	0	2	2 cos 72°	2 cos 144°	0	(R_x, R_y)	(xz, yz)
E_{2g}	2	2 cos 144°	2 cos 72°	0	2	2 cos 144°	2 cos 72°	0		(x^2-y^2, xy)
A_{1u}	1	1	1	1	−1	−1	−1	−1		
A_{2u}	1	1	1	−1	−1	−1	−1	1	z	
E_{1u}	2	2 cos 72°	2 cos 144°	0	−2	−2 cos 72°	−2 cos 144°	0	(x, y)	
E_{2u}	2	2 cos 144°	2 cos 72°	0	−2	−2 cos 144°	−2 cos 72°	0		

S_4	E	S_4	C_2	S_4^3	Linear Functions, Rotations	Quadratic Functions
A	1	1	1	1	R_z	x^2+y^2, z^2
B	1	−1	1	−1	z	x^2-y^2, xy
E	1	i	−1	$-i$	$x+iy, R_x+iR_y$	
	1	$-i$	−1	i	$x-iy, R_x-iR_y$	(xz, yz)

T_d	E	$8C_3$	$3C_2$	$6S_4$	$6\Sigma_d$	Linear Functions, Rotations	Quadratic Functions
A_1	1	1	1	1	1		$x^2+y^2+z^2$
A_2	1	1	1	−1	−1		
E	2	−1	2	0	0		$(2z^2-x^2-y^2, x^2-y^2)$
T_1	3	0	−1	1	−1	(R_x, R_y, R_z)	
T_2	3	0	−1	−1	1	(x, y, z)	(xy, xz, yz)

O	E	$8C_3$	$6C_2'$	$6C_4$	$3C_2 = C_4^2$	Linear Functions, Rotations	Quadratic Functions
A_1	1	1	1	1	1		$x^2 + y^2 + z^2$
A_2	1	1	−1	−1	1		
E	2	−1	0	0	2		$(x^2 - y^2, 2z^2 - x^2 - y^2)$
T_1	3	0	−1	1	−1	$(x, y, z)(R_x, R_y, R_z)$	
T_2	3	0	1	−1	−1		(xy, xz, yz)

O_h	E	$8C_3$	$6C_2$	$6C_4$	$3C_2 = C_4^2$	I	$8S_6$	$6\Sigma_d$	$6S_4$	$3\Sigma_h$	Linear Functions, Rotations	Quadratic Functions
A_{1g}	1	1	1	1	1	1	1	1	1	1		$x^2 + y^2 + z^2$
A_{2g}	1	1	−1	−1	1	1	1	−1	−1	1		
E_g	2	−1	0	0	2	2	−1	0	0	2		$(2z^2 - x^2 - y^2, x^2 - y^2)$
T_{1g}	3	0	−1	1	−1	3	0	−1	1	−1	(R_x, R_y, R_z)	
T_{2g}	3	0	1	−1	−1	3	0	1	−1	−1		(xz, yz, xy)
A_{1u}	1	1	1	1	1	−1	−1	−1	−1	−1		
A_{2u}	1	1	−1	−1	1	−1	−1	1	1	−1		
E_u	2	−1	0	0	2	−2	1	0	0	−2		
T_{1u}	3	0	−1	1	−1	−3	0	1	−1	1	(x, y, z)	
T_{2u}	3	0	1	−1	−1	−3	0	−1	1	1		

$C_{\infty v}$	E	$2C_\infty$	$\infty\Sigma_v$	Linear Functions, Rotations	Quadratic Functions
$A_1 = \Sigma^+$	1	1	1	z	$x^2 + y^2, z^2$
$A_2 = \Sigma^-$	1	1	−1	R_z	
$E_1 = \Pi$	2	$2 \cos \varphi$	0	(x, y) (R_x, R_y)	(xz, yz)
$E_2 = \Delta$	2	$2 \cos 2\varphi$	0		$(x^2 - y^2, xy)$
$E_3 = \Phi$	2	$2 \cos 3\varphi$	0		
…	…	…	…		
E_n	2	$2 \cos n\varphi$	0		

$D_{\infty h}$	E	$2C_\infty^\varphi$	$\infty\Sigma_v$	I	$2S_\infty^\varphi$	∞C_2	Linear Functions, Rotations	Quadratic functions
$A_{1g} = \Sigma_g^+$	1	1	1	1	1	1		$x^2 + y^2, z^2$
$A_{2g} = \Sigma_g^-$	1	1	−1	1	1	−1	R_z	
$E_{1g} = \Pi_g$	2	$2 \cos \varphi$	0	2	$-2 \cos \varphi$	0	(R_x, R_y)	(xz, yz)
$E_{2g} = \Delta_g$	2	$2 \cos 2\varphi$	0	2	$2 \cos 2\varphi$	0		$(x^2 - y^2, xy)$
$A_{1u} = \Sigma_u^+$	1	1	1	−1	−1	−1	z	
$A_{2u} = \Sigma_u^-$	1	1	−1	−1	−1	1		
$E_{1u} = \Pi_u$	2	$2 \cos \varphi$	0	−2	$2 \cos \varphi$	0	(x, y)	
$E_{2u} = \Delta_u$	2	$2 \cos 2\varphi$	0	−2	$-2 \cos 2\varphi$	0		

Solutions to Study Questions

CHAPTER 1

1.1

Right \ Left	E	S_4^1	C_2	S_4^3
E	E	S_4^1	C_2	S_4^3
S_4^1	S_4^1	C_2	S_4^3	E
C_2	C_2	S_4^3	E	S_4^1
S_4^3	S_4^3	E	S_4^1	C_2

1.2 One fourfold rotation axis perpendicular to the molecular plane; four twofold rotation axes: two containing Cl–Au–Cl (opposite Cl) and two bisecting Cl–Au–Cl (Cl–Au–Cl at 90°); one horizontal (molecular) plane; four planes: two vertical planes containing Cl–Au–Cl (opposite Cl) and two dihedral planes bisecting Cl–Au–Cl groups (Cl–Au–Cl at 90°); one fourfold improper rotation axis perpendicular to the molecular plane (coincides with the principal rotation axis); one inversion center.

1.3

		Cl4				Cl2				Cl3		
$C_4 \times \Sigma_v(xz)$	Cl1	Au	Cl3	$= C_4$	Cl1	Au	Cl3	$=$	Cl2	Au	Cl4	$= \Sigma_d(xy)$
		Cl2				Cl4				Cl1		

		Cl4				Cl3				Cl1		
$\Sigma_v(xz) \times C_4$	Cl1	Au	Cl3	$= \Sigma_v(xz)$	Cl4	Au	Cl2	$=$	Cl4	Au	Cl2	$= \Sigma_d(x, -y)$
		Cl2				Cl1				Cl3		

		Cl4				Cl4				Cl3		
$C_4 \times \Sigma_h(xy)$	Cl1	Au	Cl3	$= C_4$	Cl1	Au	Cl3	$=$	Cl4	Au	Cl2	$= S_4$
		Cl2				Cl2				Cl1		

		Cl4				Cl3				Cl3		
$\Sigma_h(xz) \times C_4$	Cl1	Au	Cl3	$= \Sigma_h(xz)$	Cl4	Au	Cl2	$=$	Cl4	Au	Cl2	$= S_4$
		Cl2				Cl1				Cl1		

1.4

	E	C_2	Σ_v	Σ_v'
E	E	C_2	Σ_v	Σ_v'
C_2	C_2	E	Σ_v'	Σ_v
Σ_v	Σ_v	Σ_v'	E	C_2
Σ_v'	Σ_v'	Σ_v	C_2	E

1.5 c_6 (perpendicular to the molecular plane) is the principal axis; the axes c_3 and c_2 are coaxial to the principal axis; three axes c_2' pass through two opposite carbon atoms; three axes c_2'' pass between carbon atoms.

1.6 PCl_5: c_3 (Cl–P–Cl is the main axis), $3c_2$ (P–Cl), σ_h, s_3 (coaxial to the main axis), and $3\sigma_v$.

1.7 The symmetry operations of PCl_5 are E, C_3, C_3^2, C_2', C_2'', C_2''', Σ_h, S_3, S_3^5, Σ_v', Σ_v'', Σ_v'''. They belong to the $\boldsymbol{D}_{3h}$ point group.

1.8 The point group of difluoroethylene is $\boldsymbol{C}_{2v} = \{E, C_2, \Sigma_v, \Sigma_v'\}$.

1.9 The point group of ethane is $\boldsymbol{D}_{3d} = \{E, C_3, C_3^2, C_2', C_2'', C_2''', I, S_6^1, S_6^5, \Sigma_{d'}, \Sigma_{d''}, \Sigma_{d'''}\}$.

1.10 The point group of hydrogen chloride is $\boldsymbol{C}_{\infty v} = \{E, C_\infty, \infty\Sigma_v\}$.

1.12 Three classes $\rightarrow$ three IRs: Γ_1, Γ_2, Γ_3. Γ_1 is the totally symmetric representation A_1 (characters 1, 1, 1). Rule 1: $1 + (l_2)^2 + (l_3)^2 = 6 \rightarrow l_2 = 1, l_3 = 2$. Rule 3: $1 \cdot 1 \cdot 1 + 2 \cdot 1 \cdot \chi_{22} + 3 \cdot 1 \cdot \chi_{23} = 0 \rightarrow \chi_{22} = 1, \chi_{23} = -1$. Again, $1 \cdot 1 \cdot 2 + 2 \cdot 1 \cdot \chi_{32} + 3 \cdot 1 \cdot \chi_{33} = 0 \rightarrow \chi_{32} = -1$, $\chi_{33} = 0$. So, the IRs are A_1 (1,1,1); A_2 (1, 1, –1); E (2, –1, 0).

1.13 The unit cell of the CsCl crystal is BCC with Cs^+ ion at the center and $8 \cdot 1/8 = 1$ Cl^- ion. Each ion placed at a vertex cube is shared by 8 cells that are found in a node.

CHAPTER 2

2.1 Co^{2+} has seven d electrons (9 – 2 = 7).

2.2 Mn^{3+}: $1s^22s^22p^63s^23p^63d^4$; Fe^{3+}: $1s^22s^22p^63s^23p^63d^5$; Fe^{2+}: $1s^22s^22p^63s^23p^63d^6$; Co^{3+}: $1s^22s^22p^63s^23p^63d^6$; Ni^{2+}: $1s^22s^22p^63s^23p^63d^8$; Cu^+: $1s^22s^22p^63s^23p^63d^{10}$; Zn^{2+}: $1s^22s^22p^6 3s^23p^63d^{10}$.

2.3 The iron ion is in the 2+ oxidation state, both in the first and in the second complex. The electronic configuration as a free ion is d^6, and then ion has a high-spin configuration (2 paired electrons and 4 unpaired electrons $\uparrow\downarrow, \uparrow, \uparrow, \uparrow, \uparrow$). All tetrahedral complexes are high spin. The configuration in a tetrahedral field is $e^3t_2^3$ ($\Delta_T < P$), therefore $\uparrow\downarrow\uparrow, \uparrow\uparrow\uparrow$ $\begin{array}{c}|\uparrow|,|\uparrow|,|\uparrow|\\ |\uparrow\downarrow|,|\uparrow|\end{array}$. Thus, the LFSE of the $[FeCl_4]^{2-}$ complex is $(-0.6 \cdot 3 + 0.4 \cdot 3)\ \Delta_T + 1P = -0.6\ \Delta_T + 1P$. The configuration in an octahedral field is $t_{2g}^4e_g^2$, therefore $\begin{array}{c}|\uparrow|,|\uparrow|\\ |\uparrow\downarrow|,|\uparrow|,|\uparrow|\end{array}$. Thus, the LFSE of the $[Fe(OH_2)_6]^{2+}$ complex is $(-0.4 \cdot 4 + 0.6 \cdot 2)\ \Delta_O + 1P = -0.4\ \Delta_O + 1P$.

The iron ion is in the 3+ oxidation state in the $[Fe(CN)_6]^{3-}$ complex. The electronic configuration as a free ion is d^5. Thus, the ion could have in the octahedral field either a low-spin configuration (4 paired electrons and 1 unpaired electron $\uparrow\downarrow, \uparrow\downarrow, \uparrow$,–,–) or a high-spin configuration (5 unpaired electrons $\uparrow, \uparrow, \uparrow, \uparrow, \uparrow$). In the $\boldsymbol{O}_h$ field ($\Delta_O > P$), we expect a low-spin configuration $t_{2g}^5e_g^0$ $\begin{array}{c}|-|,|-|\\ |\uparrow\downarrow|,|\uparrow\downarrow|,|\uparrow|\end{array}$. Hence, the LFSE of the $[Fe(CN)_6]^{3-}$ complex is $(-0.4 \cdot 5 + 0.6 \cdot 0)\ \Delta_O + 2P = -2\ \Delta_O + 2P$.

2.4 $(CN)^-$ is a stronger ligand than Cl^-. Thus, the splitting of d orbitals of the Co^{2+} ion is much larger in $[Co(CN)_6]^{4-}$ than in $[CoCl_6]^{4-}$ So Co^{2+} enters as a low-spin ion in the first complex, whereas in the second complex it is a high-spin ion.

2.5 Fifteen microstates as shown in the table below. Free ion terms are: ^3P, ^1S, ^1D.

		M_S			Term
		−1	0	1	
	+2		1↑1↓		^1D
	+1		1↑0↓		^1D
		1↓0↓	1↓0↑	1↑0↑	^3P
M_L	0		1↑−1↓		^1D
			0↑0↓		^1S
		1↓−1↓	1↓−1↑	1↑−1↑	^3P
	−1		−1↑0↓		^1D
		0↓−1↓	0↓−1↑	0↑−1↑	^3P
	−2		−1↑−1↓		^1D

2.6 Highest.

2.7 From Exercise 6.2.2 and Table 2.6 of the book, the following ground terms are: Mn^{3+}: d^4: ^5D; Fe^{3+}: d^5: ^6S; Fe^{2+}: d^6: ^5D; Co^{3+}: d^6: ^5D; Ni^{2+}: d^8: ^3F.

2.8 Wrong! The pure crystal can absorb light because of the transition from the valence band to the conduction band, induced excitons, and color centers (defects of lattice).

2.9 It depends on the position of the valence electrons. The d electrons of transition metal ions are very sensitive to the environment field while the f electrons of rare-earths ions are less sensitive.

2.10 The electronic configuration in the octahedral weak field: Mn^{3+}: d^4: $t_{2g}{}^3e_g{}^1$; Fe^{3+}: d^5: $t_{2g}{}^3e_g{}^2$; Fe^{2+}: d^6: $t_{2g}{}^4e_g{}^2$; Co^{3+}: d^6: $t_{2g}{}^4e_g{}^2$; Ni^{2+}: d^8: $t_{2g}{}^6e_g{}^2$.

2.11 Partially correct. The tetrahedral field strength is 4/9 times lower than that of the octahedral field due both to 4 anions being present instead of 6 and to greater distances from the central ion.

2.12 Inspect the Orgel diagram from Figure 2.52. Ni^{2+} has a d^8 configuration. In the octahedral field, three transitions (spin allowed) must be in the ascending order of their energies: $T_{2g} \leftarrow A_{2g}$, $T_{1g} \leftarrow A_{2g}$, $T_{1g} \leftarrow A_{2g}$. In the tetrahedral field, three transitions (spin allowed) must also be: $T_2 \leftarrow T_1$, $T_1 \leftarrow T_1$, $A_2 \leftarrow T_1$. Their order on the energy scale depends on the tetrahedral field strength.

2.13 The larger the energy splitting of the orbitals, the shorter is the wavelength of the absorption corresponding to that transition. The largest splitting is for the CN ligand that is the highest as shown in the spectrochemical series. Thus, the complex with the shortest wavelength absorption is $[Ti(CN)_6]^{3-}$.

CHAPTER 3

3.1 Atomic orbitals.

3.2 sp^2—trigonal planar; sp^3—tetrahedral; sp^3d^2—octahedral.

3.3 Wrong! The chemical bond is a pure quantum effect that can be very limited if described by means of classical physics. The covalent bond is just a better model than the ionic one. It is based on the overlapping of electron wave functions because they can be localized but they are not limited. The interaction between shared electrons is called the "exchange effect"; this brings about an increased localization of paired electrons between the bonded atoms. Another important feature of bonding is that all components of the molecule (nuclei and electrons) never rest, but their movements are somewhat correlated.

3.4 Any molecular bond is formed by atomic orbitals with proper orientation. The σ bonds are formed by any atomic orbitals oriented closer to the interatomic axis. The π bonds are formed by atomic orbitals oriented perpendicular to the interatomic axis and if they are still available (not used to the σ bond).

3.5 The number of electrons is not important, though the number of available orbitals is. A metal ion has five *d* orbitals, three *p* orbitals, and one *s* orbital in the valence shell compared to four orbitals of carbon. So it can use them to contribute to several molecular orbitals in order to accommodate all the electrons needed to bind the ligands.

3.6 Correct.

3.7 sp^3d.

3.8 Alkanes (methane, butane, *propane*, butane) are hydrocarbons where all carbon atoms are sp^3 hybridized and all carbon bonds are tetrahedral. The carbon chain has a zig-zag structure due to the 109.5° angle between C–C bonds.

3.9 Alkenes (ethene $H_2C{=}CH_2$, *propene* $H_2C{=}CH\text{-}CH_3$, butene $H_2C{=}CH\text{-}CH_2\text{-}CH_3$) are hydrocarbons that contain several σ bonds and at least one π bond as part of the molecular structure. Each alkene molecule has a double bond carbon–carbon: one σ and one π bond. The σ bond is formed by the overlapping of two sp^2 hybrid orbitals. The π bond is formed by the overlapping of two p_z atomic orbitals.

3.10 Triatomic molecules that possess 16 electrons in the outer shells are linear, for example, CO_2, whereas the 17th electron present determines the bending of the molecule. The structure of NO_2 is similar to that of the OH_2 molecule (nonlinear) and belongs to the symmetry $\boldsymbol{C}_{2v}$, due to the presence of nonbonding electrons (2 × 2 of O_a, 2 × 2 of O_b, 2 + 1 of N). There are 12 valence orbitals: 2*s*, $2p_x$, $2p_y$, and $2p_z$ of each of the three atoms N, O_a, and O_b. Thus, there will be 12 molecular orbitals. Only orbitals which transform alike under symmetry operations of the $\boldsymbol{C}_{2v}$ group can be combined. The MOs result from the overlapping between both atomic orbitals of nitrogen and a linear combination of atomic orbitals of both oxygen atoms with the same symmetry: A_1, A_2, B_1, and B_2. The atomic orbitals of oxygen are located lower on the energy scale than the orbitals of nitrogen due to the higher electronegativity of oxygen. It is convenient to consider linear combinations of oxygen atomic orbitals before combining with nitrogen. Also, the orbitals 2*s* and 2*p* of nitrogen will make hybrid orbitals before bonding to oxygen. The molecular orbitals will be: 2 σ-bonding and their correspondent 2 σ-antibonding; 1 π-bonding and 1 π-antibonding as a result of the combination of p_z orbitals (perpendicular to the molecule plane) of N, O_a, and O_b; 3 σ and 1 π-nonbonding due to both oxygen atoms and 2 σ-nonbonding due to nitrogen. The total of 17 electrons fills molecular orbitals from lower energy to higher energy. The last partly filled orbital is a nonbonding one. The MOs are sketched in the figure below. Note that the MO positions on the energy scale were not calculated. For a detailed treatment, see also Ballhausen, C. G. and H. B. Gray, 1965, *Molecular Orbital Theory*, W. A. Benjamin Inc., New York, Chapter 6-3.

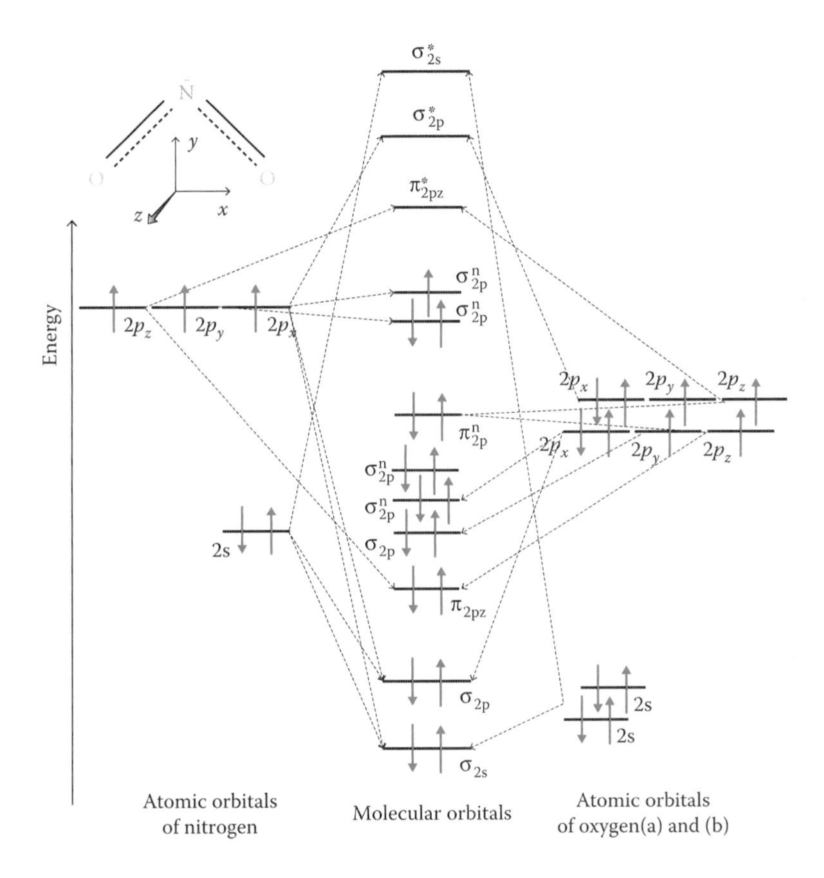

CHAPTER 4

4.1 CO_2 no, H_2O yes, H_2S yes, CH_4 no, NH_3 yes, BF_3 no (all three bonds are polarized, but they are equally spaced out to form an equilateral triangle, and thus the total molecular dipole is zero!).

4.2 Yes, but not too much.

4.3 The first sentence is correct, but the second one is wrong. Therefore, "the transition is forbidden if the direct product does not contain the totally symmetric representation."

4.4 If the dipole moment of benzene is zero results in the transition probability is zero, no light is absorbed. However, the benzene absorption bands in UV have been known since the 1880s! In fact, we draw a ring in order to represent delocalization of the double bond C=C, but really it is a two-state system (see Feynman, R. P., Leighton, R. B., Sands, M. 1970. The *Feynman Lectures on Physics*, Vol. 3, Addison-Wesley); therefore, the molecule is not centrosymmetric. The electronic absorption is due to transitions involving π electrons. The representation generated by π orbitals under operations of the $\boldsymbol{D}_{6h}$ group

D_{6h}	E	$2C_6(z)$	$2C_3$	C_2	$3C_2'$	$3C_2''$	I	$2S_3$	$2S_6$	$\sigma_h(x,y)$	$3\Sigma_d$	$3\Sigma_v$
B_{2g}	1	−1	1	−1	−1	1	1	−1	1	−1	−1	1
E_{1g}	2	1	−1	−2	0	0	2	1	−1	−2	0	0
A_{2u}	1	1	1	1	−1	−1	−1	−1	−1	−1	1	1
E_{2u}	2	−1	−1	2	0	0	−2	1	1	−2	0	0
Γ_π	6	0	0	0	−2	0	0	0	0	−6	0	2

will be reduced to $\Gamma_\pi = B_{2g} + E_{1g} + A_{2u} + E_{2u}$ using the above table. The order of corresponding levels is: A_{2u}, E_{1g}, E_{2u}, and B_{2g}. By filling with 6 electrons the first two levels as shown in the figure below, one obtains the ground state and then the first excited state.

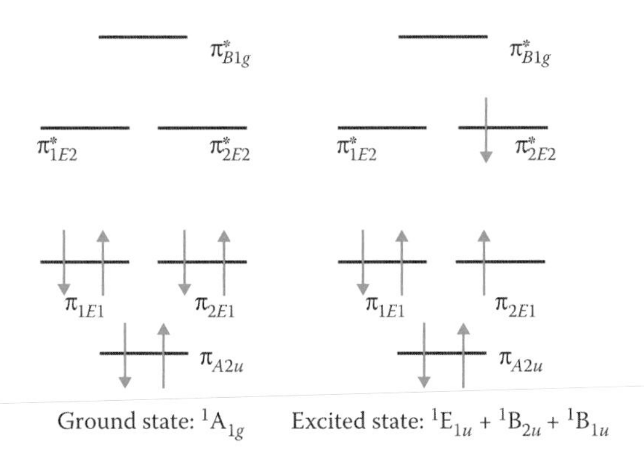

The ground state is a singlet and the excited state will be a singlet too as allowed by the spin selection rule. Their symmetries will be given by the direct products:

$$(A_{2u})^2 \otimes (E_{1g})^4 = 3A_{1g} + \cdots$$

$$(A_{2u})^2 \otimes (E_{1g})^3 \otimes (E_{2u})^1 = 3B_{1u} + 3B_{2u} + 5E_{1u}$$

Thus, three transitions are expected to occur: $^1B_{2u} \leftarrow {}^1A_{1g}$, $^1B_{1u} \leftarrow {}^1A_{1g}$, and $^1E_{1u} \leftarrow {}^1A_{1g}$ (written in order of increasing energy).

If we take the direct product of the state symmetries and the dipole operator (A_{2u} for z and E_{1u} for the x, y pair) for each of these states given above, we find that only the $^1E_{1u} \leftarrow {}^1A_{1g}$ transition is allowed by symmetry due to the product $E_{1u} \otimes E_{1u} \otimes A_{1g} = A_{1g} + \cdots$ containing the total symmetric representation. Thus, this transition presents the most intense absorption band.

Similarly, the $^1B_{2u} \leftarrow {}^1A_{1g}$ and $^1B_{1u} \leftarrow {}^1A_{1g}$ transitions are symmetry forbidden, and thus they have a lower probability. They can occur if they are properly coupled to vibrational transitions:

$^1B_{2u} \leftarrow {}^1A_{1g}$ coupled to B_{1g} or E_{2g} vibrations
$^1B_{1u} \leftarrow {}^1A_{1g}$ coupled to B_{2g} or E_{2g} vibrations.

These are called vibronic coupled electronic transitions.

4.5 We suppose that the HCl molecule is a linear harmonic oscillator, and thus the oscillation frequency is given by

$$\nu = \frac{1}{2\pi}\sqrt{\frac{k}{m_r}}$$

The force constant k depends on the bond strength between H^+ and Cl^-. Isotopic substitution of Cl^{35} by Cl^{37} (different mass but same charge) results in the same force constant. The reduced mass is calculated as in the two-body problem of Newtonian mechanics:

$$m_r = \frac{m_H \cdot m_{Cl}}{m_H + m_{Cl}}$$

Therefore, the ratio between the wavenumbers of two peaks can be calculated as follows (c is the speed of light):

$$\frac{\tilde{\nu}_{37}}{\tilde{\nu}_{35}} = \frac{\nu_{37}/c}{\nu_{35}/c} = \sqrt{\frac{m_{r35}}{m_{r37}}} = \sqrt{\frac{35(1+37)}{37(1+35)}} \Rightarrow \tilde{\nu}_{37} = 2883.83\ \text{cm}^{-1}$$

Thus, the separation is roughly 2.17 cm^{-1}.

4.6 The spectral range of the IR instrument is limited at 400 cm^{-1} by the detector sensitivity. It means that it could measure only 454 and 770 cm^{-1} vibrations. There are two IR active vibrations $\tilde{\upsilon}_3$ (antisymmetric stretch) and $\tilde{\upsilon}_4$ (bend). The stretching mode needs more energy, so a 770 cm^{-1} high peak should be assigned to the $\tilde{\upsilon}_3$ stretch. The 454 cm^{-1} vibration does not appear in the IR spectrum, and thus it will be assigned only to Raman $\tilde{\upsilon}_1$ stretching (higher energy). The present data are not enough to assign 220 and 312 cm^{-1} peaks either to $\tilde{\upsilon}_2$ or $\tilde{\upsilon}_4$. The measurement of the IR spectrum by a longer wavelength range instrument will help to assign 312 cm^{-1} frequency to the $\tilde{\upsilon}_4$ bending mode. What about the small peak around 1550 cm^{-1}? It is certainly the overtone of the $\tilde{\upsilon}_3$ antisymmetric stretching mode.

T_d	Frequency (cm^{-1})	Normal Modes	Activity
A_1	454	$\tilde{\upsilon}_1$	*Raman*, symmetric stretch
E	220	$\tilde{\upsilon}_3$	*Raman*, bend
T_2	770	$\tilde{\upsilon}_3$	*IR* and *Raman* antisymmetric stretch
	312	$\tilde{\upsilon}_3$	*IR* and *Raman* bend

4.7 The oscillation frequency depends on the atomic mass as $1/\sqrt{m}$. Therefore, the frequency increases in the following order: C–Br, C–Cl, C–O, C–C, C–H.

4.8 The asymmetric stretching vibration and both bending vibrations will induce an electric dipole (see figure below) responsible for infrared light absorption. The stretching vibration corresponds to higher frequency. The molecule bend can produce in two perpendicular planes and they are equivalent (degenerated) corresponding to the same frequency.

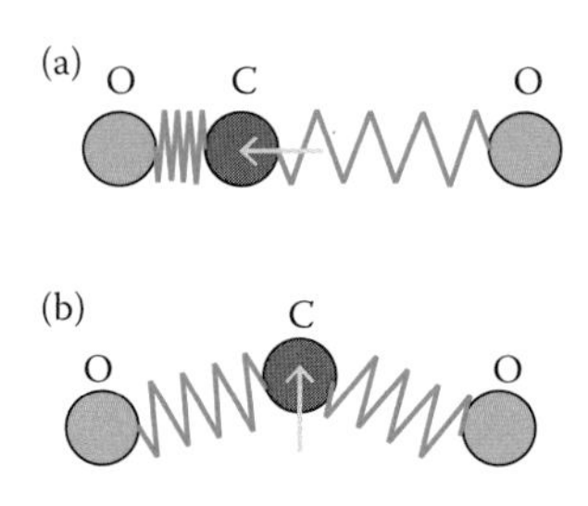

4.9 The energy of excitation is 33,333 cm^{-1}. The energy of emission peak is 25,839 cm^{-1} and 29,940 cm^{-1}, respectively. The difference between excitation energy and emission energy is 3393 cm^{-1} for the small peak. The Raman shift in water is approximately 3400 cm^{-1}. The Raman peak has a low intensity and is often overlapped by the fluorescence. Raman scattering occurs from all solvents and the Raman shift peak is a common presence of the dilute solution emission spectrum.

4.10 The same functional group will show the same absorption features wherever the molecule belongs. The position and amplitude of the infrared absorption peaks slightly depend on the molecule. Thus, each functional group absorbs in a range depending on the molecule (see table below). The infrared absorption is therefore useful to quickly identify the functional groups that are present in a molecule.

Functional Group	Spectral Range (cm^{-1})	Assignment
Hydroxyl (–OH)	3000–3600	Stretching vibration of the O–H bond
Amino ($–NH_2$)	3000–3600	Stretching vibration of the N–H bond
Methyl ($–CH_3$)	2800–3100	Stretching vibration of the C–H bond; CH_3 (lower frequencies), CH_2 (middle frequencies), CH (higher frequencies)
Carbonyl (–CO)	1600–1800	Stretching vibration of the C=O bond
Phosphate ($–PO_4$)	900–1200	Antisymmetric stretching vibration

CHAPTER 5

5.1 The bands become thinner at lower temperature and separate better. The intensities of the absorption bands forbidden by Laporte's rule are strongly dependent on temperature in an octahedral environment and they are relatively easy to recognize when temperature decreases.

5.2 Fortunately not! We can use several types of chemical bonds to represent any bond in the molecule and the infrared spectrum will be reduced to a few frequencies. As an example, each C–H bond in a given molecule has the same value of the stretching frequency that can roughly vary from 2800 to 3300 cm^{-1} from alkanes to alkynes.

5.3

	← UV →	← Vis →	← NIR →		← IR → Far IR →	
Wavelength	200 nm	400 nm	750 nm	2000 nm	3 μm	25 μm
Wavenumber	50,000 cm^{-1}	25,000 cm^{-1}	13,333 cm^{-1}	5000 cm^{-1}	3333 cm^{-1}	400 cm^{-1}

5.4 To convert half-width from nm to cm^{-1}, you should use the following formula:

$$\Delta\tilde{v} = \frac{10^7}{\lambda - (\Delta\lambda/2)} - \frac{10^7}{\lambda + (\Delta\lambda/2)} = 10^7 \frac{\Delta\lambda}{\lambda^2 - (\Delta\lambda/2)^2} \approx 10^7 \frac{\Delta\lambda}{\lambda^2}$$

Find the corresponding formula to convert half-width from cm^{-1} to nm.

Band Position	Wavelength	200 nm		3 μm		
	Wavenumber (cm^{-1})		25,000		5000	400
Half-width	$\Delta\lambda$	0.24 nm	0.24 nm	1 nm	6 nm	6.25 μm
	$\Delta\tilde{\upsilon}$ (cm^{-1})	60	15	1.11	15	1

5.5 The NaCl (and all alkali halides) material window has a high transmittance in the IR range until longer wavelength, but it is water soluble so it is not recommended for aqueous samples (biological samples). Water-resistant windows made of CaF_2 or ZnSe are used.

5.6 Convert λ_{ex} into the corresponding wavenumber $\tilde{\upsilon}$ = 28,571 cm^{-1}. Then the energy of Raman scattered radiation will be $\tilde{\upsilon}$ = (28,571 – 3400) cm^{-1} = 25,171 cm^{-1}. Convert back to wavelength λ_{em} = 397 nm.

5.7 c. The transmittance is not proportional to the sample concentration. Therefore,

$$\frac{c_2}{c_1} = \frac{\ln(1/T_2)}{\ln(1/T_1)} = \frac{\ln(1/0.64)}{\ln(1/0.80)} = 2$$

5.8 There are spectrophotometers that can really measure beyond 6, or even 8, absorbance. However, measurement with a very high absorbance value is not recommended, since it will lead to more error. The most accurate quantitative analyses should be performed at roughly $A = 1$ or less and thus dilute the sample in a controlled manner until the absorbance is close. Using a narrower cell (2–3 times shorter) could also be useful if it is not the way to make a precise dilution of the original sample. Therefore, the most recommended answer, although not mandatory, is (b). Do not forget that the sample absorbance must be in the range of standards that you are using.

5.9 Stoke and anti-Stokes Raman lines are symmetric with respect to the Rayleigh scattering peak. Stokes lines are always much stronger than anti-Stokes lines. The Rayleigh peak is never measured (shutter closed) in order to protect the photodetector. Its position lies midway between the paired Raman lines (Stokes and anti-Stokes).

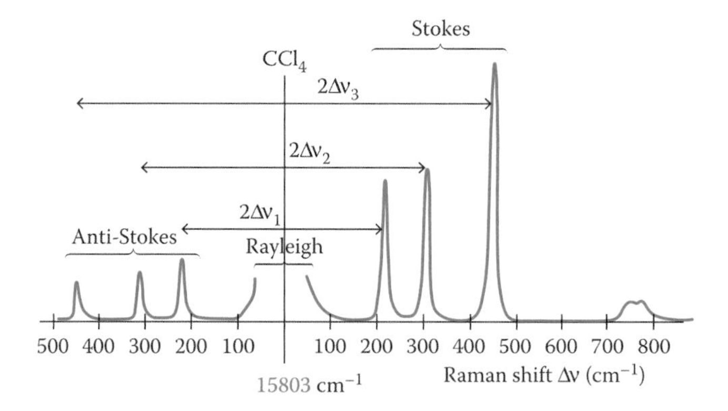

5.10 First, you should determine the reciprocal dispersion of your instrument:

$$\frac{\text{Spectral range}}{\text{Number of pixels}} = \frac{(1050 - 350)\,\text{nm}}{2048\ \text{pixels}} = 0.34\ \frac{\text{nm}}{\text{pixel}}$$

Because the pixel resolution is 3.2 pixels for a 10 μm slit width, you can determine the full-width at half-maximum (FWHM), that is, the optical resolution of the spectrograph (see below figure left):

$$FWHM = 0.34\ \frac{\text{nm}}{\text{pixel}} \times 3.2\ \text{pixel} \approx 1\ \text{nm}$$

We note that when calculating the spectral resolution in order to determine the FWHM of a band, a minimum of three pixels is required.

Second, you should evaluate data from the journal following the illustration on the right side of the figure. Draw an arbitrary axis x, where you should mark the position and FWHM of each band. Let us suppose the distance between the centers of two bands measured on this axis is 25 mm.

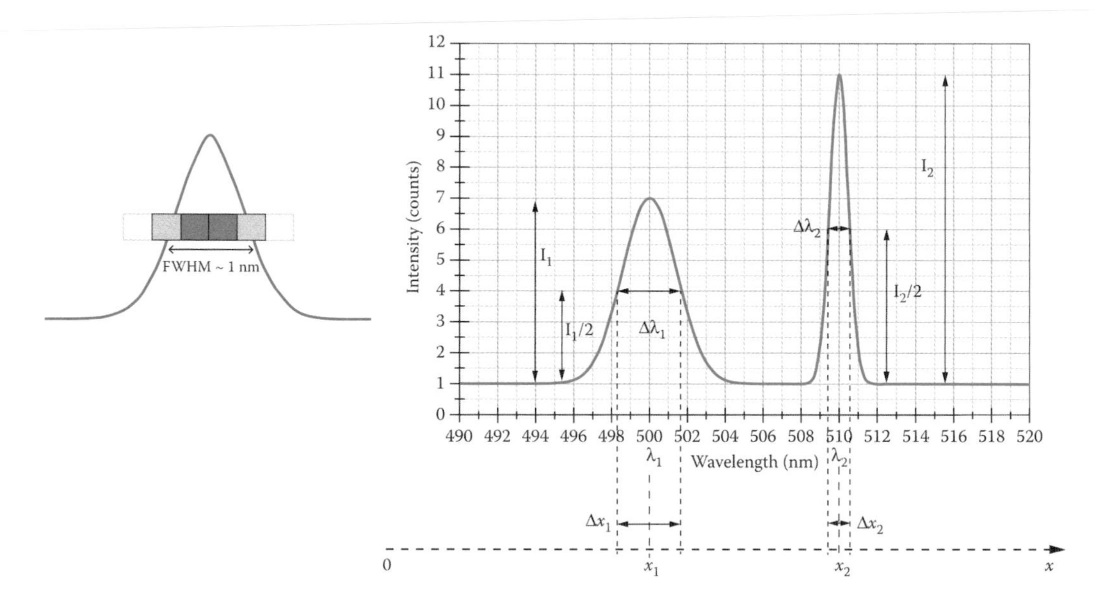

Thus, the recorded reciprocal dispersion is given by

$$\frac{\lambda_2 - \lambda_1}{x_2 - x_1} = \frac{510 - 500}{25} = 0.4 \frac{\text{nm}}{\text{mm}}$$

Measure the distance in millimeters at the FWHM of each band. If $\Delta x_1 = 6$ mm, it results in bandpass of the instrument of 6 mm × 0.4 nm/mm = 2.4 nm for the first band. If $\Delta x_2 = 2$ mm, the bandpass of the instrument is then 2 mm × 0.4 nm/mm = 1.2 nm for the second band.

Therefore, your instrument is at its limit for measuring the second band, but it is good enough to measure the first band.

Important note: Always think of both spectral resolution and throughput because of the trade-off between them. The answer was positive due to the slit width taken being very narrow. For a 10-μm slit, you should take into consideration only strong light sources and/or very sensitive detectors. Otherwise, a good resolution is useless.

Index

D

E

N

O

P

T